KU-575-376

A Guide to the

FRESHWATER LIFE

of Britain and North-West Europe

Field Guide to the

FRESHWATER LIFE

of Britain and North-West Europe

Richard Fitter · Richard Manuel

Collins
Grafton Street, London

William Collins Sons & Co Ltd
London · Glasgow · Sydney · Auckland
Toronto · Johannesburg

ISBN 0 00 219143-1

Filmset by Ace Filmsetting Ltd, Frome, Somerset
Colour reproduction by Adroit Photo Litho Ltd, Birmingham
Printed and bound by Wm. Collins Sons & Co. Ltd, Glasgow

Contents

Acknowledgments

The authors wish to thank the following colleagues who have read and made helpful comments on the sections indicated: Dr A. H. Fitter (Bryophytes), Dr R. Ingle (Decapoda), Dr P. Macauley (Hydra), Dr P. G. Willmer (Algae and Invertebrates generally), Mr J. Wright (Mollusca, Tardigrada); and Mr A. C. Jermy, who supplied information for identifying *Potamogeton*.

Colour photographs provided by Oxford Scientific Films are credited (OSF). Others were taken by the authors (vascular plants, Richard Fitter; invertebrates, and algae, Richard Manuel) except for the following: M. Amphlett, Pl. 320; T. Halliday, Pls. 313, 314, 315; RSPB, Pls. 326, 327, 329.

The line drawings for chapters 1–24, with the exception of chapter 5 (Flowering Plants), and for the Collection and Study section are by Richard Manuel. The publishers are grateful to Bell & Hyman Ltd for permission to reproduce material from Stella Ross-Craig's Drawings of British Plants (Figs. 5.3, 4, 5, 6, 7, 8, 9, 10, 12, 13, 14, 15, 16, 17, 18, 19, 20, 21, 22, 23, 24, 26, 27, 29, 30, 31, 32, 33, 34, 40, 41, 42, 43, 44, 45, 46, 56, 57).

Figs. 5.1, 2, 11, 25. 28, 35, 36, 37. 38, 39, 47. 48. 49. 50. 51. 52. 53. 54. 55. 58. 59, 60. 61, 62, 63, 64, 65, 66; 25.1, 19, 20, 26, 28, 29, 30 and 31 are by Andrew Robinson.

The illustrations for chapters 26–29 are by Norman Arlott. The balance of illustrations of fishes (chapter 25) are taken from British Freshwater Fishes by C. Tate Regan (Methuen, 1911).

About the Guide

This is a field guide to the plants and animals of all freshwater ecosystems of the north-western quadrant of Europe. All possible groups of organisms, including microscopic ones, have been included, even those that seem to be rare or obscure. Some are thought to be rare just because people do not know how or where to look for them. Marine and brackish-water species are not included unless they extend into fresh water for a significant part of their life-cycle.

In short, the book seeks to enable readers to identify freshwater animals and plants with reasonable ease and reasonable certainty to a reasonable level, beyond which would-be specialists will wish to consult more advanced works. For those who do wish to study any group in greater detail, we supply a brief specialist reference (if one exists) after each group, full references being given on *p. 356*.

The Area Covered

As we are dealing with freshwater ecosystems, we have based our area on north- and north-west-flowing rivers and their catchment areas on the mainland. The area stretches southwards from Iceland, Ireland, Scotland and the Arctic Circle in Norway, to the mouth of the Loire on the Atlantic coast of France. Thence the boundary runs eastwards as far south as Basel, along the watersheds of the Loire, the Seine, the Rhine, and the other rivers that run into the North and Baltic Seas as far east as the Vistula, and back up through the Baltic Sea and the Gulf of Bothnia to the Arctic Circle in Sweden. Thus south- and south-east-flowing streams, such as those that fall into the Rhône or the Danube, are excluded, as these would take us into the largely different warm-temperate and Mediterranean climatic regions.

Despite the foregoing, the guide is applicable well beyond these somewhat arbitrary boundaries. With some minor subtraction and addition of 'fringe' species and groups, it effectively covers the whole of Europe except the lands bordering the Mediterranean and Black Seas, and the Iberian peninsula. This includes the Baltic States, Poland, Hungary, Czechoslovakia, East and West Germany, as well as providing adequate coverage of the alpine regions of France, Switzerland and Austria. Although these regions may differ slightly in the species that are present, our policy of describing mainly groups and genera of organisms, with less emphasis on species, increases the guide's value outside the designated area.

How to Use the Guide

For most groups we provide identification keys (see *p. 12*), but most readers will doubtless flick through the illustrations first, to try and find a quick 'good match', especially if they already have some idea of the type of organism they are trying to identify. We have therefore placed the line drawings as close to their descriptions as practicable. NB The keys and descriptions have been compiled in the context of fresh water only, and ignore terrestrial or marine members of the same group.

For lower plants and invertebrates, descriptions of genera and species are headed by the currently accepted scientific name followed by the vernacular name (for higher plants and vertebrates the vernacular name comes first). In some cases, commonly-used, but out-of-date or invalid scientific names are also given in brackets, together with a second vernacular name if this is also commonly used. All descriptions and keys assume that the relevant introductory passage has been read.

Measurements are all in the metric scale. The standard microscopic unit is the micrometre or micron (1μm) which is 1/1000th of a millimetre. Measurements of total length do not include any appendages.

The treatment of each group varies according to the relative ease of identification, the physical size or conspicuousness of both macroscopic and microscopic individuals and their frequency. Roundworms (or nematodes), for example, are abundant and easily recognizable, with very characteristic, uniform appearance and style of movement; but their very uniformity makes it difficult (even for professional taxonomists who are not nematode specialists) to identify them to genus, family, or even order. So these mostly inconspicuous or microscopic animals are treated briefly. Snails, on the other hand, being relatively large, conspicuous and abundant, are keyed to species, because most people, with a hand lens and a little practice, can identify the individual species. Between these two extremes most groups of invertebrates and lower plants are identified to genus or family only, except for a few distinctive or monotypic species, i.e., those which are the only members of their family or genus.

Many of the higher plants and vertebrates in the book are more marginal to freshwater ecosystems than most of the lower plants and invertebrates. Therefore some groups, such as birds (which are covered in detail in specialist field guides) are dealt with fairly cursorily. Others, notably the flowering plants, have both fully and marginally aquatic species, and we have treated the marginal ones fairly arbitrarily – to include them all would have overloaded the text.

Identifying Plants and Animals
Before attempting to identify an organism it is necessary to understand something of its basic biological features and characteristics. These will mostly be familiar to readers with some biological knowledge or training, to others many of the terms and concepts will be new or at best hazy in meaning. Little previous knowledge is required for the identification of most of the larger organisms – vertebrates and higher plants, for instance – but many invertebrates, algae, and all microscopic organisms are more demanding. Terms and concepts such as the cell, photosynthesis and photosynthetic pigments, sexual and asexual reproduction, flagella and cilia, colonies and coloniality, and soft-bodiedness are therefore explained at length in the glossary (*p. 361*).

The descriptions in this book apply largely to living organisms. Many of the characters can be seen only in life; preservation may cause contraction and distortion that obscures important detail, especially in soft-bodied invertebrates. However, some creatures, such as small arthropods, are too lively to observe in sufficient detail and it may be necessary to preserve these, after they have been studied alive, in order to obtain an accurate identification. Methods for preservation are given on *p. 346*.

Identification of algae and invertebrates is rarely taken below the level of genus. The characters defining species in these groups are almost invariably obscure and difficult to determine and, besides, the number of species involved would be overwhelming.

The major groups of organisms, kingdoms or phyla, are summarized on *pp. 22–32*. This outlines the characteristics of each group in turn. Illustrations of typical organisms are included when these are likely to be of assistance. From this the reader is led to the relevant section where a more complete description of, and introduction to the group concerned will be found, together with keys or other means of identification.

Using Identification Keys A biological key is a step-by-step elimination procedure, using the physical characteristics of an organism to decide between two or more alternative descriptions at each step in the key – ultimately arriving at an identification of the organism or the group to which it belongs (see *p. 180* for a typical example). Each step of the key is numbered on the left side of the page, and the descriptions within it are indicated by letters, i.e., 1. a) b) etc. With the subject before you, study the alternative paragraphs in step 1, and decide which of them best fits your specimen. Then refer to the right-hand column to find the number that indicates the next step to be consulted. Continue in this way until you reach a taxon and page reference to lead you to the more detailed description in the main text.

We have not used formal identification keys in all groups, because they are not always easy to apply, or are unsuitable for the group concerned. Some groups are so large that we can only describe a few of their commoner representatives.

Alternative means of identification include the use of tables, or splitting large groups into smaller units by means of an informal 'key' integrated with the descriptive text. The use of tables is self explanatory, and is most suited for separating closely-related organisms such as plant species. In such cases, it is necessary to compare a number of characters which individually are not diagnostic, but which together enable a positive identification. In 'integrated keys' each step is defined by capital letters (not numbers as in formal keys). Thus **A** denotes the first unit and **AA** the remaining units; **B** the next unit (within **AA**) and **BB** the remainder; and so on, e.g., *p. 249*. This reduces a large number of taxa to manageable units within which individual descriptions can be directly compared.

In any group it is possible that the user will find an organism not included in this book: a locally-abundant population of an otherwise rare species, for example. In such cases it will be necessary to consult more specialized literature to achieve an identification, so wherever possible we have provided references to such works at the end of each group.

Naming and Classifying Plants and Animals

In everyday speech or writing we apply names to objects, people, or places, in order to identify them instantly without having to go into a detailed description of their salient features. Scientific names serve the same purpose but attempt to be more precise than 'common' names; to eliminate confusion and ambiguity by ensuring that a different name is used for every one of the three million or so known species. These names have the same meaning throughout the world, regardless of national languages. The use of Latin, Greek, or 'Latinized' words in these names stems from the time when Latin was the language of scholars the world over. At one time all learned texts were written in Latin and it is only in the present century that the practice has finally died out.

It is a sad fact that the prospect of having to use scientific terminology fills many a would-be naturalist with nervous apprehension or even horror. Yet many other activities, such as computing, car-mechanics, or photography, all entail the use of their own particular jargon which must be mastered before one can become proficient. In natural history, as in these other subjects, many creatures and concepts have no 'common names' and there is thus no alternative to the jargon of their scientific names.

Before embarking on a description of how classification and scientific names are used it is necessary to define the subjects involved. Classification itself is the arranging of living organisms into designated groups (called taxa, singular taxon) usually in a manner intended to reflect their relationships; Nomenclature is the application of names to the organisms and groups of organisms concerned according to prescribed Codes of Nomenclature (see Jeffrey, 1977); and the study and practical application of classification, which necessarily involves nomenclature as well, is known as Taxonomy. All this may be a little confusing at first but the essential point is that the whole is a filing system for organisms, a tool to be used and not an end in itself.

Scientific classification is traditionally a 'family-tree' arrangement known formally as a taxonomic hierarchy. Starting from the main trunk (kingdom), each branching point on the tree is a rank and these ranks are formally known as, variously, classes, orders, families, etc., the final twig on each branch being a species. There is no limit to the number of branches springing from each rank.

From this it follows that the basic unit of classification is the species (a word that is both singular and plural), all higher taxa being groups of species. Closely related species are placed together in a group called a genus. A species' name consists of two words, the generic name, which is common to all species in that genus, and the specific name, which is different for each species in that genus (though the same specific name may be used for species in different genera). Thus the common bird known as the house sparrow bears the scientific name *Passer domesticus*. *Passer* is the generic name, the initial letter of which is always capitalized; *domesticus* is the specific name and is uncapitalized. There are several other species in the genus *Passer*, for instance *Passer montanus* (the tree sparrow) but no other species in the genus can bear the name *domesticus*. This method of using two words to name each species is known as the binomial system and was first introduced by an 18th-century Swedish naturalist, Carl Linnaeus.

Scientific names of species are always printed in italics. When the context is clear it is customary to abbreviate the generic name to its initial letter to avoid endless repetition: e.g., *P. domesticus*. If the genus is known but the species is uncertain it is written *Passer* sp.; if a number of species are involved then it becomes *Passer* spp. Sp. and spp. are also used as abbreviations for species (singular) and species (plural) respectively, in any context.

Genera that are closely related are grouped into families, families into orders, orders into classes, and classes into phyla. When necessary, intermediate ranks are inserted at the appropriate level, e.g. order, suborder, infraorder, superfamily, tribe, family. Names of families and higher ranks are not italicized. (Family names, which are often derived from the principal genus in that family, must end in the suffix -idae in zoology, -aceae in botany; other ranks are not usually so strictly governed.)

A phylum is a major taxonomic group in which all the included species are clearly related to each other by basic similarities in their structure, development, and other criteria, and yet just as clearly different from those belonging to other phyla. Although most of the vital characters which determine phyla are often internal it is not difficult, with experience of studying and handling a variety of animals and plants, to gain a 'feel' for the phylum concept, and eventually to recognize an animal's relationships by its general appearance, type of movement, life-style, etc. This is the main aim of the 'Recognition features' paragraph at the beginning of a group introduction.

The phyla are traditionally grouped into two great kingdoms, Plantae and Animalia, but many modern taxonomists tend to increase this number to four or five. In this book we recognize four kingdoms: Monera (bacteria and blue-green algae), Fungi, Plantae and Animalia. Some authorities place all single-celled organisms in a separate kingdom, Protista.

As an example of a taxonomic hierarchy we have chosen the family tree of the most familiar of all animal groups, the phylum Chordata, to which we ourselves belong. Only the branches leading to our original example, the house sparrow, are shown complete, *p. 16*.

In a book such as this it is customary to arrange the phyla in a sequence leading from the 'lower' groups (the least specialized or exhibiting the simplest grade of structure) to the 'higher' ones (more specialized or complex in structure). This does not imply that 'lower' animals are in any way inferior to 'higher' ones; they are simply different. Nevertheless man, being an egocentric animal, tends to place his own group in the highest position! The sequence of phyla used in this book is shown on *p. 17*.

'Common names' derived from formal scientific terminology are frequently used and many have become part of the naturalist's everyday language. Thus we may say 'mollusc', 'oligochaete', or 'hydra', to refer to members of the phylum Mollusca, the subclass Oligochaeta, or the genus *Hydra*, respectively.

Apart from the formal grouping of phyla into kingdoms, there are several useful informal groupings that are widely used:

Vertebrates are animals which possess backbones, and usually other bones too, such as fishes, amphibians, reptiles, birds and mammals

Invertebrates is a negative term for all animals that are not vertebrates

Protozoa refers to all the single-celled (unicellular) animals

Metazoa is the collective term for all the other, multicellular animals (except the sponges, phylum Porifera – an odd, in-between group)

Arthropods are animals that possess a hardened skin (cuticle) jointed for movement, and paired, jointed limbs, during at least part of their lives, and include Crustacea, Chelicerata (spiders and mites), and Uniramia (centipedes, millipedes and insects)

Similarly the plant kingdom is divided into 'higher' or vascular plants (those with provision for sap flow); bryophytes, mosses and liverworts; and algae, which includes all the remaining phyla of lower plants, many of which are unicellular.

An Example of Taxonomic Hierarchy

Phylum CHORDATA (Animals which, during at least part of their lives, possess a dorsal nerve cord running the length of their bodies.)

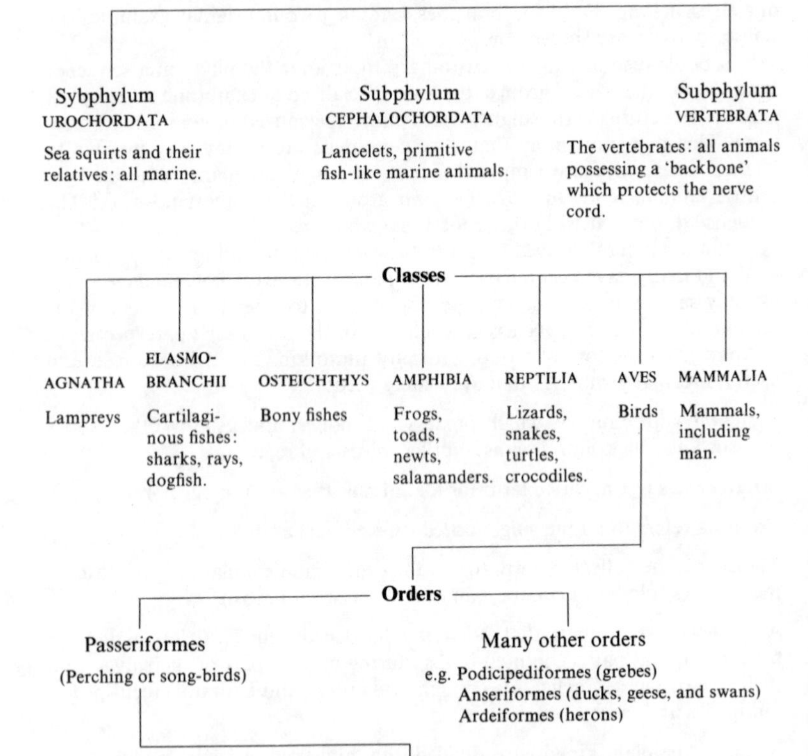

Families

Passeridae

(Sparrows and others)

Many other families:

e.g. Sylviidae (warblers)
Corvidae (crows, rooks, jays, magpies)
Cinclidae (dippers)

Genus and species
Passer domesticus
(The house sparrow)

The Sequence of Phyla Used in this Guide

Kingdom	Group	Phylum	Common name
Kingdom MONERA		Phylum various	Bacteria
		Phylum CYANOPHYTA	Blue-green Algae
Kingdom FUNGI		Phylum various	Fungi and Moulds
Kingdom PLANTAE	Algae (lower plants)	Phylum EUGLENOPHYTA	Green Algae
		CHLOROPHYTA	Green Algae
		XANTHOPHYTA	Yellow-green Algae
		PRASINOPHYTA	
		CHRYSOPHYTA	Golden-brown Algae
		CRYPTOPHYTA	
		DINOPHYTA	Dinoflagellates
		BACILLARIOPHYTA	Diatoms
		CHAROPHYTA	Stoneworts
		RHODOPHYTA	Red Algae
	Bryophytes	BRYOPHYTA	Mosses and Liverworts
	Higher (vascular) plants	PTERIDOPHYTA	Ferns and Horsetails
		ANGIOSPERMOPHYTA	Flowering Plants
Kingdom ANIMALIA	Protozoa	Phylum SARCOMASTIGOPHORA	Amoebae, Flagellates
		MICROSPORIDA	
		CILIOPHORA	Ciliated Protozoans
		PORIFERA	Sponges
		CNIDARIA	Hydroids, Jellyfishes
		PLATYHELMINTHES	Flatworms, Flukes, Tapeworms
	Metazoa	NEMERTEA	Proboscis Worms
		NEMATODA	Roundworms
		NEMATOMORPHA	Horsehair Worms
		ACANTHOCEPHALA	Spiny-headed Worms
		ROTIFERA	Rotifers, Wheel Animalcules
		GASTROTRICHA	Hairybacks
		ENTOPROCTA	Entoprocts
		ANNELIDA	Segmented Worms, Leeches
		BRYOZOA	Moss Animals
		MOLLUSCA	Snails, Mussels
	Arthropods	TARDIGRADA	Water Bears
		CHELICERATA	Spiders, Mites
		CRUSTACEA	Waterfleas, Shrimps, Crayfishes
		UNIRAMIA	Insects
		CHORDATA	All Vertebrates

Freshwater Habitats

Ecosystems in the freshwater environment are as complex and varied as those on land. Those on land are influenced by three main factors – climate, soil, and the activities of man, whereas in fresh water, of course, water largely replaces soil and atmosphere, and has a smaller temperature range than the atmosphere, so climate is less important. But the influence of man is just as pervasive in the water as on land.

There is a major division between still and moving waters; between lakes, ponds and canals on the one hand, and rivers and streams on the other. Each of these contains a number of distinct habitats which depend on such factors as depth, speed and chemical composition of the water (*Pls. 1–8*).

Animals and plants living wholly in water can satisfy all their needs for air, food and shelter within it, since water contains solutions of many nutrients, in addition to the life-giving oxygen. But many aquatic organisms spend part of their lives on land, are emergent and breathe air from the atmosphere, or have roots that not only hold them fast but obtain nutriment from the bottom mud. By no means all freshwater organisms depend wholly on the water, and this further complicates the definition of a freshwater organism. There are similar complexities in defining freshwater habitats, for even such an apparently straightforward distinction as that between lakes and ponds still has no scientific basis.

The simplest freshwater habitat is a rainwater puddle, which rarely has time to be colonized by any plants or animals before it dries up. In deep shade, however, a puddle may last long enough to acquire microscopic animals, algae, and plants which live or grow in wet mud rather than actually in the water. Some larger pools regularly dry out in summer and have a very specialized flora and fauna. However, once a pool holds even a little water throughout a normal season, we are at the beginning of a continuum that only ends with Lake Baikal – the world's largest freshwater body. There is no point at which a definitive line can be drawn between a pond and a lake or even between a puddle and a pond. One can only indicate characteristics common to most ponds, as compared with most lakes, and vice versa.

A pond, then, is likely to be a small body of stagnant fresh water, shallow enough for plants rooted on the bottom to grow all over it (though this also depends on the clarity of the water), and to ensure a fairly uniform temperature throughout. A lake is a larger body of still water, often too deep for emergent plants to grow in the middle, and with marked differences between the temperature of the surface and the bottom. A lake is often large enough to have some streams flowing in to it and/or out of it. Numerous exceptions can be found to these rough and ready definitions, but they do at least make it clear that many so-called lakes in both urban and rural parks are in fact large ponds. Similarly, canals and river backwaters, which share many of the same characteristics, can be regarded as linear ponds. Indeed ponds sometimes originate from cut-off oxbows of rivers.

Layers of warm and cold water are a striking feature of larger lakes in summer. Water is heaviest at 4°C and hence in winter the chilled surface layers tend to sink, and wind currents mix the water until it is an even 4°C throughout. In spring and summer, however, the sun warms the surface water, forming a warmer, lighter layer called the epilimnion. This is separated by the thermocline – a narrow band where the temperature changes rapidly – from the cold water at the bottom called the hypolimnion. In British lakes the thermocline is often at a depth of about 9–10 m. In some lakes, such as Lac Leman (Geneva), a remarkable phenomenon known as the *seiche* occurs, when a strong wind piles the epilimnion up against one side of the lake. When the wind drops, the warm surface water slops back to the other side, and continues to oscillate over the colder water for some time, just as if a tilted bath had been righted.

Lakes and ponds both have five distinct habitat zones, starting on the edge or shore, where all the plants are aerial, with only their roots in the water. Many animals derive food and shelter from the submerged stems and roots of these plants, but need to retreat into deeper water in summer, when the edge of the water is likely to dry out. Next comes a deeper zone, which rarely, if ever, dries out, enabling plants rooted on the bottom to push their flowering spikes up into the air, often to be pollinated by insects or the wind. Many animals also take advantage of the plants in this zone to live a partly or fully aquatic existence. The third and deepest zone contains only submerged plants, such as stoneworts. Finally, the surface and the bottom each provide a distinct habitat. Some animals, such as water skaters and whirligig beetles, take advantage of the surface tension to spend much of their time skating about on the top of the water, and duckweeds and other plants may pass their whole lives there. On the bottom, the nutrient-rich mud provides a habitat for many animals and a resting place for some plants, such as the water soldier, which though normally floating, has a rooted stage as a seedling.

Another classification of fresh waters is based on the chemistry of the water. Oligotrophic waters have few nutrients, the lack of calcium in particular making them 'soft', and usually have comparatively little animal or plant life. Such waters are usually found in hill districts, and have notably cold, clear water, except sometimes on peat, where the water may become quite brown, and these waters are termed dystrophic. Waters with many nutrients, usually warmer because they are found in the lowlands, are called eutrophic, and are rich in animal and plant life; in chalk and limestone districts the calcium content makes the water 'hard'.

Nowadays many lakes, ponds and rivers are so contaminated by chemicals drained off highly-fertilized farmland and become so eutrophic that most of the vegetation is swamped by the algae stimulated by excessive quantities of nutrients. The decomposition of the algae removes oxygen from the water and may make it more sterile than oligotrophic water. This process, known as eutrophication, has virtually destroyed the once rich animal and plant life of the Norfolk Broads.

Rivers and streams too have no scientific definition, except that rivers are usually larger than streams, which in turn are larger than brooks, becks or burns; in fact, as with lakes and ponds, we stay with the commonsense perceptions of our Anglo-Saxon ancestors. Rivers and streams share all five of the habitat zones of lakes and ponds, but can be further classified according to the speed of the water, the several zones being named after their most characteristic fishes. The speed of the water is a most important ecological factor, for just as terrestrial organisms have to avoid being blown away by the wind, so the inhabitants of flowing water must adapt themselves to avoid being swept away by the current.

All rivers and streams originate either in ponds (including the mountain ponds known as tarns) or in springs or flushes, which rise out of a subterranean aquifer of some kind. These sources share the characteristics of a pond, but as they have an outflow, they cannot be wholly stagnant. Immediately below the outflow, a rivulet is too small for fish, but in hills or mountains it soon develops into a troutbeck. It is fairly shallow and fast-flowing because of the steep gradients; the water is clear and becomes well oxygenated by splashing over stones, rocks and waterfalls. This is the trout zone, where the bottom is usually gravelly or pebbly, there are few if any higher plants, and mosses and liverworts predominate.

Next comes the grayling or minnow zone, where the water gradually slows down as it approaches the lowlands, and enough silt accumulates to allow some higher plants to grow on the bottom. In these first two zones, the water is oligotrophic and salmonid fishes are characteristic. In the next two the so-called 'coarse fish' of the carp and other families predominate in the mainly eutrophic water.

As the stream debouches into the lowlands and becomes a river, it slows down still more and builds up increasing amounts of silt and mud, and much more vegetation. This is the barbel zone, transitional to the lowland rivers. When it finally reaches the flat lowlands, the river slows right down, accumulates much more mud and vegetation, and is termed the bream zone. In the lower reaches of rivers, many animals adapted to still water may be found living along the margins, where the flow is minimal or non-existent. The pollution by eutrophication, which has harmed many lakes, also affects the lower reaches of rivers – the final repository of the excessive amounts of fertilizer applied to modern farmland.

The river reaches the sea in an estuary, where the water becomes brackish, if not actually salt, and few freshwater species can survive. Equally, few of the primarily saltwater species can penetrate above it. The flounder and the freshwater crustacean *Gammarus duebeni* are among those that do.

Conservation Problems

Unpolluted fresh water is one of the most endangered habitats in Britain and north-western Europe. So all naturalists and others interested in freshwater wildlife should be concerned that as many as possible of the freshwater habitats in their area remain unspoiled. Some threats, such as the eutrophication of ponds due to overuse of agricultural fertilizers, are so general that the individual can do little about them, except join local county conservation trusts or a national body, such as the Royal Society for the Protection of Birds or Friends of the Earth, which specialize in grappling with this kind of environmental problem.

What the individual can do is to refrain both from any polluting acts, either as a householder or as a visitor to the countryside, and from collecting any freshwater animal or plant to the point where its local population is threatened. Education is important, and is more effective if children can themselves catch and handle specimens; but wherever possible these should be returned to the water and not taken back to the classroom. It only needs an enthusiastic class of 8- or 9-year olds to eliminate all the local sticklebacks or frogspawn. Moreover the future prospects of an individual fish are not improved by the trauma of being netted, examined and thrown back into the water, and even less so if it has been caught on a hook.

Summary of the Major Groups of Freshwater Organisms

The following guide contains a brief synopsis of the main features of each kingdom and the major divisions within them, illustrated with typical examples where this is helpful. Terms printed in **heavy type** are defined in the Glossary (*p. 361*).

Kingdom MONERA

All monerans are **procaryotes**; all remaining plants and animals are **eurar-yotes**.

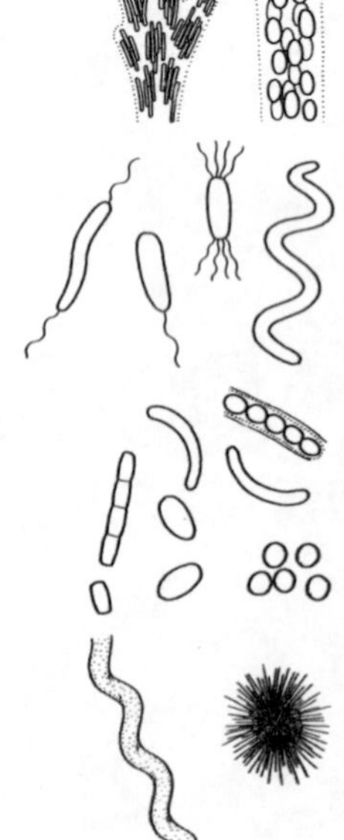

BACTERIA (*Pl. 9*) All phyla of Kingdom MONERA, except Cyanophyta (*p. 33*).

Microscopic unicellular organisms; solitary or **colonial**; individual **cells** very small, typically up to 10 μm, rarely more than 25 μm, but colonies may be much larger; individuals rod-shaped, spherical, oval, or spiral; immobile or moving by **gliding** or **flagellar** action; colourless or pigmented.

Bacterial colonies may form slimy sheets, irregular masses, or bunches of fine filaments, often easily visible to the naked eye. Accumulations of bacteria are often accompanied by characteristic and unpleasant smells – ammonia, hydrogen sulphide, etc.

BLUE-GREEN ALGAE (*Pls. 10–13*) Phylum **Cyanophyta**, Kingdom MONERA (*p. 35*).

General features as in bacteria above but **flagella** are never present. **Photosynthetic pigments**, characteristically deep blue-green in colour, are present, diffused throughout the cell, not contained in **plastids**. Most forms are **colonial**, forming regular or irregular slimy masses of varied texture (sometimes quite hard) or mats of long filaments; colonies often large enough to be visible to the eye, some may become several centimetres (or more) across, typically appearing blue-green or blackish, but some are yellowish, brownish, or plain green to the naked eye.

Kingdom FUNGI

FUNGI Macroscopic freshwater fungi are immobile **multicellular** organisms forming masses of fine branching filaments, usually colourless, growing over or within the food matter (any dead organism, less often living ones unless in wounds) – (*Pl. 14*). Numerous **microscopic** forms occur in fresh water but these are not described in detail in this book.

Kingdom PLANTAE

Plants are organisms that possess **photosynthetic pigments** (typically green but may be other colours) contained in **plastids**, and obtain their food by **photosynthesis** (exceptions on *p. 24*); only certain **microscopic** forms are mobile.

ALGAE (*Pls. 15–30*) Ten phyla of Kingdom PLANTAE (*p. 42*).

Simple plant organisms that are not bryophytes or **vascular plants** (see below); a difficult group to characterize briefly as it contains such a wide variety of forms.

Micro- or **macroscopic**; **unicellular** (solitary or **colonial**) or **multicellular**; cells typically more than 25 μm (cf. Cyanophyta, *p. 22*); sometimes forming plant bodies of substantial size. **Microscopic** algae are often green but may be brown, yellowish, or red (**plastid** colour); fixed or free; immobile or mobile, locomotion by **gliding** over a substratum or swimming with **flagella**. **Macroscopic** forms are nearly always green (otherwise red); fixed or free, never mobile; forming long, hair-like filaments, spherical or irregular growths such as hollow tubes, cushions or thin encrustations on rocks, etc.; texture usually soft and slimy but some are hard due to accumulation of calcium. In practice few algae are likely to be confused with higher plants except for stoneworts and some 'red' algae.

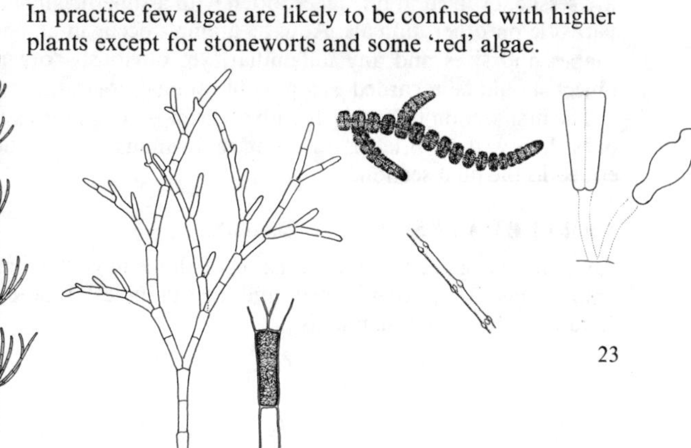

MOSSES and LIVERWORTS (*Pls. 31–34*) Phylum **Bryophyta** (*p. 64*)

Non-**vascular**, non-flowering plants typically comprising a fixed stem with two or more ranks of simple (unveined) leaves arranged more or less regularly along it. Some liverworts form flattish, irregularly divided or lobed fronds, encrusting or floating free. Reproduction by **spores** borne in cup-shaped, stalked or sessile capsules.

VASCULAR ('HIGHER') PLANTS (*Pls. 35–100*) Phylum **Pteridophyta** (Ferns) and Phylum **Angiospermophyta** (Flowering plants) (*p. 69*)

Green plants with an internal fluid transport (**vascular**) system which is largely responsible for their crisp texture. All flowering and seed-producing plants, and the few aquatic ferns and horsetails, belong to this group. Nearly all are 'typical plants' with roots, stems and leaves, and should be easily recognized as such. Possible exceptions are the tiny rootless duckweed *Wolffia* (*Pl. 91*) and perhaps the other duckweeds and the floating fern *Azolla* (*Pls. 37–38*). Also, beware of stoneworts (*Pls. 26, 27*) – algae which strongly resemble vascular plants.

Kingdom ANIMALIA

Animals are organisms that feed by eating other organisms. They never possess **photosynthetic pigments** (but some are green due to algae living in their tissues). Most familiar animals, mainly terrestrial ones, are mobile – walking, running, crawling, flying or swimming – but many aquatic animals are **sessile**, or fixed in one place, often with a superficially plant-like form, or **parasitic** on other animals. Aquatic animals occur in an enormous range of shapes and sizes and any unfamiliar, yet obviously 'organized' non-green object should be regarded as a possible animal organism.

The first section below deals with all groups except the exclusively **parasitic** ones, followed by brief notes on **epizooic** animals; parasitic forms are described in the final section.

VERTEBRATES Phylum **Chordata**

An easily-recognized group consisting of all the most familiar animals: fishes, amphibians (frogs, toads, newts, and their tadpoles), reptiles (lizards, snakes, terrapins), birds and mammals.

INVERTEBRATES Numerous phyla

All animals that are not vertebrates (above). This is an extremely diverse group including all worms, snails, mussels, water fleas, shrimps, spiders, mites, insects and their larvae, and a host of less well-known creatures such as sponges, hydras, rotifers and moss animals. The rest of this section is devoted to the invertebrate phyla.

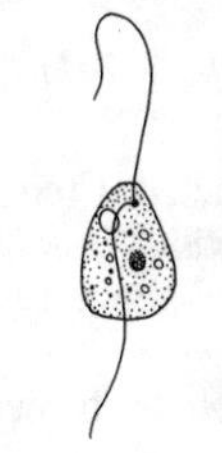

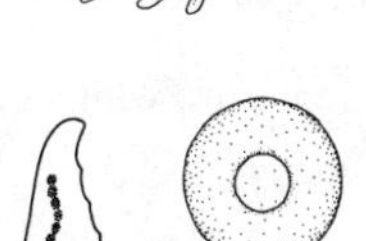

PROTOZOA (*Pls. 101–114*) Three phyla in fresh water (*p. 108*)

Microscopic, occasionally **macroscopic**, up to 5 mm; solitary or **colonial**, **unicellular** animals; fixed, **sessile**, free-swimming or **parasitic**; locomotion by flowing (**amoeboid** motion), **flagellar** or **ciliary** swimming or gliding; **cilia** or **flagella** may be present; mostly colourless (the few green protozoans possess cilia, never flagella, which distinguishes them from algae). The **single-celled** nature of a protozoan is not always obvious to the beginner, but becomes relatively easy to recognize with practice; reference to the plates should help.

NB All the multicellular animals, including the vertebrates (but excluding the sponges, which are nowadays regarded as huge colonies of unicells) are collectively known as METAZOA (see *p. 17*).

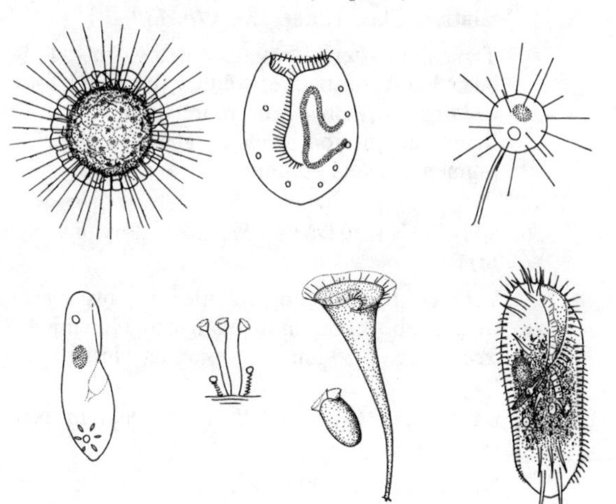

SPONGES Phylum **Porifera** (*Pls. 115–116, p. 124*)

Fixed, immobile, often fragile, irregular spongy growths up to 10 cm or more across; encrusting or forming erect 'fingers'; surface finely perforated and slightly rough to the touch – often green, but may be dirty yellowish, brown or whitish. Common in clean waters (not winter).

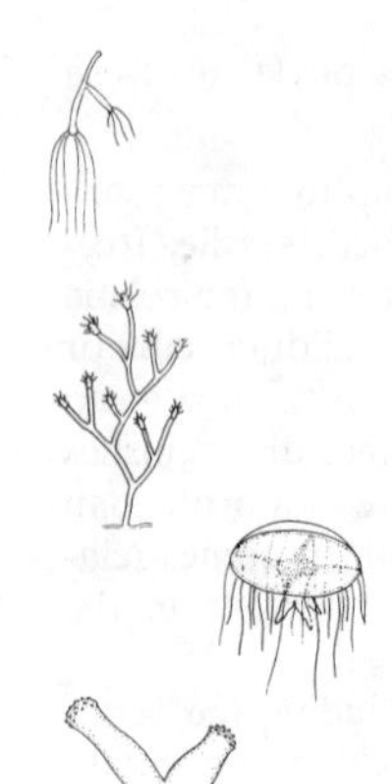

HYDRAS and JELLYFISHES (*Pls. 117–119*)
Phylum **Cnidaria** (*p. 128*)

Soft-bodied, contractile, radially symmetrical (circular in a section across the **longitudinal axis**), usually with slender, very contractile tentacles; four distinct freshwater types:

- Hydras: about 10 mm, **sessile**, green or brown, very common
- Colonial hydroid: colony several cm high, fixed, branching, individuals pink, uncommon
- Jellyfish: 2 cm diameter, colourless, free-swimming, very rare
- Hydroid stage of jellyfish: 2 mm, no tentacles, whitish, fixed, very rare

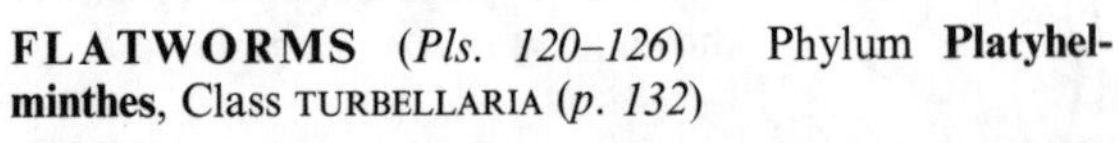

FLATWORMS (*Pls. 120–126*) Phylum **Platyhelminthes**, Class TURBELLARIA (*p. 132*)

Typically flattened, unsegmented worms; 1–35 mm; **soft-bodied** and contractile; solitary (but some form chains by **budding**); **eyespots** 2 or more than 10, or absent; free-living, locomotion by **ciliary gliding**; colourless, white, or pigmented; very common.

RIBBONWORMS (*Pl. 128*) Phylum **Nemertea** (*p. 141*)

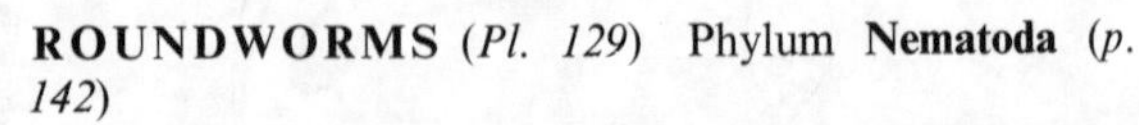

Very contractile, unsegmented worms resembling flatworms (above) but more elongated and with 4–8 **eyespots**; free-living; 1–30 mm; red, pink or whitish; uncommon.

ROUNDWORMS (*Pl. 129*) Phylum **Nematoda** (*p. 142*)

Non-contractile but flexible, cylindrical, unsegmented worms, usually pointed at one or both ends; colourless/whitish; free-living (or **parasitic**), locomotion by strongly characteristic sinuous writhing movements; 1–30 mm; very common.

HORSEHAIR WORMS (*Pl. 130*) Phylum **Nematomorpha** (*p. 144*)

Similar to Nematoda but very elongated, up to 500 × 1 mm, and blunt-ended; pigmented, reddish or brown; local and seasonal.

ROTIFERS or WHEEL ANIMALCULES (*Pls. 132–136*) Phylum **Rotifera** (*p. 147*)

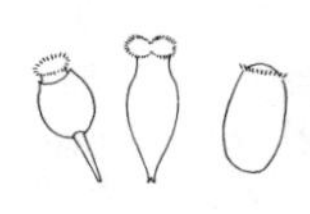

Microscopic – 2 mm; solitary or aggregating; attached, **sessile**, or free-living, locomotion by **looping**, levering with foot, or **ciliary** swimming; body shape very varied, basically worm-like and contractile, but often with a **lorica** or hardened **cuticle** protecting it, some sessile species inhabit tubes of mucus or particulate matter; colourless or yellowish, translucent. *Special feature*: a prominent patch or ring(s) of large **cilia** on retractile head region (absent in a few sessile forms – see figs). Very common.

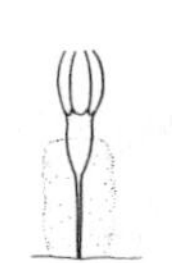

HAIRYBACKS (*Pl. 137*) Phylum **Gastrotricha** (*p. 156*)

Microscopic; solitary; free-living; locomotion by **ciliary gliding**; body shape fairly constant and characteristic, scarcely contractile; colourless; common.

ENTOPROCTS Phylum **Entoprocta** (*p. 158*)

Stalked **colonial** animals up to 2.5 mm tall, attached beneath stones; stalks bead-like; **zooids** cup-shaped with about 15 non-retractile tentacles; very rare.

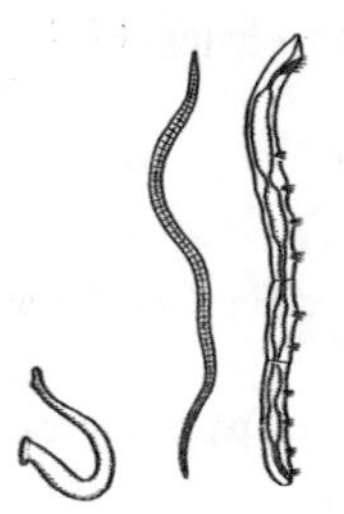

SEGMENTED WORMS and LEECHES (*Pls. 138–147*) Phylum **Annelida** (*p. 159*)

Worms with **sub**-cylindrical, contractile bodies divided transversely into many segments; 1–300 mm or more; solitary (some form chains by **budding**); free-living (but some inhabit fixed tubes); locomotion by **cilia** on head (rare), **looping**, or earthworm-like stretching/contracting movements; colourless or pigmented, often red; very common.

MOSS ANIMALS (*Pls. 148–151*) Phylum **Bryozoa** (*p. 173*)

Colonial, colonies up to 50 mm or more, forming attached masses of fine (up to 1 mm) branching tubes or structured masses of translucent jelly; **zooids** 1–3 mm with retractile crown of 12–100 colourless tentacles; locally common in clean waters.

SNAILS and MUSSELS (*Pls. 152–172*) Phylum **Mollusca** (*p. 178*)

Soft-bodied contractile animals protected by a hard shell; 1–150 mm; solitary; free-living (one attached sp.), locomotion by crawling on or levering with fleshy foot; shell either a single coiled structure (snails), single and conical (limpets), or in two matching oval halves hinged together (mussels); very common.

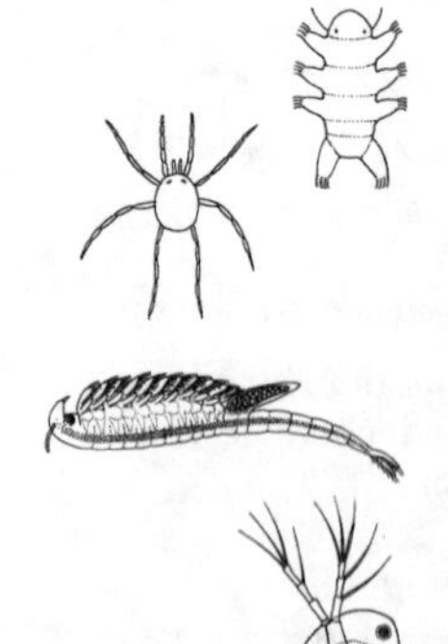

ARTHROPODS Four phyla (*p. 193*)

Animals with paired and (usually) jointed legs and other appendages; **microscopic** – 150 mm; solitary; free–living of **parasitic**. Body typically segmented but this is not always apparent; skin usually more or less hardened (**sclerotized**) to form a jointed 'shell' in many types, in others only the head is sclerotized and the rest of the body soft. Arthropods are extremely common animals and easily outnumber all other animal species together.

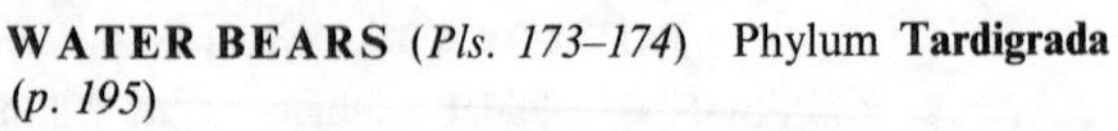

WATER BEARS (*Pls. 173–174*) Phylum **Tardigrada** (*p. 195*)

Microscopic; 4 pairs unjointed legs; characteristic shape.

SPIDERS and MITES (*Pls. 175–179*) Phylum **Chelicerata** (*p. 198*)

Microscopic – 20 mm; 4 pairs jointed legs.

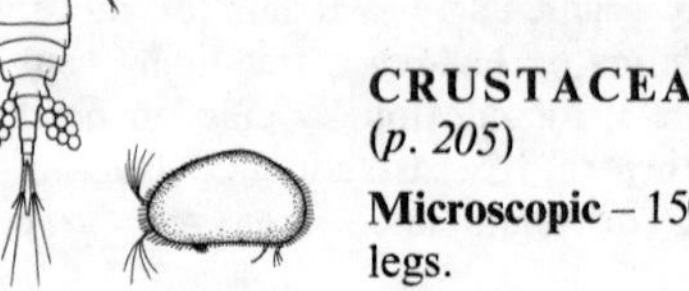

CRUSTACEANS (*Pls. 180–196*) Phylum **Crustacea** (*p. 205*)

Microscopic – 150 mm; typically more than 4 pairs jointed legs.

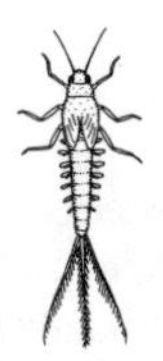

INSECTS (*Pls. 197–305* Phylum **Uniramia**, Subphylum **Hexapoda** (*p. 233*)

Microscopic – 80 mm; 3 pairs jointed legs (absent in some larvae, such as 'maggots', but **prolegs** may be present instead); adults usually winged; juveniles very variable in degree of **sclerotization**, often entirely soft-bodied.

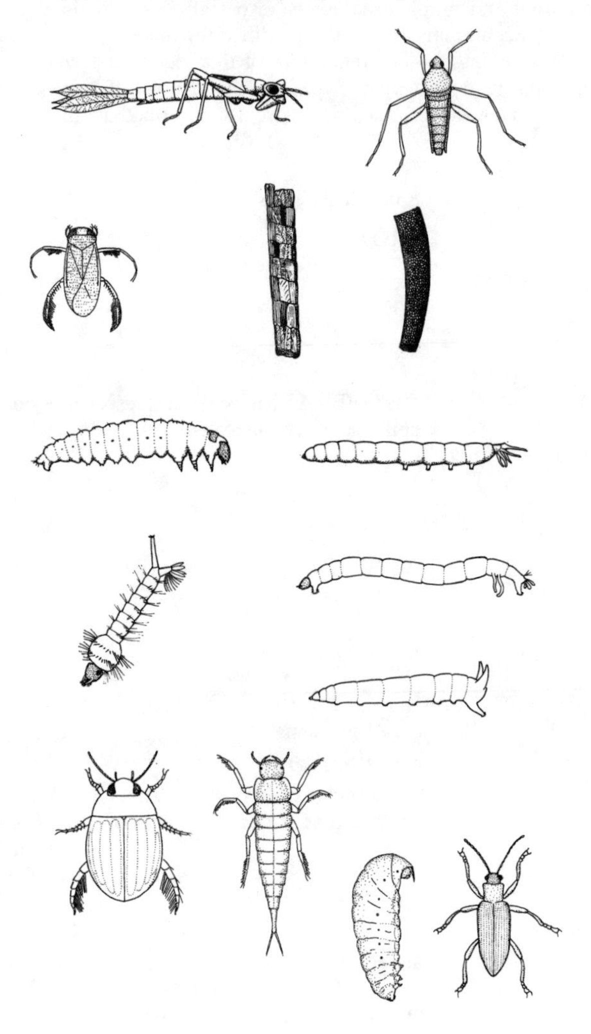

ANIMALS LIVING ON OTHER ANIMALS

In the aquatic environment, it is common to find small or microscopic animals living attached – temporarily or permanently, internally or externally – to other larger animals. Some of these are **parasites**, which derive nourishment by feeding directly on the **host** animal (e.g., sucking its blood); others are **epizooic** – and simply use the host (especially arthropods) as a site for attachment. Such animals are greatly modified for this way of life, and thus may be difficult to classify. The remainder of this section describes all such animals occurring in fresh water. Types which do not differ from the descriptions already given earlier (marked *), are not described again.

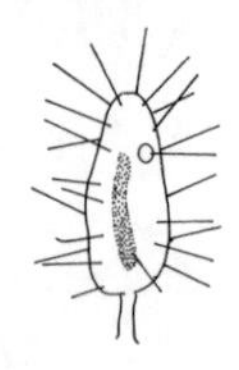

Epizooic Animals

PROTOZOA* = Suctorians (*p. 119*)
= Peritrichs (*p. 122*)

ROTIFERA* Many species, especially bdelloids (*p. 147*) epizooic on crustaceans.

ANNELIDA* Small, colourless worm living on soft parts of snails = *Chaetogaster* (*p. 159*)

Parasitic Animals

PROTOZOANS*

Sub-circular whitish spots, about 1 mm across, on or in the skin of fishes = White spot disease. *Ichthyphthirius*, Phylum **Ciliophora** (*p. 120*)

Baggy whitish cysts or raised boils in skin or flesh of fishes. Probably caused by sporozoans — Phylum **Microsporida** (*p. 116*)

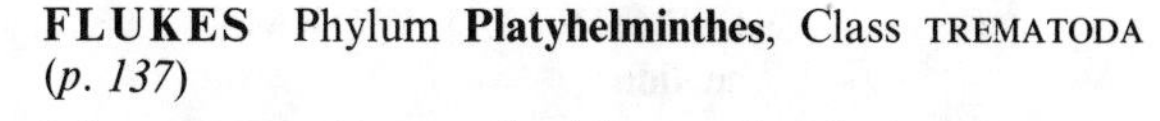

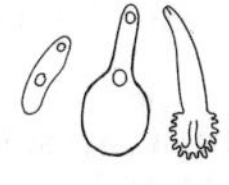

FLUKES Phylum **Platyhelminthes**, Class TREMATODA (*p. 137*)

Microscopic – 30 mm, flattish to cylindrical, **soft-bodied**, unsegmented worms living in or on **vertebrates**; often with hooks or suckers for attachment to host.

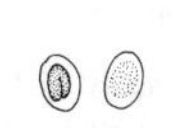

Metacercaria larvae often encyst on or in the skin and flesh of fishes, snails, worms and other hosts, forming tiny (up to 1 mm) oval or circular translucent cysts.

Cercaria larvae are free-living, not parasitic: **microscopic** but usually visible to eye; colourless/whitish; swimming by vibrating tails or crawling by looping.

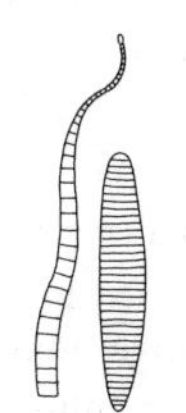

TAPEWORMS Phylum **Platyhelminthes**, Class CESTODA (*p. 139*)

Macroscopic (up to 10 m!), distinctly flattened, **soft-bodied** 'worms' comprising numerous **sub**-rectangular 'segments' tapering to a narrow 'head' which is attached (often by hooks or suckers) to the gut-lining of the **host** (a **vertebrate**).

Later larval stages also found in vertebrates, loose in the body cavity or embedded in the flesh: relatively featureless worms, with or without apparent 'segmentation'; may cause gross distortion of host (see *Pl. 127*).

ROUNDWORMS* (*Pl. 129*) Phylum **Nematoda** (*p. 142*)

Common internal parasites of many types of aquatic organisms; also free-living.

HORSEHAIR WORMS* (*Pl. 130*) Phylum **Nematomorpha** (*p. 144*)

Internal parasites of insects; also free-living.

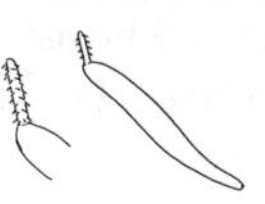

SPINY-HEADED WORMS (*Pl. 131*) Phylum **Acanthocephala** (*p. 146*)

Sub-cylindrical, unsegmented worms up to 50 mm long, with a retractile spiny **proboscis**; exclusively parasitic.

Adult worms are intestinal parasites of vertebrates; orange or white.

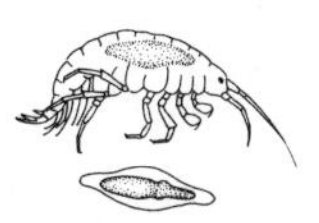

Larvae (*cystacanths*) are found occupying thin-walled cysts in the body cavity of **arthropods** (esp. *Gammarus*); orange or whitish; up to 5 mm long; **proboscis** usually kept withdrawn. Larvae often visible from outside host (*Pl. 131*)

SEGMENTED WORMS* (*Pls. 144–145*) Phylum **Annelida**

External parasites of crayfishes; up to 10 mm; small posterior sucker. = *Branchiobdella*, Class OLIGOCHAETA (*p. 166*).

External parasites of fishes, molluscs, ducks and other hosts; suckers present at both ends (anterior one sometimes obscure); 5–150 mm; free-living for much of the time; many can swim. Leeches, Class HIRUDINEA (*p. 166*).

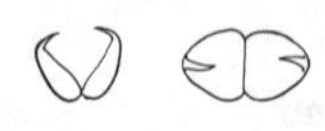

MUSSELS* Phylum **Mollusca** (*p. 178*)

Tiny (up to 1 mm) **bivalved** larvae embedded in skin or fins of fishes. = *Glochidia* larvae of mussels.

ARTHROPODS*

CHELICERATES* Phylum **Chelicerata** (*p. 198*)

Tiny (up to 2 mm) oval or globular larvae fixed to cuticle of insect host by **rostrum**; usually red or orange; may have *three* pairs of legs (*Pl. 179*). Larvae of **water mites**, Suborder HYDRACARINA (*p. 199*).

Adult mites, with 4 pairs of legs, living inside freshwater mussels. = *Pentatax* spp., Suborder HYDRACARINA (*p. 199*).

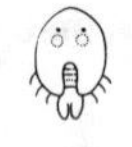

CRUSTACEANS* Phylum **Crustacea** (*p. 205*)

External parasites of fishes, temporarily attaching (not fixed) to skin or fins, can glide over host or swim free; 1–8 mm; shape very characteristic. = Fish louse, *Argulus*, Class BRANCHIURA *Pl. 191* (*p. 222*).

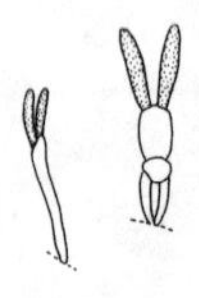

External parasites of fishes, often fixed to skin, fins, or gills of host, but some can glide over host; others (cyclopoid in form, Fig. 5.28) may be capable of swimming free. 1–30 mm; ♀♀ of some forms greatly modified with reduced or absent limbs and segmentation, recognizable by pair of oval or cylindrical egg sacs. = Parasitic COPEPODA (*p. 220*).

CHAPTER 1 Bacteria and Blue-green Algae

Bacteria

The organisms collectively known as bacteria are all, individually, very small, rarely exceeding five microns (0.005 mm) in length. But in spite of their diminutive size, bacteria are of vital importance to all other life forms. Most of them obtain their nourishment by chemically breaking down organic remains (dead plants and animals) or inorganic substances, converting them to a form which, when released into the environment, can be assimilated by plants and animals. They thus introduce into the food web the basic raw materials required by higher life forms, many of which cannot be obtained by any other means.

In the past bacteria have been variously classified as primitive plants, 'microbes', or 'protists', usually in a single phylum of their own. Many modern scientists now consider, however, that all procaryotic organisms should be treated as a kingdom, Monera, containing about sixteen phyla. Most of these phyla can only be identified and classified by chemical tests which determine their mode of nutrition and chemical products. One group, the blue-green algae (phylum Cyanophyta), is very distinctive and generally larger than other monerans, and is dealt with separately (*p. 35*).

Structurally, bacteria are simple procaryotic cells. Some are immobile, some can creep slowly by exuding slime (mucilage), and others possess flagella which enable them to swim. Bacterial flagella are remarkable in that they appear to revolve in a kind of ball and socket joint; they bear no relationship to the flagella and cilia of eucaryotic organisms. Bacteria reproduce by simple cell division and frequently form large aggregations or colonies. Most species can form resistant spores to survive unfavourable conditions and aid dispersal. The shapes of bacteria are often distinctive: rod-shaped bacilli, spherical cocci and spiral spirochaetes (*Fig. 1.1*). Some are very colourful as they contain various pigments of red, blue, brown, purple, etc; others derive their colours from the presence of mineral salts.

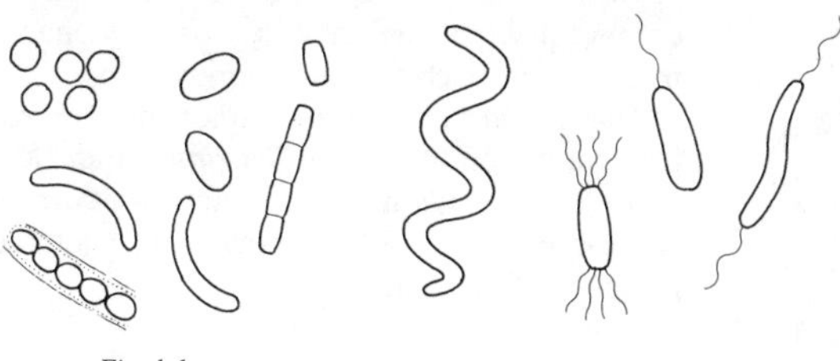

Fig. 1.1

Bacteria are probably best known to the layman as the 'germs' that cause disease in plants and animals. But their beneficial effects far outweigh this notorious reputation, earned for the whole group by relatively few species. Some of them provide the basis for modern antibiotics which cure disease. Most 'higher' animals utilize bacteria to aid digestion, in the breakdown of cellulose in ingested plant material, for instance. Bacteria are present in enormous numbers in the alimentary tracts of all animals.

The free-living bacteria feed on many different substances. Some absorb various inorganic elements and compounds, converting them to a soluble form which can be utilized by higher plants. Probably the most important of these are the nitrogen-fixing bacteria, which make atmospheric nitrogen available to green plants. Many protozoans and small metazoans utilize bacteria more directly by simply eating them.

In clean fresh water the presence of bacteria is usually only evident where there are dead remains of plants or animals. The unpleasant smells and blackish discolourations that often advertise such decaying material are caused by gases such as ammonia, methane, or hydrogen sulphide, which are released by the bacteria. In marshy areas such gases may become trapped in the composting layer of dead vegetation on the bottom. This is the aptly named black layer which is caused by exclusion of oxygen (anoxia). Will-o-the-wisps, the ghostly lights that can sometimes be seen hovering in the air above marshy places, are caused by the gradual release of incandescent methane (marsh gas) resulting from bacterial action.

In small stagnant waters rich in dissolved iron salts, iron-reducing bacteria signal their presence as a precipitate that rapidly stains the water a deep rust colour. Many bacteria frequent mineral-rich springs where they metabolize dissolved sulphur or other salts.

A characteristic bacterial regime often develops in waters contaminated with sewage effluent. This results in the formation of unpleasant slimy mats, dirty white or greyish-brown in colour, growing over the substratum. *Sphaerotilus* (*Fig. 1.2*) is the best-known cause of these and, although a bacterium, is often called the sewage 'fungus'. It forms branching filaments of mucilage, about 10 μm wide, containing numerous, small, rod-shaped cells aggregated into bundles.

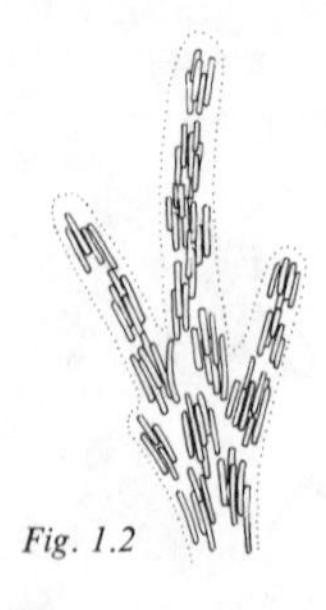

Fig. 1.2

Sphaerotilus frequently occurs in rivers in the vicinity of sewage outfalls and is the basis of a characteristic sewage community that includes various oligochaete worms and insect larvae such as chironomids.

Two genera that are sometimes encountered in 'normal' freshwater conditions are *Beggiatoa* and *Thiothrix* (*Pl. 9*), relatively large filamentous bacteria that metabolize hydrogen sulphide, but unlike many other bacteria, require the presence of oxygen.

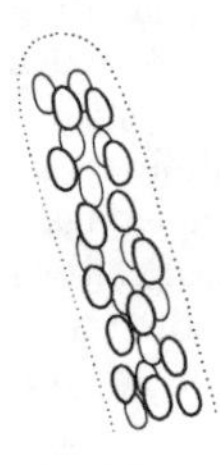

Fig. 1.2

Beggiatoa (*Fig. 1.2*) forms colourless mucilaginous filaments up to about 10 μm diameter and several millimetres long. It lives where the black layer, rich in hydrogen sulphide, is near to the surface of the mud, forming slimy and unsightly mats of aggregated filaments over the surface of the substratum. These mats may attain a size in excess of 1 sq m.

Thiothrix (*Pl. 9*) is similar in form to *Beggiatoa* but its filaments are fixed by one end to a substratum such as filamentous algae, usually in small bunches. These genera may be said to be non-photosynthesizing parallels of the filamentous blue-green algae.

Reference: Edmondson, 1966.

Blue-green Algae
Phylum CYANOPHYTA (MYXOPHYCEAE)

Recognition features Individual cells are microscopic, but most species form chains (filaments) or aggregations embedded in masses of jelly large enough to be visible to the naked eye; colour typically blue-green to blackish.

The blue-green algae, so called because of the characteristic colouration of most species, are the simplest known plants. Their procaryotic cell-type means that they are closely related to the bacteria (above) although until recently they have been classified together with the (eucaryotic) algae (*p. 42*). They are, in effect, bacteria that function as plants. Blue-green algae are amongst the earliest life-forms known on earth and were responsible for the production of curious fossil structures known as stromatolites. Some fossil stromatolites are believed to be more than 2500 million years old, yet their descendants are still to be found living in shallow tropical seas. Blue-green algae occur in all wet or aquatic environments ranging from the polar ice-caps to thermal springs with temperatures up to 90°C. They are probably most commonly encountered as slimy, dark green encrustations on wet or submerged rocks, or as the occasional 'water-blooms' which discolour the surface of still waters, resembling spilt green paint.

Cyanophyte cells are very small, rarely being more than about 25 μm in diameter and often much less. They never possess flagella. Under the microscope, their contents are relatively featureless, except for an apparently granular texture. The photosynthetic pigments are diffused throughout the cell, not contained in discrete packages (plastids) as in eucaryotic plants, and although the typical colour is a rich blue-green, yellow, olive, brown, red or violet forms may occur. The cells often contain numbers of tiny blackish gas bubbles which cause many species to appear very dark, almost black in dense aggregations.

Apart from the normal vegetative cells, some genera develop differentiated cells, spores or heterocysts. Spores (*Fig. 1.8*) are dense, often elongated, thick-walled bodies; heterocysts (*Fig. 1.8*) are hollow, thick-walled cells with internal nodular thickenings. Reproduction is by cell division, spore production, or fragmentation in filamentous types. Sexual reproduction is unknown in this group.

The cells are either separate or joined end-to-end to form a filament or trichome, each cell or trichome being enveloped by a sheath of jelly-like mucilage, which can be very thick. Free-living filamentous species can move about by a gliding motion which is thought to be caused by mucilage secretion. Many species form aggregations or colonies of separate cells or trichomes, usually contained in a common mass of mucilage, formed by the mucilage sheaths coalescing. The texture of these colonies varies from soft and slimy to firm and leathery, or even stony. Colonies of most species are large enough to be seen with the unaided eye and may reach several millimetres or even centimetres across. They may be fixed to a substratum or free, benthic or planktonic. Some planktonic species are notorious for causing water blooms, a brief, seasonal superabundance of one species which, by depleting the oxygen content of the water, or releasing toxins, may poison aquatic life and endanger animals that come to drink. Water blooms are most likely to occur in late summer and are often a result of eutrophication.

About 1500 living species of Cyanophyta are known. The three main freshwater orders and some of the more important genera are described below.

Order CHROOCOCCALES

Cells separate, solitary or forming aggregations; not producing spores.

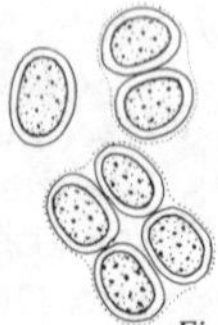

Fig. 1.3

Chroococcus *Pl. 10, Fig. 1.3*
Free-living, single cells or clusters of two or four, with a thick mucilage sheath. Cells often large, up to about 30 μm; colour variable, blue-green, brown or bluish. Common, usually found in small pools or ditches.

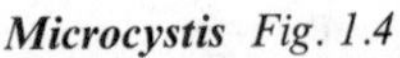

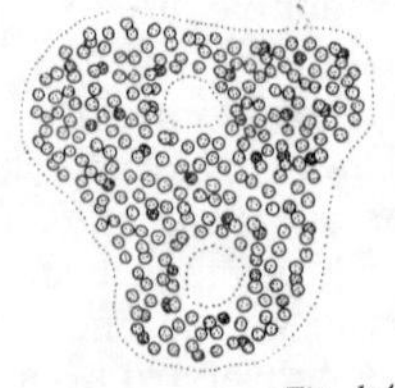

Fig. 1.4

Microcystis *Fig. 1.4*
Small, spherical cells, about 5 μm diameter, closely packed in irregular masses of mucilage. Common in still waters, in the plankton, or lying free on the substratum; sometimes forms blooms.

***Aphanothecae* (*Coccochloris*)**
Small, oval or elongated cells, up to 8 μm long, loosely aggregated in a soft mass of mucilage. In marshy pools or on wet rocks.

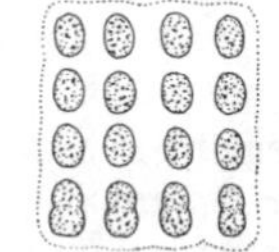

Fig. 1.5

Merismopedia *Fig. 1.5*
Flat, regularly-shaped colonies of small oval cells up to 10 μm long, formed by the cells dividing in strictly one plane. In the plankton of still waters.

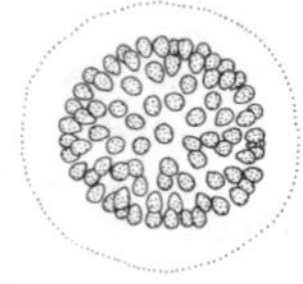

Fig. 1.6

***Gomphosphaeria* (*Coelosphaerium*)** *Fig. 1.6*
Cells spherical or pear-shaped, aggregated into a ball and embedded in a thick layer of mucilage; large colonies often become irregular in shape; whole colony up to 100 μm or more. Common in the plankton of lakes, sometimes forming blooms.

Order CHAMAESIPHONALES

Cells usually solitary, dividing internally to produce numerous, small, rounded spores.

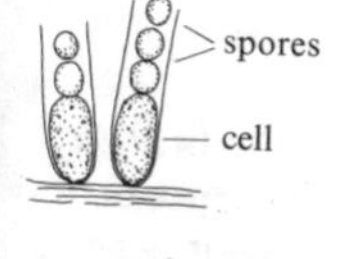

Fig. 1.7

Chamaesiphon *Fig. 1.7*
Single elongated cells up to 10 μm long in a thin tubular sheath, attached by one end to a substratum; many small spherical spores are released from the distal end, temporarily accumulating in the sheath. Large numbers often build up to form a substantial, blackish, slimy encrustation on rocks or plants such as *Cladophora*.

Order OSCILLATORIALES

Cells joined end to end (with no intervening mucilage) forming cylindrical filaments (trichomes); spores and heterocysts often present.

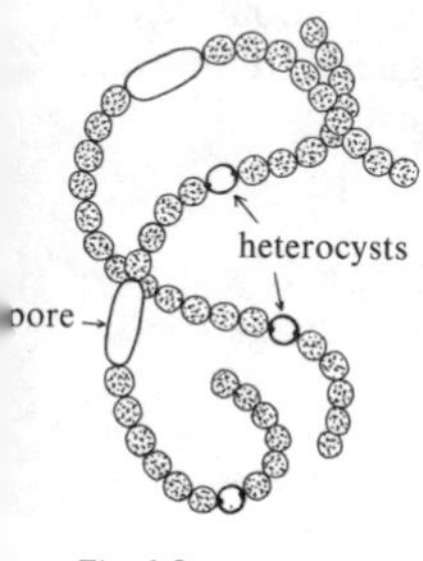

Fig. 1.8

Anabaena *Pl. 11, Fig. 1.8*
Free-living, unbranched filaments composed of rounded cells up to 10 μm diameter, resembling a string of beads; elongated spores and heterocysts (never adjacent) are usually present. A common planktonic form in still waters, with many different species; may form blooms.

Cylindrospermum
Similar to *Anabaena* but spores are formed adjacent to heterocysts.

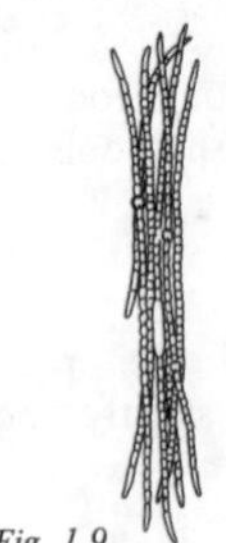
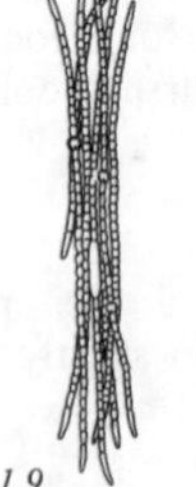

Fig. 1.9

Nostoc
Trichomes similar to *Anabaena* but aggregated in a mass of mucilage; colonies may exceed 1 cm across, are usually attached to a substratum, and vary in colour from yellowish to deep blackish green. On stones in rivers, or in damp places out of the water.

Aphanizomenon *Fig. 1.9*
Stiff-looking filaments, resembling *Anabaena* but with more elongated spores, gathered into loose bundles. A common planktonic form, often forming blooms.

Fig. 1.10

Gloeotrichia *Pl. 12, Fig. 1.10*
Each filament tapers to a point at the distal end and bears a heterocyst at the other; these aggregate to form roughly spherical, fuzzy-looking colonies several millimetres across. A common form in the surface plankton of still waters; may cause blooms.

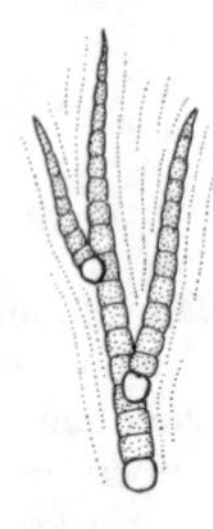

Fig. 1.11

Rivularia *Fig. 1.11*
Filaments similar to *Gloeotrichia* but frequently branching; aggregated filaments are arranged in a firm, globular mucilage matrix several millimetres across. Lives attached to stones or similar substrata, usually in running waters, particularly in upland streams where it may become encrusted with lime-scale.

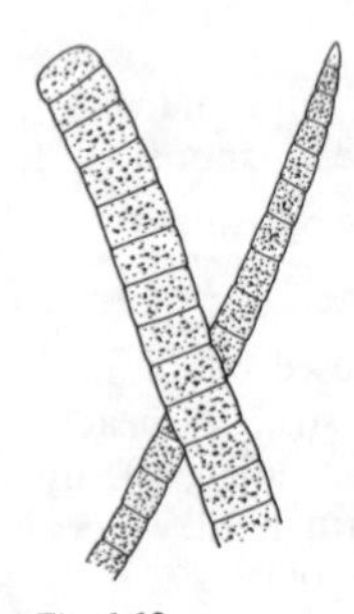

Fig. 1.12

Oscillatoria *Pl. 13, Fig. 1.12*
Plain, unbranched filaments without spores or heterocysts, no obvious sheath of mucilage, up to about 20 μm diameter. This and the following two genera are difficult to distinguish; each has numerous species.

Lyngbya
Resembles *Oscillatoria*, but filaments covered by a substantial (but not necessarily easily-seen) mucilage sheath.

Phormidium
Similar to *Lyngbya* but sheaths of adjacent filaments tend to coalesce.

[The above three genera of simple, filamentous blue-

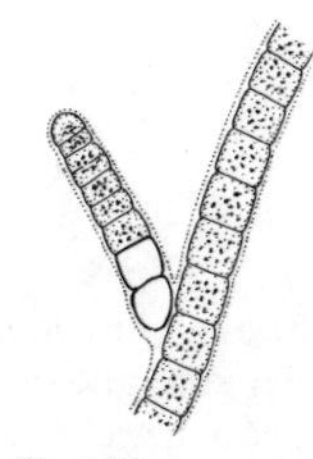

green algae are common in waters of all types, particularly in the autumn when they may become abundant enough to form mats covering the substratum. If disturbed (e.g., by a collector's net) they may dissociate and rise into suspension in the water, but they soon re-settle and spread out evenly, by creeping, to form a mat again. This strange phenomenon is easily observed in collected samples.]

Fig. 1.13

Tolypothrix *Fig. 1.13*
Cylindrical filaments, up to 10 μm diameter, with frequent branching, each branch with one or two heterocysts at its base. Usually found growing on other vegetation in ponds and streams.

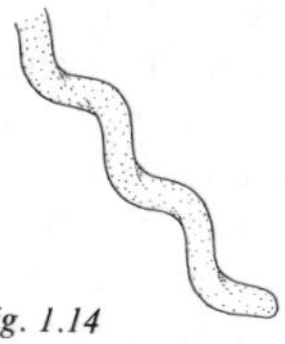

Fig. 1.14

Spirulina *Fig. 1.14*
Slender filaments, 1–5 μm diameter, wound into a fixed spiral shape, variable according to species; divisions between cells obsolete and not visible. *Arthrospira* is similar, but cross-divisions between cells are present. Both are planktonic forms of still waters.

References: Round, 1981. Edmondson, 1966.

CHAPTER 2 Fungi and Moulds

Fungi are non-motile organisms that reproduce by spores. Although bearing a superficial resemblance to plants, they lack photosynthetic pigments and do not photosynthesize. Instead, fungi are either parasitic – feeding on the cells or tissues of living plants and animals, or saprophytic – feeding on dead organisms; some are both. More than 80,000 species of fungi are known, most of which occur in terrestrial environments, but a few are always aquatic.

The complete fungus body is called the thallus. It consists of a three-dimensional network (the mycelium) of fine, usually microscopic, branching filaments (hyphae), which grows on and through a substratum that is usually the food organism. The hyphae secrete enzymes that break down the food externally and absorb the nutrients directly through their walls. When the fungus is mature, erect fruiting bodies grow out from the mycelium, borne on hyphal stalks. These give rise to sexually or asexually produced spores, often in great quantities. The best-known examples of fruiting bodies are the mushrooms and toadstools of our fields and woods. But most are much smaller than these; the colourful speckles that announce the presence of mildews and moulds on rotting foodstuffs or damp walls are also the fruiting bodies of fungi. Fruiting bodies are only present for part of the time, and when absent, fungi are remarkably inconspicuous, in spite of their abundance and diversity.

Apart from their greater size, fungi have many parallels with the bacteria. Many forms are injurious to man, causing disease, sometimes to man himself (ringworm, athletes' foot, etc.) but more often attacking his crops, domestic animals, and other possessions. These are the various rots, mildews, blights, moulds, rusts, smuts, etc. But not all fungi are bad. Many provide man with the means to combat disease, providing antibiotics such as penicillin. Yeasts are fungi that change sugars to alcohol by fermentation, a property whose importance requires no elaboration! And of course yeasts have many uses in cookery. Mushrooms and other large fungi are delicious to eat, although some have potent effects on man's nervous system, being poisonous or hallucinogenic. Some fungi are involved in symbiotic relationships with plants, aiding the plant body to absorb nutrients from the soil, a function that is vitally important to some plant groups such as the orchids. Lichens are 'plants' formed by a symbiotic association between various species of fungi and algae; no lichens are aquatic.

Three groups of fungi are common in fresh water but accurate identification is very difficult and will not be attempted here.

The chytrids (phylum Chytridiomycota) are obscure, microscopic parasites of algae and other fungi. The thallus consists of the original spore body, which is usually spherical and may be situated inside or outside the host cell,

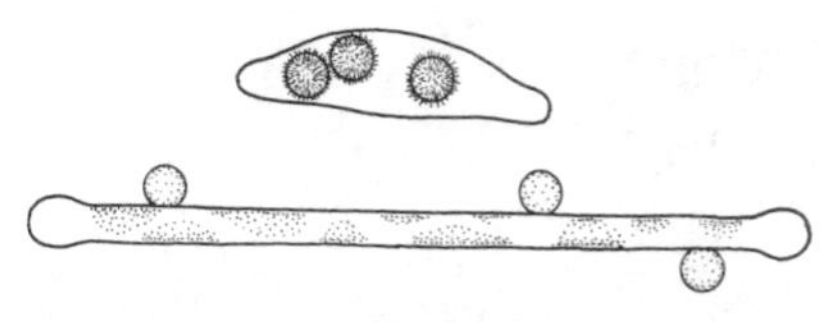

Fig. 2.1

from which a mycelium of fine hyphal filaments invades the host cell (*Fig. 2.1*). Chytrids can be found by carefully examining suitable algal hosts – diatoms and desmids are particularly liable to infection – under the microscope. It is thought that chytrids are probably an important controlling factor in populations of planktonic algae such as Asterionella.

Fungi belonging to the phylum Deuteromycota are called imperfect (*fungi imperfecti*) because they have no known sexual phase. They reproduce asexually by means of spores borne in fruiting bodies called conidia. Imperfect fungi of the order Hyphomycetes are very common saprophytes in freshwaters where they are usually found in decaying leaves, especially in running water. They are all effectively microscopic and the only visible parts are the conidiophores – the hyphal stalks that bear the conidia. The mycelium is hidden within the substratum. Conidiophores vary greatly in form but most bear three or four radial branches at their tips, each branch terminating in a conidium. The entire structure is no more than about 100 μm across and less than 1 mm tall. (*Fig. 2.2*).

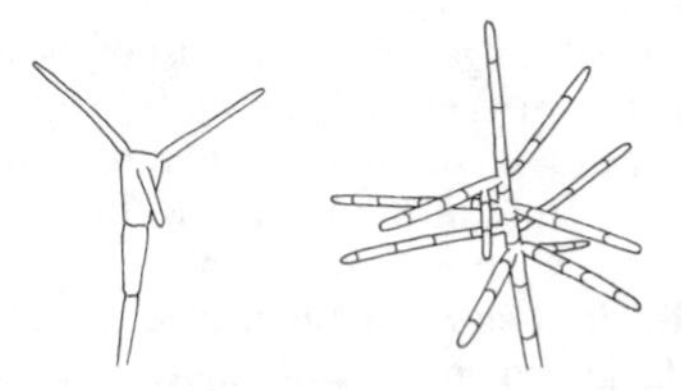

Fig. 2.2

The water moulds, *Saprolegnia* and related genera belonging to the order Saprolegniales (phylum Phycomycota), are the only conspicuous freshwater fungi. They appear as fluffy, cotton-wool-like growths, on dead animals (*Pl. 14*) and plants, and may also occur as secondary infections on diseased or wounded animals, particularly vertebrates. Spores are produced asexually in the cylindrical sporangia that appear at the tips of the hyphal stalks. These are 'perfect' fungi and reproduce sexually by a kind of conjugation between hyphal tips of different mycelia, resulting in the production of flagellated swimming zoospores. The occurrence of flagella in this group has led some authors to remove it from the Kingdom fungi altogether, but it seems more likely that this is simply an adaptation to aquatic life.

References: Edmondson, 1966. Ingold, 1975.

CHAPTER 3 Algae
(10 Phyla)

The term algae covers a miscellaneous collection of all the relatively simple (eucaryotic) plant organisms that do not belong to the groups of 'higher' plants. Bryophyta (liverworts and mosses), Pteridophyta (ferns and horsetails), and Angiospermophyta (flowering plants) are the phyla of higher plants found in fresh water; all other aquatic plants are algae, which range from numerous microscopic forms to simple macroscopic 'pondweeds' and the large seaweeds.

Algae abound in all aquatic or wet environments. They are the dominant plants in the sea – all 'seaweeds', with a few exceptions, are algae, and here they often reach a great size, some of the giant kelps attaining a length of 50 m or more. In fresh water most of the algal species are microscopic, unicellular or colonial organisms, but a fair proportion form plant-bodies large enough to be seen with the naked eye. Some, such as various filamentous algae and the stoneworts, grow in substantial masses and may, for a short period at least, dominate the flora in some waters. Algae can be found in almost any place that holds water, however small or impermanent: puddles, gutters and water butts, bird baths, abandoned crockery, and similar places are all likely to contain an algal flora.

The single-celled algae include genera from most of the ten or so freshwater algal phyla, and exhibit a great range of shapes, colours and sizes. Some are mobile, swimming by means of flagella, or 'gliding' over the substratum; others are non-motile – drifting in the plankton, fixed to a substratum, or lying free. Colonial algae also show a remarkable range of forms: globular clusters, often embedded in a sphere of mucilage; simple or branched filaments of cylindrical cells joined end to end; encrustations on rocks or other plants; tufts, skeins, or cushions of more complex filaments; gelatinous green blobs; irregular green tubes; and many others.

Because they embrace such a mixture of types, algae are difficult to define simply as a group, but in practice they are rarely likely to be confused with many higher plants. The unicellular, colonial and filamentous types are distinct enough, but stoneworts, which are highly organized algae, resemble higher plants in many respects. Their whorls of cylindrical 'leaves' are superficially similar to those of some higher plants such as *Myriophyllum* (milfoil) and *Ceratophyllum* (hornwort). But close examination will reveal that stoneworts consist of relatively few very large cells (often several millimetres long), in contrast to the numerous very much smaller cells of the higher plants. Some of the more simply-structured liverworts (e.g., *Riccia*) might be mistaken for algae, as might the tiny floating fern *Azolla* and the almost microscopic rootless duckweed *Wolffia*. But if studied under the microscope, or even with a good hand lens, these will be seen to comprise solid blocks or sheets of cells, a type of structure that rarely occurs in the algae.

The cell walls of most algae are rigid and hence their shape is fixed, but a few genera, notably *Euglena*, are plastic and can change their shape. Many of the unicellular and colonial forms possess flagella, one, two or four being present. These are a useful identification feature, but unfortunately are not always easy to see under the microscope. (If you wait for the specimen to become moribund, the flagella will eventually slow down enough to be visible before it dies.) Many algae secrete a slimy substance, mucilage, which makes them slimy to the touch. Although it is virtually invisible under the microscope, mucilage can be revealed by introducing a very dilute solution of indian ink to the water. This will tint the water but not the mucilage, and the latter shows up as a colourless zone around the subject.

An algal cell contains many different structures, but few of these can be observed consistently without the use of special preparatory techniques. The most obvious and important feature is the plastid or chromatophore, which contains the photosynthetic pigments. (Plastids are often called chloroplasts but strictly, this term should only be applied to green-pigmented ones.) Plastids vary in number, shape and colour in different algae. The plastid colour is often characteristic of a phylum but may not be reflected in the colour of the whole plant: the 'red' algae *Lemanea* and *Batrachospermum*, for instance, are usually dark green or greyish-purple to the eye. The outlines of plastids are not always clear, as in many Chlorophyta, but in this case, the bright green colour alone is diagnostic of the phylum. Plastids in the golden-brown algae (Chrysophyta) are often inconspicuous in living specimens, which thus appear to be colourless, and suggest that one is dealing with a flagellated protozoan (*p. 111*). Some algae possess an eyespot or stigma, usually bright red and easily seen if present. Pyrenoids – structures concerned with starch production – usually appear as colourless spots in the plastids, but cannot always be seen.

Some algae do not obtain food by photosynthesis and these are a potential source of confusion. Genera such as *Astasia* and *Chilomonas* are clearly very closely related, by their structure, to photosynthesizing forms – *Astasia* to *Euglena*, *Chilomonas* to *Cryptomonas* – but they are unpigmented and feed on other organisms. Despite this animal-like mode of nutrition, their affinities clearly lie with the plant kingdom and they are classified as such.

Algae reproduce asexually by various forms of division, fragmentation, or spore production. Sexual reproduction occurs either by conjugation or production of separate male and female cells, antheridia and oogonia respectively. Apart from the stoneworts, the sexual cells of algae are small, inconspicuous and short-lived; they are not mentioned further in the text.

Algae are predominantly seasonal organisms. Many species can only be found at certain times of the year, others are present for much of the time but their numbers fluctuate, with massive abundances interspersed by periods of relative scarcity. Some species may cause the water to turn green (or other colours) by their sheer numbers, especially in small or confined waters, but these plagues of 'green water' rarely last long. Planktonic species are more seasonal than, say, some of the filamentous algae, and their seasons may last only a week or two in each year, with a succession of species waxing and waning throughout the year. There are characteristic spring forms which appear early in the year, thriving in cold-water conditions but declining as warmth returns. But some, such as *Batrachospermum*, may persist all year round in cold spring outlets. Late spring and summer see the greatest abundance and diversity of species, particularly green algae which require a high light level. In autumn the blue-green algae (*p. 35*) tend to come into their own, and often cover the formerly green and lush underwater landscape with a sombre inkiness that lingers into the winter.

Collecting algae presents few problems and the methods described on *p. 346* will produce a good variety. A useful way to concentrate a selection of different forms is to squeeze out a mass of filamentous algae, or other fine-leaved plant life over a small container. This will often yield a rich harvest of algae and many other small aquatic organisms. Collections made specifically for algae should be transported carefully. Leave a large air space in the container, keep it cool and do not overcrowd the specimens. Many algae deteriorate rapidly and die unless great care is taken over handling them. Remember, also, to remove any animals (small crustaceans in particular) that might eat them.

Unlike the other major groups in this book we have not treated the algae phylum by phylum. Accurate identification of algae depends largely on the chemistry of their pigments and details of internal structure that would not be practical to use here. Instead we have grouped them according to their gross external appearance into six groups, characterized on *p. 46*. Each group contains a selection of the more common genera which can be identified with reasonable certainty from the drawings and descriptions. It is emphasized that this selection is only a small proportion of the total number of algal genera it is possible to find in fresh water, so identifications made this way should be treated with caution. Further, it should be pointed out that many genera contain large numbers of species which may vary in appearance from the examples illustrated.

The classification of the algal genera in this book is shown below; this follows Round (1981).

Phylum EUGLENOPHYTA — Unicellular, with many green plastids, or unpigmented; one main flagellum.
Euglena, Astasia, Phacus, Trachelomonas.

Phylum CHLOROPHYTA — Unicellular, colonial, filamentous, or plant-like; plastids bright green, variable in number, often single and large; some genera are unpigmented; flagella present or absent.

usually two or four if present. This is by far the largest phylum of algae occurring in fresh water.
Haematococcus, Chlamydomonas, Pandorina, Eudorina, Volvox, Gonium, Chlorella, Ankistrodesmus, Actinastrum, Micractinium, Coelastrum, Pediastrum, Scenedesmus, Dictyosphaerium, Gonatozygon, Pleurotaenium, Closterium, Euastrum, Micrasterias, Xanthidium, Cosmarium, Staurodesmus, Staurastrum, Desmidium, Cosmocladium, Ulothrix, Microspora, Draparnaldia, Stigeoclonium, Aphanochaete, Oedogonium, Cladophora, Mougeotia, Spirogyra, Zygnema, Cylindrocapsa, Chaetophora, Tetraspora, Hydrodictyon, Coelochaete, Enteromorpha.

Phylum XANTHOPHYTA

Unicellular, colonial, or filamentous, with numerous small, yellowish-green plastids; flagella, if present, single or two unequal.
Tribonema, Vaucheria, Botrydium.

Phylum PRASINOPHYTA

Unicellular, with a single, large green plastid divided into four lobes; four equal flagella.
Pyramimonas.

Phylum CHRYSOPHYTA

Unicellular or colonial; usually with two golden-brown plastids; flagella, if present, two unequal.
Ochromonas, Mallomonas, Synura, Uroglena, Dinobryon.

Phylum CRYPTOPHYTA

Unicellular, with one or two usually brown plastids, sometimes unpigmented; two unequal flagella.
Cryptomonas, Chilomonas.

Phylum DINOPHYTA (Dinoflagellates)

Unicellular, with numerous tiny brown or reddish plastids, or unpigmented; two flagella, originating from a groove around the cell's equator, one lies in this groove, the other trails from a longitudinal groove.
Gymnodinium, Peridinium, Ceratium.

Phylum BACILLARIOPHYTA (Diatoms)

Unicellular or colonial, with two large or numerous small brown plastids; flagella absent; cell wall made of silica in two close-fitting halves.
Cyclotella, Stephanodiscus, Melosira, Asterionella, Diatoma, Meridion, Fragilaria, Synedra, Tabellaria, Cocconeis, Cymbella, Gomphonema, Gyrosigma, Navicula, Nitzschia, Pinnularia, Attheya, Rhizosolenia.

Phylum CHAROPHYTA (Stoneworts)	Green algae which form large plants consisting of jointed stems with whorls of cylindrical branchlets. *Chara, Nitella, Tolypella.*
Phylum RHODOPHYTA (Red algae and green seaweeds)	Unicellular, colonial, encrusting, or plant-like; pigments red but some freshwater plants are green; flagella absent. *Batrachospermum, Lemanea, Hildenbrandia.*

Definitions of the Artificial Groups of Algae Used in this Book

1. Flagellated cells, usually free-swimming, solitary or colonial; plastids usually green or some shade of brown; mostly microscopic but some colonies may exceed 1 mm. *(p. 46)*
2. Non-flagellated, non-motile green cells, solitary or colonial; not fixed to a substratum; not forming filaments (excepting certain desmids: desmid cells are divided into two mirror-image halves, which in filamentous species are separated by a circular constitution; true filamentous algal cells, groups 4 below, are not constricted like this); mostly microscopic. *(p. 50)*
3. Diatoms (phylum Bacillariophyta only). Non-flagellated, solitary or colonial brown cells; cell wall made of silica, in two parts that fit together like a pill box and its lid; some can move by 'gliding': microscopic. *(p. 54)*
4. Filamentous green algae. Chains of cylindrical cells joined end-to-end, green or yellow-green, branched or unbranched, attached or free, never with flagella; macroscopic or microscopic. (NB Some desmids form filaments – see group 2 above; filamentous diatoms are brown.) *(p. 57)*
5. Stoneworts (phylum Charophyta only). Macroscopic plants up to 50 cm tall, consisting of very large cylindrical cells (several millimetres long) which form a jointed stem with a whorl of cylindrical branchlets arising from each joint. *(p. 59)*
6. Other forms that do not fit the categories above. These are mostly macroscopic plants with a wide range of growth forms, fixed, encrusting, or free-floating; usually some shade of green or red in colour. *(p. 61)*

Descriptions of the Genera of Freshwater Algae

Group 1 Flagellated unicells or colonies, usually green or brown. Members of six phyla are represented in this group.

Phylum EUGLENOPHYTA	***Euglena*** *Pl. 15, Fig. 3.1* Typically elongated single cells with one long flagellum visible, many grass-green plastids, and a prominent red stigma; up to about 100 μm long but

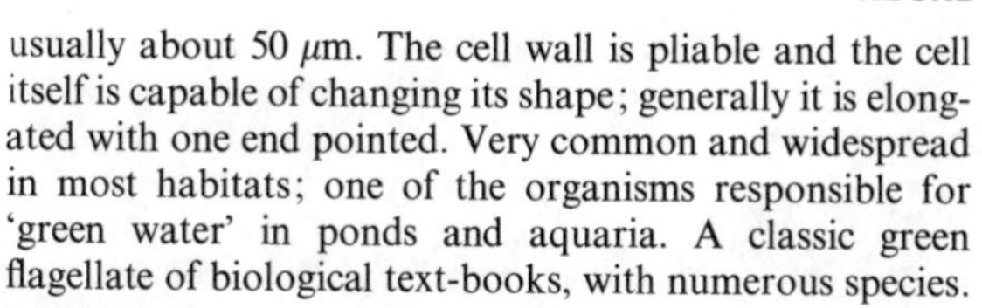

usually about 50 μm. The cell wall is pliable and the cell itself is capable of changing its shape; generally it is elongated with one end pointed. Very common and widespread in most habitats; one of the organisms responsible for 'green water' in ponds and aquaria. A classic green flagellate of biological text-books, with numerous species.

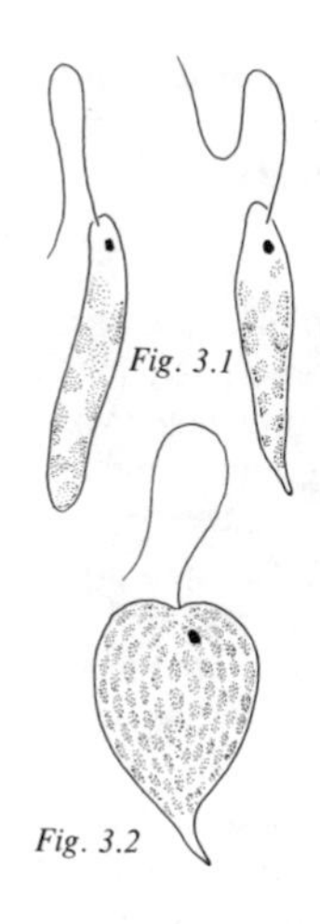

Fig. 3.1

Fig. 3.2

Astasia This is structurally almost identical to *Euglena* but lacks photosynthetic pigments, hence it is colourless and functions as an animal. Usually found in small bodies of water.

Phacus *Fig. 3.2* Green solitary cells with one flagellum. Cell wall rigid, the cell itself flattened with the posterior end drawn out to a point. Several species: in some the cell is twisted, in others it is divided longitudinally into lobes which may be twisted. Up to about 50 μm long. Frequent in still waters, bogs and marshes.

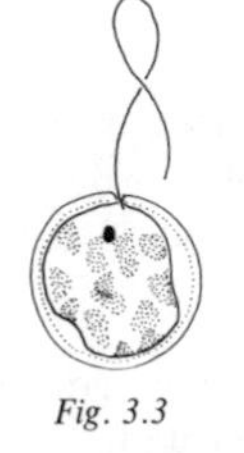

Fig. 3.3

Trachelomonas *Fig. 3.3* A small spherical green cell, up to 30 μm long, with one long flagellum. The cell is enclosed in a brownish outer 'shell' which usually obscures the green pigments.

Phylum CHLOROPHYTA

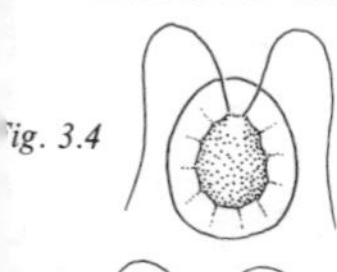

Fig. 3.4

Haematococcus *Fig. 3.4* A small solitary cell, up to 25 μm long, with two long flagella; cell contents separated from the wall by a clear layer. Normally green but in impoverished habitats, e.g., clean rainwater, may become bright red. In small standing waters, sometimes tinting the water green or red.

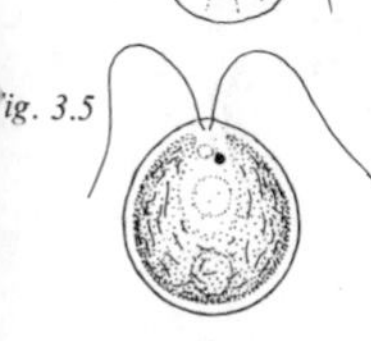

Fig. 3.5

Chlamydomonas *Pl. 16, Fig. 3.5* A solitary oval or spherical cell with two long flagella, a single large green plastid and a red stigma. Up to 20 μm diameter. Very common in standing waters amongst decaying vegetation. Like *Euglena*, it is a 'text-book' green flagellate.

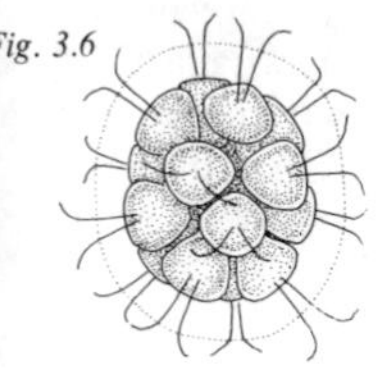

Fig. 3.6

Pandorina *Fig. 3.6* A colony of green, *Chlamydomonas*-like cells, usually sixteen in number, clustered centrally within a ball of mucilage. Colonies up to 50 μm across. Frequent in still waters.

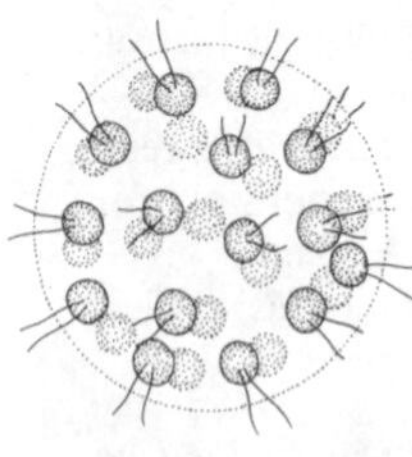

Fig. 3.7

Eudorina *Fig. 3.7* A colony of green, *Chlamydomonas*-like cells, usually thirty-two, spaced out on the periphery of a sphere of mucilage. Colony up to 100 μm or more in diameter. Common in still waters.

Volvox *Pl. 17* Spherical colonies consisting of upwards of 500 small, green, bi-flagellate cells more or less equally spaced on the surface of a hollow globe of mucilage. Most colonies contain daughter, or even grand-daughter, colonies within them. Large colonies may exceed 1 mm diameter. Frequent in still or slow flowing waters. *Volvox* is one of the most beautiful objects to be found in fresh water and is large enough to be easily visible to the naked eye.

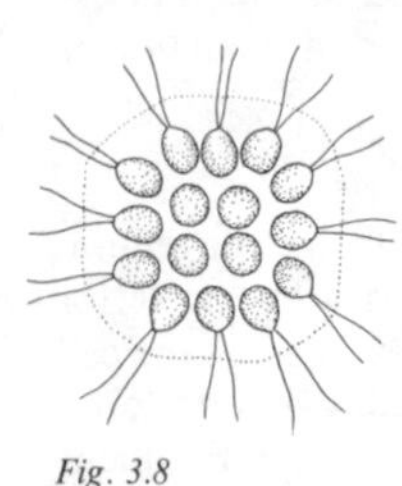

Fig. 3.8

Gonium *Fig. 3.8* Colonies of four to sixteen green, bi-flagellated cells arranged in a flat square in a mucilage matrix. Colonies up to 100 μm across. Fairly common in still or slow flowing waters.

Phylum
PRASINOPHYTA

Pyramimonas *Fig. 3.9* A solitary, green, pear-shaped cell divided into lobes by four longitudinal grooves; with four flagella; up to 30 μm long. Locally frequent in small standing waters.

Phylum
CHRYSOPHYTA

Ochromonas *Fig. 3.10* Solitary, with one or two golden-brown plastids and two flagella of unequal length; up to 25 μm long. Frequent in ponds and small standing waters.

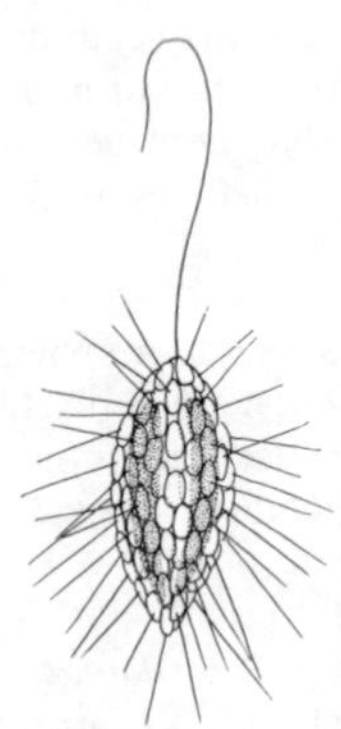

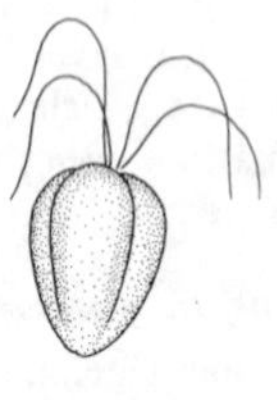

Fig. 3.9

Fig. 3.10

Fig. 3.11

Mallomonas *Fig. 3.11* A solitary, elongated-oval cell with one flagellum and one or two golden-brown plastids; up to 30 μm long. The cell wall is covered with numerous tiny scales of silica; in some species these bear long siliceous needles. Found in the plankton of still waters of all sizes.

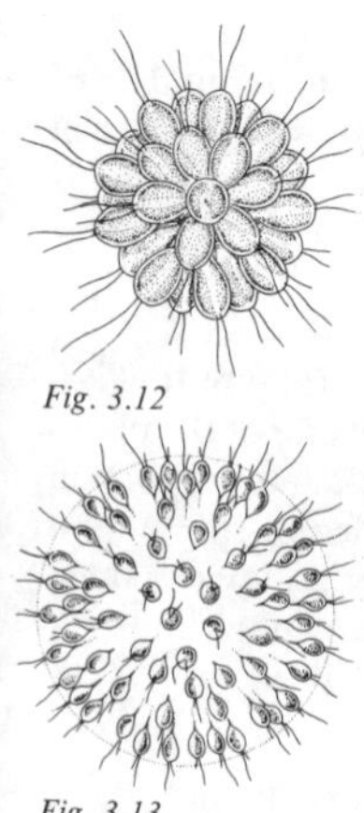
Fig. 3.12

Fig. 3.13

Synura *Fig. 3.12* A colony of elongated pear-shaped cells, each with two golden-brown plastids and two unequal flagella, joined by their narrow ends to form a ball up to 400 μm diameter. Common in shallow ponds and ditches in the summer.

Uroglena *Fig. 3.13* Globular colonies consisting of many small, golden-brown, unequally bi-flagellated cells spaced out at the surface of a ball of mucilage. Individual cells about 10 μm long; whole colony up to at least 500 μm across. Frequent, but rarely abundant, in the plankton of ponds and lakes.

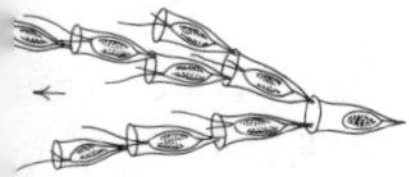

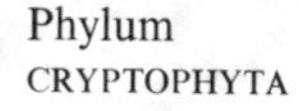
Fig. 3.14

Dinobryon *Fig. 3.14* Distinctive, irregularly branching colonies of clear 'shells', each containing a golden-brown cell with two unequal flagella. Shell shape varies according to species but they are usually in the form of a tapering cylinder resembling a stemless wine glass. Shells up to 40 μm long; colonies may consist of 20 or more shells. A free-swimming planktonic form (that moves, somewhat surprisingly, 'blunt' end first), frequent in still waters.

Phylum
CRYPTOPHYTA

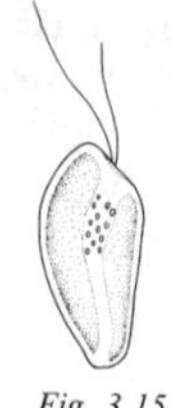
Fig. 3.15

Cryptomonas *Fig. 3.15* A solitary, asymmetrical cell with two unequal flagella and a distinctive gullet-like invagination at the anterior end; two olive or brownish plastids are present; up to 50 μm long. Common in ponds and ditches.

Chilomonas Similar to *Cryptomonas* but colourless, lacking photosynthetic pigments. A very common genus in waters rich in decaying organic matter.

Phylum
DINOPHYTA
(Dinoflagellates)

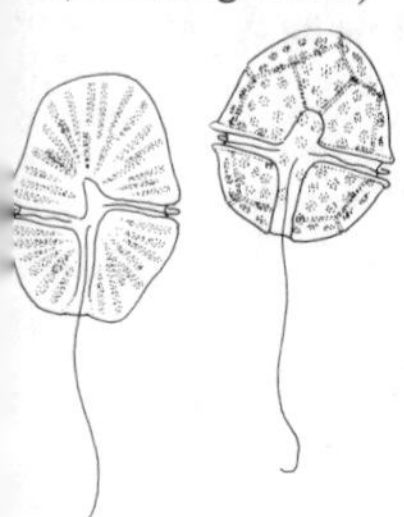
Fig. 3.16 Fig. 3.17

Gymnodinium *Fig. 3.16* Solitary, oval cells with two flagella attached to one side: one emerges posteriorly from a longitudinal groove, the other lies in a groove around the middle of the cell. The cell's surface is smooth and unfaceted (see genera below). Numerous species, most possessing many small brown plastids, but some are colourless; 10–300 μm long. Common in the plankton of ponds, small pools and ditches.

Peridinium *Pl. 18, Fig. 3.17* Solitary, pear-shaped or roughly spherical cells with two flagella arranged as in

Gymnodinium. The cell wall is divided into a number of plates with conspicuous joints, so that the whole cell appears faceted. Numerous species, with brown plastids or colourless; up to about 50 μm long. Common in lakes and ponds.

Fig. 3.18

Ceratium *Fig. 3.18* Basically similar in structure to *Peridinium* but differs dramatically in that several of the plates are drawn out to form spikes. The length and proportion of these spikes vary greatly between different species and also from locality to locality. Several species, up to 500 μm long including spikes, usually with brown plastids. A common genus in the plankton of large lakes.

Group 2 Non-flagellated, non-motile green cells, solitary or colonial, never fixed to a substratum.

Phylum
CHLOROPHYTA

Fig. 3.19

Chlorella *Fig. 3.19* Small solitary cells, spherical or oval, each with a single large cup-shaped green plastid. Following division, two or four daughter cells may be temporarily imprisoned within the wall of the parent cell. Cells up to about 15 μm diameter. Numerous species occur in all freshwater habitats; may be responsible for 'green water'.

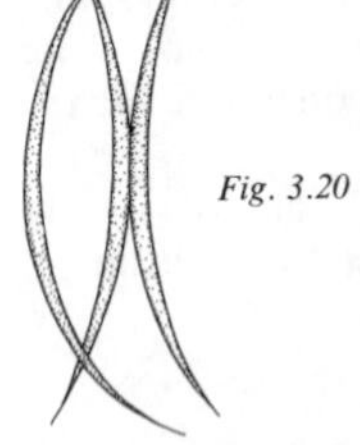
Fig. 3.20

Ankistrodesmus *Fig. 3.20* Slender, needle-like green cells drawn out to a sharp point at each end, straight or variously curved. Tend to form clusters, sometimes becoming twisted together. Up to at least 70 μm long. Widespread and common in small standing waters, may contribute to 'green water'.

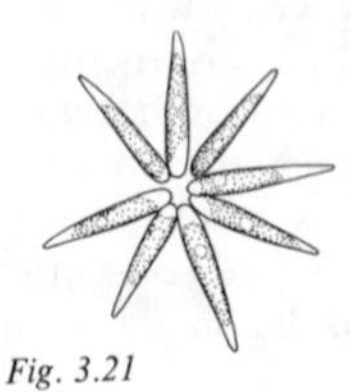
Fig. 3.21

Actinastrum *Fig. 3.21* Small star-like colonies of four or eight elongated cigar- or carrot-shaped cells joined loosely together by one end to a central point. Colonies about 50 μm across. Frequent in still or slow-flowing waters.

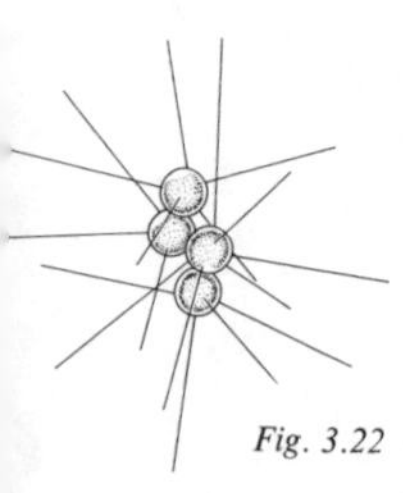
Fig. 3.22

Micractinium *Fig. 3.22* Small spherical green cells, up to 20 μm diameter, solitary or forming clusters, each one bearing several long, stiff hairs. Frequent in still or slow-flowing waters.

Coelastrum *Fig. 3.23* Spherical colonies, up to 150 μm diameter, consisting of many spherical or oval green cells joined together by their cell walls or thread-like struts. A planktonic form, mostly found in still waters.

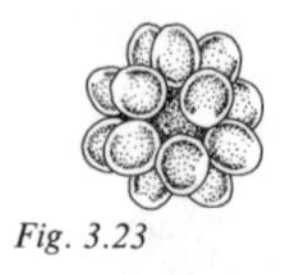
Fig. 3.23

Pediastrum *Pl. 19, Fig. 3.24* Colonies of irregular polygonal green cells united to form a flat disc. The peripheral cells are usually deeply notched so that the colony has a toothed outline. Colonies up to 50 μm or more across; widespread and common.

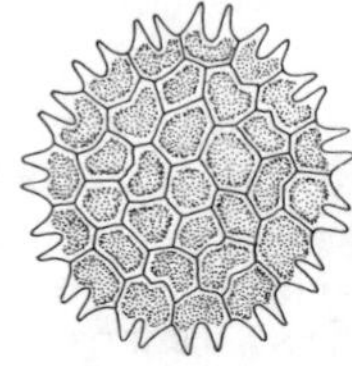
Fig. 3.24

Scenedesmus *Fig. 3.25* Small colonies of four or eight cells stacked together edge to edge, often with spines which are typically longer on the end cells. Numerous species that vary greatly in shape; individual cells up to 20 μm long. Very common in most habitats.

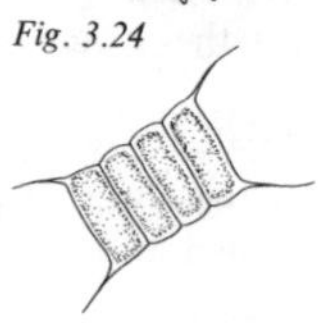
Fig. 3.25

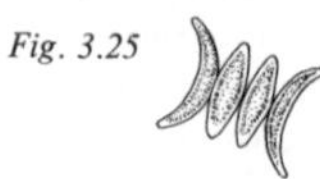

Dictyosphaerium *Fig. 3.26* Globular or irregular colonies of small spherical green cells in a very transparent mucilaginous matrix. The cells are joined to a common central point by very fine threads. Cells up to 20 μm diameter, colonies up to 200 μm or more; common in still or slow-flowing waters.

The remaining algae in this group are known as **desmids**. These are green unicells with rigid cell walls, widely varied in shape, that are divided into two equal halves (semicells) by a more or less distinct median constriction (sometimes absent) or a gap in the plastid(s). The two semicells are mirror images of each other. Most are solitary but a few genera are colonial.

Desmids are typical of neutral or acid-water areas with only a few genera being found in alkaline waters. They occur in all types of habitat from moorland pools and bogs to large lakes or rivers.

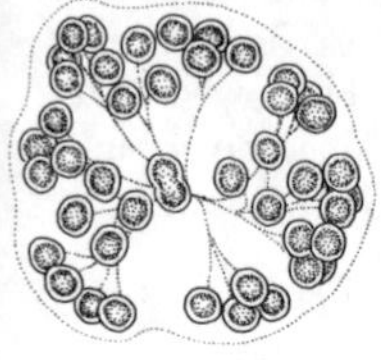
Fig. 3.26

Gonatozygon *Pl. 20, Fig. 3.27* An elongated, cylindrical desmid, often slightly bent in the middle which is not constricted; up to 300 μm long.

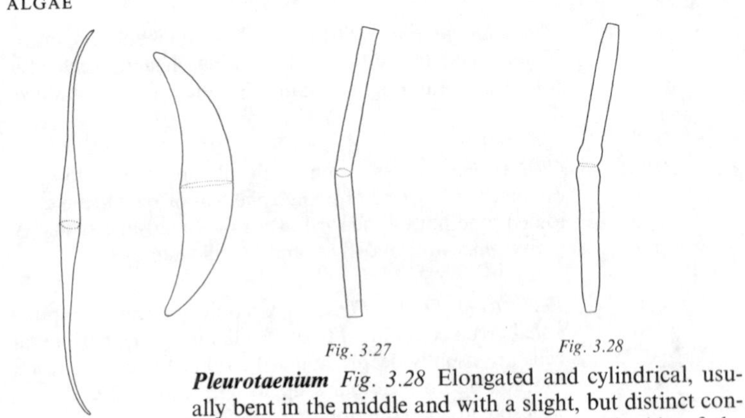

Fig. 3.27 Fig. 3.28

Fig. 3.29

Pleurotaenium *Fig. 3.28* Elongated and cylindrical, usually bent in the middle and with a slight, but distinct constriction; often with a slight bulge on each side of the middle; up to 700 μm long.

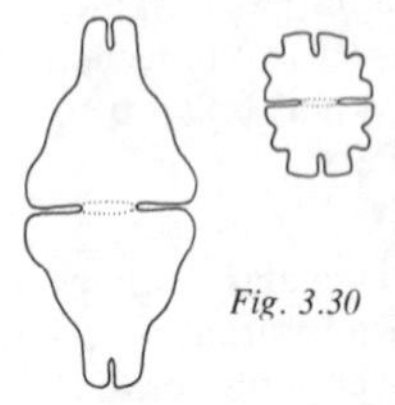

Fig. 3.30

Closterium *Pl. 21, Fig. 3.29* Cells are usually curved, but sometimes straight, and tapering to each end; no constriction between semicells; most species are crescent-shaped. This is a very large genus containing numerous species which vary greatly in size and shape; the largest may attain 800 μm. Very common, some of the species being found in alkaline waters.

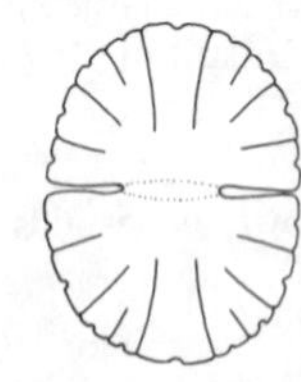

Fig. 3.31

Euastrum *Pl. 22, Fig. 3.30* Cell typically in the form of an elongated hexagon, always with a deep median constriction and variously ornamented with notches, tubercles, etc. The apex of each semicell is either cleft or concave. A large, very variable genus whose many species rarely exceed 200 μm in length, most being less than 50 μm. Found mostly in small standing waters.

Micrasterias *Fig. 3.31* Markedly flattened cells, rounded or oval in outline, with a variable number of deep notches or incisions; median constriction always deep. Never with tubercles, as in many *Euastrum* spp., but some species bear small spines. Numerous species, encompassing a wide range of shapes, up to 400 μm long. Common in small pools, bogs, etc.

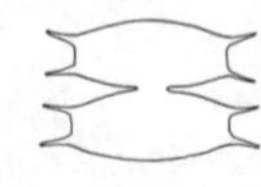

Fig. 3.32

Xanthidium *Fig. 3.32* The semicells are much broader than long, and separated by a deep constriction. The lateral lobes are armed with spines in various arrange-

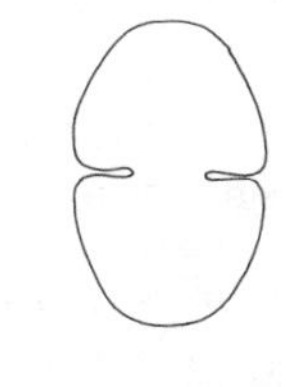

ments according to species and a tubercle is often present between the lobes. Whole cell up to 100 μm long. Common in the plankton of lakes but also in small standing waters.

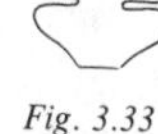

Fig. 3.33

Cosmarium *Fig. 3.33* Outline of cell variable: oval, rounded, hexagonal, lobed, etc., but never notched or incised, apart from the median constriction which is usually deep. This is the largest desmid genus with over 300 British species and many more European ones. The largest are up to 200 μm long but most are less than 50 μm. Found in most habitats, planktonic or benthic; a few species occur in alkaline waters.

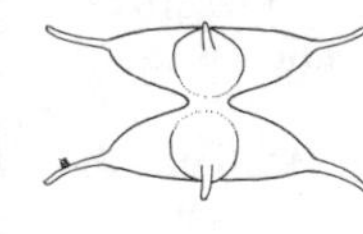

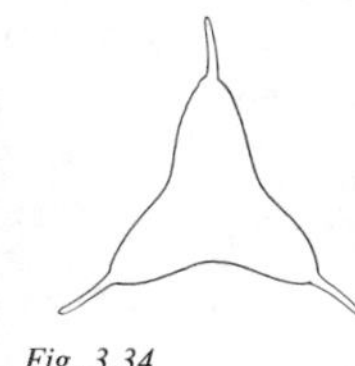

Fig. 3.34

Staurodesmus *Fig. 3.34* Each semicell is triangular or three-lobed in end view, each lobe being simply rounded or terminating with a nipple or spine. The cell wall is typically smooth and never spiny or granular (but it may be finely punctured). Many species are known, up to about 70 μm long; most occur in the plankton of large lakes.

This and the following genus sometimes occur in two-lobed or multi-lobed forms.

Staurastrum *Fig. 3.35* Basic form similar to *Staurodesmus* but the surface is always ornamented with spines (sometimes very long) or granulations. There are numerous species, ranging from those with plain, rounded lobes, to others in which the lobes are drawn out into long projections, usually one but sometimes several projections on each lobe. Up to 100 μm long but mostly less than 50 μm. A very common genus in the plankton of lakes.

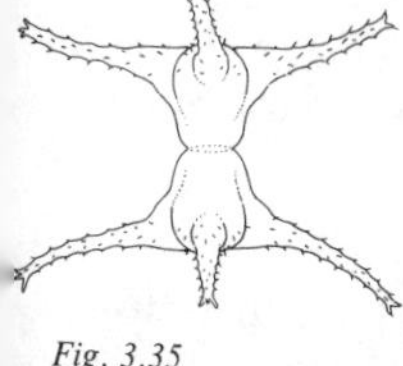

Fig. 3.35

Desmidium *Fig. 3.36* This is typical of the several desmid genera that form chain-like colonies. Each cell is broader than long with only a slight median constriction. Viewed from the end the outline may be elliptical, triangular, or rectangular, according to species. The whole chain is enclosed in a sheath of mucilage and is often twisted. Width of chain up to 50 μm. Filamentous desmids occur in the plankton or amongst other plant life in the shallows.

Cosmocladium Colonial: loose clusters of *Cosmarium*-like cells are embedded in a ball of soft mucilage, the cells being interconnected by strings of denser mucilage. Colonies up to 150 μm across; individual cells to 30 μm. Found in the plankton or amongst marginal plants.

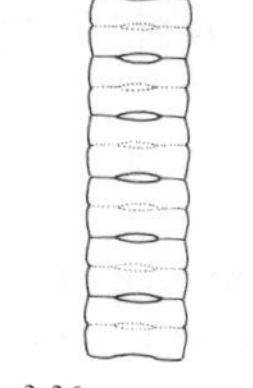

Fig. 3.36

Reference: Lind & Brook, 1980.

Group 3 Diatoms
Phylum BACILLARIOPHYTA

Unicellular, solitary, or colonial algae with two large, or many small brownish plastids. Each cell is enclosed within a capsule made of silica – the frustule – which consists of two halves or valves, which have rims that fit together like a pill box and its lid (*Fig. 3.37*). The region where the rims overlap is the girdle; nearly all diatoms (unless stated otherwise) are rectangular in girdle view. The frustule is very varied in shape, although constant within species, and its two valve faces are intricately and minutely sculptured. This sculpturing forms the basis of the classification of diatoms but most genera can be recognized by their overall shape. The drawings are valve-face views unless indicated by **g** (girdle view). Most diatoms secrete a sheath or stalk of mucilage, the latter for attachment to a substratum, and some are capable of a gliding movement when in contact with a solid surface.

Diatoms are very common in most freshwater habitats. At times they become so abundant as to form a dense, brownish encrustation over plants and other objects; those that secrete mucilage stalks may form, *en masse*, a slimy layer of soft jelly up to 1 cm thick over the substratum; and the planktonic forms occasionally discolour the water by their numbers.

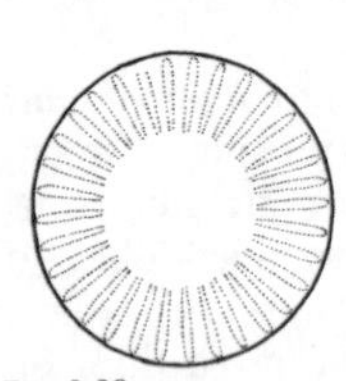

Fig. 3.37

Fig. 3.38

Cyclotella *Fig. 3.38* Pill-shaped diatoms up to 50 μm diameter. The sculpturing on the marginal parts of the valves differs from that of the central region (which may be plain), usually consisting of radially-arranged perforations. Common in the plankton of lakes.

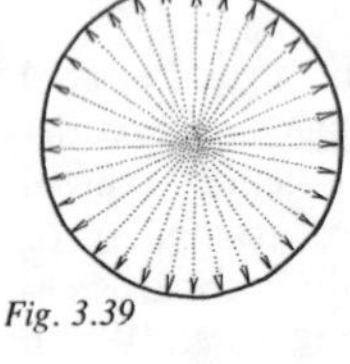

Fig. 3.39

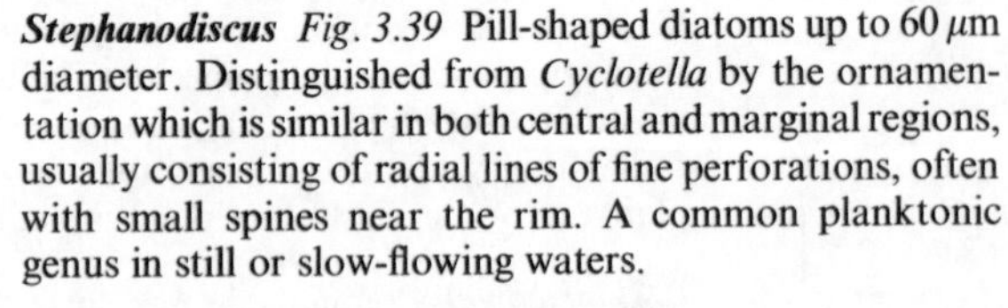

Stephanodiscus *Fig. 3.39* Pill-shaped diatoms up to 60 μm diameter. Distinguished from *Cyclotella* by the ornamentation which is similar in both central and marginal regions, usually consisting of radial lines of fine perforations, often with small spines near the rim. A common planktonic genus in still or slow-flowing waters.

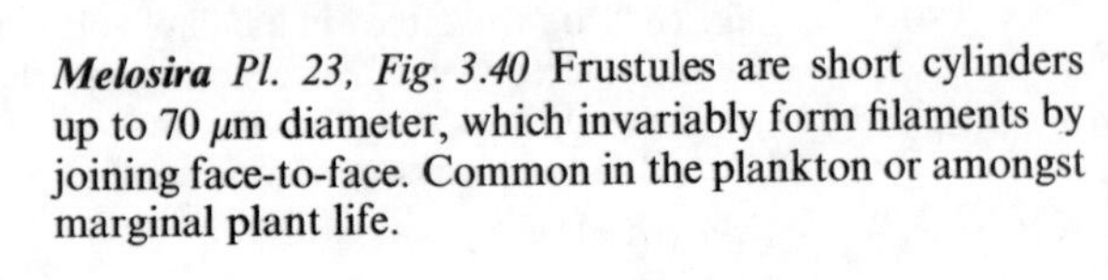

Fig. 3.40

Melosira *Pl. 23, Fig. 3.40* Frustules are short cylinders up to 70 μm diameter, which invariably form filaments by joining face-to-face. Common in the plankton or amongst marginal plant life.

Asterionella *Fig. 3.41* Frustule elongated, swollen a little at each end in valve view, up to 100 μm long. Characteristically forms star-like colonies by joining at the corners. A well-known and seasonally abundant genus in the plankton of large lakes. (Some other genera, e.g., *Tabellaria, Synedra*, may form star-shaped colonies but can be distinguished by their individual shapes.)

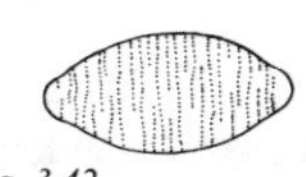
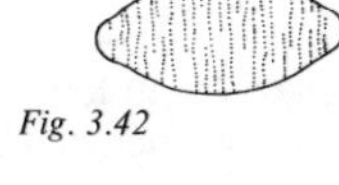

Fig. 3.42

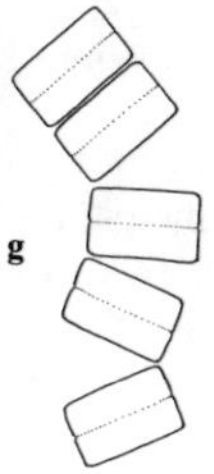

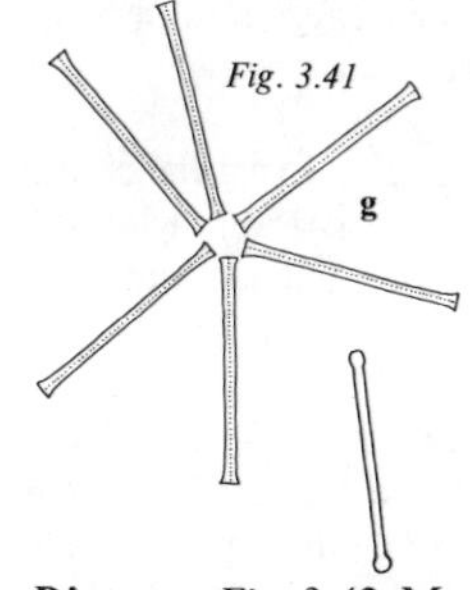

Fig. 3.41

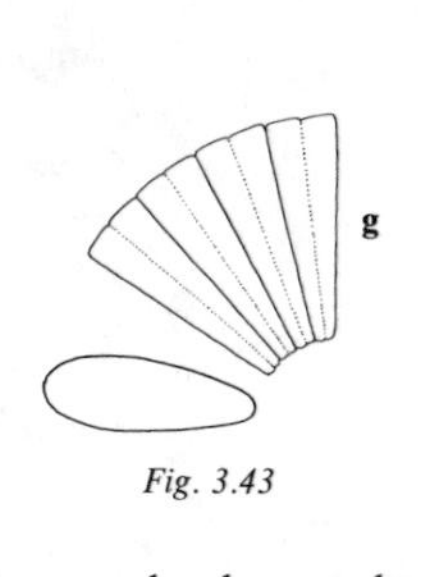

Fig. 3.43

Diatoma *Fig. 3.42* Most species are oval or lemon-shaped in valve view, but some are elongated and may resemble *Asterionella*, from which they are distinguished by their strong transverse sculpturing; up to 80 μm long. Forms irregular chains or zig-zag colonies by joining at valves or corners. Planktonic or benthic; common.

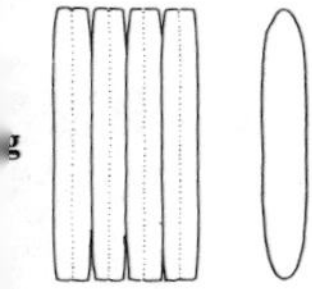

Fig. 3.44

Meridion *Fig. 3.43* Wedge-shaped in girdle view, club-shaped in valve view, up to 70 μm long. Typically forms fan-shaped colonies by joining at the valves. Frequent amongst plant life at lake or river margins.

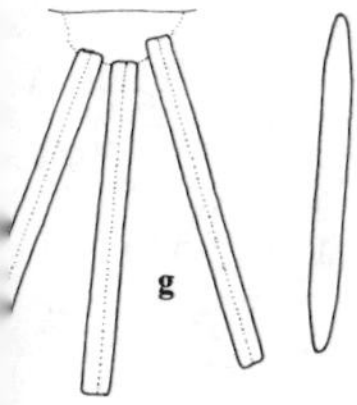

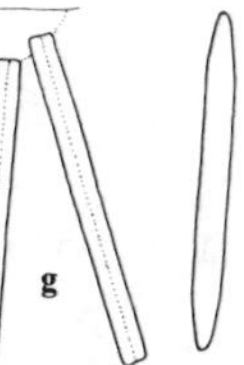

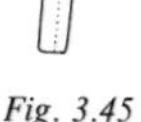

Fig. 3.45

Fragilaria *Fig. 3.44* Elongated diatoms, narrowing at the ends in valve view, up to 100 μm long. Forms broad, tape-like filaments by joining valve to valve. Common.

Synedra *Fig. 3.45* Resembles *Fragilaria* in frustule shape but is solitary, planktonic or attached to a substratum by one end, usually stuck in a pad of mucilage which it shares with other individuals, forming a rayed cluster. Rarely forms free, star-shaped clusters like *Asterionella*. Frustule up to 400 μm long. Very common.

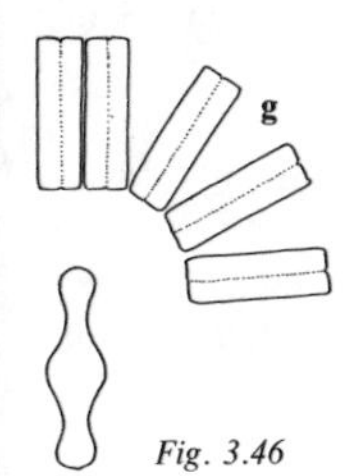

Fig. 3.46

Tabellaria *Fig. 3.46* In valve view, the frustules are swollen at each end and in the middle, different species

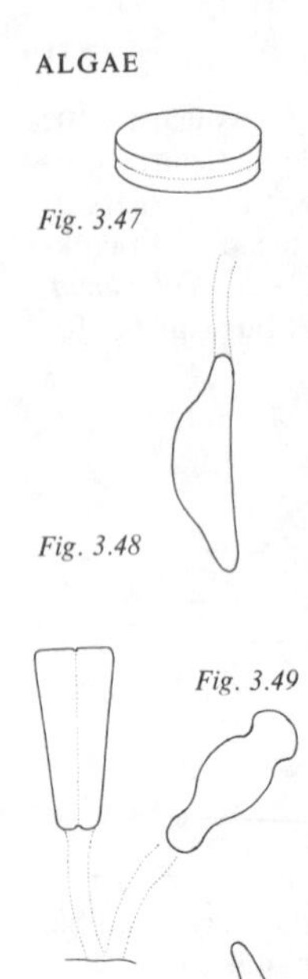
Fig. 3.47

Fig. 3.48

Fig. 3.49

varying in proportion; up to 100 μm long. Forms irregular chains or zig-zags by joining at the corners; sometimes in stars like *Asterionella*. Very common.

Cocconeis *Fig. 3.47* Small oval diatoms up to 25 μm long. Solitary, usually attached to other plant life, particularly *Cladophora*, lying flat on one valve. Very common, sometimes present in such large numbers that the substratum is completely obscured.

Cymbella *Fig. 3.48* Frustule characteristically plano-convex in shape, up to 150 μm long. Solitary or in small groups, attached to a substratum by a stalk of mucilage. Common.

Gomphonema *Fig. 3.49* Wedge-shaped in girdle view, swollen in the middle and at one (rarely both) ends in valve view, up to 120 μm long. Attaches to a substratum by a stalk or pad of mucilage; very common.

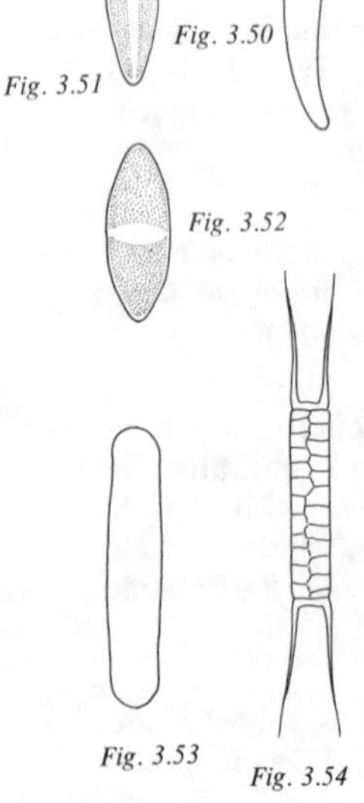
Fig. 3.50

Fig. 3.51

Fig. 3.52

Fig. 3.53

Fig. 3.54

Gyrosigma *Pl. 24, Fig. 3.50* A distinctive diatom with an elongated frustule curved in a graceful S-shape; up to 200 μm long. A common benthic form.

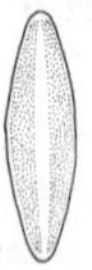

Navicula *Fig. 3.51* Spindle-shaped diatoms with two large plastids arranged longitudinally, side-by-side. Numerous species, up to 200 μm long. Very common in most habitats, creeping around on the substratum.

Nitzschia *Fig. 3.52* Spindle- or boat-shaped, with two large plastids occupying opposite ends of the cell. Numerous species, which vary greatly in their proportions; up to 200 μm long, but most are much smaller. Very common in many habitats.

Pinnularia *Fig. 3.53* Elongated diatoms with rounded ends and plastids arranged as in *Navicula*. Several species up to 300 μm long. Common.

Attheya *Fig. 3.54* and ***Rhizosolenia*** These are planktonic diatoms of remarkable shape: the girdle region is greatly elongated, forming a tube consisting of overlapping scales; each valve bears one (*Rhizosolenia*) or two (*Attheya*) long spines. Up to 100 μm long excluding spines.

Reference: Barber & Haworth, 1981.

Group 4 Filamentous green algae.

These are mostly formed of cylindrical cells joined end-to-end, but in one genus, *Vaucheria*, the internal cell walls have disappeared. Some genera are simple (unbranched), others are branched. Some grow free but most are attached to a substratum, especially in rivers where they may form long trailing skeins, or grow in masses on the surface of rocks where they build up into dense cushions containing entrapped sediment and are often several centimetres thick. This group includes representatives of the phyla Chlorophyta and Xanthophyta, but some desmids (group 2) and diatoms (group 3) also form filamentous chains.

Phylum
CHLOROPHYTA

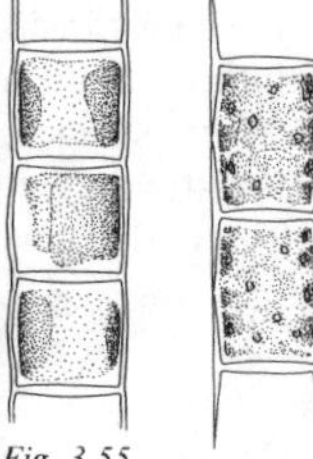

Fig. 3.55 *Fig. 3.56*

Ulothrix *Fig. 3.55* Filaments free or attached by one end, not branching, up to 50 μm diameter. Each cell contains a single, broad plastid which is wrapped around the inner circumference of the cell wall but does not form a complete ring. Frequent in many habitats.

Microspora *Fig. 3.56* Filaments unbranched and usually unattached, up to 25 μm diameter. Cells not much longer than wide, with a single large plastid which encircles the inside of the cell wall, often appearing to be broken up. Cell walls consist of H-shaped structures, one of these overlapping half of two adjacent cells (cf. *Tribonema*). Locally common in ponds and streams.

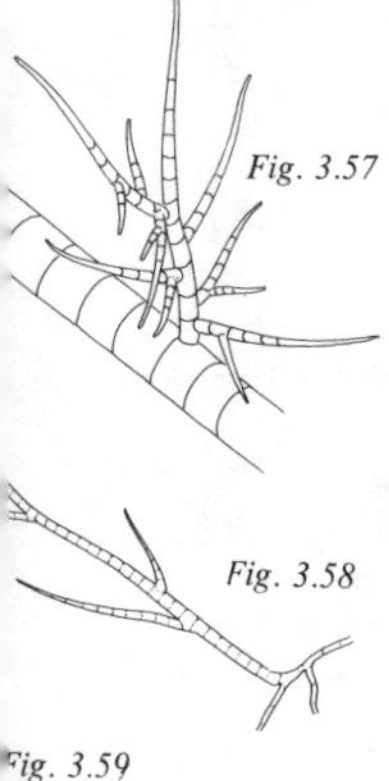

Fig. 3.57

Fig. 3.58

Draparnaldia *Fig. 3.57* Filaments attached by one end, up to 70 μm diameter, with feathery tufts of much finer filaments branching off laterally; whole plant may be several centimetres long. The bright green cells resemble those of *Ulothrix*. Occurs in many habitats growing on firm substrata, often in dense tufts.

Stigeoclonium *Fig. 3.58* Slender filaments, usually not more than about 25 μm diameter, fairly regularly branched and tapering gently towards the tips; mucilaginous and slimy to the touch. Common in many habitats, usually growing attached by one end but may also creep along the substratum.

Fig. 3.59

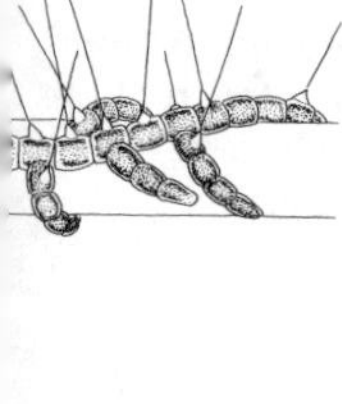

Aphanochaete *Fig. 3.59* Slender filaments of somewhat bulbous cells, up to 15 μm diameter, with numerous side branches, which grow along the substratum attached by one side. Some of the cells give rise to long hairs with bulbous bases. Typically grows on other plants, especially other filamentous algae such as *Cladophora*.

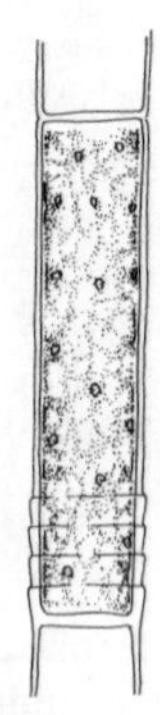
Fig. 3.60

Oedogonium *Fig. 3.60* Filaments free or attached, unbranched, up to 50 μm diameter. Cells much longer than wide, each with a single diffuse (net-like) plastid; sometimes the cell wall is formed into several encircling ridges at one end. Common in still waters, often forming floating masses.

Cladophora *Fig. 3.61* Filaments up to 100 μm diameter, attached at one end, profusely and irregularly branched, forming tufts or skeins which may exceed one metre in length. Texture not unlike fine human hair: definitely not mucilaginous and often rough to the touch. Cells elongated, with a single diffuse plastid, varying from medium to dark green in colour.

This is probably the most common and widespread genus of filamentous algae. It is abundant in both marine and freshwater habitats, attached to any firm surface. Its rough texture makes it an ideal site for the attachment of other organisms and it is usually a rich source of diatoms and other algae, protozoans, rotifers, etc.

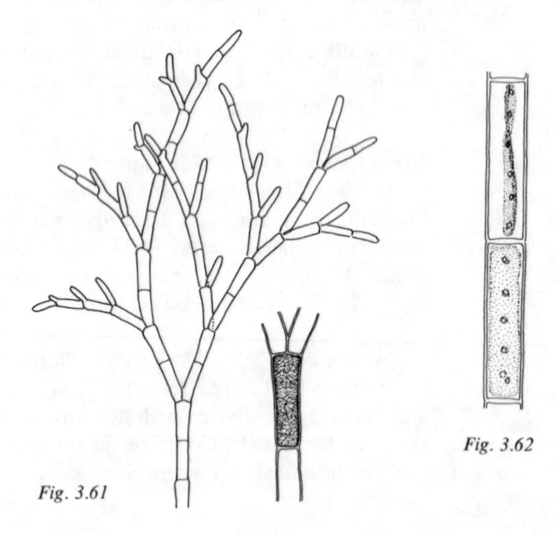
Fig. 3.61

Fig. 3.62

Mougeotia *Fig. 3.62* Unbranched filaments up to 50 μm diameter, mucilaginous and slimy to the touch. Cells elongated, with one flat band-like plastid which lies along the mid-line of the cell. Frequent in still waters.

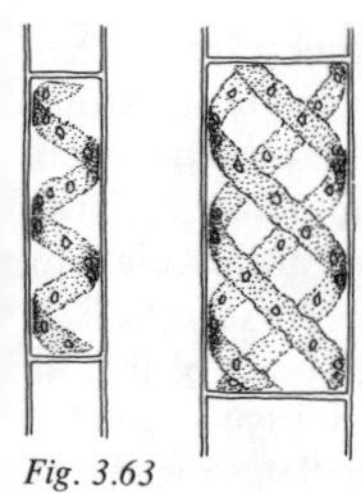
Fig. 3.63

Spirogyra *Pl. 25, Fig. 3.63* Long, usually free, unbranched filaments up to 150 μm diameter, with a mucilage sheath which makes it slimy to the touch. Cells usually elongated, with one or more narrow, ribbon-like plastids, bright grass-green in colour, wound in a neat spiral along the cell. Common in still or slow-flowing waters, often forming floating rafts or tangled masses amongst other vegetation. This genus is commonly used as a text-book example of a filamentous alga.

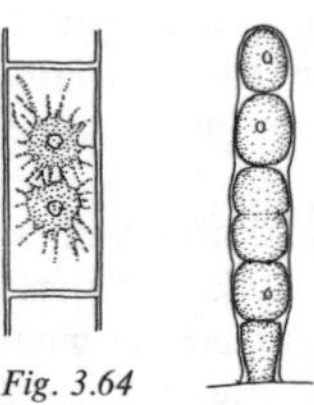
Fig. 3.64

Fig. 3.65

Zygnema *Fig. 3.64* Unbranched filaments up to 50 μm diameter, mucilaginous and slimy to the touch. Each cell has two star-like plastids. Frequent in both still and flowing waters.

Cylindrocapsa *Fig. 3.65* Small, unbranched filaments of oval cells, typically attached to a substratum by one end and growing perpendicularly; plastid single, large and filling whole cell, bright green. Common on stones, glass walls of aquaria, etc., and on other plants, often growing with *Coelochaete*.

Phylum XANTHOPHYTA

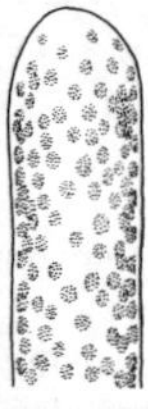

Tribonema *Fig. 3.66* Slender, unbranched filaments up to 20 μm diameter; cells often a little swollen, with many, roughly disc-shaped, pale yellowish-green plastids lying close to the cell wall; cells made of H-shaped pieces like those of *Microspora*. Frequent in small, standing waters.

Vaucheria *Fig. 3.67* Stout filaments up to 200 μm diameter, with occasional branching. Internal cross walls between cells are lacking and the filament often appears to be hollow; plastids small and numerous, bright pale yellowish-green in colour. Common in many habitats, usually attached to a substratum.

Fig. 3.67

Fig. 3.66

Group 5 Stoneworts or brittleworts

The stoneworts are macroscopic green plants possessing a higher level of organization than any other freshwater algae. Each plant consists of a stem of elongated internodal cells (many millimetres long) alternating with short nodal cells. From each node arises a whorl of lateral branchlets, themselves consisting of nodes and internodes; secondary stems may also originate at the stem nodes. In some species of *Chara* the internodal cells are covered by a layer (cortex) of protective (cortical) cells. Stoneworts vary in appearance from low bushy tufts to tall, delicate and slender plants which may reach a height of 50 cm or more. They are anchored to soft, muddy or sandy, substrata by a mass of root-like rhizoids, from which new plants may arise by budding.

The prominent reproductive cells (*Pl. 26*) appear at the nodes or at forks in the branchlets; these are the female oogonia and the male antheridia. An oogonium (*Fig. 3.68*) is a single egg cell surrounded by a protective sheath of five spirally-wound cells. At the apex of this structure is a circlet of five or ten crown cells. Each antheridium is spherical, often bright orange in colour, and constructed from eight umbrella-shaped cells (*Fig. 3.69*). Plants may be of one sex (monoecious) or both (dioecious).

Stoneworts are so-called because they have a strong affinity for lime. In calcium-rich waters, their cells soon become covered by a crisp surface layer of 'scale' which eventually becomes quite dense, marring their natural translucency and attractive bright green colour. It also makes the plants quite brittle – hence their alternative name.

Stoneworts typically are found in shallow clear waters, although they sometimes inhabit deep, clear lakes and often occur in brackish localities. They usually grow in dense beds, but invariably suffer in competition with filamentous algae or higher plants. Five genera, two of which are very rare, occur in north-west Europe; the three principal genera are described below.

Phylum
CHAROPHYTA

Chara *Pl. 26, Figs. 3.70, 71* Cortex present in most species; branchlets unforked, typically about equal length in each whorl; not more than one secondary stem arising from each node. Oogonia with five crown cells, borne above antheridia at nodes of branchlets (*Fig. 3.71*). This is the largest genus of stoneworts, with many common and widespread species.

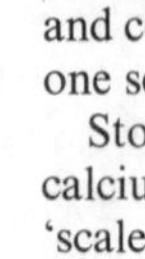

Nitella *Figs. 3.68, 69, 72* Cortex never present; branchlets forked, often repeatedly, the forking generally equal in each whorl; not more than two secondary stems arising from each node. Oogonia compressed, with ten crown cells; antheridia borne *in* the forks of the branchlets, with oogonia lateral to them. Typically a tall, slender, delicate plant in habit, the lime encrustation usually thin, sometimes absent altogether; several species with a widespread distribution.

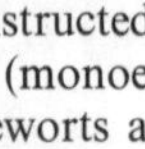

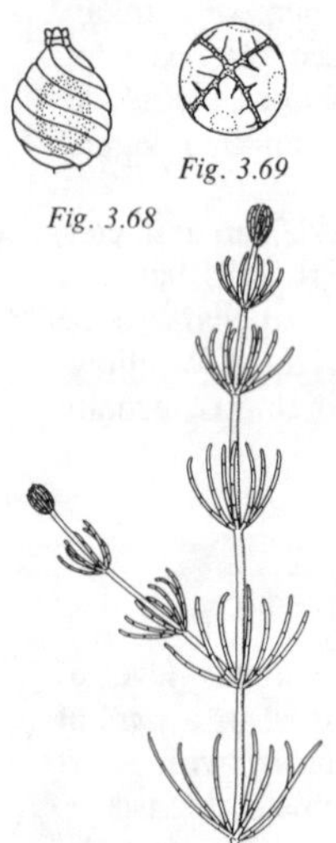

Fig. 3.68

Fig. 3.69

Fig. 3.70

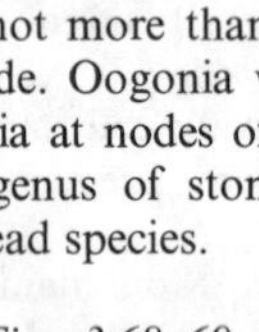

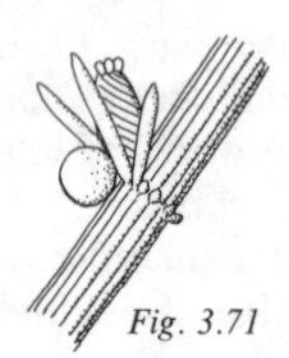

Fig. 3.71

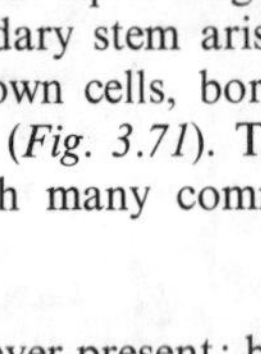

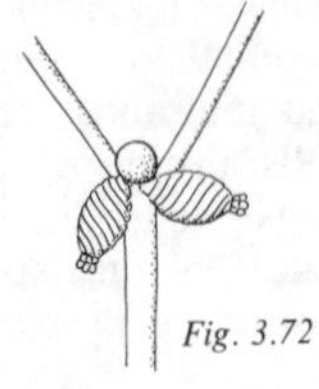

Fig. 3.72

Fig. 3.73

Tolypella *Pl. 27, Fig. 3.73* Cortex never present; fertile branchlets (those bearing reproductive cells) forked, typically with dense tufts of branchlets forming around the reproductive cells; sterile branchlets long, irregular in length, and not usually forked; more than two secondary stems may arise from a node. Oogonia not compressed, with ten crown cells; antheridia borne laterally at nodes with oogonia lateral to them. Typical growth habit low and bushy, somewhat untidy, with stouter stems than *Nitella*, from which it usually differs greatly in overall appearance. Several species, widespread, but local in distribution.

References: Allen, 1950. Groves & Bullock-Webster, 1920 & 1924.

Group 6 Other algal forms

These are mostly macroscopic plants of varied appearance: encrusting, fixed, or free-floating in habit; green, red, grey or brownish in colour. They include genera from three phyla.

Phylum
CHLOROPHYTA

Chaetophora *Pl. 28* A filamentous alga that forms mucilaginous colonies of bright green, branched filaments. These radiate from a central point to produce balls up to several millimetres in diameter, or branching growths up to 10 mm or more long, with fine, translucent threads projecting from the surface. Colonies are attached to other plants, twigs, stones or other substrata, and are common in many habitats in spring and summer.

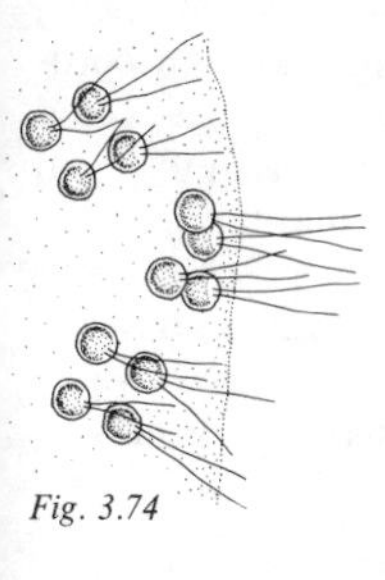
Fig. 3.74

Tetraspora *Fig. 3.74* This forms elongated, irregular masses of soft mucilage, up to 25 mm long, in which are imbedded numerous tiny, spherical green cells, usually in groups of four. Two very fine hairs, like motionless flagella, project from each cell. Occasional in ponds or slow rivers, attached to various substrata.

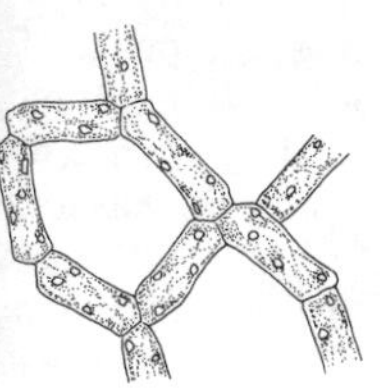
Fig. 3.75

Hydrodictyon *Fig. 3.75* Distinctive colonies of large green cells, up to 10 mm long when fully grown, that link together to form a two- or three-dimensional network. Its occurrence is somewhat erratic, growing on various substrata in ponds or slow-moving waters.

Coelochaete *Fig. 3.76* Colonies form flat encrustations one cell thick and up to 1–2 mm across on any firm substratum.

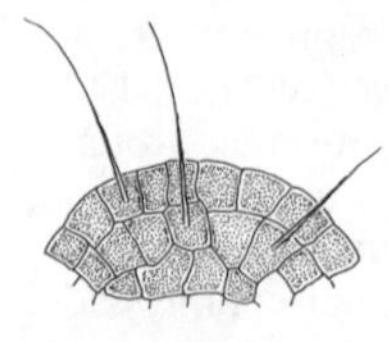
Fig. 3.76

The green cells are irregular in shape and some have long hairs arising from their surface. Very common.

Enteromorpha *Pl. 29* Bright green tubular colonies of soft texture and very irregular shape, usually about 1–20 mm across and up to at least 20 cm long. Very common; attached or free, often floating due to gas bubbles trapped in the tube, in still or slow-flowing waters. Sometimes blown into dense masses.

Phylum
XANTHOPHYTA

Botrydium *Fig. 3.77* Dark green globules up to 2.5 mm diameter, consisting of a mass of wall-less cells (*cf. Vaucheria*) with numerous small plastids. Found on mud, each globule anchored by a colourless, root-like rhizoid. It occurs just above or below the water level, often appearing when the level drops in summer, and only lives for a few days.

Fig. 3.77

Phylum
RHODOPHYTA
(red algae)

Batrachospermum Frogspawn alga *Pl. 30, Fig. 3.78* The whole plant consists of profusely-branched strands composed of a central filament of small, bead-like cells from which arise, at close intervals, dense whorls of fine lateral branches; pale to dark greenish, sometimes blackish. The texture of the whole plant is very soft and slimy, being covered with a mucilage layer varying in thickness according to species and season; it may become very thick and gelatinous – just like frogspawn. Whole plant up to 10 cm or more long, each strand up to about 2 mm thick.

This distinctive alga occurs in cold water (mostly a winter or spring form) in shaded places: in fast streams and rivers, spring outlets, or occasionally in ponds or lakes where conditions are suitable; usually attached to a firm substratum or anchored in sand or gravel.

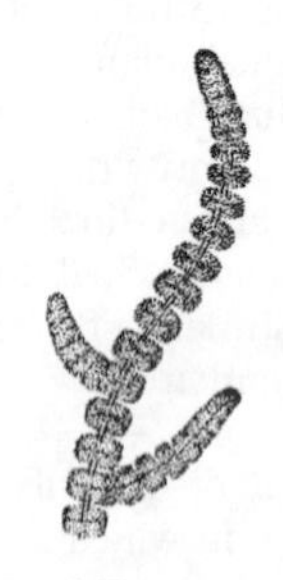
Fig. 3.78

Lemanea *Fig. 3.79* Long, gently-tapering filaments up to 1 mm diameter, with occasional branching, composed of cylindrical (multicellular) segments about 5–10 mm long with knobbly joints. The colour is drab olive-green, brownish or greyish-purple. Its texture is firm but still slimy to the touch. It forms tufts up to 25 cm long, attached firmly to rocks in the lower reaches of fast-flowing rivers and streams.

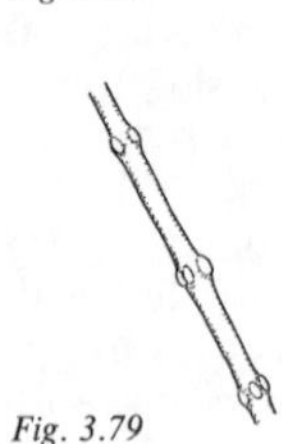
Fig. 3.79

Hildenbrandia This alga forms thin encrustations, bright crimson to blackish in colour (like drying blood!), on stable rocks. It prefers shaded conditions but may be found in open stretches of water growing under boulders; if too much light is available it becomes overgrown by more vigorous plants such as *Fontinalis* or *Chamaesiphon*. The only freshwater species in north-west Europe is *H. rivularis* (there are several marine species) which is characteristic of the upper, torrential, reaches of cold rivers originating in mountain or moorland areas.

References to Algae (general): Belcher & Swale, 1978, 1979. Pascher, 1913, 1914, 1915, 1927. Round, 1981.

CHAPTER 4 Mosses and Liverworts
(Phylum Bryophyta)

Mosses (Musci) and Liverworts (Hepaticae) are two groups of green lower plants which comprise the phylum Bryophyta. The mosses always have distinctly differentiated leaves and stems, and fall into two broad groupings, the acrocarps, whose stems are mainly upright, and the pleurocarps, which are usually creeping or spreading. Liverworts may or may not have such leaves and stems. Unlike higher plants, bryophytes do not reproduce by means of flowers and seeds, but have special organs which contain the male and female gametes respectively. These eventually unite to produce a fruit from which spores are shed as a fine dust. Almost all bryophytes grow in damp places, but not many actually grow in the water.

The drawings in this section are of a growing shoot with a close-up of a single leaf. Overall size range and typical size (in brackets) of leaf is provided. Important features include centre vein (nerve) present or absent, and its relative length; edges toothed or entire, sometimes with differentiated rim; basal auricles of enlarged cells present in some (e.g., *Brachythecium*).

Acrocarpous Mosses

The undecayed remnants of bog mosses ***Sphagnum*** (Sphagnaceae) make up the peat found both in blanket bogs on moorland, and in valley bogs forming in shallow depressions that may later give rise to raised bogs. Uniquely among the mosses, their branches are in whorls, and bear tiny leaves that differ from those actually on the stems. Only two widespread species are actually aquatic, growing in the acid water of bog pools: ***S. cuspidatum*** and ***S. auriculatum***. Both form loose, yellowish-green mats, and have long trailing stems up to 40 cm, with very narrow, feathery, submerged leaves quite distinct from the broader aerial ones. *S. cuspidatum* may turn red or orange and has toothed leaves, broadly triangular on the stems, but longer and narrower on the branches. *S. auriculatum* var. *auriculatum* is often coppery-red and has less obviously hanging branches and less triangular leaves.

Fig. 4.1

Fissidens crassipes (Fissidentaceae) *Pl. 31, Fig. 4.1*
Lanceolate, dark-green, untoothed leaves arranged in two rows up the short (to 5 cm) stems. Rocks in calcareous streams.

Fig. 4.2

Fig. 4.3

Dicranella palustris (Dicranaceae) *Fig. 4.2*
Best identified by bright golden-green tufts or mats by mountain streams and flushes, together with *Philonotis fontana.* Broad, blunt-tipped leaves spread well out from 2–20-cm stems.

Cinclidotus fontinaloides (Pottiaceae) *Fig. 4.3*
Dark-green, tongue-shaped, untoothed, 4-mm leaves, closely ranked up long (up to 18 cm) stems. Rocks or submerged logs in streams, rivers and lakes.

Grimmia alpicola var. ***rivularis*** (Grimmiaceae)
Like the last species, but with leaves only 2 mm long and growing in mountain streams.

Rhacomitrium aciculare (Grimmiaceae)
Dark-green, blunt-tipped, tongue-shaped leaves, which crowd up 2–8-cm stems. Rocks, and fast non-calcareous streams in hill districts. *R. aquaticum* has usually yellowish- or olive-green leaves with more pointed tips. Often at waterfalls.

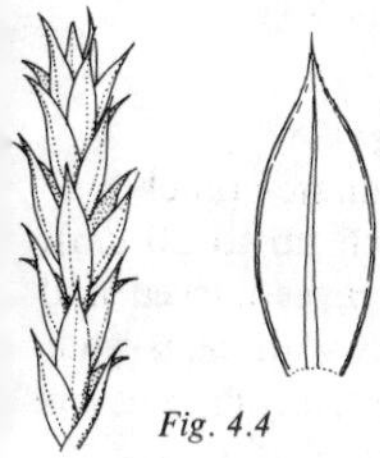

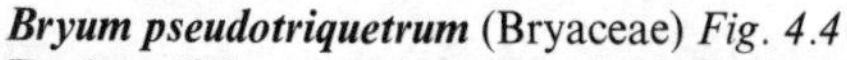

Bryum pseudotriquetrum (Bryaceae) *Fig. 4.4*
Dark or light green leaves, usually slightly tinged purple, lanceolate and pointed, crowded up 2–8-cm stems, which are covered with a brownish down. Mostly in bogs, marshes and dune slacks, but often in springs and wet flushes on mountains with *Philonotis fontana* and its associates.

Fig. 4.4

Philonotis fontana (Bartramiaceae) *Fig. 4.5*
Distinctive, bright golden-green or glaucous moss with clumps of upright stems to 12 cm, covered at the base with red-brown hairs, and small, tightly packed, triangular leaves. Especially characteristic of mountain springs and flushes with acid water, along with the last species, *Brachythecium rivulare* and the liverwort *Jungermannia cordifolia. P. calcarea* is a very similar plant of calcareous flushes.

Fig. 4.5

Fig. 4.6

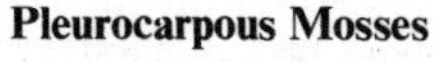

Pleurocarpous Mosses

Fontinalis antipyretica (Fontinalaceae) *Pl. 32, Fig. 4.6*
One of the few mosses with a vernacular name – Willow Moss. One of the commonest and best-known aquatic mosses, growing on submerged stones, logs and tree-roots

in fast, slow-moving and still water, and in both calcareous and acid water. The dark green, sometimes reddish, keeled leaves are arranged in 3 rows along the very long (50–70 cm) stems. *F. squamosa* has leaves not keeled and grows only in fast-flowing, non-calcareous streams.

Crateroneuron commutatum
Golden-green, sometimes turning orange-brown, with the shoots pinnately branched and leaves curved. Springs on mountains.

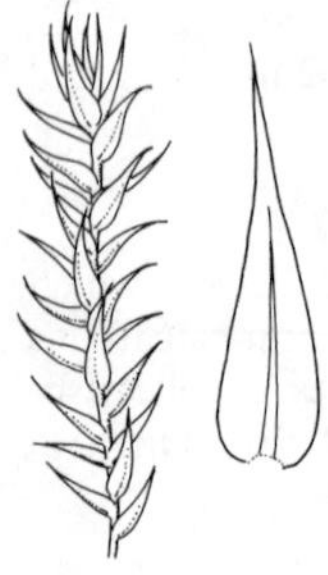

Fig. 4.7

Leptodictyon (formerly ***Amblystegium***) ***riparium*** (Hypnaceae) *Fig. 4.7*
Bright green with ovate-lanceolate leaves tapering to a very fine point, spreading from the stems; capsule with a red stalk. Stones and submerged logs in lowland streams or ponds, especially with calcareous water. *A. fluviatile* is deep green, with many long branches held almost parallel to the stem and stiff, untoothed leaves. Stones in fast-flowing calcareous streams. *A. tenax* has the stiffer toothed leaves.

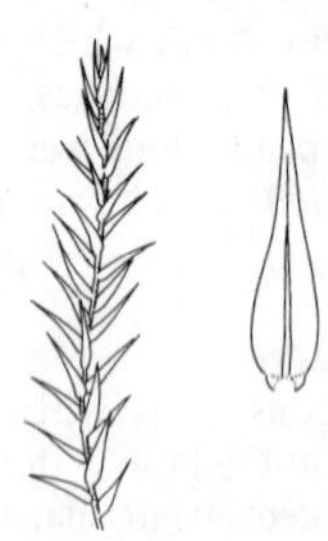

Fig. 4.8

Drepanocladus aduncus (Hypnaceae) *Fig. 4.8*
Green, varying both in the number of pinnate branches up the stems, and in length (which may be up to 30 cm); shoots usually with straight tips, but leaves curved and tapering to a fine tip. Pools and ditches with calcareous and neutral water. ***D. fluitans*** may be green, yellow-green or rufous, but is best identified by the hooked tips of the shoots. Prefers acid pools. Cf. *Hygrohypnum luridum*. ***D. revolvens*** is yellowish-green, often tinged orange or purple, with stems irregularly-branched and leaves markedly curved, almost into a circle; bog or fen pools.

Hygrohypnum luridum (Hypnaceae)
Forms loose, dull yellow or brownish-green patches, the branches curved at the tip, and leaves slightly concave, with short blunt tips (unlike the somewhat similar *Drepanocladus fluitans*). Rocks or submerged wood in streams, especially fast-flowing waters in hills and with basic water. ***H. ochraceum*** has laxer patches with longer, dull yellow shoots. ***H. eugyrium*** is more compact and glossier, with small orange-brown patches on the leaves. The two latter species are only in mountain streams.

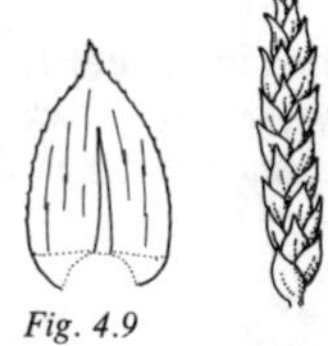

Fig. 4.9

Brachythecium rivulare (Hypnaceae) *Fig. 4.9*
Bright golden-green, creeping but with upright branches, leaves pointed, oval, tinged orange at the base. In and by streams and waterfalls, often in mountain flushes along with *Philonotis fontana* and its associates.

. 4.10

Eurhynchium riparioides (Hypnaceae) *Fig. 4.10*
Bright or deep green, patch-forming and long-stemmed, leaves pointed, oval and minutely-toothed. One of the commoner mosses of fast-flowing water, especially by waterfalls, but also occasionally in still or slow-flowing waters.

Hyocomium armoricum (Hypnaceae) *Fig. 4.11*
Golden-green or deep green, with the stems, to 20 cm, often pinnately branched, leaves very short, heart-shaped, minutely-toothed and with a pointed tip. Streams and waterfalls in hill districts, preferring acid water.

Fig. 4.11

Liverworts (Hepaticae)
Liverworts differ from mosses in various ways. All mosses, but only some liverworts, have distinct stems and leaves. Only in liverworts are any leaves deeply-lobed or cut. All leafy liverworts have their leaves in either two or three ranks up the stem.

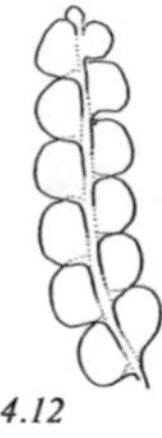

Fig. 4.12

Chiloscyphus polyanthus (Jungermanniaceae) *Fig. 4.12*
Grows in patches with almost-rounded, overlapping greenish or brownish leaves, the aquatic form being dark green and well-branched. On rocks and stones in springs and streams.

Fig. 4.13

Solenostoma (formerly ***Jungermannia***) ***cordifolium*** (Jungermanniaceae) *Fig. 4.13*
Clumps of upright stems to 10 cm, with oval to heart-shaped leaves as long as they are broad. Mountain springs along with the moss *Philonotis fontana* and its associates, and rocks in mountain streams along with the liverwort *Scapania undulata*.

Nardia compressa
Like the last species, with large, purplish tufts, but leaves are concave and lie pressed together up the stems. Mountain streams.

Marsupella emarginata (Marsupellaceae)
Forms greenish-brown or red-brown patches, its roundish leaves with a notched tip inserted transversely up erect 3–10-cm stems. In and by mountain streams and lakes, sometimes well submerged.

Fig. 4.14

Scapania undulata *Pl. 33, Fig. 4.14*
Two-lobed, folded leaves, which are either bright green and untoothed, or purple-red and finely toothed; stems up to 12 cm in the water. Fast streams especially in hill districts, often with *Solenostoma cordifolia*; sometimes in the lowlands.

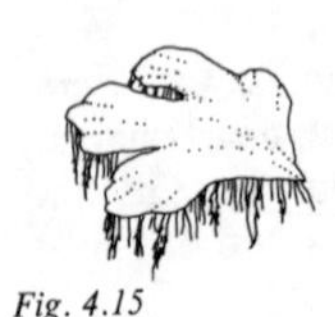
Fig. 4.15

Ricciocarpus natans (Ricciaceae) *Fig. 4.15*
Free-floating, like duckweeds (Lemnaceae) or water-fern *Azolla*, with much broader leaves than the following species and much less common. Distinguished also by the long violet scales that hang downwards like duckweed roots. In still, basic water.

Fig. 4.16

Riccia fluitans *Pl. 34, Fig. 4.16*
Two aquatic forms – one in floating masses of narrow, repeatedly-forked leaves, the other with broader leaves and submerged on the bottom mud. Ponds and ditches with alkaline water.

CHAPTER 5 Ferns and Flowering Plants

(Phyla Pteridophyta and Angiospermae)

All flowering plants, ferns and fern allies that grow actually in the water, as distinct from on the bank or shore, are included. This inevitably raises problems with plants that normally grow on the margins of fresh water, sometimes in and sometimes out of the water. In general, species that are primarily land or marsh plants but sometimes just overlap into the water, e.g., when water levels are unusually high, are either omitted or treated fairly briefly. This applies especially to trees and shrubs.

No European trees are genuinely aquatic, as the swamp cypress *Taxodium distichum* (sometimes planted in Europe) is in North America. However, some willow trees and shrubs *Salix* spp. and alder *Alnus glutinosa* do grow in very wet conditions in swamps and along freshwater margins, and contribute to aquatic ecosystems in three important ways. They restrict the amount of light that reaches the water surface; they add their dead leaves to the store of nutrients in the mud on the bottom; and their roots provide shelter and a substratum for many aquatic animals, enabling them to survive in fast-flowing streams where they would otherwise be swept away.

Tabular Key to Vascular Plant Leaves

F = Floating Plant   T = also Terrestrial Plant

Leaves Submerged Only

Flowers Submerged:	quillworts, Pillwort (T), hornworts (F), Awlwort, waterworts, Marsh Pennywort (T), water starworts (*Callitriche brutia, hamulata, hermaphroditica, truncata*), Horned Pondweed, naiads
Flowers Floating:	water-thymes, Canadian Waterweed, Tape-grass
Flowers Aerial:	water crowfoots (*Ranunculus circinatus, fluitans, penicillatus, trichophyllus* and rarely also *aquatilis and peltatus*), Awlwort, water milfoils, Water Violet, bladderworts (F), Water Lobelia, Ribbon-leaved Water Plantain, pondweeds (*Potamogeton acutifolius, alpinus, berchtoldii, compressus, crispus, epihydrus, filiformis, friesii, gramineus, lucens, obtusifolius, pectinatus, perfoliatus, praelongus, pusillus, rutilus, trichoides, vaginatus, × zizii*), Opposite-leaved Pondweed, Pipewort
No Flowers:	Shoreweed (T), Ivy-leaved Duckweed (F), Water Soldier (F), Needle Spike-rush (T)

Leaves Submerged and Floating

Flowers Submerged: Floating Fern (F), Blinks (T)

Flowers Floating: Blinks (T), Water Chestnut, water starworts (T) (*Callitriche brutia, cophocarpa, hamulata, obtusangula, palustris, platycarpa, stagnalis*).

Flowers Aerial: Yellow Water-lily, Least Water-lily, water crowfoots (T) (*Ranunculus aquatilis, baudotii, ololeucos, peltatus, penicillatus* (rarely), *tripartitus*) pondweeds (*Potamogeton alpinus, epihydrus, gramineus, natans, nodosus, polygonifolius* and rarely ×*zizii*), Common Club-rush, Floating Club-rush

Leaves Submerged and Aerial:

Flowers Aerial: New Zealand Water Stonecrop, Marsh St. John's Wort (T), Hampshire Purslane (T), Marestail, water parsnips, water dropworts, Fool's Watercress (T), Creeping Marshwort (T), Lesser Marshwort, Monkey Flower (T), Brooklime (T), water speedwells (T), Bulbous Rush (T), Ribbon-leaved Water-plantain, Water Soldier (F), Arrowhead

Leaves Floating Only

Flowers Floating: water ferns (F), White Water-lily, Shining Water-lily, Frogbit (F), Fat Duckweed (F), Common Duckweed (F), American Duckweed (F), Great Duckweed (F)

Flowers Aerial: Amphibious Bistort (T), water crowfoots (T) (*Ranunculus aquatilis, baudotii, hederaceus, omiophyllus, peltatus*), Floating Arrowhead, Star-fruit (T), Cape Pondweed, bur-reeds (*Sparganium angustifolium, glomeratum, gramineum, hyperboreum, minimum*)

No Flowers: Rootless Duckweed

Leaves Floating and Aerial

Flowers Aerial: arrowheads, Floating Water-plantain, Heart-shaped Water-plantain, sweet-grasses, Branched Bur-reed (rarely)

Leaves Aerial Only

Flowers Aerial: Arrowhead, water-plantains (lesser, common, narrow-leaved), Flowering Rush, Water Soldier (F), Bog Arum (T), cotton-grasses (T), Bottle Sedge (T), Bladder Sedge (T), Water Sedge (T).

No Leaves

Flowers Aerial: Water Horsetail, Common Club-rush, Common Spike-rush, Needle Spike-rush

No Flowers: Rootless Duckweed

FERNS and FERN ALLIES (Phylum Pteridophyta)

QUILLWORT FAMILY (Isoetaceae)

Quillworts are the only true flowerless plants with tufts of submerged leaves, but although they can always be told by having spore-cases with accompanying ligules instead of flowers, their leaves in the barren state are so similar to those of several tufted flowering plants that they are all set out in the table below. The presence of these underwater species in a lake or tarn can often be ascertained by finding their leaves, detached by storms, along the drift-line on the shore.

Tabular Key to Plants with Submerged Quill-like Leaves

	Leaf tip	*Leaf length*	*Leaf section*	*Other characters*
Quillwort	bluntly pointed	8–25 cm, to 60 cm	rounded	leaves dark green, stiff, sometimes recurved
Spring Quillwort	pointed	5–15 cm	flattened	leaves pale green, flaccid, recurved
Awlwort (*p. 77*)	pointed	2–7 cm	rounded	leaves straight
Lobelia (*p. 89*)	blunt	2–4 cm	hollow	leaves recurved
Shoreweed (*p. 89*)	pointed	1.5–10 cm	semi-cylindrical	has runners, forms sward
Pipewort (*p. 99*)	pointed	3–10 cm	flattened	leaves translucent, with jointed sections
Needle Spike-rush (*p. 105*)	pointed	to 50 cm	rounded	leaves thread-like, forms sward

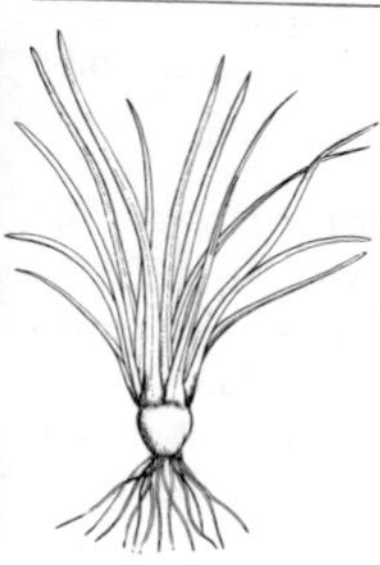

Fig. 5.1

Quillwort *Isoetes lacustris* (*Fig. 5.1*). Leaves quill-like, stiff, dark green, sometimes recurved, abruptly and bluntly pointed, with a tiny flap-like ligule above the spineless spore-cases at their base; spores ripe May–July. Completely submerged in lakes, tarns and pools in hill districts on acid soils.

Spring Quillwort *I. echinospora* has shorter, more flaccid, pale grass-green leaves, tapering more evenly to the sharp point; female spores spiny. Sometimes in lowland districts. **Slender Quillwort** *I. tenuissima* is similar, but has bulbous-looking leaf-bases and triangular ligules; France only.

HORSETAIL FAMILY (Equisetaceae)
Leafless and flowerless plants with their spores in blackish terminal cones, ripening in June. They are sometimes semantically confused with the flowering plant Marestail (*p. 80*).

Water Horsetail *Equisetum fluviatile* (*Pl. 35*) may have short branches, but often has unbranched stems only, with or without terminal cones. Stems have a central hollow and 10–30 obscure ridges, and die down in winter. Ponds, tarns and lakes, growing right in the water; also in swamps.

Marsh Horsetail *E. palustre* is usually branched, the stems with a very small central hollow and 6–10 conspicuous ridges; usually in marshes, rarely right in the water. In their early stages, before any flowerheads appear, the stems of the club-rushes (*p. 104*) and spike-rushes (*p. 105*) are superficially similar to these two horsetails.

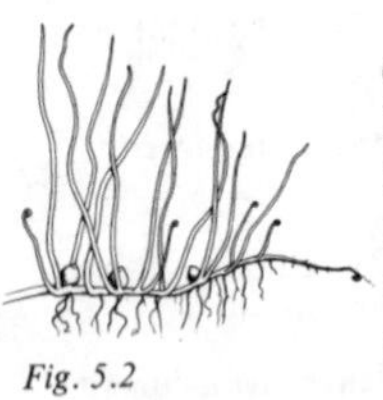
Fig. 5.2

CLOVER FERN FAMILY (Marsileaceae)
Pillwort *Pilularia globulifera* (*Fig. 5.2*). Has a mat of 3–8-cm yellow-green, thread-like leaves, coiled like fern fronds when young, with small blackish pill-like spore-cases at their base; spores ripe June–September. In or by shallow muddy ponds; an endangered European plant.

FLOATING FERN FAMILY (Salviniaceae)
Floating Fern *Salvinia natans* (*Pl. 36*). Has two oblong floating leaves, 10–14 mm × 6–9 mm, and a third one submerged and so finely divided as to appear root-like, with the spore-cases at its base; spores ripe August–October. Muddy ponds, ditches. Germany and Netherlands only.

WATER FERN FAMILY (Azollaceae)
Two species, both introduced from North America. The more frequent is **Water Fern** *Azolla filiculoides* (*Pls. 37, 38*), 1–10 mm across, whose minute two-lobed leaves overlap along the stems, the upper lobes blunt with a broad pale margin, and redden conspicuously in autumn. Spore-cases on the lower lobe of the first leaf of each branch, ripening August–October. Ponds, canals and other still water. **Carolina Water Fern** *A. caroliniana* differs

mainly in its pointed leaf-lobes having narrower pale margins. See *p. 101* for the differences between the aquatic ferns and the duckweeds.

FLOWERING PLANTS (Phylum Angiospermophyta)

1. DICOTYLEDONES
(Plants with two seed-leaves or cotyledons)

WILLOW FAMILY (Salicaceae)
In view of their marginal impact on freshwater ecosystems (see *p. 69*), only a few of the commoner willows and sallows are mentioned here.

Crack Willow *Salix fragilis* (*Pls. 39, 40*) and **White Willow** *S. alba* are two of the commonest riverside trees, often pollarded (beheaded to produce a crop of thin branches) in southern England. They have long catkins, appearing with the lanceolate leaves. Crack Willow has twigs that break easily, and April catkins. White Willow has May catkins, and silky white hairs under the leaves that make the whole tree look silvery at a distance. There is also a range of hybrids between the two.

Grey Willow or **Grey Sallow** *S. cinerea* is a shrub whose short catkins appear in late March and April before the shorter, broadly oval leaves, which often have rusty hairs beneath. Its twigs have raised striations under the bark. **Goat Willow** or **Goat Sallow** *S. caprea*, which is less frequent by water but commoner away from it, lacks both the twig striations and the rusty hairs.

BIRCH FAMILY (Betulaceae)
Alder *Alnus glutinosa* (*Pls. 42, 43*) is the tree most likely to be found growing actually in the water. Its catkins appear in February and March, well before the distinctive rounded leaves.

Fig. 5.3

BISTORT FAMILY (Polygonaceae)
Amphibious Bistort *Polygonum amphibium* (*Pl. 44*). Has oblong, floating leaves and conspicuous pink flowering spikes, appearing June–September. It grows in still or slow-flowing water and also has a completely, often barren, terrestrial form.

Water-pepper *P. hydropiper* (*Fig. 5.3*), with narrow, peppery-tasting leaves and much slenderer flowering

Fig. 5.4

Fig. 5.5

spikes, not appearing till July, is a bare-mud species that overlaps into shallow water.

Water Dock *Rumex hydrolapathum* (*Pl. 45*). A very tall (to 2 m) marginal plant that grows 'with its feet wet'. It has very long, fairly broad leaves and a typical dock-type flower-spike, appearing July–September. The very local **Scottish Dock** *R. aquaticus* (*Fig. 5.4*) has broader, triangular leaves.

PURSLANE FAMILY (Portulacaceae)

Blinks or Water Blinks *Montia fontana* (*Fig. 5.5*). A versatile plant that grows submerged, emergent, on wet mud or in bare dry places. Its leaves are narrow, often broader at the tip, and in opposite pairs. Its forking clusters of tiny white, five-petalled, stalked flowers, appearing April–October, are cleistogamous when submerged, i.e., the seeds ripen without the flowers opening (see also *p. 79*).

WATER-LILY FAMILY (Nymphaeaceae)

The water-lilies are the classic rooted plants with floating leaves and emergent or floating flowers. Four species occur in our area, two with white and two with yellow flowers, all flowering June–August.

White Water-lily *Nymphaea alba* (*Pl. 46*). Has all leaves floating, almost circular and 10–30 cm across, the basal lobes not overlapping. Its many-petalled floating flowers are 50–200 cm across and its fruits carafe-shaped. The non-British, **Shining Water-lily** *N. candida* has overlapping basal leaf-lobes and smaller flowers. Both grow in lakes, ponds and other still water.

Yellow Water-lily *Nuphar lutea* (*Pl. 47*) has both oval floating leaves, 12–40 cm by 9–30 cm, and large cabbagey submerged leaves. Its five- to six-petalled flowers, held a few inches above the surface, are much smaller, only 6 cm across, than White Water-lily (but larger than those of **Fringed Water-lily**, *p. 82*), and smell faintly of brandy, whence the folk-name Brandybottle; fruits globular. It grows in both still and slow-moving water. **Least Water-lily** *N. pumila* is smaller still, with flowers only half the size and wider gaps between the petals, growing only in upland lakes and tarns. The two yellow water-lilies hybridize and the hybrid may occur in the absence of Least Water-lily.

HORNWORT FAMILY (Ceratophyllaceae)

Hornworts *Ceratophyllum* (*Pl. 49*). Submerged, rootless plants of still or slow-moving water, with whorls of forked leaves, whose linear segments are toothed. Flowers minute, solitary, unstalked, with eight or more linear petals, the whitish males and green females at the base of separate leaf-whorls; appearing July–September. **Rigid Hornwort** *C. demersum* has rather stiff, dark green leaves and three spines on its nut-like fruit. **Soft Hornwort** *C. submersum* has laxer, paler green leaves and fruits with one or no spine; sometimes in brackish water.

BUTTERCUP FAMILY (Ranunculaceae)

Water Crowfoots *Ranunculus* subgenus *Batrachium* (*Pl. 48*). Eleven species, with floating and/or submerged leaves, and five-petalled white buttercup-like flowers, appearing April to September. They are all rather variable, depending on the depth and speed of the water in which they grow, but unless otherwise stated, the untoothed floating leaves have three lobes broadest at the tip, the submerged leaves are much divided into hairlike segments, the petals have a yellow claw, the nectary at the base of each petal is half-moon-shaped, the fruiting receptacle is hairy and the achenes have their beak near, but not at the tip.

Fig. 5.6

Ivy-leaved Crowfoot *R. hederaceus* (*Fig. 5.6*) has floating leaves only and is the only species whose leaf-lobes are broadest at the base; petals 2.5–3.5 cm and receptacle hairless. Very shallow, still water or streams, more often on wet mud. **Round-leaved Crowfoot** *R. omiophyllus* has longer petals (5–6 mm) and sometimes five leaf-lobes, and is the only species with the achene beak actually at the tip.

Three-lobed Water Crowfoot *R. tripartitus* always has floating leaves and usually also submerged ones; petals 1.25–4.5 mm and sepals blue-tipped. Very shallow, still water on acid soils. Round-leaved and Three-lobed Water Crowfoots hybridize. **All-white Crowfoot** *R. ololeucos* has larger petals (7–15 mm) with a white claw and sepals not usually blue-tipped, Western Europe, not Britain.

Brackish Water Crowfoot *R. baudotii* usually has floating leaves and always has submerged ones; petals 5–10 mm, sepals blue-tipped. The only species with elongated receptacle, winged achenes and confined to brackish or saline water.

Pond Water Crowfoot *R. peltatus* may lack either floating or submerged leaves, and has five-lobed floating leaves, petals 12–15 mm and pear-shaped nectaries. One of the two commonest water crowfoots, it grows in still or slow-flowing water, rarely more than 1 m deep.

Common Water Crowfoot *R. aquatilis* usually has floating leaves and always has submerged ones; petals 5–10 mm. The only species whose five-lobed floating leaves are toothed and with circular nectaries. Generally the commonest water crowfoot; habitat as Pond Water Crowfoot.

Thread-leaved Water Crowfoot *R. trichophyllus* has submerged leaves only and petals 3.5–5.5 mm. Habitat as Pond Water Crowfoot. **Fan-leaved Water Crowfoot** *R. circinatus* also has no floating leaves and is the only species whose submerged leaves lie stiffly all in the same plane; petals 4–10 mm. Still and slow-flowing, occasionally slightly brackish water.

Tabular Key to Water Crowfoots

	Floating leaves	*Submerged leaves*	*Petal length (mm)*	*Other characters*
R. hederaceus Ivy-leaved	Yes	No	2.5–3.5	Leaf-lobes broadest at base; receptacle hairless
R. omiophyllus Round-leaved	Yes	No	5–6	Achene beak at tip
R. tripartitus Three-lobed	Yes	Usually	1.25–4.5	Sepals blue-tipped
R. ololeucos All-white	Yes	Usually	7–15	Petal claw white
R. baudotii Brackish	Usually	Yes	5–10	Sepals blue-tipped; receptacle elongated; achenes winged; brackish water
R. peltatus Pond	Usually	Usually	12–15	Leaves 5-lobed, nectaries pear-shaped
R. aquatilis Common	Usually	Yes	5–10	Leaves 5-lobed, toothed; nectaries circular
R. trichophyllus Thread-leaved	No	Yes	3.5–5.5	
R. circinatus Fan-leaved	No	Yes	4–10	Leaves all in same plane: sometimes slightly brackish
R. fluitans River	No	Yes	7–13	Very long stems and leaves; nectaries pear-shaped; receptacle hairless
R. penicillatus Stream	Sometimes	Yes	10–15	Very long stems and leaves; floating leaves 5-lobed

Fig. 5.7

River Water Crowfoot *R. fluitans* grows in slow- to fast-flowing, non-calcareous rivers and streams, and so has very long stems, to 6 m, and leaves, to 8 cm. It has submerged leaves only, petals 7–13 mm, nectaries pear-shaped and receptacles hairless. **Stream Water Crowfoot** *R. penicillatus* is its calcareous-water counterpart, with petals 10–15 mm, hairy receptacles and occasionally five-lobed floating leaves.

Marsh Marigold or **Kingcup** *Caltha palustris* (*Pl. 50*). Has very large, buttercup-like, yellow flowers and large, kidney-shaped leaves, and can grow in very wet places, but is really a marsh or swamp plant. Two plants with yellow buttercup-like flowers and narrow leaves, **Greater Spearwort** *Ranunculus lingua*, tall, to 120 cm, with flowers 30–50 cm across, and **Lesser Spearwort** *R. flammula* (*Fig. 5.7*) shorter, to 60 cm, are also really marsh or fen plants that sometimes stray into shallow water.

CABBAGE FAMILY (Cruciferae)

A family characterized by its flowers all having four petals arranged crosswise.

Great Yellowcress *Rorippa amphibia* (*Pl. 51*). A bank plant that often extends into nearby shallow, still or slow-moving water. It is tall, to 120 cm, with lanceolate toothed or pinnately lobed leaves; panicles of small (6 mm) yellow flowers, appearing June–August; and long-stalked ovoid or oblong fruits.

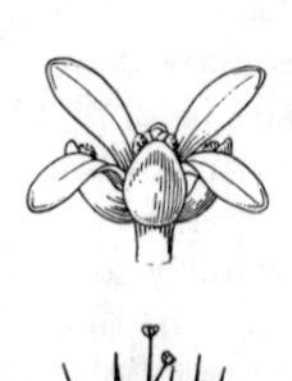

Watercress *Nasturtium officinale* (*Pl. 52*). The well-known salad plant, usually growing half-immersed in shallow, flowing water. Its peppery-tasting leaves are pinnate and remain green in winter; flowers white, appearing May–October; pods with two rows of seeds. **One-rowed Watercress** *N. microphyllum* flowers a fortnight later and has leaves turning purplish-brown in autumn and longer pods with only one row of seeds. The watercress most often cultivated, also frequently growing wild, is the hybrid between these two, whose leaves turn purplish-brown and whose pods are usually deformed.

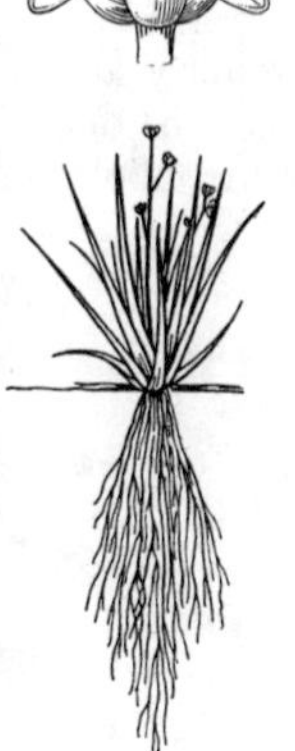
Fig. 5.8

Awlwort *Subularia aquatica* (*Fig. 5.8*). One of the plants with submerged tufts of narrow leaves tabulated on *p. 71*. Its leaves are pointed and its short spikes of narrow-petalled white flowers, appearing June–September, may also be submerged. Lakes and tarns on acid soils in hill districts.

STONECROP FAMILY (Crassulaceae)

New Zealand Water Stonecrop *Crassula helmsii* (*Pl. 53*). A perennial, creeping and rooting at the nodes; has sterile submerged shoots and emergent and terrestrial flowering ones. Leaves somewhat fleshy, opposite, joined at the base, linear to narrowly oval. Flowers stalked, solitary at the base of the leaves, four-petalled, white or pale pink, with filaments and black anthers; appearing July–September. Ponds, introduced by aquarists. In Europe two other species growing in damp muddy and sandy places on acid soils that are flooded in winter could be confused with the terrestrial form of *C. helmsii*. **Northern Water Stonecrop** *C. aquatica* is annual, with pointed linear leaves and solitary, unstalked, white flowers. **Southern Water Stonecrop** *C. vaillantii* differs from the northern species in its blunt-tipped leaves and forking clusters of stalked flowers.

ST. JOHN'S WORT FAMILY (Guttiferae)

Bog St. John's Wort *Hypericum elodes* (*Pl. 54*). Grows half immersed in shallow water at the edges of pools and lakes, and on nearby damp mud. Its roundish, opposite leaves are conspicuously grey with hairs, and its five-petalled yellow flowers open much less widely than other St. John's worts.

WATERWORT FAMILY (Elatinaceae)

Waterworts *Elatine* (*Fig. 5.9*) are small, creeping plants, rooting at the nodes, growing in or by shallow water, usually on sandy or peaty soils. Leaves fresh green, often reddening, usually in opposite pairs, narrowly oval or, if submerged, more or less linear or strap-shaped. Flowers pink or whitish-pink, usually unstalked, solitary at the base of the leaves appearing July–September. The differences between the four species in the region are tabulated below, and compared with Blinks (*p. 74*).

Fig. 5.9

Tabular Key to Waterworts

	Annual/ Perennial	*Flowers stalked*	*No. of petals*	*No. of stamens*	*Leaves*
*Whorled *E. alsinastrum*	A or P	No	4	8	Whorled
Six-stamened *E. hexandra*	A	Yes	3	6	Opposite
Eight-stamened *E. hydropiper*	A	No	4	8	Opposite

*Three-stamened *E. trianda*	P	No	3**	3	Opposite
Blinks *Montia fontana*	A or P	Yes	5***	3 or 5	Opposite

* non-British
** sometimes petal-less when submerged
*** flowers white, cleistogamous when submerged

LOOSESTRIFE FAMILY (Lythraceae)

Purple Loosestrife *Lythrum salicaria* (*Pl. 55*). A tall, to 150 cm, bank plant that extends into nearby shallow water. It has unstalked, lanceolate leaves in opposite pairs or whorls of three, and conspicuous spikes of bright red-purple six-petalled flowers, appearing June–August.

Fig. 5.10

Water Purslane *L. portula* (*Fig. 5.10*), as inconspicuous as its relative is conspicuous, is a mud plant that also grows in nearby shallow water. A more or less prostrate, creeping annual, its stems and opposite oval leaves are both often reddish, and its obscure small pinkish flowers with six petals or green flowers with no petals, appearing July–August, are solitary at the base of the leaves.

WATER CHESTNUT FAMILY (Trapaceae)

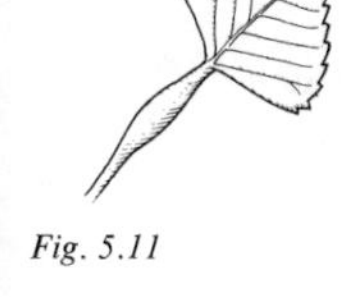

Fig. 5.11

Water Chestnut *Trapa natans* (*Fig. 5.11*) has toothed, rhomboid floating leaves, 1–4 cm long, and untoothed, linear submerged ones that soon fall. Its small, white, four-petalled flowers, appearing June–August, are at the base of the floating leaves. The nut-like fruits have two to four horns. In both still and flowing water, neither strongly acid nor strongly calcareous. Non-British.

EVENING PRIMROSE FAMILY (Onagraceae)

Fig. 5.12

Hampshire Purslane *Ludwigia palustris* (*Fig. 5.12*). Grows in and by fresh water. A creeping perennial, to 50 cm, it has reddish stems; glossy, pointed oval, red-veined, opposite leaves; and pairs of small, often red-tinged, green flowers, with no petals, at the base of the upper leaves. Rare in Britain.

WATER MILFOIL FAMILY (Haloragaceae)

Water Milfoils *Myriophyllum* (*Pls. 56, 57*) grow largely submerged in still and slow-moving water, often with long trailing stems, to 200–300 cm, and whorls of feathery leaves with hair-like segments. The whorled spikes of four-petalled flowers project above the surface, male flowers above female, with a few hermaphrodite whorls between, all with whorls of leaf-like bracts at their base; appearing June–September. There are three species, **Whorled Water Milfoil** *M. verticillatum*, **Spiked Water Milfoil** *M. spicatum* and **Alternate Water Milfoil** *M. alternifolium*, whose main characters may be summarized as follows:

Tabular Key to Water Milfoils

Leaves	Whorls of five:	Whorled
	of four:	Spiked, Alternate
	Longer than internodes between whorls:	Whorled
	Equalling the internodes:	Spiked, Alternate
Bracts	Pinnate, longer than flowers:	Whorled, Alternate (female)
	shorter than flowers:	Alternate (male)
	Undivided, shorter than flowers:	Spiked
Flowers	Whorls of five:	Whorled
	of four:	Spiked
	of 1–4:	Alternate (female)
	of 1–2:	Alternate (male)
	Hermaphrodites always present:	Whorled, Spiked
	rare:	Alternate
Petals	Red:	Spiked male
	Yellow streaked red:	Alternate male and female
	Yellow-green:	Whorled (male)
	Greenish (small):	Spiked (female)
	None:	Whorled (female)
Fruits	Smooth:	Whorled
	Dotted:	Spiked, Alternate
Habitat	Sometimes in brackish water:	Spiked
	Preferring acid water:	Alternate

MARESTAIL FAMILY (Hippuridaceae)

Marestail *Hippuris vulgaris* (*Pl. 58*). Grows in both still and slow-moving water, when it is erect and emergent, to 60 cm or more, and in fast streams, when it trails in the water like water milfoils. Its narrow, pointed leaves are in whorls of 6–12, stiff and dark green in the air, laxer and paler in the water. The tiny pink male and female flowers, with no petals, appearing June–July, are at the base of the leaves; fruit a smooth nut. Not to be confused with Horsetails (*p. 72*), which have neither leaves nor flowers.

CARROT FAMILY (Umbelliferae)

A family characterized by its small, usually white, five-petalled flowers being grouped in flat heads or umbels, which are usually in turn arranged umbrella-wise in a main compound umbel. Key characters for the aquatic species (except Marsh Pennywort) are the number of divisions in the pinnate leaves; whether the main umbels are terminal or opposite the leaves; and the presence or absence of bracts at the base of the main umbel. These are tabulated below.

Tabular Key to White-flowered Aquatic Umbellifers

	Habit	*Aerial leaves*	*Submerged leaves*	*Umbels terminal or leaf-opposed*	*Bracts*
Great Water Parsnip	Erect	1-pinnate	2–3-pinnate	T	2–6
Lesser Water Parsnip	Erect/ sprawling	1-pinnate	3–4-pinnate	L–O	many
Fine-leaved Water Dropwort	Erect/ floating	3-pinnate	3–4-pinnate	T/L–O	0–1
River Water Dropwort	Usually floating	1–2-pinnate	2-pinnate	L–O	0–1
Fool's Watercress	Creeping	1-pinnate	1-pinnate	L–O	0(1–2)
Creeping Marshwort	Creeping	1-pinnate	1-pinnate	L–O	3–7
Lesser Marshwort	Floating	1-pinnate	1-pinnate	L–O	0
Cowbane	Erect	2–3-pinnate	2–3-pinnate	T	0

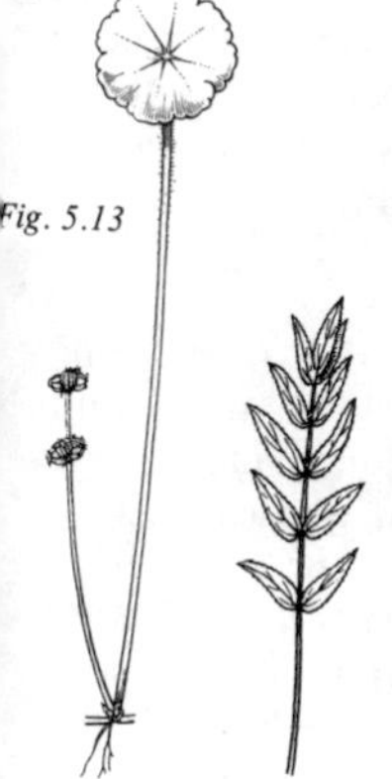

Fig. 5.13

Fig. 5.14

Marsh Pennywort *Hydrocotyle vulgaris* (*Fig. 5.13*) is atypical, having tiny, pink, single umbels at the base of almost circular leaves, which are held aloft from its creeping stems; flowering June–August. Wet marshes and very shallow water.

Great Water Parsnip *Sium latifolium* (*Fig. 5.14*). Tall, to 150 cm, the submerged leaves with linear lobes and the bracts undivided; flowering June–September. In and by still water. **Lesser Water Parsnip** *Berula erecta* (*Pl. 59*) is usually much shorter, 30–150 cm, with more sharply toothed aerial leaves, bracts more or less pinnate and flowers appearing a fortnight later.

Fine-leaved Water Dropwort *Oenanthe aquatica.* Stems to 150 cm, stem-leaves three-pinnate, lower leaves with linear or hair-like lobes, flowers appearing June–September and flower-stalks not thickened in fruit. More often in than by still or slow-moving water. **River Water Dropwort** *Oe. fluviatilis* has stem-leaves only 1–2-pinnate and flower-stalks thickened in fruit, and usually floats on the surface.

Fool's Watercress *Apium nodiflorum* (*Pl. 60*). Has roots at most leaf-junctions, scarcely or unstalked umbels and usually no bracts; flowering June–September. By, or immersed in, very shallow, still or flowing water. The rather rare **Creeping Marshwort** *A. repens* roots at all leaf-junctions and has umbels always well stalked and 3–7 bracts. The smaller and slenderer **Lesser Marshwort** *A. inundatum* is usually largely submerged in deeper water, and its lower leaves, even if not submerged, have linear or hairlike segments; flowering June–July.

The very poisonous **Cowbane** *Cicuta virosa* grows to 120 cm and has almost triangular leaves, their lobes sharply toothed, flowering July–August.

Fig. 5.15

PRIMROSE FAMILY (Primulaceae)

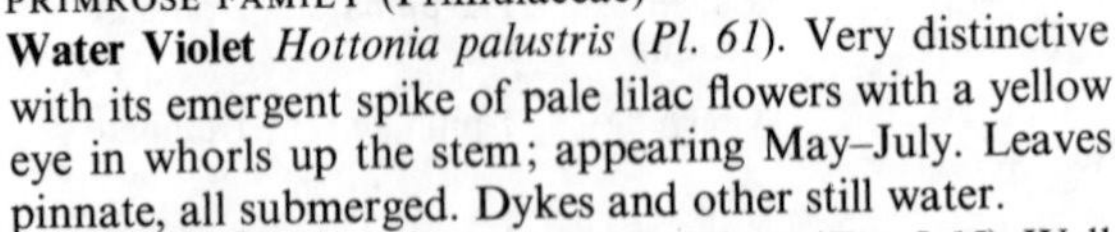

Water Violet *Hottonia palustris* (*Pl. 61*). Very distinctive with its emergent spike of pale lilac flowers with a yellow eye in whorls up the stem; appearing May–July. Leaves pinnate, all submerged. Dykes and other still water.

Creeping Jenny *Lysimachia nummularia* (*Fig. 5.15*). Well known in gardens for its yellow, bell-like flowers, appearing June–August. Stems creeping, with opposite rounded leaves. Also grows quite wild by water and may stray into it, but is then usually barren.

Tufted Loosestrife *L. thyrsiflora* (*Fig. 5.16*). A fen and marsh plant which may also fringe the shallow margins of northern lakes. Its small yellow flowers, arranged in pairs of rounded heads at the base of the upper leaves (but not right at the top), appear June–July, but it is a shy flowerer and the barren leafy stems can be puzzling.

Fig. 5.16

BOGBEAN FAMILY (Menyanthaceae)

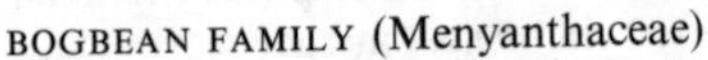

Bogbean *Menyanthes trifoliata* (*Pl. 62*). A creeping perennial, colonizing shallow pools and wet marshes with large trefoil leaves held conspicuously aloft and handsome spikes of pink flowers, often white within, whose petals are fringed with long white hairs, appearing April–June.

Fringed Water-lily *Nymphoides peltata* (*Pl. 63*). Leaves roundish to kidney-shaped floating, purplish beneath and sometimes spotted purple above, which at 3–10 cm across are much smaller than those of the true water-lilies (*p.*

Fig. 5.17

Fig. 5.18

74). Its deep yellow flowers, with five fringed petals, appearing June–September, are likewise smaller and more star-like than the yellow water-lilies. Still and slow-moving water.

BORAGE FAMILY (Boraginaceae)

Water Forget-me-nots *Myosotis* (*Pl. 64*). Bank plants that overlap slightly into the water. They have oblong to lanceolate leaves and sky-blue flowers similar to those of garden forget-me-nots, appearing June–September. The following table shows the main differences between the five species in our area, viz., **Water Forget-me-not** *M. scorpioides* (*Fig. 5.17*), **Creeping Forget-me-not** *M. secunda* (*Fig. 5.18*), **Pale Forget-me-not** *M. stolonifera*, **Tufted Forget-me-not** *M. laxa* and **Jersey Forget-me-not** *M. sicula*.

Tabular Key to Water Forget-me-nots

Hairs on stems and sepals closely appressed:	Water, Pale, Tufted, Jersey
spreading on lower stem:	Creeping
sometimes none:	Tufted
Leaves short and bluish-green:	Pale
Flowers pale blue:	Pale, Jersey
4–10 mm:	Water
4–6 mm:	Creeping
4–5 mm:	Pale
2–4 mm:	Tufted
2–3 mm:	Jersey
Petals notched:	Water, Creeping, Pale
not notched:	Pale, Tufted, Jersey
Sepals cut to $\frac{1}{3}$:	Water
to $\frac{1}{2}$:	Creeping, Jersey
to $\frac{2}{3}$:	Pale
to $\frac{3}{4}$:	Tufted
blunt:	Jersey
Fruiting stalks 1–2 times as long as sepal-tube:	Water
1–3 times:	Jersey
2–3 times:	Tufted
3–5 times:	Creeping, Pale

WATER STARWORT FAMILY (Callitrichaceae)

Water Starworts *Callitriche* (*Pl. 65*). A difficult group, their leaf-shape varying with the presence or depth of water and the ripe fruit, often needed for certain identification, being sometimes hard to find. Leaves opposite, oval or rhomboid to linear, often forming a rosette on the surface or on wet mud. Flowers tiny, green, with no petals, male (with yellow anthers) and female separate and usually solitary at the base of the leaves; appearing April–September. Fruits usually brown, more or less rounded or globular, not or

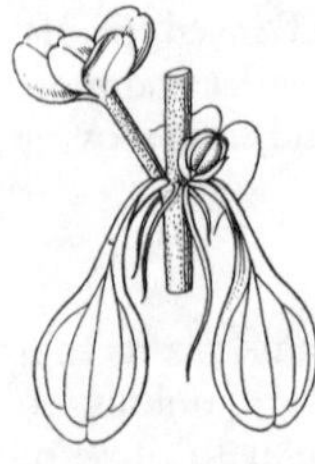
Fig. 5.19

scarcely stalked, the styles spreading or erect and the segments (mericarps) winged (an important character). Still and flowing water, some species always submerged, others either submerged, partly emergent or terrestrial on wet mud.

They are here arranged in the three groups suggested by the Dutch botanist H. D. Schotsman. Principal characters only are given in the species descriptions, all leaf and fruit characters being tabulated below.

1. Plant always submerged, with transparent linear leaves.

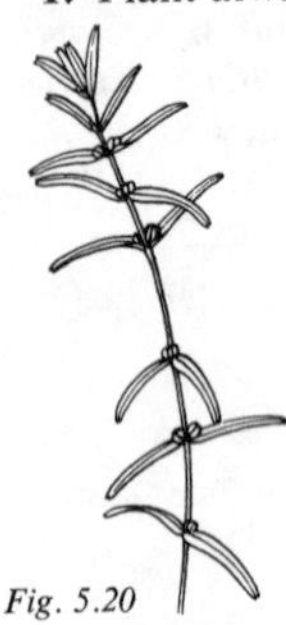
Fig. 5.20

Autumnal Water Starwort *C. hermaphroditica* (*Fig. 5.20*) has leaves widest at the base, tapering to a distinctly notched tip, and fruits unstalked or almost so, broadly-winged and the styles often turned down. Often in larger waters or canals.

Short-leaved Water Starwort *C. truncata* has leaves shorter than the last species and only shallowly notched or even square-tipped, and fruits wider than long, distinctly stalked and not or only narrowly winged. In shallower water.

2. Plant with three forms: submerged and sterile, aquatic with floating rosette, and terrestrial. Submerged leaves not transparent.

Common Water Starwort *C. stagnalis* (*Fig. 5.19*) is generally the commonest species, and the only one growing regularly on shady damp ground away from water. Rosette and terrestrial leaves varying from broadly elliptical to spathulate or almost rounded. Male and female flowers sometimes paired at base of same leaf. Fruits pale brown, broadly winged; in terrestrial form the styles recurved.

Fig. 5.21

Blunt-fruited Water Starwort *C. obtusangula* has rosette leaves markedly fleshy, rhomboidal and more numerous (12–20) than the other species, making it the easiest water starwort to recognize in the vegetative state. Fruits elliptical, unwinged. Often in slow-moving, especially nutrient-rich, water.

Blunt-winged Water Starwort *C. cophocarpa* has rosette leaves elliptical to slightly rhomboid, and fruits not winged but with a distinct blunt margin. Often in calcareous waters on sand and mud.

Various-leaved Water Starwort *C. platycarpa* (*Fig.*

5.21) has dark green elliptical rosette and terrestrial leaves, and narrowly winged fruits. Prefers nutrient-rich waters.

Marsh Water Starwort *C. palustris* has rosette leaves elliptical to almost rounded, and blackish oval fruits, winged at the tip. Shallow, still, nutrient-poor water, not in marshes.

3. Plant with three forms, all fruiting: submerged, aquatic with floating rosette, and terrestrial. Submerged leaves not transparent.

Hooked Water Starwort *C. hamulata* has submerged leaves linear with a broader, deeply notched tip and rosette leaves elliptical to slightly oval. Fruits scarcely stalked, narrowly winged, with styles turned down and appressed to the fruit. Cool, nutrient-poor lakes and streams.

Intermediate Water Starwort *C. brutia* is slenderer than the last species, with the submerged leaves irregularly notched and not broader at the tip, and fruits of terrestrial plants distinctly stalked. Still, often shallow water. This and *C. hamulata* used to be lumped together as *C. intermedia*, and small aquatic forms are hard to separate.

Tabular Keys to Water Starworts

1. Submerged Leaves

	Shape	*Tip*	*Green colour*
C. hermaphroditica Autumnal	widest at base	distinctly notched	transparent
C. truncata Short-leaved	widest at base	shallowly notched or square	transparent
C. stagnalis Common	narrowly elliptical	notched	mid
C. obtusangula Blunt-fruited	almost linear	notched	mid
C. cophocarpa Blunt-winged	linear	notched	mid
C. platycarpa Various-leaved	linear	notched	mid
C. palustris Marsh	linear	notched	mid
C. hamulata Hooked *C. brutia* Intermediate	linear	broad, deeply notched	mid

2. Rosette Leaves

	No. of leaves	Shape	Green colour
C. stagnalis Common	c6	broadly elliptical/ spathulate/almost rounded	pale
C. obtusangula Blunt-fruited	12–20	rhomboid	mid
C. cophocarpa Blunt-winged	c10	elliptical/slightly rhomboid	mid
C. platycarpa Various-leaved	6–8	elliptical	dark
C. palustris Marsh	c10	elliptical/almost rounded	mid
C. hamulata Hooked *C. brutia* Intermediate	c10	elliptical/slightly oval	mid

3. Terrestrial Leaves

	Shape	Green colour
C. stagnalis Common	as rosette	mid
C. obtusangula Blunt-fruited	narrower	yellowish
C. cophocarpa Blunt-winged	elliptical	mid
C. platycarpa Various-leaved	as rosette	mid
C. palustris Marsh	short elliptical	mid
C. hamulata Hooked	elliptical	dark
C. brutia Intermediate	elliptical	dark

4. Fruits

	Shape	*Colour*	*Segments*	*Styles*
C. hermaphroditica Autumnal	suborbicular		broadly winged, at least near top	spreading or down-turned
C. truncata Short-leaved	wider than long		not or only narrowly winged	spreading or down-turned
C. stagnalis Common	suborbicular	pale brown	broadly winged	erect or spreading: recurved on mud
C. obtusangula Blunt-fruited	elliptical	brown	unwinged	erect or spreading
C. cophocarpa Blunt-winged	suborbicular/ oblong	brown	unwinged	erect
C. platycarpa Various-leaved	suborbicular	brown	narrowly winged	erect or spreading
C. palustris Marsh	obovate	blackish	winged at tip	erect
C. hamulata Hooked	suborbicular		narrowly winged	down-turned, appressed
C. brutia Intermediate	suborbicular		broadly winged	down-turned, appressed

MINT FAMILY (Labiatae)

Marsh Woundwort *Stachys palustris* (*Pl. 66*). A bank plant that strays into the water, is faintly aromatic and has opposite pairs of narrow leaves and spikes of pinkish-purple two-lipped flowers, appearing June–October.

Water Mint *Mentha aquatica* (*Pl. 67*). Another straying bank plant, strongly aromatic, with broader leaves and roundish heads of much smaller lilac two-lipped flowers, appearing July–September.

FIGWORT FAMILY (Scrophulariaceae)

Monkey Flower *Mimulus guttatus* (*Pl. 68*). Often overlaps from the bank into the water by streams and pools. It has opposite pairs of pointed oblong leaves and very showy, large, red-spotted, bright yellow flowers, appearing June–September. Two less common relatives are **Blood-drop Emlets** *M. luteus* (*Pl. 69*), smaller with large red blotches (and often hybridizing with Monkey Flower), and **Musk** *M. moschatus*, stickily hairy all over and with smaller all-yellow flowers. All three are introduced from the Americas.

Water Speedwells *Veronica* Three mud plants that often grow in nearby shallow water. They are easily told by flower colour. **Brooklime** *V. beccabunga* (*Pl. 70*) has broad leaves and bright blue flowers, appearing May–September. **Water Speedwell** (*Fig. 5.22*) *V. anagallis-aquatica* (*Pl. 71*) has narrower leaves and pale blue flowers, appearing June–August. **Pink Water Speedwell** *V. catenata* is similar but with pale pink flowers.

For **Mudwort** *Limosella aquatica* see under Shoreweed (*p. 89*).

BLADDERWORT FAMILY (Lentibulariaceae)

Bladderworts *Utricularia* are rootless insectivorous plants, their leaves having both numerous thread-like green segments, usually ending in short bristles, and tiny bladders that entrap tinier animals. Flowers yellow, two-lipped, spurred, in emergent leafless spikes. Growing in still, usually fairly deep, water, and hard to identify except in flower (see table below).

Tabular Key to Bladderworts

	Spike height in cm	*No. of flowers in spike*	*Flower size in mm*	*Shade of yellow*	*Spur shape*
Greater	10–30	4–10	12–18	deep	bluntly pointed
Southern	10–30	4–10	12–18	lemon	blunt
Intermediate	10–20	2–4	10–15	deep with red-brown lines	long, narrowly pointed
Lesser	4–15	2–8	6–8	pale	short, blunt

Greater Bladderwort *U. vulgaris* (*Pl. 72*) has all leaves pinnately divided, with both green segments and bladders, and flowers with the lower lip turned down, on stalks 2–3 times as long as their bract. **Southern Bladderwort** *U. australis* is very similar, but has the lower lip flat with a wavy margin and stalks 3–5 times as long as the bract.

Intermediate Bladderwort *U. intermedia* has two kinds of leaves: green, palmately lobed, bladderless and floating; and colourless, with bladders and more or less sunk in the bottom mud. Usually in shallow, peaty water. **Pale Yellow Bladderwort** *U. ochroleuca* is intermediate between this and the next species and may be their hybrid. **Lesser Bladderwort** *U. minor* has similar leaves but without bristles, and grows especially in shallow peaty moorland pools.

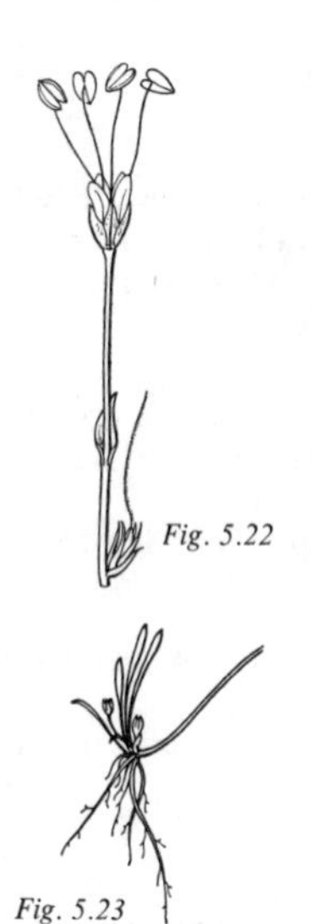

Fig. 5.22

Fig. 5.23

Fig. 5.24

PLANTAIN FAMILY (Plantaginaceae)

Shoreweed *Littorella uniflora* (*Fig. 5.22*). One of the plants with submerged rosettes of strap-shaped leaves tabulated on *p. 71*. It never in fact flowers when wholly submerged, but the barren plants with their creeping runners may form an underwater sward. Its solitary greenish-white four-petalled flowers, the stalked males with prominent stamens and the females unstalked, appear on sand or gravel by shallow non-limy water June–September. The semi-cylindrical linear leaves then also distinguish it from **Mudwort** *Limosella aquatica* (Scrophulariaceae) (*Fig. 5.23*), which grows on mud, not normally submerged, and has long-stalked elliptical leaves with white or pinkish flowers, all stalked and with both stamens and styles, and also appearing June–September.

BELLFLOWER FAMILY (Campanulaceae)

Water Lobelia *Lobelia dortmanna* (*Fig. 5.24*). Another plant with submerged rosettes of strap-shaped leaves (see *p. 71*), which are blunt-tipped and often recurved. Its pale lilac, two-lipped flowers are in a spike on stems leafless except for a few obscure flap-like leaves, and appear in July–August, in lakes and tarns on acid soils in hill districts.

2. MONOCOTYLEDONES

(Plants with one seed-leaf or cotyledon)

WATER-PLANTAIN FAMILY (Alismataceae)

Perennials with three-petalled flowers. Most grow in and by still and slow-flowing water. Some leaf and flower characters are tabulated below.

Tabular Key to Water Plantains and Arrowheads (Alismataceae)

Aerial Leaves	
arrow-shaped:	Arrowhead, Duck Potato, Stiff Arrowhead
heart-shaped:	Heart-shaped Water-plantain
ovate/oblong:	Duck Potato, Floating Arrowhead, Stiff Arrowhead. Common Water-plantain, Star-fruit
lanceolate:	Lesser Water-plantain, Narrow-leaved Water-plantain. Ribbon-leaved Water-plantain
linear:	Floating Arrowhead, Stiff Arrowhead

Floating Leaves

heart-shaped: Heart-shaped Water-plantain
kidney-shaped: Frogbit (*p. 91*)
ovate/oblong: Arrowhead, Floating Arrowhead (often with 2 basal lobes), Duck Potato, Stiff Arrowhead, Ribbon-leaved Arrowhead, Floating Water-plantain, Star-fruit
linear: Floating Arrowhead, Stiff Arrowhead

Flowers

all white: Floating Arrowhead, Duck Potato, Stiff Arrowhead. Ribbon-leaved Arrowhead, Ribbon-leaved Water-plantain, Heart-shaped Water-plantain
white with yellow spot: Stiff Arrowhead, Floating Water-plantain, Star-fruit. Frogbit (*p. 91*)
white with purple spot: Arrowhead
lilac: Common Water-plantain, Ribbon-leaved Water-plantain
pink: Lesser Water-plantain, Narrow-leaved Water-plantain

Arrowhead *Sagittaria sagittifolia* (*Pl. 74*) grows to 90 cm, and has large aerial leaves shaped like an arrowhead, as well as lanceolate floating ones and ribbon-like submerged ones. The showy whorled racemes of separate male and female white flowers, 20 mm across, usually have a purple patch on each petal, and purple anthers; appearing July–August.

Fig. 5.25

Floating Arrowhead *S. natans* (*Fig. 5.25*), a Scandinavian species, has all leaves linear to oblong, often with two basal lobes, floating and sometimes also aerial; flowers all white with yellow anthers; appearing July–August.

Three North American arrowheads with all-white flowers are occasionally naturalized. **Duck Potato** *S. latifolia* has aerial leaves only, arrow-shaped or ovate; anthers yellow. **Stiff Arrowhead** *S. rigida* has ovate aerial leaves, flower-spike often bent near the lowest whorl and petals sometimes with a yellow patch. The much shorter **Ribbon-leaved Arrowhead** *S. subulata* has all leaves ribbon-like or the floating ones lanceolate.

Lesser Water-Plantain *Baldellia ranunculoides* is smallish, to 20 cm, with narrow lanceolate aerial leaves, and whorls of long-stalked pale pink flowers, 10–15 mm across, appearing June–September. A creeping variety has rooted tufts of leaves and flowers sometimes solitary. Prefers shallow, acid, often peaty water.

Fig. 5.26

Floating Water-Plantain *Luronium natans* (*Fig. 5.26*) grows only in the water and has both elliptical floating

Fig. 5.27

leaves and linear submerged ones. Flowers solitary, with a yellow spot on each white petal, 12–15 mm across, appearing May–August.

Common Water-Plantain *Alisma plantago-aquatica* (*Pl. 73*) grows to 100 cm and has broad lanceolate to ovate leaves, rounded or heart-shaped at the base. Flowers in whorled panicles, long-stalked, pale lilac, 8–10 mm across, appearing June–September. Fruits with a long beak arising at or near the middle. **Narrow-leaved Water-Plantain** *A. lanceolatum* (*Fig. 5.27*) has narrower leaves, tapering to the stalk, pinker flowers and fruits with a short beak arising near to tip.

Ribbon-leaved Water-Plantain *Alisma gramineum* (incl. the Scandinavian *A. wahlenbergii*), 5–30 cm high, has long-stalked, narrowly lanceolate aerial leaves and ribbon-like floating ones. Flowers white or lilac, 5–7 mm across, appearing July–September.

Fig. 5.28

Heart-shaped Water-Plantain *Caldesia parnassifolia* (*Fig. 5.28*) grows to 90 cm and has long-stalked, heart-shaped leaves, both aerial and floating. Flowers white, 10–15 cm across, appearing July–September. Non-British.

Star-fruit *Damasonium alisma* (*Fig. 5.29*). Only 5–30 cm high, it has bluntly oval leaves floating, submerged or, when growing on mud, aerial. Flowers usually in small whorls, about 6 mm across, petals white with a yellow spot, appearing June–September. Fruits spreading like a six-pointed star.

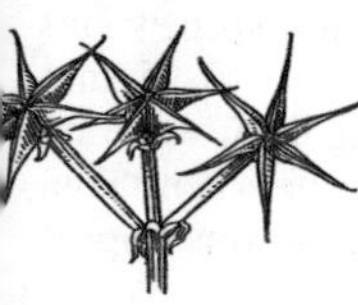

Fig. 5.29

FLOWERING RUSH FAMILY (Butomaceae)

Flowering Rush *Butomus umbellatus* (*Pl. 75*). Tall, to 150 cm, with long three-cornered grass-like or rush-like leaves, and conspicuous umbels of bright pink three-petalled flowers, appearing July–August. In and by still and slow-moving water.

FROGBIT FAMILY (Hydrocharitaceae)

Frogbit *Hydrocharis morsus-ranae* (*Pl. 76*). Free-floating in still water, with bronzey-green kidney-shaped floating leaves; flowers about 2 cm across with a yellow spot on each of their three white petals, appearing July–August.

Water Soldier *Stratiotes aloides* (*Pl. 77*). Rests on the bottom of ponds and other still water for most of the year. but in late spring the whole plant rises to the surface, so that both its lanceolate spine-toothed leaves and its 3–4-cm three-petalled white flowers, appearing June–August, emerge above the surface.

Water-thymes are submerged rooted perennials of still and slow-moving water, mostly aliens to Europe, introduced, either accidentally or carelessly, by aquarists during the present century, although Canadian Waterweed was first seen in Ireland in 1836. They are best distinguished by their leaves (see table below), which are all minutely toothed. Their three-petalled flowers float on the surface, male and female on separate plants; females float at the end of long thread-like stalks, but males, which are uncommon in Europe, often break off and float free.

Tabular Key to Water-thyme Leaves

Tip pointed:	Nuttall's, Greater, Curly, Esthwaite Waterweed
Tip bluntly pointed:	Canadian Waterweed, large-flowered
No. in whorls:	Curly, alternating spirally; Canadian water 2–3, Greater 3, Nuttall's 3–4, Large-flowered 3–5, Esthwaite Waterweed 3–8.
max length (mm):	Canadian Waterweed 15, Nuttall's 15, Esthwaite Waterweed 20, Greater 20, Curly 30, Large-flowered 40.
max width (mm):	Nuttall's 1.8, Canadian waterweed 2, Esthwaite waterweed 2, Greater 2.5, Curly 3, Large-flowered 5.

Canadian Waterweed or Water-thyme *Elodea canadensis* (*Pl. 78*), until recently much the commonest species, has oblong, bluntly pointed, dark green leaves, 10–15 mm long and 2 mm wide, either in overlapping whorls of 3 up the stem or paired at its base; not collapsing when taken from the water. The white or lilac-tinged flowers appear May–September.

Nuttall's Water-thyme *Elodea nuttallii*, also from North America, is now supplanting Canadian waterweed in some areas. It has longer, narrower, more lanceolate and pointed, paler green leaves, in more widely-spaced whorls of 3–4 that collapse out of the water, and smaller, white flowers.

Greater Water-thyme *Elodea ernstiae* (*E. callitrichoides*) from South America is larger than Canadian Waterweed, with longer leaves, to 25 mm, that taper to a fine point, in whorls of three; flowers larger, always white.

Large-flowered Water-thyme *Egeria densa*, also from South America, has still longer and broader but bluntly pointed leaves, to 40 mm long and 5 mm across, more

Fig. 5.30

densely packed in whorls of 3–5; flowers white, 15–20 mm across, the males in heads of 2–4, appearing May–August.

Curly Water-thyme *Lagarosiphon major* from South Africa has its narrow, markedly curved leaves, densely crowded spirally up the stem; flowers pinkish, with many male flowers to a head.

Esthwaite Waterweed *Hydrilla verticillata* (*Fig. 5.30*) is the only member of the group native to Europe, where it is rare, extinct in Britain and in only one lough in Ireland. Its linear leaves are more distinctively toothed, in whorls of 3–8, and its transparent, red-streaked flowers appear July–August.

Tapegrass *Vallisneria spiralis* has long ribbon-like leaves coming direct from its roots, 3–5-veined and slightly toothed at the tip. Flowers pinkish-white, male and female on separate plants; males many in a head, often breaking off and floating free; females solitary, the fruiting stalks twisting; appearing June–October. Native as far north as central France, introduced further north.

CAPE PONDWEED FAMILY (Aponogetonaceae)

Cape Pondweed or Water Hawthorn *Aponogeton distachyos*. Has long-stalked, oblong, shining floating leaves and emergent forked spikes of fragrant white flowers with dark purple anthers, appearing April–October. A South African plant sometimes naturalized in ponds and other still water.

PONDWEED FAMILY (Potamogetonaceae)

Pondweeds *Potamogeton* grow in still or slow-moving water and have floating and/or submerged alternate leaves, all very variable in size, the submerged ones and their stipules translucent, the floating ones usually green. Flowers small, green, four-sepalled, with no petals, in spikes, appearing May–August. There are numerous hybrids, but only *P.* × *zizii*, is described here, as it is the only one to produce fruits.

The first twenty of the twenty-three species described have their flower spikes projecting above the surface. The first seven species have floating leaves.

Broad-leaved Pondweed *P. natans* (*Pl. 79*), the commonest and most conspicuous species with floating leaves, has these ovate-elliptical, 3–12 cm × 1–7 cm, and (unlike all other pondweeds) jointed at their junction with a long stalk. Submerged leaves linear, 15–30 cm × 1–3 mm; stipules large, 5–12 cm. Still water only.

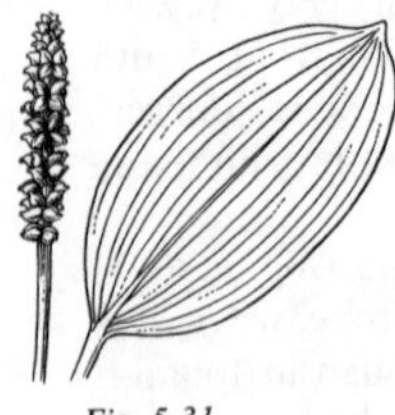

Fig. 5.31

Bog Pondweed *P. polygonifolius* (*Fig. 5.31*) has floating leaves ovate-lanceolate, often reddish, 2–6 cm × 1–4 cm; submerged leaves stalked, narrower, 8–20 cm × 1–3 cm; stipules blunt, 2–4 cm; and fruits red-brown. Shallow, usually acid water and bogs, often almost terrestrial. Its calcareous-water counterpart is **Fen Pondweed** *P. coloratus*, with floating leaves translucent, net-veined and 2–7 cm × 2–5 cm; submerged leaves longer and narrower; stipules joined near the base; and fruits greenish.

Loddon Pondweed *P. nodosus* has floating leaves elliptical or ovate-elliptical, 6–15 cm × 3–6 cm; submerged leaves longer, narrower, net-veined, stalked, and 10–20 cm × 2–4 cm; and stipules conspicuous, pointed, 7–10 cm and joined right at the base. Slow-flowing rivers.

Fig. 5.32

American Pondweed *P. epihydrus* has floating leaves oblong-elliptical, 4–7 cm × 1–2 cm; submerged leaves ribbon-like, 5–7-veined, unstalked, 8–20 cm × 3–8 mm; and stipules very short, ovate. Native in the Hebrides, introduced elsewhere, including canals in West Yorkshire.

The next two species may or may not have floating leaves.

Various-leaved Pondweed (*Fig. 5.32*) *P. gramineus* is often tinged red, and may have a few floating leaves, elliptical to ovate-elliptical, long-stalked, 3–7 cm × 1–2.5 cm; submerged leaves narrower, unstalked, 3–8 cm × 3–12 mm; stipules conspicuous, green, pointed, 2–5 cm; and flower stalks thickened upwards. *P.* × *zizii* is its hybrid with Shining Pondweed and the only European pondweed hybrid to produce fruits. It rarely has floating leaves, but its submerged leaves are broader and more tapered to the base than *gramineus* and differ from *lucens* in being very shortly or unstalked; stipules two-keeled like *lucens*. **Reddish Pondweed** *P. alpinus* is often all tinged red and may have long-stalked, elliptical floating leaves, 3–8 cm × 1–2 cm; submerged leaves lanceolate to linear-lanceolate, net-veined, short-stalked, 6–15 cm × 1–2 cm; stipules conspicuous, blunt, 2–6 cm; and flower-stalks not thickened upwards. Mainly in non-calcareous water.

The remaining sixteen species have submerged leaves

only, all except the first four narrowly ribbon-like or thread-like, and all except Shining Pondweed unstalked.

Fig. 5.33

Shining Pondweed *P. lucens* has leaves, broad, more or less elliptical, wavy-edged, shortly stalked, 10–20 cm × 3–5 cm; and stipules large, bluntly pointed, conspicuously two-keeled, 3–8 cm. Usually in calcareous water.

Long-stalked Pondweed *P. praelongus* has leaves broadly strap-shaped but pointed, wavy-edged, half-clasping the stem, with a hooded tip, to 20 cm × 4.5 cm; stipules long, blunt, not ridged or keeled, 1–6 cm; and flower spike rather loose.

Perfoliate Pondweed *P. perfoliatus* (*Fig. 5.33*) has leaves dark green, ovate-lanceolate, clasping the stem, 2–6 cm × 1–4 cm; and stipules small, to 1 cm, blunt, not keeled and not persistent.

Fig. 5.34

Curled Pondweed *P. crispus* has leaves often reddish, strap-shaped, markedly wavy, toothed, 3–9 cm × 8–15 mm; stipules short, 1–2 cm, blunt and triangular; flowers sometimes reddish; and fruits with a long curved beak.

Opposite-leaved Pondweed *Groenlandia densa* (*Fig. 5.34*) has distinctive opposite ovate-lanceolate leaves, to 4 cm × 1.5 cm, no stipules and very short flower spikes.

A group of eleven pondweeds without floating leaves have their submerged ones narrow and grass-like, the first five 2–4 mm wide, the rest usually less than 2 mm. The leaves are either sharply pointed, bluntly pointed, or blunt with a minute point at the tip.

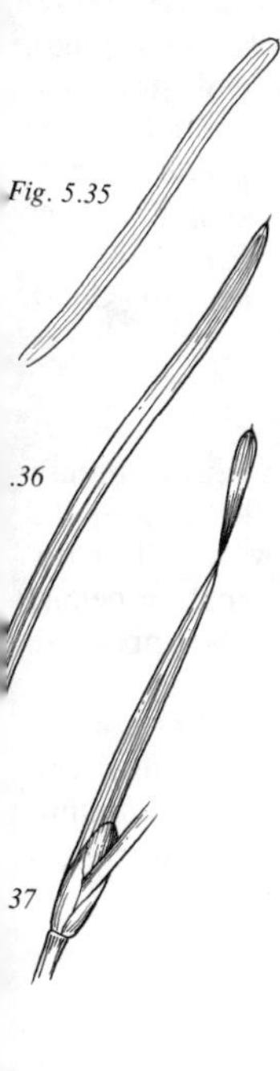
Fig. 5.35

.36

37

Blunt-leaved Pondweed *P. obtusifolius* (*Fig. 5.35*) is one of the two with the broadest leaves, 2–4 mm, three-veined, rounded and minutely pointed at the tip, with open and transparent stipules. The other, **Sharp-leaved Pondweed** *P. acutifolius* (*Fig. 5.36*), with leaves usually 2–3 mm, many-veined and sharply pointed at the tip and similar stipules, is confined to calcareous water. **Grass-wrack Pondweed** *P. compressus* (*Fig. 5.37*) differs from Blunt-leaved Pondweed in its markedly compressed stems, many-veined leaves and many-flowered spikes.

Flat-stalked Pondweed *P. friesii* also has flattened stems, with leaves 2–3 mm, five-veined and either blunt and minutely pointed at the tip, or tapering to a sharp point; stipules whitish and tubular at the base at first; and flower spikes with 3–4 separate whorls.

Fig. 5.38

Fig. 5.39

Small Pondweed *P. berchtoldii* (*Fig. 5.38*) has leaves 1–2 mm, three-veined, blunt and minutely pointed at the tip; stipules open, blunt; and flower spike 2–8 mm. Has a wide tolerance of both calcareous and acid water. The somewhat similar **Lesser Pondweed** *P. pusillus*, has narrower, bluntly pointed leaves, rarely more than 1 mm; stipules tubular to above half-way; and flower spike 6–12 mm. Calcareous water only.

Hair-like Pondweed *P. trichoides* (*Fig. 5.39*) also has very narrow, thread-like leaves, up to 1.5 mm but usually less than 1 mm, which are three-veined, gradually tapering to a fine point, and open stipules; flower spike 10–15 mm, longer than the two previous species. The very similar northern and far western **Shetland Pondweed** *P. rutilus* has tubular stipules, pointed and markedly veined, and shorter flower spikes, 5–10 mm.

Three pondweeds have their flowers as well as their leaves submerged; see also **Horned Pondweed** *Zannichellia* below.

Slender-leaved Pondweed *P. filiformis* has very narrow, almost hairlike leaves, usually less than 1 mm, blunt and yellowish-green; stipules tubular; and flower spikes with widely spaced whorls.

Fennel Pondweed *P. pectinatus* (*Fig. 5.40*) has leaves usually less than 2 mm, but pointed and with open stipules, the flower spikes with the whorls fairly close together. Base-rich and brackish water, often in rivers and canals; common in parts of the Thames. **Sheathing Pondweed** *P. vaginatus*, local in brackish water in Scandinavia, has some leaves broader and blunter than either of the last two species, the stipules open and the flower spikes with only short gaps between the whorls.

Fig. 5.40

TASSELWEED FAMILY (Ruppiaceae)

Tasselweeds *Ruppia* are submerged waterweeds of brackish or occasionally freshwater pools and ditches near the sea. Leaves thread-like, up to 1 mm wide, alternate. Flowers hermaphrodite, greenish, with no sepals or petals, in pairs, floating on the surface on long stalks; appearing July–September.

Beaked Tasselweed *R. maritima* (*Fig. 5.41*) has pointed leaves and somewhat recurved fruiting stalks, to 6 cm. **Spiral Tasselweed** *R. cirrhosa* has darker green, blunt-tipped leaves, and longer, to 8 cm, spirally coiled fruiting

Fig. 5.41

Fig. 5.42

stalks. Sometimes in slow-flowing water, and in brackish pools inland.

HORNED PONDWEED FAMILY (Zannichelliaceae)

Horned Pondweed *Zannichellia palustris* (*Fig. 5.42*). A wholly submerged waterweed of still or slow-moving fresh or brackish water. Leaves thread-like, up to 2 mm wide, usually opposite, with stipules clasping the stem but soon falling. Flowers green, with no sepals or petals, the male stalked and the female unstalked, on the same plant, solitary or in small clusters at the base of the leaves; appearing May–August.

NAIAD FAMILY (Najadaceae)

Naiads *Najas* are wholly submerged waterweeds of lakes and tarns. Leaves narrow, opposite or in whorls of three. Flowers green, unstalked, 1–3 together at the base of the leaves; the two-lipped males with a conspicuous spathe-like bract, the females without sepals, petals or bract; appearing August–September.

Fig. 5.43

Holly-leaved Naiad *N. marina* (*Fig. 5.43*) has distinctive broad leaves, to 6 mm wide, with conspicuous, holly-like teeth, and male and female flowers on different plants. Fresh or brackish water; in Britain confined to the Norfolk Broads.

Fig. 5.44

Slender Naiad *N. flexilis* (*Fig. 5.44*) has thread-like leaves, less than 1 mm wide, with more than twenty tiny teeth; male and female flowers on the same plant. Fresh water in the north. The non-British Lesser Naiad *N. minor* has broader leaves, to 3 mm, with fewer than seventeen small teeth.

Tabular Key to Pondweeds and Other Submerged Waterweeds with Grass-like or Thread-like Leaves (cf Water Milfoils, *p. 80*)

Leaves:	alternate:	*Potamogeton, Ruppia*
	opposite:	*Zannichellia, Najas*
	whorls of three:	*Ceratophyllum* (*p. 74*) *Najas*
	forked:	*Ceratophyllum* (*p. 74*)
	5 mm or more wide:	*Potamogeton epihydrus, P. obtusifolius, P. acutifolius, P. compressus, P. friesii, Najas marina, N. minor*
	2–5 mm wide:	*Potamogeton berchtoldii, P. pectinatus, P. vaginatus, Zannichellia*

	up to 1 mm wide:	*Ceratophyllum* (*p. 74*), *Potamogeton pusillus, P. trichoides, P. rutilus, P. filiformis, Ruppia, Naias flexilis, Scirpus fluitans* (*p. 104*)
Flowers:	emergent spikes:	*Potamogeton* (except three species below)
	floating on surface:	*Ruppia*
	submerged spikes:	*Potamogeton filiformis, P. pectinatus, P. vaginatus*
	submerged at base of leaves:	*Ceratophyllum* (*p. 74*), *Zannichellia, Ruppia*
	hermaphrodite:	*Potamogeton, Ruppia*
	monoecious (male & female separate on same plant:	*Ceratophyllum* (*p. 74*), *Zannichellia, Najas flexilis, N. minor*
	dioecious (male and female on different plants):	*Najas marina*

NB Most of the vernacular names in this group are so artificial and rarely used by those interested in the plants, that only the scientific names are given in this key.

IRIS FAMILY (Iridaceae)

Yellow Iris or Yellow Flag *Iris pseudacorus* (*Pl. 80*). A marsh or swamp plant that sometimes gets right into the water. Its broad, sword-shaped leaves and conspicuous yellow flowers, appearing May–August, are unmistakable.

RUSH FAMILY (Juncaceae)

Bulbous Rush *Juncus bulbosus* (incl. *J. kochii*) (*Fig. 5.45*). The only truly aquatic rush of the region, having a form with long floating stems. Leaves hollow, grass-like, slightly flattened or grooved, often jointed and sometimes bronzy or reddish in the water. Flowers small, with six petals (or sepals), green or brown, in a forking cluster of up to twenty small heads, often replaced by green shoots; appearing June–August.

Fig. 5.45

Several other rush species overlap into the water, but are not true aquatics. The most frequent of these are three that have no leaves, but clusters of flowers near the top of a stiff stem: **Hard Rush** *J. inflexus* (*Pl. 81*) with ridged greyish stems and a fairly loose flower cluster; *Soft Rush J. effusus* (*Pl. 82*) with smooth or faintly ridged green stems and flower clusters tight or loose; and **Compact Rush** *J. conglomeratus* (*Pl. 83*) with conspicuously ridged green stems and flowers in a tight cluster.

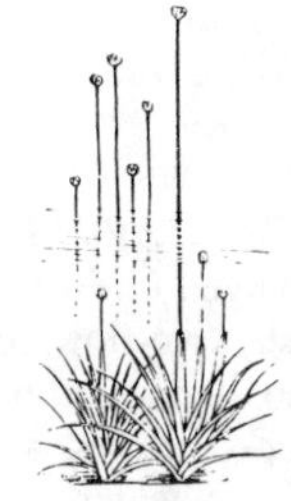
Fig. 5.46

PIPEWORT FAMILY (Eriocaulaceae)

Pipewort *Eriocaulon septangulare* (*Fig. 5.46*). One of the plants with submerged rosettes of strap-shaped leaves tabulated on *p. 71*. Its leaves are translucent and pointed, and its flat button-like heads of two-petalled white flowers, usually male in the middle and female on the edge, are borne in leafless emergent spikes; appearing July–September. Shallow tarns, pools and lake margins on peaty soils in western Scotland and western Ireland; otherwise only in North America.

GRASS FAMILY (Gramineae)

Only eight species of grass are in any true sense aquatic, and these can be neatly divided into four tall species, attaining 200 cm or more, and four shorter species, not exceeding 120 cm. The four tall species all make patches or more extensive beds in and by still and slow-moving fresh water, and have their flowers in panicles with unawned spikelets. Their distinctive features are as follows:

Common Reed: very tall, no ligule, silky hairs on spikelets, late flowering.
Reed Canary-grass: ligule torn, panicle compressed.
Reed Sweet-grass: short pointed ligule, brown spot at leaf junctions on stems.
Swamp Grass: narrow leaves, long pointed ligule.

Fig. 5.47

Fig. 5.48

Common Reed *Phragmites australis* (*Pl. 84*). The tallest grass, capable of reaching 350 cm, and forming the most extensive beds in and by both fresh and brackish water. Its stems persist through the winter as hard canes, but its broad, to 5 cm, greyish, smooth-edged leaves shrivel up in winter. It is the only aquatic grass with a ring of hairs instead of a ligule. Panicle spreading, the spikelets dark purple with long silky hairs; appearing August–October.

Reed Canary-grass *Phalaris arundinacea* (*Fig. 5.47*). Leaves green, rough-edged, to 18 mm wide, persisting through the winter when dead; ligules long, torn. Panicle compact, lobed, pale purple, white or pale yellow; appearing June–August.

Reed Sweet-grass *Glyceria maxima* (*Pl. 85*). Leaves bright green, smooth-edged, not overwintering, to 18 mm wide and having a brown mark at their stem-junction, and a short rounded ligule ending in a sharp point. Panicle spreading, green, sometimes tinged purple or yellow; appearing June–August.

Swamp Grass *Scolochloa festucacea* (*Fig. 5.48*). Leaves

Fig. 5.49

Fig. 5.50

narrower, to 12 mm, rough-edged, with long pointed ligules. Panicles spreading, green; appearing June–July. Non-British.

The four shorter species all have unawned spikelets.

Whorl Grass *Catabrosa aquatica* (*Fig. 5.49*). The branches of its spreading panicle are arranged in distinctive alternate half-whorls, the purplish two-flowered spikelets appearing May–July. It grows to 70 cm and has creeping runners, sweet-tasting blunt-tipped leaves, and long blunt ligules. In and by shallow fresh water, and often on wet mud. On wet mud it could be confused with **Creeping Bent** *Agrostis stolonifera* (*Fig. 5.50*), with even longer runners, pointed ligules, complete whorls, one-flowered spikelets and panicles contracted in fruit.

The three smaller sweet-grasses *Glyceria* are loosely tufted and grow by and very often right in shallow water, their green or greyish leaves floating on the surface. Their flower clusters are narrow green panicles, and they are best distinguished by their ligules and lemmas, as summarized below:

Tabular Key to Smaller Sweet-grasses (*Glyceria*)

	Ligule	*Lemma*	*Spikelet Length*
Floating	sharply pointed	pointed	18–35 mm
Hybrid	oblong	bluntly pointed	15–35 mm
Plicate	bluntly pointed	blunt	10–25 mm
Small	abruptly pointed	toothed	15–25 mm

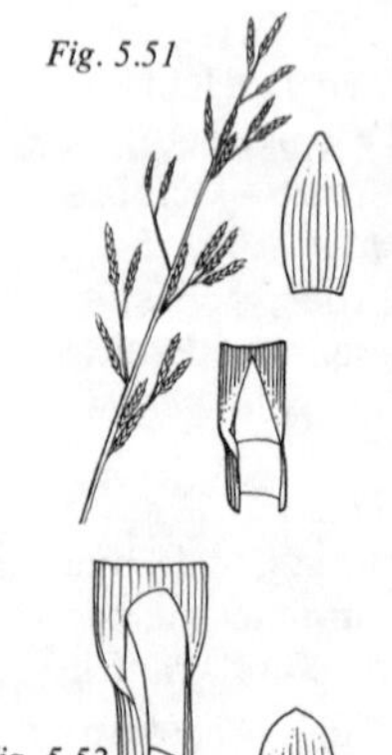
Fig. 5.51

Fig. 5.52

Floating Sweet-grass *G. fluitans* (*Fig. 5.51*) is both the commonest and the tallest, to 120 cm, and has the longest spikelets; appearing May–August. **Plicate Sweet-grass** *G. plicata* (*Fig. 5.52*) is shorter, to 80 cm, and has a more branched and spreading panicle; June–August. **Hybrid Sweet-grass** *G.* × *pedicillata* (*Fig. 5.53*), the hybrid between the last two species, is intermediate in most respects, and is not uncommon; as it spreads vegetatively, it may occur in the absence of its parents. **Small Sweet-grass** *G. declinata* (*Fig. 5.54*), grows only to 50 cm, with stems often curved and leaves usually greyish; flowering June–August.

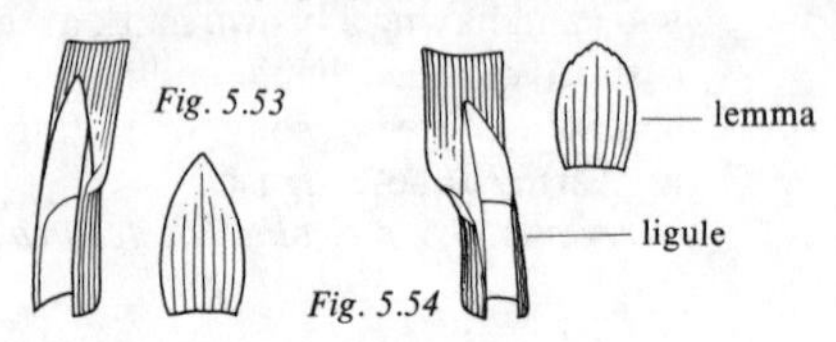

Fig. 5.53

Fig. 5.54

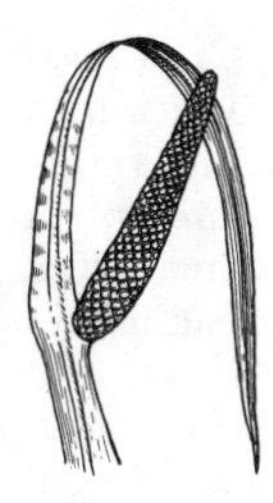

Fig. 5.55

ARUM FAMILY (Araceae)

Sweet Flag *Acorus calamus* (*Fig. 5.55*). Another plant of shallow water along the margins of lakes, ponds and slow-moving rivers. It grows to 125 cm and has sword-shaped leaves like those of Yellow Iris (*p. 98*), but with crinkled edges and a sweet smell when crushed. The tiny green flowers are tightly packed into a phallic spike at the base of a leaf-like spathe, which equals or exceeds the leaves; appearing June–July.

Two other aroids are planted as ornamentals and occasionally naturalized in shallow water in Britain.

Bog Arum *Calla palustris* (*Pl. 86*) has long-stalked, heart-shaped leaves and tight spikes of tiny green flowers (producing red berries) that have a spathe similar to the leaves but white inside; appearing April–August. **Skunk Cabbage** *Lysichiton americanum* (*Pl. 90*) is taller and has a large clump of broadly oval unstalked leaves, which appear after the foetid flowers, whose stout green spike (producing green berries) has a conspicuous, bright yellow, boat-shaped spathe; appearing April–May.

DUCKWEED FAMILY (Lemnaceae)

The duckweeds are the smallest European flowering plants, sometimes so numerous that they carpet water surfaces with green. They grow only in still water, usually ponds and ditches, but sometimes the backwaters of rivers and streams. They have no stems and their minute flowers have no petals.

Tabular Key to Duckweeds and Water Ferns

Duckweeds	*Leaf shape*	*Leaf width*	*Other leaf features*	*Roots*
Common	rounded	1.5–5 mm	3-veined	single
American	rounded	1.5–3 mm	1-veined	single
Fat	rounded	2–5 mm	swollen	single
Ivy-leaved	lanceolate	5–15 mm long	translucent, submerged, some at right angles	single
Great	rounded	4–10 mm		several
Rootless	oval	0.5–1 mm		none
Aquatic Ferns (*p. 72*)				
Floating Fern	oblong	10–14 mm		several
Water Fern	pinnate	1–10 mm	reddening	several
Carolina Water Fern	pinnate	1–10 mm	reddening	several

Common Duckweed *Lemna minor* (*Pl. 87*). Leaves rounded, three-veined, 1.5–5 mm across, each with a single root. **American Duckweed** *L. minuscula*, a very similar recent invader, is usually appreciably smaller, but can only be certainly distinguished by its one-veined leaves. **Fat Duckweed** *L. gibba* is also similar, but usually has markedly swollen leaves.

Fig. 5.56

Ivy-leaved Duckweed *Lemna trisulca* (*Pl. 88*) has more lanceolate, translucent leaves, 5–15 mm long, at right angles to each other, and floats just below the surface until the plant actually flowers.

Great Duckweed *Spirodela polyrhiza* (*Pl. 89, Fig. 5.56*) has larger, rounded leaves, 4–10 mm across, often purplish beneath and with a tuft of roots.

Rootless Duckweed *Wolffia arrhiza* (*Pl. 91*), the smallest European flowering plant, has single egg-shaped leaves only 0.5–1 mm across and without roots.

BUR-REED FAMILY (Sparganiaceae)

Bur-reeds *Sparganium* have linear leaves and rounded, male above female, heads of small green flowers; fruits conspicuously beaked, making the heads bur-like. Only the first two species have, or may have, emergent leaves.

Tabular Key to Bur-Reeds

Some leaves emergent:	*erectum, emersum*
Leaves keeled:	*erectum, emersum, glomeratum*
translucent:	*minimum*
Flower-spike branched:	*erectum*
Only one male flower-head:	*minimum, hyperboreum, glomeratum*
Style straight:	*erectum, emersum, angustifolium, glomeratum*
bent or down-turned:	*minimum, gramineum*
none:	*hyperboreum*

Branched Bur-reed *S. erectum* (*Pl. 92*), generally the commonest species, is the least exclusively aquatic, preferring shallow water near the banks of ponds, dykes and slow-moving rivers. It grows to 150 cm or even taller, and has stiff iris-like leaves, three-sided and keeled, especially near the base, nearly always emergent and only rarely floating. Flowers in leafy branched spikes; appearing June–August.

Unbranched Bur-reed *S. emersum* (*Pl. 93*) grows more in flowing than in still water, and has numerous long, ribbon-like floating leaves as well as emergent ones similar to Branched Bur-reed. It has an unbranched spike of 3–10 male flower-heads above 3–6 female, the lower often stalked, with the lowest leaf-like bract longer than the top head; appearing July–August. Fruits with a straight style. In Scandinavia cf. Grass-like Bur-reed below.

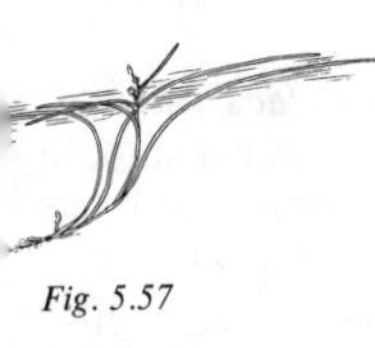
Fig. 5.57

Five more bur-reeds have floating leaves only.

Floating Bur-reed *S. angustifolium* (*Fig. 5.57*) has long flat, ribbon-like leaves, not keeled and with inflated sheaths. Flower-heads two male close together above 2–4 spread-out female, the lower usually stalked, and with the lowest leaf-like bract much longer than the male heads; appearing July–September. Fruit brown, with a straight style. Pools and tarns, preferring acid water. **Least Bur-reed** *S. minimum* (*Fig. 5.58*) has thinner, translucent leaves, only slightly inflated sheaths, one male head above 2–3 all unstalked female, the lowest bract scarcely longer than the male head, and the fruits paler, with a bent style; prefers calcareous water.

Fig. 5.58

The remaining species are confined to Scandinavia.

Boreal Bur-reed *S. hyperboreum* has thicker leaves than Floating Bur-reed, one male above 2–3 female heads, and yellower fruits with one style. **Grass-like Bur-reed** *S. gramineum* is more like Unbranched Bur-reed, but has all its very long leaves floating, and neither keeled nor inflated at the base. Its flowers are in 2–6 male above 3–7 female heads, the lower often stalked, and dark brown fruits with a down-turned style. **Northern Bur-reed** *S. glomeratum* has keeled leaves, and a more compact spike than other bur-reeds, the 1–2 male being immediately above the 3–5 crowded female heads, and the lowest bract at least three times as long as the spike; fruits waisted, with a short straight style.

REED-MACE FAMILY (Typhaceae)

Bulrush or Cattail *Typha latifolia* (*Pl. 94*). An aggressive plant that invades shallow water and causes it to silt up, growing in large patches and beds. Stems to 200 cm or more; leaves linear, 10–20 mm wide. Male flowers yellowish-buff in a tight spike immediately above the cigar-shaped spike of chocolate-brown female flowers; appearing July–August. **Lesser Bulrush** *T. angustifolia* is less common and has narrower, about 5 mm, leaves and a gap between the male and female spikes.

SEDGE FAMILY (Cyperaceae)

A very varied family in appearance, six of the first eleven species looking superficially like rushes, while the last eight (*Carex*) are more confusable with grasses.

Common Club-rush or Bulrush *Scirpus lacustris* (*Pl. 95*). Best known for its tall, to 300 cm, rounded smooth green leafless stems, emerging from slow-moving rivers and streams, lakes and larger ponds. Its strap-like leaves are all submerged. Flowers in loose clusters of egg-shaped red-brown spikelets near the top of the stems; appearing June–August. Its ssp. *tabernaemontani*, **Grey Club-rush**, has glaucous stems, red-dotted flowers and two instead of three styles. **Triangular Club-rush** *S. triqueter* is much less common and has a single strap-like leaf at the base of its three-sided stems; June–September.

Floating Club-rush *Scirpus fluitans* is confusingly like some of the smaller pondweeds (see *p. 96*). It forms a mat of floating and largely submerged stems and narrowly linear leaves in still and slow-moving, usually acid water. Flowers in a single, long-stalked, emergent greenish spikelet, differing from all pondweeds in lacking petals and sepals; appearing May–July.

Cotton-grass *Eriophorum* (*Pl. 97*). Two species are especially likely to be found in bog pools. **Common Cotton-grass** or **Bog Cotton** *E. angustifolium* has stems three-sided at the top, leaves 3–5 mm wide, and flowers in a loose, drooping cluster of 3–7 smooth-stalked spikelets. These look very different in April and May, when they have bright yellow anthers, than later, when the fruits develop their long white cottony hairs. The much less common, but more often actually in the water, **Slender Cotton-grass** *E. gracile* has stems all three-sided, leaves only 1–2 mm and spikelets with rough three-sided stalks. Often grows with Slender Sedge (below).

SPIKE-RUSHES (*Eleocharis*)

Flowers in single spikelets at the top of leafless, usually rounded stems, the leaves being replaced by brown sheaths. Most of them grow in and by shallow water and have three styles and the lowest glume completely encircling the base. Other stem and flower characters are tabulated below. Only the four species most likely to be found growing in the water are described.

Tabular Key to Spike-rushes

	Sheath Truncated	*Lowest Glume*
Common	Squarely	Short
Many-stalked	Obliquely	¼ length of spikelet
Needle	Squarely	½ length of spikelet
Dwarf	Obliquely	¾ length of spikelet

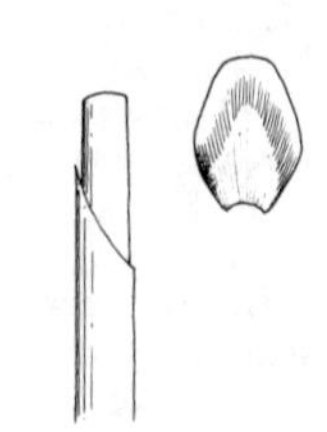

Fig. 5.59

Common Spike-rush *E. palustris* (*Pl. 96*) grows in small tufts, to 60 cm, in both basic and slightly acid water. Flowers yellowish– to dark brown, with only two styles and the two lowest glumes each only half-encircling the base; appearing May–August.

Many-stalked Spike-rush *E. multicaulis* (*Fig. 5.59*) grows in large tufts, to 30 cm, in bog and other pools on acid soils. Flowers brown, often with small green shoots instead of flowers; appearing June–August.

Needle Spike-rush *E. acicularis* makes bright green submerged 'lawns' in seasonal shallow water, only flowering as the water recedes. Stems distinctively four-angled, to 50 cm when submerged, but usually to 10 cm when flowering. Flowers dark brown; appearing June–October.

Dwarf Spike-rush *E. parvula* grows on tidal mud or by saline pools inland. Its short, to 8 cm, thread-like stems have underground runners that end in whitish tubers. Flowers greenish; appearing June–July.

Fig. 5.60

Great Fen Sedge *Cladium mariscus* (*Fig. 5.60*). A most distinctive patch-forming, tall, to 250 cm. plant, growing in and by shallow, usually basic or calcareous water, especially in fens and swamps. Its long (to 200 cm), broad, 10–15 mm, saw-edged leaves are characteristically bent in a V-shape. Flowers in a clustered panicle of pale brown egg-shaped spikelets; appearing July–August.

True Sedges (*Carex*)

None are really aquatic, but several are freshwater margin plants that regularly grow in the water. This selection is inevitably somewhat arbitrary, but includes those species mot often found growing away from the bank. All *Carex* species can be told from grasses by their solid stems and lack of joints at the leaf-junctions. They also have separate male and female flowers which, in the species described below (except Greater Tussock Sedge) are in separate spikes, the male flowers produce conspicuous yellow anthers in May and June. Fruits are often important, and sometimes essential, for identification.

All species described have sharply three-sided stems, which are rough to the touch, except Bottle Sedge, which has bluntly three-sided stems, and Slender and Water Sedges, which are both bluntly three-sided and smooth. The lowest bract is longer than the whole spike in all but Greater Tussock and Tufted Sedges. Some stem, leaf and flower characters are tabulated below.

Tabular Key to Sedges

	Stems 3-sided	*Leaf width* ×	*Leaf colour*	*Female spike*
Greater Tussock	sharply, rough	5–7 mm	dark green	ovoid, unstalked
Slender	bluntly	1–2 mm	grey-green	very narrowly cigar-shaped, short-stalked
Lesser pond	sharply, rough	7–10 mm	glaucous when young	narrowly cigar-shaped, un- or short-stalked
Greater pond	sharply, rough	6–15 mm	glaucous when young	narrowly cigar-shaped, un- or short-stalked
Cyperus	sharply, rough	5–12 mm	yellow-green	shortly and broadly cigar-shaped, long-stalked
Bottle	bluntly, rough	2–7 mm	greyish	shortly cigar-shaped, short-stalked
Bladder	sharply, rough	4–8 mm	yellow-green	almost ovoid, long-stalked
Water	bluntly, smooth	3–5 mm	glaucous above, green below	narrowly cigar-shaped, un- or short-stalked
Tufted	sharply rough	3–6 mm	glaucous	cigar-shaped, unstalked
Lesser Tufted	sharply, rough	3–7 mm	glaucous	narrowly cigar-shaped, un- or short-stalked

Fig. 5.61

Greater Tussock Sedge *C. paniculata* (*Fig. 5.61*) is easily told by its substantial tussocks, up to 250 cm high and 100 cm across, with leaves exceedingly rough to the touch. Flowers in unstalked spikes, male and female mixed.

Slender Sedge *C. lasiocarpa* (*Fig. 5.62*) grows in bog and other acid pools, often with Slender Cotton-grass (above). It is loosely tufted, with very narrow leaves and flower-spikes.

Fig. 5.62
Fig. 5.63
Fig. 5.64
Fig. 5.65
Fig. 5.66

Lesser Pond Sedge or **Marsh Sedge** *C. acutiformis* (*Pl. 100*), one of the commonest waterside sedges, forms patches in and by many ponds and rivers. It has 2–3 male spikes with bluntly pointed glumes. **Great Pond Sedge** *C. riparia* (*Pl. 99, Fig. 5.63*), almost as common, grows in similar places, but has broader leaves and 3–6 male spikes with the glumes sharply pointed.

Cyperus Sedge or **Hop Sedge** *C. pseudocyperus*, growing by and in still and slow-moving water, is loosely tufted with only one male and 3–5 long-stalked, drooping, sausage-shaped female spikes, the lowest bract much longer than the top spike.

Bottle Sedge *C. rostrata* often grows in a band round the shallower edges of upland lakes and tarns. Stems bluntly three-sided, leaves overwintering; and the lowest bract sometimes only equalling the top spike. The rather similar **Bladder Sedge** *C. vesicaria* (*Fig. 5.64*) has sharply three-sided stems, no overwintering leaves, and a longer lowest bract.

All the above species have three styles; the remaining three have only two.

Water Sedge *C. aquatilis* (*Fig. 5.65*) grows in swamps and in and by still and slow-moving water, mainly in hill districts. It is loosely tufted, with smooth, bluntly three-sided stems. **Tufted Sedge** *C. elata* (*Fig. 5.66*) makes small tussocks in fens, swamps and shallow lakes or pools. Its lowest bract is much shorter than the top spike. **Slender Tufted Sedge** *C. acuta* (*Pl. 98*) often grows alongside both Tufted Sedge and the two Pond Sedges. It is tufted and its stems are both sharply three-sided and rough to the touch.

CHAPTER 6 Protozoa

(Several Phyla of Unicellular Animals)

Protozoans are complete animal organisms contained within a single cell. They include solitary and colonial, free-swimming, sessile, parasitic, and symbiotic forms. The early microscopists called them animalcules (little animals) along with all other micro-organisms, or infusorians, because they often appeared, apparently spontaneously, in infusions of vegetable matter (see *p. 346*). Later, all single-celled animals were placed in a single phylum, Protozoa, but modern protozoologists recognize several phyla and the formal term Protozoa has lapsed (some now use it to designate a sub-kingdom); it is now used as a 'common name' for these animals in general.

The freshwater species range in size from a few microns to about 3 or 4 mm in length, but some marine forms can become much larger than this, up to about 20 mm. Because of their small size, protozoans are generally restricted to wet or moist environments – they would be too liable to desiccation in a dry habitat. (Most species are capable of forming resistant cysts in dry conditions, but they cannot function as living animals.) They therefore abound in marine and freshwater habitats and many species live inside other animals or plants as parasites or symbionts: malaria, sleeping-sickness and amoebic dysentery are diseases caused by protozoans that afflict man and other animals. Symbiotic forms are common in the guts of higher animals, where they perform such tasks as helping in the digestion of otherwise indigestible substances such as cellulose.

Protozoans occur in a remarkable range of shapes. Most of these are fairly constant – although the animals are usually flexible and slightly contractile – but the amoebae and some flagellates can vary their shapes almost infinitely. Most have definite anterior and posterior ends, and often recognizable dorsal and ventral surfaces. As in the algae, some species form colonies of essentially similar cells joined together in some way.

Internal structures in the protozoa are usually easier to observe than in the algae but exactly how much detail can be seen depends largely on the quality of the microscope used. The two most obvious structures are the nucleus and contractile vacuoles. The nucleus contains the chromosomes or genetic material of the cell; it is present in all eucaryotes (but not always easily seen, as in the algae). In most protozoans it is a dense, globular structure situated near the middle of the cell, but in some genera it is filamentous or formed like a string of beads (moniliform). In the drawings it is shown as a dotted area. Contractile vacuoles are cavities in the cytoplasm that are concerned with maintaining the water-balance of the cell. They are spherical or star-shaped and pulsate with a regular rhythm. Various food products may be stored as granular or crystalline matter within the cytoplasm (the semi-liquid substance that constitutes the cell) and often affect its appearance by adding colour or texture, as may ingested food organisms or particles. A few genera, mainly amoebae, construct protective cases or shells out of secreted material, sand particles, or other matter; such cases are a useful diagnostic feature of these animals.

Most free-living protozoans feed by engulfing small, living or dead organisms – bacteria, algae, rotifers, other protozoans, etc. – which are imprisoned within a food vacuole in the cytoplasm while digestion takes place. Ciliates often possess distinct mouths or food-grooves combined with complicated food-capturing structures of modified cilia. The curious suctorians have sticky, hollow tentacles which adhere to the food organism and suck out its body-contents.

Several methods of locomotion are employed by protozoans: swimming by means of flagella; swimming, walking, or gliding over a substratum by means of cilia or ciliary organelles – bunches of fused cilia which function, in effect, as legs or paddles; or amoeboid motion. Amoeboid motion is a flowing movement of the cytoplasm – the animal literally flows forward over a substratum, controlling speed, direction and overall shape by varying the viscosity of the cytoplasm in different parts of the cell. Pseudopodia (literally 'false feet') are particularly associated with amoeboid motion. They are protrusions of the cytoplasm that can take any form, although generally elongated. A typical amoeba may produce several pseudopodia during locomotion, but usually only one leads, losing its identity as the rest of the cell flows into it. Pseudopodia are also used as food-catching devices, surrounding and engulfing the food organism. In the heliozoa (sun animalcules) the 'rays' are long, slender, stiff-looking pseudopodia radiating outwards from the spherical body.

Sexual and asexual reproduction occur, with many variations: sexual reproduction takes place by conjugation, or by the fusion of differentiated sexual cells, gametes, produced by division from 'parent' cells. Division or budding are the usual means of asexual multiplication. Multiple division often occurs, a whole protozoan demolishing itself by dividing into numerous small fragments. These may simply grow up into new protozoan cells, or become gametes, or may form resistant spores which are the main distributive stage of protozoans. Spores can withstand desiccation and extreme variations in temperature; being so minute they are easily broadcast by wind or water, and have enabled many, perhaps most, species to distribute themselves worldwide.

The free-living protozoa can be divided into three groups: those bearing flagella; those bearing cilia; and those which lack cilia (athough flagella are sometimes present) and move by amoeboid motion, or do not move at all. Unfortunately such clear-cut definitions are not always appropriate. At this end of the scale of life, exceptions to any rule can be found and the parameters of any group are bound to be hazy. Even the distinction between plants and animals is not always clear: many single-celled 'plants' (organisms that are clearly structurally similar to undoubted plant cells) feed and behave like animals. However, for general purposes of identification the tripartite division above is as good as any other method, with the reservation that any colourless flagellate should be viewed with suspicion as it may turn out to be a plant!

Classification of the Major Groups of Protozoa

Phylum SARCOMASTIGOPHORA		
Subphylum MASTIGOPHORA	The flagellates	(*p. 111*)
Subphylum SARCODINA		
Superclass RHIZOPODA	Amoebae	(*p. 112*)
Superclass ACTINOPODA	Sun animalcules	(*p. 115*)
Phylum MICROSPORIDA	Sporozoans	(*p. 116*)
Phylum CILIOPHORA	The ciliates	(*p. 116*)

Key to the Major Groups of Freshwater Protozoans

---

1. a) Internal parasites, common in fishes, symptomized by the presence of cysts, boils, ulcers, etc., on the external surface of the host. Phylum MICROSPORIDA (*p. 116*)
(Similar symptoms could also be caused by other organisms, e.g., bacteria)
b) Not as above: free-living, sessile, epizooic, or externally parasitic protozoans. . . . 2

2. a) Protozoans partly or wholly covered with cilia, which may be fused to form ciliary organelles. Phylum CILIPHORA (part) (*p. 116*)
b) Cilia never present; flagella present or absent. . . . 3

3. a) Solitary or colonial protozoans with flagella, not normally progressing by amoeboid motion. Subphylum MASTIGOPHORA (*p. 111*)
b) Solitary protozoa normally progressing by amoeboid motion, or sessile; sometimes inhabiting a shell (or test); flagella rarely present. . . . 4

4. a) Unattached amoeboid forms inhabiting a test made of secreted material, sand grains, etc. . . . 5
b) Unattached, sessile, or fixed forms lacking a test (except one genus which is anchored by a stalk). . . . 6

5. a) Pseudopodia few, broad, or finger-like Subclass TESTACEALOBOSIA (*p. 114*)
b) Pseudopodia numerous, fine and thread-like. Class FILOSA (*p. 114*)

6. a) Shape variable and typically irregular; pseudopodia not radiating; typical amoebae. Subclass GYMNAMOEBIA (*p. 112*)
b) Shape more-or-less fixed and regular; pseudopodia or tentacles radiate from the body. . . . 7

7. a) Free (rarely stalked), more-or-less spherical protozoans with long, stiff pseudopodia radiating from their entire surface (not epizooic). Superclass ACTINOPODA (*p. 115*)

b) Attached protozoans, usually on short stalks: shapes various: with contractile, sometimes branched tentacles (usually epizooic).
Phylum CILIOPHORA, Class SUCTORIA (*p. 119*)

Phylum SARCOMASTIGOPHORA (Flagellates and Amoebae)
Identification of many of the protozoans in this phylum is notoriously difficult, and further complicated by the awkward fact that some amoeboid types have flagellate stages and some flagellates exhibit amoeboid characteristics at some stage in their lives. Only the larger, naked and testate amoebae and the heliozoans can be determined with any confidence; for the remainder we give only a few typical examples.

Subphylum MASTIGOPHORA (Flagellates)
Protozoans possessing one or more permanent flagella during their free-living stages. The plant or phytoflagellates, which possess photosynthetic pigments, and their related unpigmented forms, are described on *p. 46*, Algae, Group 1. The remainder are distinguished by calling them animal or zooflagellates.

Most zooflagellates are very small, rarely longer than about 25 μm and usually less than 10 μm. They live free or attached, and are solitary or colonial in habit. Many genera secrete a thin, transparent case or lorica, in which they live with the flagella protruding; others are naked.

Relationships within the zooflagellates are very uncertain and will not be discussed here; superficially, three main types can be recognized.

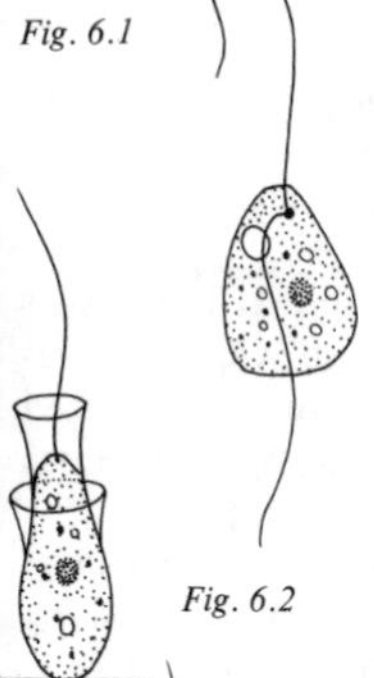

Fig. 6.1

Fig. 6.2

Fig. 6.3

Monads This group includes numerous genera of solitary, free-swimming, naked zooflagellates. Many of these produce slender pseudopodia which are not used in locomotion; up to eight flagella may be present. *Bodo* (*Fig. 6.1*) is a typical example. It usually occurs in foetid waters containing large amounts of decaying matter, farmyard effluent, etc.

Choanoflagellates Distinctive zooflagellates with a single flagellum and characterized by a flask-shaped lorica with a prominent, flared collar. Most of the numerous genera are colonial. *Salpingoeca* (*Fig. 6.2*) is the largest genus of solitary choanoflagellates, with numerous species varying greatly in shape and size (5–30 μm long), and usually fixed to a substratum.

Codonosiga (*Fig. 6.3*) forms small colonies consisting of a

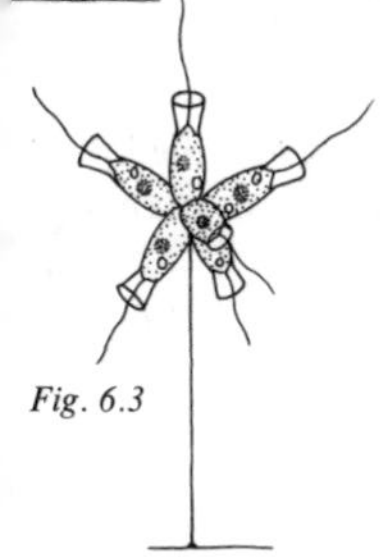

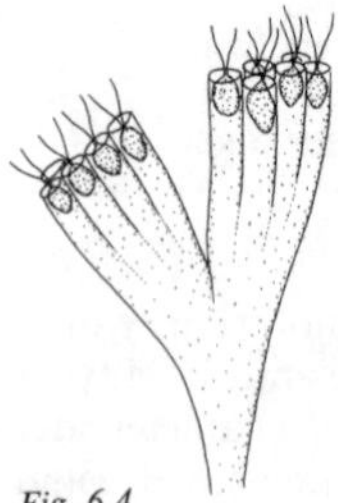
Fig. 6.4

cluster of individuals joined by their bases atop a slender, fixed stalk.

Codonocladium is similar but each individual has a short stalk connecting it to a common point on the main stalk.

Sphaeroeca forms globular, *Uroglena*-like (*Fig. 3.13, p. 49*) colonies consisting of numerous individuals embedded in the surface of a sphere of jelly. Colonies are free-swimming and up to 200 μm diameter.

Other colonial forms Some monad types build colonial structures of flimsy jelly.

Rhipidodendron (*Fig. 6.4*) makes fixed colonies up to 1 mm tall, consisting of numerous, gelatinous tubes joined side-by-side in flattened, branching bunches. It occurs mostly in acidic waters in bogs and marshes.

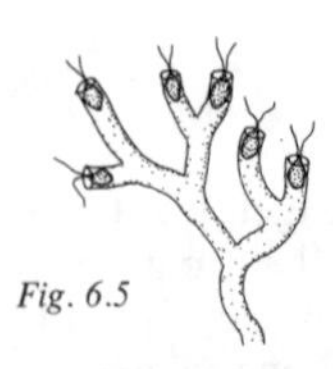
Fig. 6.5

Cladomonas (*Fig. 6.5*) makes irregularly-branched colonies of single gelatinous tubes up to 100 μm tall; frequent in small standing waters.

Reference: Pascher, 1914.

Subphylum SARCODINA

Protozoans progressing by amoeboid motion, or sessile; pseudopodia usually evident; flagella occur rarely.

Superclass RHIZOPODA

Class LOBOSEA

Pseudopodia form broad lobes or finger-like protrusions, never fine and thread-like.

Subclass GYMNAMOEBIA

The naked amoebae will be familiar to anyone who has studied biology at almost any level, but the common concept of amoebae as primitive, structureless blobs of living matter is far from true. Amoebae are highly organized cells with a definite orientation and a complex internal structure. Amoeboid motion is also far from simple and is an ideal subject for microscopical study. When first spotted under the microscope, an amoeba appears as a somewhat granular, immobile, translucent mass. After a while, when it has settled, a streaming motion may be noticed in the contents of the cell; this may be the first indication that a living organism is being observed.

Two general categories of naked amoebae occur: the typical, more-or-less flattened type with a number of relatively slender pseudopodia, and the limax (slug-like) amoebae, which are roughly cylindrical, with a single, broad (body-width) pseudopod at the leading end. Most species are very widely distributed, probably cosmopolitan; in fresh water they are common but rarely abundant.

Fig. 6.6

Amoeba *Pl. 101, Fig. 6.6* *A. proteus* is the standard, large amoeba, but the genus contains several other smaller species. *A. proteus* is up to 700 μm long, large enough to be seen easily with the naked eye, appearing as a whitish speck against a dark background, although actually colourless. There is one main pseudopod which leads during locomotion and several lateral ones: their surfaces usually bear longitudinal ridges.

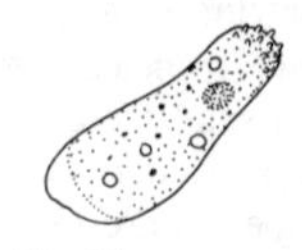
Fig. 6.7

Chaos Similar to *Amoeba* in external appearance but with numerous – often several hundred – tiny nuclei (*Amoeba* has one large one): up to 1 mm long.

Trichamoeba *Fig. 6.7* A large limax amoeba up to 500 μm long; posterior pole of body characteristically covered with small tubercles; nucleus single.

Fig. 6.8

Thecamoeba *Fig. 6.8* Body flattened, oval, with a definite thickened skin, often longitudinally ridged; nucleus single. Pseudopodia absent (or very rare), locomotion being achieved by a rolling 'track-laying' movement; several species up to 250 μm long.

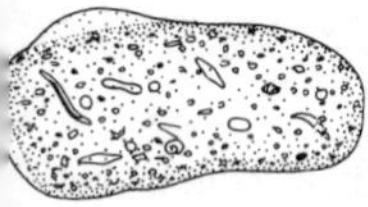

Fig. 6.9

Pelomyxa palustris *Pl. 102, Fig. 6.9* This, the sole species in the genus, is probably the largest freshwater protozoan, reaching a length of 5 mm, although 0.5–2 mm is more usual. It is a limax amoeba, dirty greyish or yellowish in colour, rather opaque and with a granular texture due to numerous sand grains and other particulate matter embedded in it. Nuclei small and numerous but usually impossible to see because of the thickness and opacity of the animal. Local, but often abundant in mud in still waters.

This amoeba possesses several unique characters (apparent only by the use of electron microscopy) which has led some protozoologists to question its relationship with other amoebae. Some have even placed it in a separate phylum of its own.

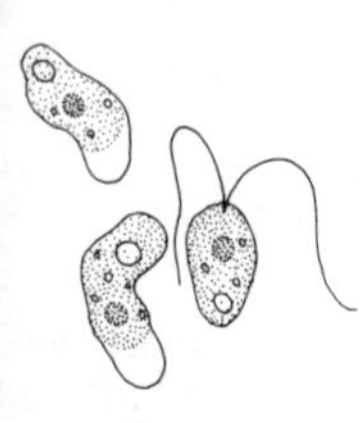
Fig. 6.10

Naegleria gruberi *Fig. 6.10* This is a common example of the numerous small limax amoebae under 50 μm long, which are frequent in soil and freshwater habitats. This species can transform temporarily into a free-swimming state possessing a flagellum, induced by sudden changes in its physical environment.

Mastigamoeba Small or medium-sized amoebae, up to 200 μm long, with a long permanent flagellum: found amongst decaying vegetation or in mud.

Reference: Page, 1976.

Subclass TESTACEALOBOSIA

Amoebae which live in shells (tests) made of a substance secreted by the animal, or of particulate material – sand grains, diatom frustules, etc. – bound together by secretion to form a rigid structure with a definite aperture from which the pseudopodia protrude. Several genera are very common and their tests may be abundant in samples of bottom debris from ponds.

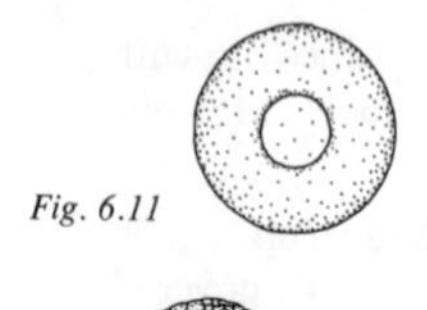

Fig. 6.11

Arcella *Fig. 6.11* Test a rounded cone, varying in height according to species, with a circular aperture in the centre of its base, made of brownish or amber material secreted by the animal. It appears disc-shaped in the normal, dorsal view seen through a microscope. Several species are very common; up to 150 μm diameter.

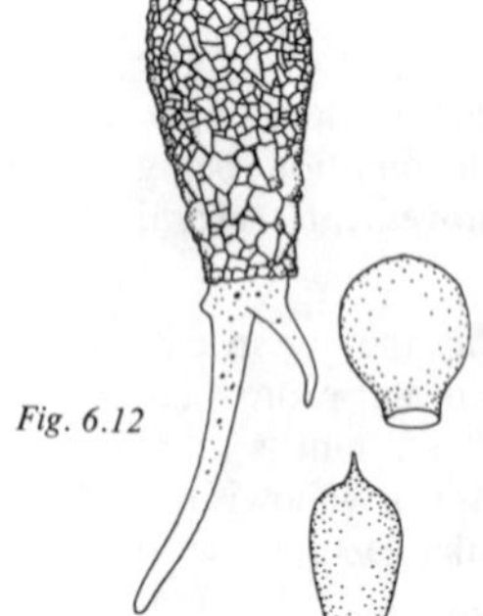

Fig. 6.12

Difflugia *Fig. 6.12* Test made of sand grains, typically vase-shaped with a circular, terminal aperture; often with a distinct neck before the aperture and occasionally bearing small spines. Numerous species are known, with tests up to 700 μm long; common and widespread.

Centropyxis *Fig. 6.13* Test of various shapes – globular, vase-like, etc. – usually with several spines; aperture ventral, never terminal as in *Difflugia*; made of sand grains embedded in a secreted matrix. Numerous species up to 400 μm long; common and widespread.

Nebela Test oval or vase-shaped, sometimes with a neck, often compressed, up to 150 μm. It is composed of secreted material covered with tiny siliceous plates obtained from its principal prey organism, *Euglypha* (below). Many species, mostly occurring in bog pools and marshes, especially amongst *Sphagnum*.

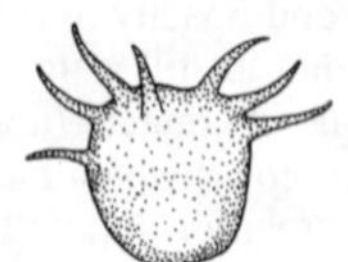

Fig. 6.13

Class FILOSA

Order GROMIIDA

Testate amoebae with very fine, thread-like branching pseudopodia. There is one main freshwater genus.

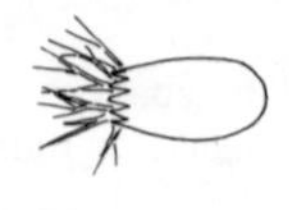

Fig. 6.14

Euglypha *Fig. 6.14* Test oval or vase-shaped, sometimes spiny, with a circular terminal aperture usually with a serrated rim; made of tiny, rounded plates of silica (may be difficult to observe); up to 100 μm long. Common in bogs and marshes.

Superclass ACTINOPODA

Sessile protozoans lacking flagella and not moving by amoeboid motion. Individuals are spherical, with radiating, hair-like pseudopodia, often supported by stiffened axial fibres. Only one class occurs in fresh water.

Class HELIOZOA (Sun Animalcules)

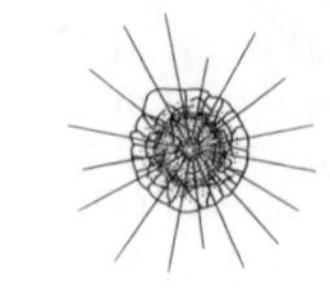

Fig. 6.15

Actinophrys *Fig. 6.15* Body with a central nucleus from which the axes of the pseudopodia radiate: usually about 50 μm diameter. Central mass of cell granular, with an outer layer consisting almost entirely of large vacuoles. Common in many habitats where there is abundant plant life, adhering to leaves or stems, occasionally free-floating.

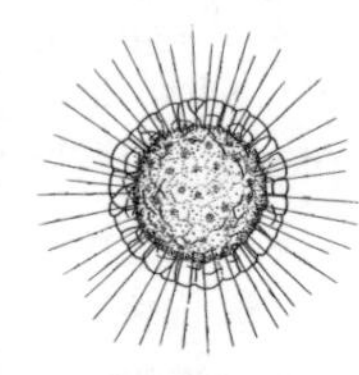

Fig. 6.16

Actinosphaerium *Pl. 103, Fig. 6.16* Similar to *Actinophrys* but much larger, usually about 0.5–1 mm diameter, sometimes larger, up to 2 mm. Nuclei small and numerous; pseudopod axes arising just beneath vacuolated layer. Optimally lit under a good microscope this is a beautiful object, and easily visible to the naked eye. Widespread and locally abundant amongst water plants, especially *Callitriche* and fine-leaved plants such as *Myriophyllum*.

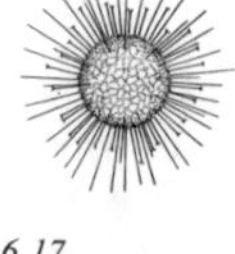

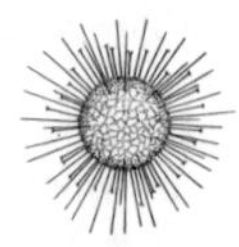

Fig. 6.17

Acanthocystis *Fig. 6.17* Small heliozoans up to 60 μm diameter, with a single, off-centre nucleus and numerous. needle-like spicules of silica radiating amongst the pseudopodia. Some species also bear tiny siliceous plates, and the tips of the spicules may be forked. Planktonic or amongst plants.

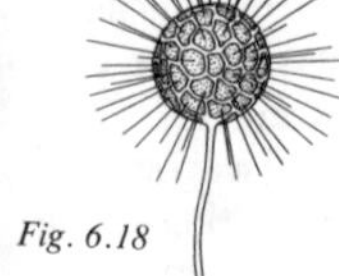

Fig. 6.18

Clathrulina *Fig. 6.18* Animal enclosed in a spherical latticework of silica, about 50–80 μm diameter, attached to a substratum by a slender stalk of similar material: pseudopodia are extended through the lattice. Widespread but rather local and sporadic in occurrence.

Phylum MICROSPORIDA (Sporozoans)

This phylum contains exclusively parasitic protozoans, many of which attack freshwater fishes and amphibians. Although the individual parasites cannot be seen or identified without making special microscope preparations, the external symptoms they cause are fairly characteristic of the phylum. Fishes become infected by ingesting spores along with their food. Once inside the host the parasites become active, multiply rapidly, and spread to various parts of the host's body where they form cysts containing numerous individual parasites. The unpleasant external boils, blisters, swellings and lesions are caused partly by the effects of the sporozoan parasites, and partly by secondary infection by bacteria or fungi. Two common types are described below.

Glugea anomala This is a common parasite of sticklebacks, *Gasterosteus* and *Pungitius*. It causes small rounded cysts, up to about 4 mm across, in the skin, on the fins, or occasionally in the eye of the fish. Other species of *Glugea* and the related *Plistophora* occur in other host species.

Myxobolus This causes large cysts to form within the flesh of the fish host, erupting as swellings up to several centimetres across on the skin, often causing scales to stand up or be displaced. There are several species which infest such hosts as barbel (*Barbus*), roach (*Rutilus*) and other members of the carp family.

Phylum CILIOPHORA (Ciliates)

The ciliates form a very distinctive group of protozoans. Except for the curious suctorians, they all possess cilia throughout their lives and are easily identified by this character alone. They are probably the lowest animals to exhibit an apparent awareness of their surroundings; their movements have a noticeably more purposeful and positive air than those of other protozoans.

The ciliate cell is usually of fairly fixed shape (although some species are contractile) due to the presence of a complex skin or pellicle, in which the cilia are imbedded, usually in longitudinal or oblique rows. The normal (macro-) nucleus may be spherical, band-like, or moniliform; it is always accompanied by one or more tiny micronuclei, a characteristic feature of the phylum but not easily seen. The position of the mouth is fixed and it is frequently surrounded by a differentiated region – the peristome – which usually bears specialized cilia. The gullet is sometimes defined by the presence of a 'basket' of rod-like trichites. Ciliary organelles occur in many ciliates. They are groups of fused cilia modified to suit various purposes (see below): membranelles, triangular flaps set in a row, and delicate undulating membranes are usually rapidly-beating devices for producing feeding currents; conical cirri, which move relatively slowly, are normally used as legs or

undulating membrane

membranelles

cirrus

paddles. In the drawings, the ciliary organelles are shown but normal body cilia are omitted.

There are very many genera of freshwater ciliates, most of which can be identified by such features as overall shape and form, shape of the nucleus (if visible), position of the contractile vacuole(s), and details of the ciliary organelles. We have included a wide selection below but as the occurrence of protozoans is very unpredictable, other genera are just as likely to be encountered.

Class KINETOPHRAGMINOPHORA

Apart from the subclass Suctoria, which have already been defined in the key (*p. 119*), members of this class are free-swimming ciliates which lack compound ciliary organelles. In some there is a specialization of the peristomial cilia but this is never conspicuous.

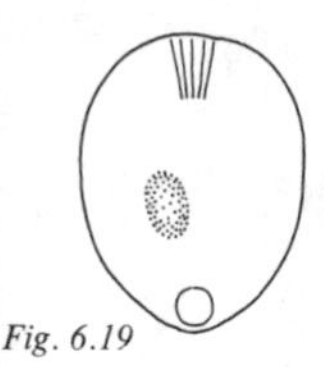

Fig. 6.19

Holophrya *Fig. 6.19* Body gooseberry-shaped with the mouth at one pole; up to 200 μm long; a common planktonic form.

Amphileptus *Fig. 6.20* Body markedly flattened, up to 250 μm long; may attach to peritrich colonies (*p. 121*), on which it feeds, by a tendril at the posterior end.

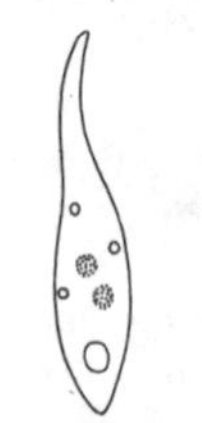

Fig. 6.20

Loxophyllum *Fig. 6.21* Body markedly flattened, up to 250 μm.

Loxodes *Fig. 6.22* Body only slightly flattened, up to 500 μm.

Dileptus *Fig. 6.23* Body cylindrical, with a long proboscis; up to 200 μm.

Lachrymaria *Fig. 6.24* Body and neck extremely contractile, neck can extend from almost nothing to several times the body length; total length to 500 μm.

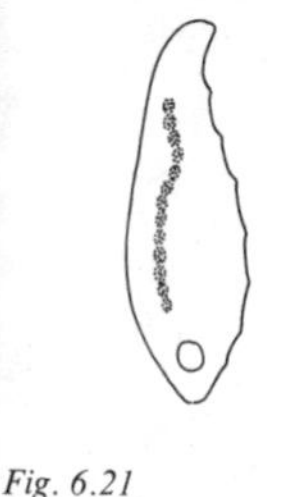

Fig. 6.21

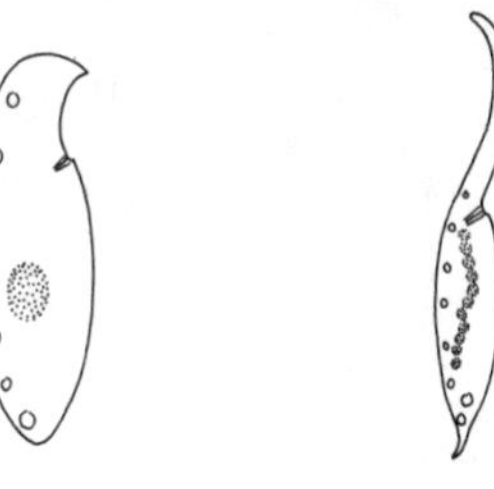

Fig. 6.22

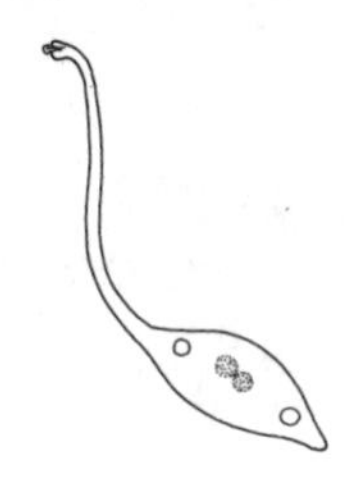

Fig. 6.23 *Fig. 6.24*

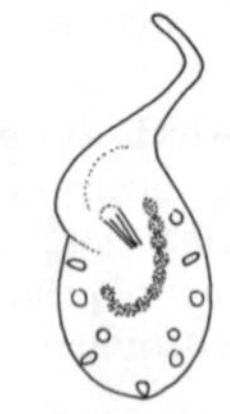

Fig. 6.25

Paradileptus *Fig. 6.25* Body cylindrical, with a spirally-twisted proboscis; up to 350 μm.

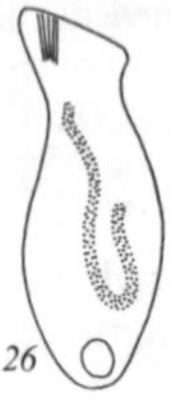

Fig. 6.26

Cranotheridium *Fig. 6.26* Body cylindrical with the anterior end pinched flat, up to 150 μm. *Spathidium* is similar in shape but lacks trichites in the gullet.

Didinium *Fig. 6.27* Body droplet-shaped with cilia reduced to two belts. Although only about 100 μm long, this protozoan is a voracious predator of much larger protozoans, such as *Paramecium*.

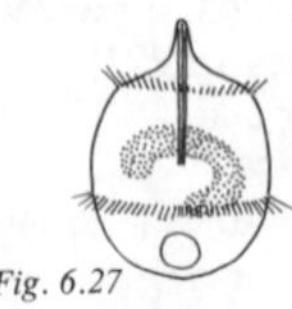

Fig. 6.27

Coleps *Pl. 104, Fig. 6.28* Body barrel-shaped, protected by many small plates which often bear spines; up to 80 μm long.

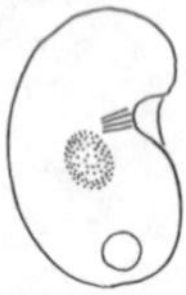

Colpoda *Fig. 6.29* Body bean-shaped with the mouth situated in a cavity on one side; 100 μm.

Bursaria *Fig. 6.30* A very large ciliate with the mouth in a deep vertical groove surrounded by enlarged cilia (these are easily mistaken for membranelles); up to 1 mm long.

Chilodonella *Fig. 6.31* Body with a distinct, flattened ventral surface on which the mouth is situated; up to 250 μm.

Spirochona *Fig. 6.32* Body urn-shaped, lacking cilia except for those on the complex spiral funnel which forms the peristome; up to 100 μm tall. This is the sole freshwater genus of the order Chonotricha; it lives epizooically, attached to the gill plates of *Gammarus* and other aquatic crustacea.

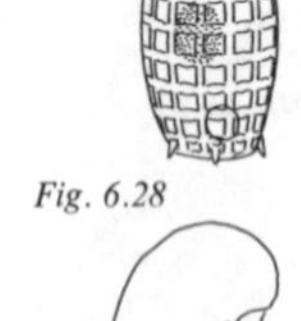

Fig. 6.28

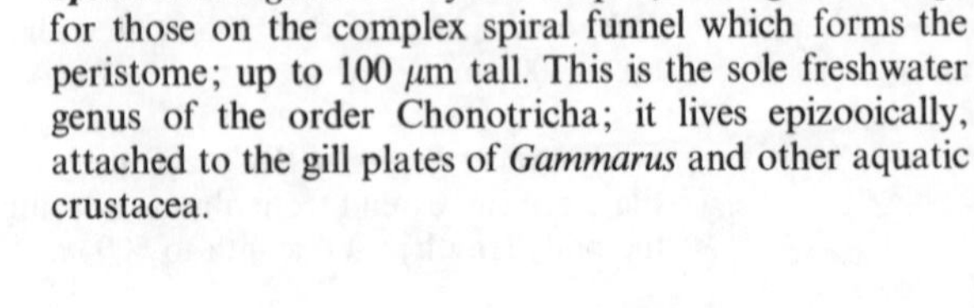

Fig. 6.29

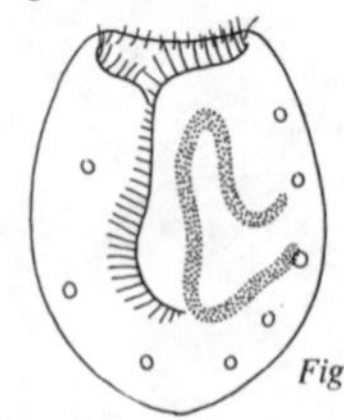

Fig. 6.30

Fig. 6.31

Fig. 6.32

Subclass SUCTORIA

Sessile ciliates which have lost all trace of cilia in the adult state (the juvenile stages are ciliated). They trap their prey, usually other ciliates, on their sticky tentacles; these are hollow, and the cell contents of the prey is sucked out through them. The tentacles are flexible and contractile, unlike the stiffer pseudopodia of the heliozoans, the only other group with which they might be confused. Suctorians are common but are easily overlooked. They are found on water plants and epizooically on other animals – crustaceans, insects, etc.

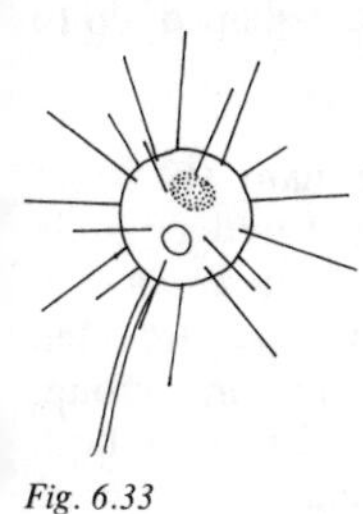

Fig. 6.33

Podophrya *Fig. 6.33* Body spherical, about 50 μm diameter, with tentacles scattered over its whole surface, and mounted on a rigid stalk.

Tokophrya *Fig. 6.34* Body shaped like an inverted cone, about 70 μm long, on a flexible stalk. Tentacles in 2–4 bunches arising from the top of the body.

Acineta *Fig. 6.35* Similar to *Tokophrya* but with a lorica. The top centre of the lorica is pinched together, forming two corner apertures from which lobes of the body emerge, with a bunch of tentacles on each lobe.

Discophrya *Fig. 6.36* Body elongated and irregular in shape, on a short stalk, with the tentacles scattered over its surface; up to 100 μm tall.

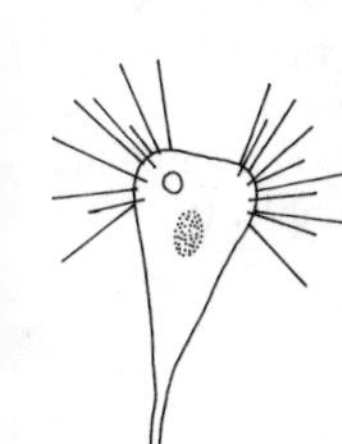

Fig. 6.34

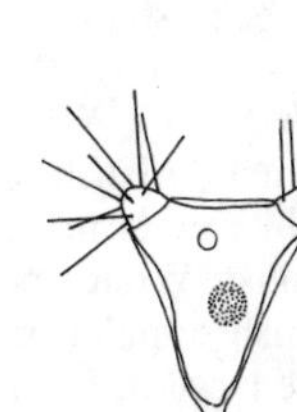

Fig. 6.35

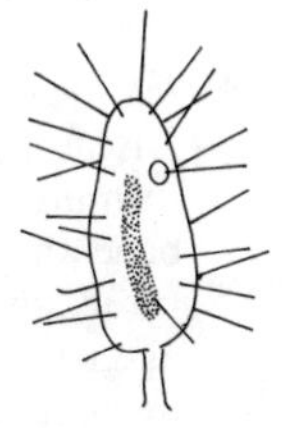

Fig. 6.36

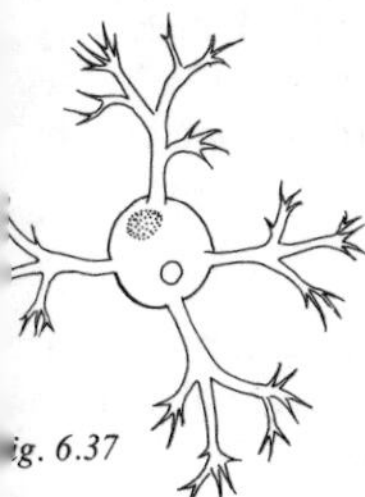

Fig. 6.37

Dendrocometes *Pl. 105, Fig. 6.37* Body globular, about 60 μm across, with several much-branched arms; the tentacles are extensions of the tips of the branches. The sole species *D. paradoxus* lives on the gill plates of *Gammarus*, often in the company of *Spirochona* (*p. 118*).

Class OLIGOHYMENOPHORA

Ciliates with normal cilia on the body (lacking in most peritrichs) and with ciliary organelles on the peristome.

Subclass HYMENOSTOMATA

Free-swimming or parasitic ciliates with normal cilia on the body and a small mouth cavity with an undulating membrane and three membranelles (not often easily observed).

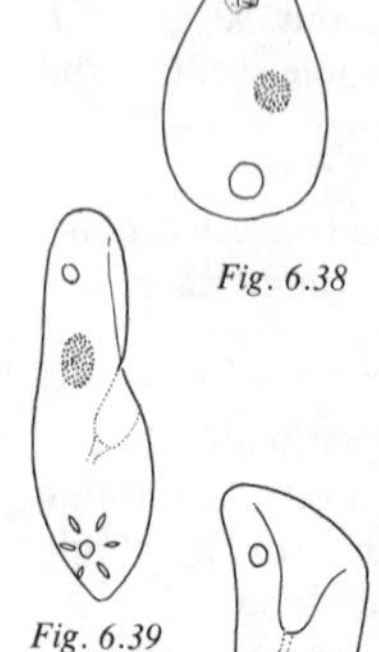

Fig. 6.38

Fig. 6.39

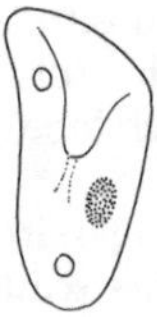

Fig. 6.40

Tetrahymena *Fig. 6.38* Body typically pear-shaped, up to 50 μm long; a rapid swimmer.

Paramecium *Pl. 106, Figs. 6.39, 40* The many species in this genus are divisible into two groups, based on body shape: the *P. aurelia* group, including *P. caudatum*, 250 μm, and *P. multimicronucleatum*, 300 μm, which are often called 'slipper animalcules'; and the *P. bursaria* group, most of which are fairly small, about 70–150 μm. In both groups, the mouth is situated near the centre of the body at the end of a long groove. The animals rotate slowly as they swim. *P. bursaria* is green, due to the presence of zoochlorellae, other species are colourless.

Paramecium is probably the best-known of all the ciliates and is commonly used as a text-book example. It is fairly common in small, standing waters.

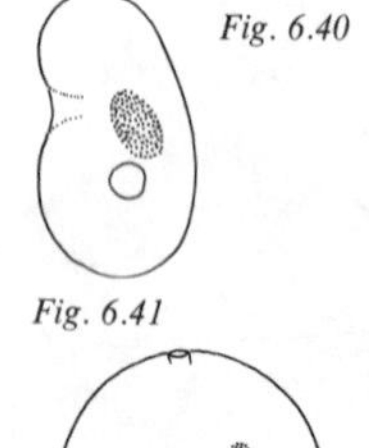

Fig. 6.41

Colpidium *Fig. 6.41* Body bean-shaped, up to 100 μm.

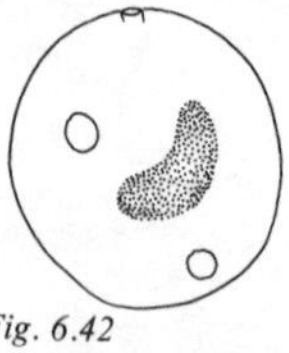

Ichthyophthirius multifiliis *Pl. 107, Fig. 6.42* (White Spot Disease) An external parasite of fishes, the adult ciliate forming characteristic white cysts up to 1 mm diameter beneath the skin of the host. White spot is frequent but rarely abundant in wild fish populations, but may reach plague proportions where large numbers of fish are kept in confinement, such as fish farms or aquaria. Unless very abundant, the parasite causes little harm to its host, but its presence encourages the growth of harmful bacteria or fungi.

Fig. 6.42

Subclass PERITRICHA

Distinctive ciliates with cylindrical or bell-shaped bodies lacking normal cilia; peristome encircled by an elaborate belt of membranelles resembling an undulating membrane; sessile (stalked or loricate) or epizooic, rarely free-swimming in adult stage, often colonial.

The stalked peritrichs, commonly known as bell animalcules, are some of the most conspicuous of the protozoa, forming colonies up to several centimetres across. They are common and often abundant on aquatic roots, plant-life, wooden pilings and similar substrata, typically appearing as a dense, fuzzy, whitish mass which contracts when disturbed. The juvenile, free-swimming stage is often common in collections of living material; they resemble the adults but lack a stalk. The loricate genera are also common but much less easy to find, apart from large colonies of *Ophrydium*. The epizooic species are usually found in close association with their hosts but can swim free.

Stalked Peritrichs

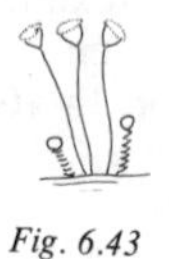

Fig. 6.43

Vorticella *Pl. 108, Figs. 6.43, 44* Solitary, but individuals are gregarious and often form densely-packed groups. Individuals bell-shaped, up to 100 μm, on a long, spirally contractile stalk; sometimes epizooic. A very common and well-known genus.

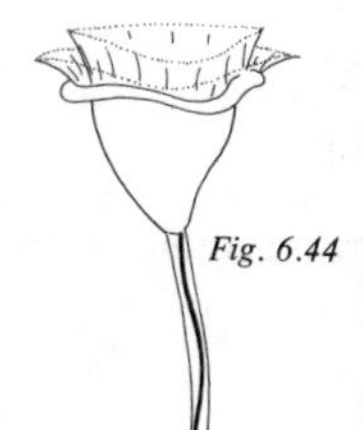

Fig. 6.44

Epistylis *Figs. 6.45, 46* Colonial, stalk branching and not contractile; individuals elongated, up to 150 μm long; mostly epizooic on crustaceans.

Carchesium *Figs. 6.47, 48* Colonies tree-like, with a main 'trunk' supporting numerous branches; trunk contracts spirally, branches contract independently; individuals bell-shaped, up to 100 μm.

Zoothamnium *Pl. 109, Fig. 6.49* Colonial, forming tree-like colonies that are smaller and more delicate than *Carchesium*; trunk contracts in a zig-zag manner, branches contract simultaneously with trunk; individuals about 50 μm long.

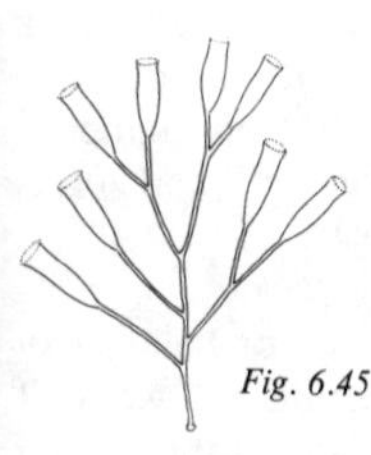

Fig. 6.45

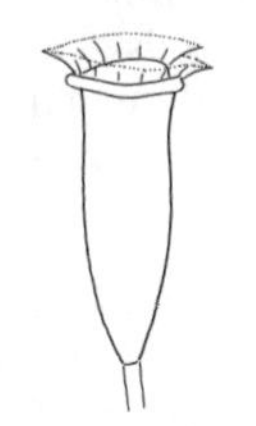

Fig. 6.46

Fig. 6.47

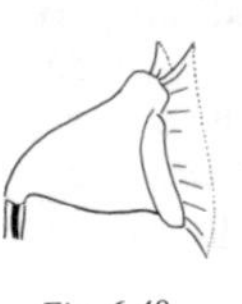

Fig. 6.48

Fig. 6.49

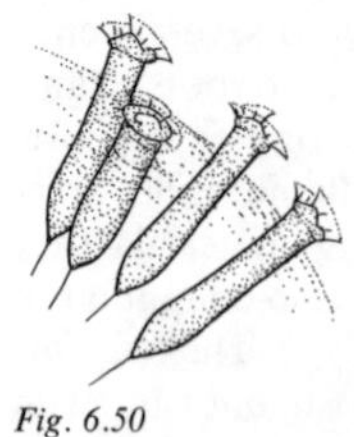

Fig. 6.50

Fig. 6.51

Loricate Peritrichs

Ophrydium *Pl. 110, Fig. 6.50* Colonial, numerous individuals embedded in an irregular or globular mass of jelly which may attain a size of 2–3 cm across. Individuals are elongated, contractile, up to 250 μm long, and usually green, due to the presence of zoochlorellae. Green colonies are very conspicuous, although easily mistaken for algae; they are usually attached to a substratum but may swim free.

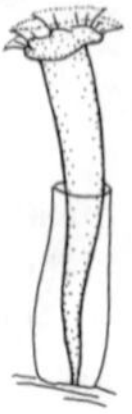

Vaginicola *Fig. 6.51* Solitary, contractile, living in a lorica which is attached to a substratum by its proximal end only, animal attached to lorica by posterior end; lorica without an internal flap. Individuals are up to 100 μm long, and occasionally two may inhabit the same lorica.

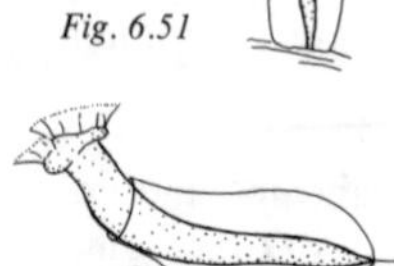

Fig. 6.52

Platycola *Fig. 6.52* Similar to *Vaginicola* but lorica is attached to substratum by one side.

Thuricola *Pl. 111, Fig. 6.53* Similar to *Vaginicola* (above) but lorica contains a lid-like internal flap.

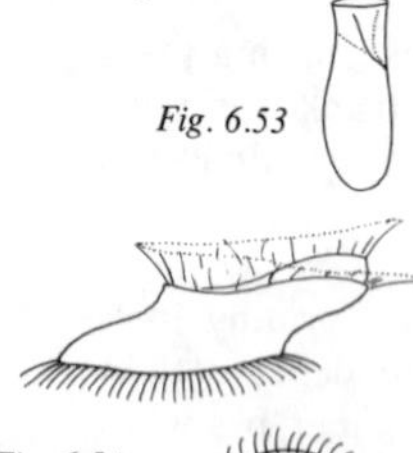

Fig. 6.53

Cothurnia Similar to *Vaginicola* but lorica is mounted on a short stalk.

Unattached Epizooic Peritrichs

Urceolaria *Fig. 6.54* Body short and obliquely depressed, with the circular peristome markedly offset from the disc-like base; base (upper half of *Fig. 6.55*) surrounded by a fringe of membranelles and with a central ring of small teeth. Several species are known, up to 100 μm diameter: *U. mitra* is common on the planarian *Polycelis* and sometimes others; another species lives on fishes.

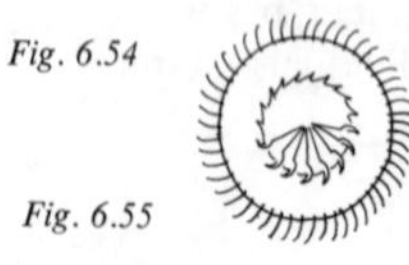

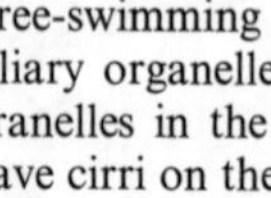

Fig. 6.54

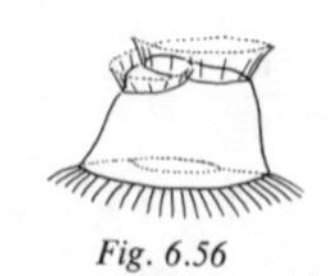

Fig. 6.55

Trichodina *Fig. 6.56* Body a short cylinder or bell-shape, with the peristome aligned with the base; base (lower half of *Fig. 6.55*) similar to that of *Urceolaria* (above) but with a ring of hooks reinforced by radial rods. Various species are epizooic on *Hydra*, fishes and amphibian tadpoles.

Fig. 6.56

Class POLYHYMENOPHORA

Free-swimming ciliates with highly-specialized ciliation consisting mostly of ciliary organelles (normal cilia often lacking); always with a row of membranelles in the peristome region, typically arranged spirally; most genera have cirri on the body.

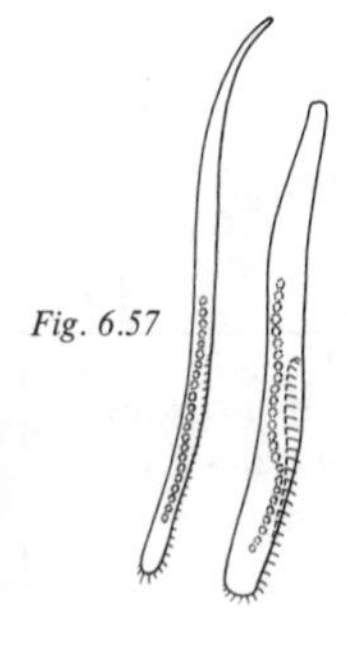
Fig. 6.57

Spirostomum *Pl.* 112, *Fig. 6.57* This elongated form is one of the largest ciliates, up to 3–4 mm long, but very contractile, often worm-like. Frequent in stagnant water.

Stentor *Pl. 113, Fig. 6.58* Semi-sessile, highly contractile ciliates, typically trumpet-shaped when attached by narrow base, but becoming oval when detached and free-swimming. Some species are colourless; one (*S. polymorphus*) is green due to zoochlorellae; and the splendid *S. coeruleus* is deep blue-green, due to pigments. The latter species is the largest and may reach 2 mm in length when extended. Some species remain attached for long periods, often in a radial cluster, and these may secrete a gelatinous mass around their bases. A well known genus, common and widespread.

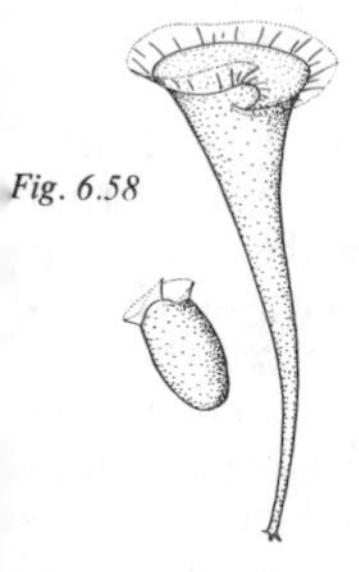
Fig. 6.58

Tintinnidium *Fig. 6.59* This and related genera are small, 50–150 μm, contractile ciliates inhabiting loricas of variable texture, often incorporating foreign particles. They live attached to various substrata, especially water plants, or swim free in the plankton.

Kerona *Fig. 6.60* The sole species *K. polyporum* is a common epizooic form found on *Hydra*; it is easily observed running up and down the stalk of the polyp; about 150 μm long.

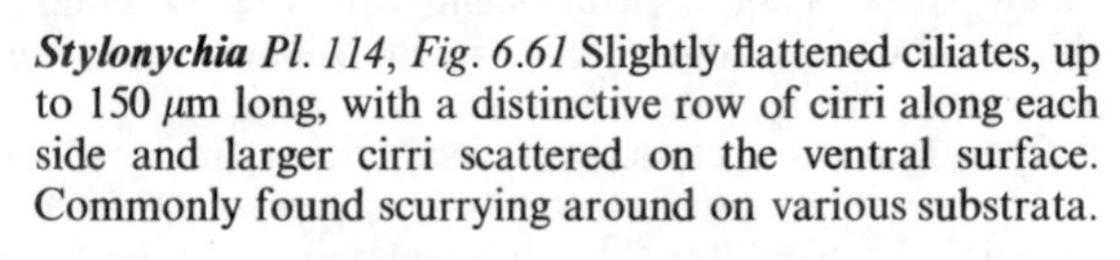

Stylonychia *Pl. 114*, *Fig. 6.61* Slightly flattened ciliates, up to 150 μm long, with a distinctive row of cirri along each side and larger cirri scattered on the ventral surface. Commonly found scurrying around on various substrata.

ig. 6.59

Euplotes *Fig. 6.62* Markedly flattened, broad-bodied ciliates, lacking the marginal rows of cirri that *Stylonychia* possesses; up to 150 μm.

References: Curds, 1982. Curds, Gates & Roberts, 1983.
Protozoa (general): Levine *et al*, 1980. Mackinnon & Hawes, 1961. Vickerman & Cox, 1967.

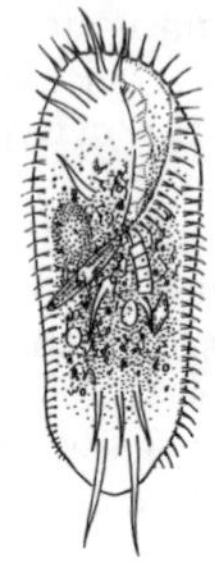
Fig. 6.61

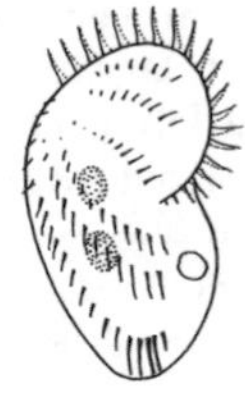
Fig. 6.60

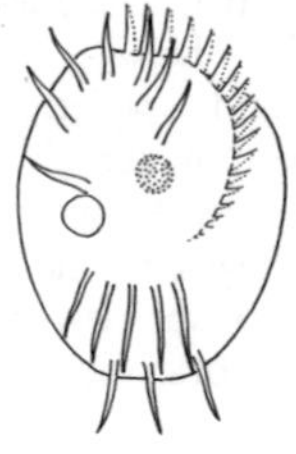
Fig. 6.62

CHAPTER 7 Sponges
(Phylum Porifera)

Recognition features Irregular, immobile, encrusting spongy growths, attached to any firm substratum, often exceeding 10 cm, usually green but may be yellow, brown or grey.

Sponges are strange, immobile organisms that are abundant in the sea, with about 10,000 species encompassing a great diversity of form, size and colour. The few freshwater species all belong to one family, Spongillidae, and by comparison are drab and uninteresting creatures, although among the largest invertebrate organisms in fresh water.

Freshwater sponges form irregular, encrusting growths of characteristic texture. The surface is slightly rough to the touch, perforated by numerous small holes and a few large ones which may be raised on cones; the whole mass is fragile, friable and definitely spongy! Shape is largely governed by conditions in the immediate environment, such as water currents, or the shape of the substratum. Common growth-forms include flat encrustations, up to about 1 cm thick, which may cover tens of square centimetres, spindle-shaped lumps around twigs, etc., and erect, branching fingers up to 20 cm long. Unfortunately, shape is not a reliable identification feature. Many species are green due to abundant zoochlorellae in the sponge, otherwise they are usually yellowish, brown, or dirty grey.

Sponges occur in clean rivers, canals, lakes and large ponds, and are likely to be encountered attached to any firm, permanent substratum. Typical sites include submerged tree roots, wooden pilings or other man-made structures, sunken branches, rocks and stones, but rarely water plants. Clean water is important and sponges are often the first organisms to die when waters become polluted.

Sponges are filter feeders. The sponge body consists of a mass of cells permeated by a labyrinth of interconnecting tunnels. These tunnels are lined with special flagellated cells resembling choanoflagellates (*p.111*) which draw currents of water through the sponge to provide a continuous supply of oxygen and food particles. The water is absorbed through numerous tiny holes in the sponge's surface and expelled through a few larger, often crater-like apertures called osculi. The sponge body is supported by a network of siliceous needles, spicules, produced by the sponge (*Fig. 7.1*). Spicules have a characteristic size and shape in each species and are valuable aids to identification.

The active cells of sponges usually die off in the winter, although fragments of the spicular network often remain. The sponge's future existence is ensured by special overwintering bodies, gemmules, which lie dormant over the winter and hatch into young sponges in the following spring. Gemmules are spherical or oval bodies about 0.5 mm diameter (*Pl. 115*) which are formed asexually in summer and autumn. They may be scattered throughout the sponge body or form a basement layer near to the substratum. They are very resistant to adverse conditions and remain viable for many years; and hence are probably the main agents in dispersal of the species. Sexual reproduction occurs in late spring and summer, resulting in tiny, ciliated larvae which swim off to settle and grow into new sponges.

Although definitely not microscopic organisms, freshwater sponges cannot be reliably identified without microscopical examination of their spicules. In most cases these can be adequately studied by simply teasing out a fragment of sponge, being sure to include a gemmule or two if these are present. Spread the fragment thinly on a microscope slide and add a drop of water and a cover-slip. Two or three kinds of spicules will be found (*Fig. 7.4*):

Megascleres (M) are the large needle-like spicules, usually about 150–400 μm long, that form the main supporting mesh of the sponge body; they may be smooth or spiny.

Microscleres (m) occur in only two genera. They are small spiny needles, about 50–150 μm long, found loose amongst the general body mass.

Gemmoscleres (G) are the most important for diagnostic purposes. They are always found in a layer surrounding the gemmules (*Fig. 7.2*). Gemmoscleres may be simple, spiny needles, or birotules, shaped like an axle with a wheel each end (*Figs. 7.3, 4*) and always less than 100 μm, often much smaller.

The following species of freshwater sponge have been reported from north-western Europe, but the group has been little-studied in the area and the distribution of most species is poorly known.

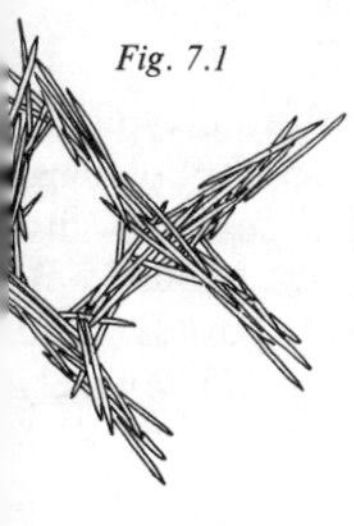
Fig. 7.1

Family Spongillidae

Spongilla lacustris *Figs. 7.1, 2, 4* Form variable, often producing erect, finger-like branches, usually green or yellowish. Spicules: **M** always smooth; **m** present, slightly curved, with tiny spines or warts; **G** curved, spiny, up to 130 μm long. A very common species occurring throughout cold-temperate regions of the Northern Hemisphere.

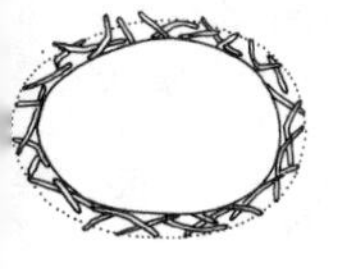
Fig. 7.2

Eunapius fragilis *Fig. 7.4* Typically flat and encrusting in form, very soft in texture, greyish or whitish, rarely green. Spicules: **M** always smooth; **m** absent; **G** straight and spiny, especially at tips, 75–140 μm long (*Fig. 17.4*). Widely distributed.

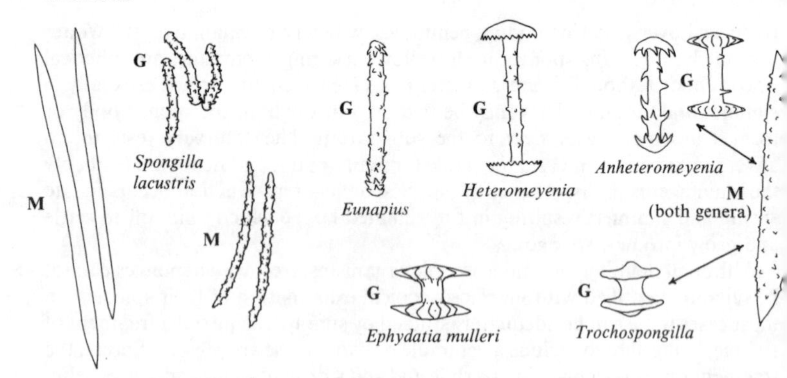

Fig. 7.4. Spicules of Freshwater sponges.

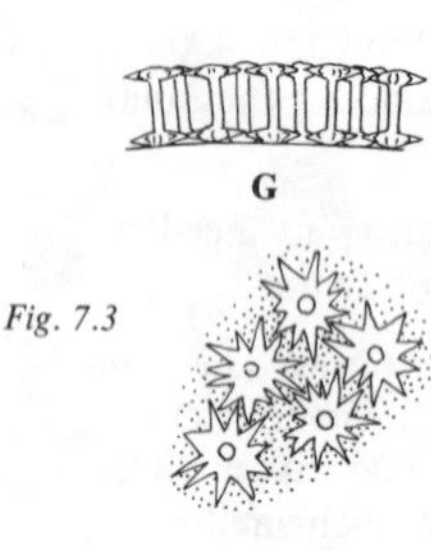

Fig. 7.3

Ephydatia fluviatilis *Pls. 115–116, Fig. 7.3* Form very variable, often forming erect fingers as in *Spongilla* but also occurs as flat encrustations, etc.; usually green but may be yellowish, greyish, or brown. Spicules: **M** usually smooth; **m** absent; **G** birotules of characteristic shape for the genus (*Fig. 7.3*), in this species they are longer than their diameter, about 25–30 μm long and 20 μm diameter. Very common and is widely distributed throughout the Northern Hemisphere.

Ephydatia mulleri Distinguished from *E. fluviatilis* by the presence of small spines on **M**, and **G** birotules being shorter than their diameter, about 12–20 μm long and 20–25 μm diameter (*Fig. 7.4*). Widely distributed throughout Northern Hemisphere.

Heteromeyenia stepanowii An encrusting form, typically forming cushion-like pads, green or yellowish. Spicules: **M** slender, with small spines; **m** present, spiny; **G** distinctive birotules (*Fig. 7.4*) in two size ranges, about 75–90 μm and 55–65 μm long. Distribution sporadic and scattered, not well known in Europe but probably widely distributed.

Anheteromeyenia ryderi Generally cushion-like, soft in texture, usually green. Spicules: **M** stout and spiny; **m** absent; **G** characteristic birotules (*Fig. 7.4*) of two types, about 50–75 μm and 30–40 μm long. A common species in eastern North America, occasionally reported in Europe; perhaps accidentally introduced.

Trochospongilla horrida Typically flat and encrusting, texture firm; yellow or brown in colour. Spicules: **M** stout and spiny; **m** absent; **G** tiny birotules (*Fig. 7.4*) about 10–12 μm long. Distribution scattered throughout Northern Hemisphere.

References: Penney & Racek, 1968. Brauer, 1909.

CHAPTER 8 Hydras and Jellyfishes
(Phylum Cnidaria (Coelenterata))

Recognition features

i) Sessile polyps. Solitary or colonial, usually with long contractile tentacles (rarely absent); green, brown, grey, pink or colourless; some types are common.
ii) Free-swimming jellyfishes; very rare.

The Cnidaria is a group of great importance in the sea, where it includes such creatures as jellyfishes, corals and sea anemones; but only a few genera, all belonging to the class Hydrozoa, occur in fresh water.

Cnidarians are animals of simple structure which occur in two basic forms: sessile polyps, which may be solitary or colonial, and free-swimming medusae (jellyfishes). Both types are represented in fresh water. Polyps are soft-bodied, contractile animals with a tubular body closed at the lower end, which forms an attachment disc, and usually with a ring of contractile tentacles arising from the upper rim, which forms the mouth. Medusae have inverted bowl-shaped, transparent bodies, with numerous contractile tentacles hanging from the rim, and a central tubular mouth; they swim by rhythmic contractions of the body or bell.

Cnidarians capture their prey, and defend themselves, with tiny, stinging organelles – cnidae or nematocysts. These are microscopic oval capsules, about 5–100 μm long, embedded in the skin of the body and tentacles in large numbers (*Fig. 8.1*). When suitably stimulated, e.g., by touching potential prey, they shoot out a long, often barbed thread. Some of the threads penetrate the victim, injecting a poison, others stick to it or tangle around hairs or other projections. Cnidae are the sole means of attack or defence available to cnidarians and thus are essential to their existence. They are fascinating objects, easily studied under the microscope by simply squashing a piece of a tentacle beneath a cover-slip on a microscope slide.

Class HYDROZOA
Family Hydridae
Hydra

Hydras are very common animals that abound in almost any permanent body of clean water. They have slender bodies and about 4–8, very contractile tentacles arranged in a single ring. Hydras are found attached to water-plants, sunken twigs and branches, rocks and stonework; in short, any firm substratum that allows them to fish a volume of open water with their tentacles. They may also hang from the surface film of the water, and wind-blown dispersal is possible by this means. Hydras will eat any small animal, e.g., a water flea, that blunders into their waiting tentacles and can be subdued by their cnidae.

Asexual reproduction occurs by budding (*Fig. 8.3*), the young polyps breaking away from the parent when fully-formed. This can build up local populations with surprising rapidity. Sexual reproduction occurs seasonally, mature polyps bearing the gonads externally on the body wall. The yellowish ovaries are spherical and the testes are roughly conical (*Fig. 8.2*); individuals may bear either or both. The fertilized eggs develop into tiny, ciliated planula larvae, which swim off to settle and develop into new polyps.

The epizooic ciliates *Kerona* and *Trichodina* (*pp. 123, 122*) are frequently found scuttling around on the outside of the hydra's body.

Fig. 8.1

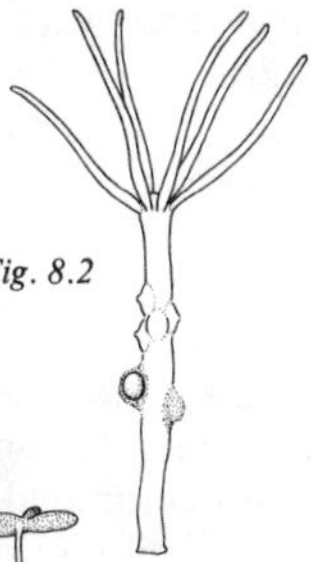

Fig. 8.2

Fig. 8.3

Hydra viridissima (Green Hydra) *Pl. 117, Fig. 8.2* The only European species that has zoochlorellae (symbiotic green algae) in its body, which are responsible for its bright green colour. Because of this it is sometimes placed in a separate genus, *Chlorohydra*. The body rarely exceeds 15 mm long and is usually about 4–10 mm. The tentacles are always shorter than the body and about 6–10 in number. Green hydras are typical of small water bodies – spring ponds, ditches, duck ponds, etc. – but are less tolerant of pollution than other species; they are common throughout Britain and Europe.

Hydra oligactis *Pl. 118, Fig. 8.3* The only non-green hydra that can be reliably distinguished from the other species without detailed study of the nematocysts and sexual stages. In mature specimens, the lower part of the body forms a slender stalk, distinct from the upper part. The whole body is up to 30 mm long and the tentacles, when fully relaxed, can become remarkably long, at least 100 mm. Colour varies from reddish- to purplish-brown. Common in running waters, lakes and large ponds; typically in places exposed to wave action.

Other brown hydras The remaining European species of *Hydra* are all brownish in colour, do not have a stalk, and the tentacles are shorter than the body which rarely exceeds 15 mm. The commonest of these is *H. attenuata* (*H. vulgaris*) but it cannot easily be distinguished from the others, which include *H. braueri*, *H. graysoni* and *H. circumcincta*.

References: Ewer, 1948. Grayson, 1971. Brauer, 1909.

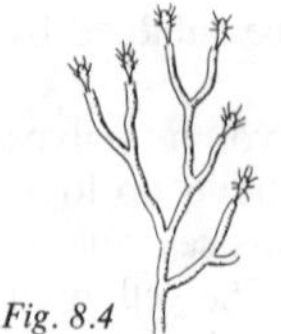

Fig. 8.4

Family Clavidae

Cordylophora lacustris *Figs. 8.4, 5* This species forms branching, shrub-like colonies up to 5–6 cm tall. The polyps are unlike those of *Hydra*, being club-shaped with about 12 tentacles arranged in several whorls. Each polyp is borne at the end of a slender stalk which is covered by a tube of chitinous material. New branches are formed from buds that arise on the old branches. Some buds give rise to male or female reproductive polyps: these are vase-shaped and lack tentacles. After fertilization, the eggs develop into planula larvae which swim off to found a new colony.

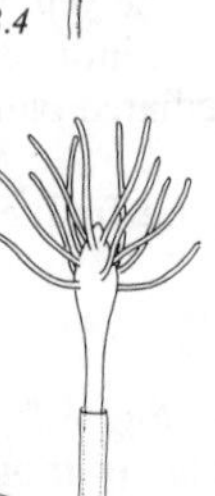

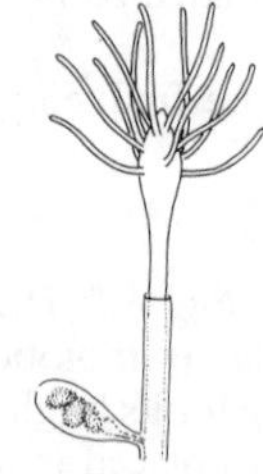

Fig. 8.5

Cordylophora is found attached to water-plants, tree roots, pilings and similar substrata, in fresh, or slightly brackish water in the lower reaches of rivers and places like the Norfolk Broads or parts of the Baltic. But it has also been found in canals and other waters having no direct connection to the sea; it is rather local and uncommon.

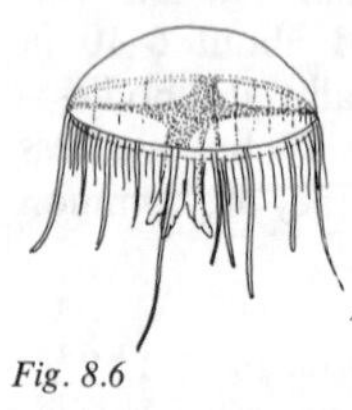

Fig. 8.6

Fig. 8.7

Family Olindiidae

Craspedacusta sowerbyi (Freshwater Jellyfish) *Pl. 119, Figs. 8.6, 7* This interesting and distinctive creature can hardly be mistaken for any other freshwater animal. Mature jellyfish may reach 20 mm across the bell but in north-western Europe are usually smaller. The medusa is the sexually-reproducing stage in the life-cycle of this species, the other stage being a tiny, tentacle-less polyp (*Fig. 8.7*) that was formerly called *Microhydra ryderi* before its connection with *Craspedacusta* was discovered. This polyp stage, or hydroid, lives on plants or amongst bottom debris, budding off other polyps which eventually separate and wander off. During warmer weather, the hydroids may produce medusa buds, which develop into tiny jellyfish only 1–2 mm across; these then break off and swim away. To complete the life-cycle, the medusae grow to maturity and propagate sexually, producing planula larvae which develop into the next hydroid generation. It is unlikely that the entire life-cycle is ever completed in temperate climates.

Craspedacusta is not native to Europe, but its true place of origin is unknown. It has been found in aquaria and hot-house ponds throughout Europe, probably being introduced with exotic water plants. It was first discovered in the Botanic Gardens in Regent's Park, London. Wild populations occasionally turn up in lakes, reservoirs and canals, but rarely persist for more than a few years; medusae only appearing in unusually long, warm spells of weather.

CHAPTER 9 Flatworms, Flukes, Tapeworms

(Phylum Platyhelminthes)

This is a large phylum of predominantly parasitic worms, although one class, Turbellaria, consists of common, free-living forms. Most platyhelminthes are contractile, very soft-bodied and easily damaged. Individual worms are unsegmented, but some types form chains of connected individuals by asexual budding. Typically, the body is flattened, often elongated, with distinct dorsal and ventral surfaces, and usually with a definite 'head-end' scarcely qualifying as a proper head. Parasitic species often bear specialized features connected with their way of life.

The three classes can be distinguished thus:

a) Free-living worms, never with rounded suckers.
Flatworms, Class TURBELLARIA below

b) Parasitic, mostly internal but some external; usually with one or two, sometimes more, small, rounded suckers; some juvenile stages are free-swimming and microscopic. Flukes, Class TREMATODA (*p. 137*)

c) Internal parasites, typically forming chains of reproductive units; juvenile stages diverse but usually elongated, relatively featureless worms unlikely to be confused with other platyhelminthes.
Tapeworms, Class CESTODA (*p. 139*)

Class TURBELLARIA (Flatworms or Planarians)

Recognition Features Free-living, more or less flattened (occasionally cylindrical) contractile worms, unsegmented but some small species form chains of several individuals joined nose to tail; colourless or pigmented; 0.5–35 mm long. Movement is by ciliary gliding over a substratum, often accompanied by wriggling movements – only a few of the very small species can swim. Out of water most species contract to an immobile oval blob of fragile jelly.

The freshwater Turbellaria can be divided into two groups, the order Tricladida (*p. 135*) which includes the larger (more than 5 mm) flatworms, and Microturbellaria (below), an assemblage of several orders of small or microscopic (mostly less than 5 mm) worms which are not easily distinguished, including the groups formerly known as rhabdocoels.

Flatworms are common freshwater animals found in most habitats, often in large numbers. They move with the body extended – the characteristic shape shown in the figures. When alarmed, they contract, twisting and distorting themselves briefly before setting off in a new direction.

Most flatworms are carnivorous, feeding on small crustaceans, insect larvae, etc.; some microturbellarians feed on algae and plant detritus.

Two or more dark eyespots are usually present at the head end, but some species are blind. The mouth varies in position: in some species it is at the anterior end, in others it is situated near the middle of the body on the ventral surface. The mouth leads to a pharynx which may be thick and muscular, or tubular; in some groups it can be everted when feeding. The gut is a blind-ended tube (there is no anus), straight and broad or branched and complex. It is usually visible, especially when filled with food, but may be obscure in the dark-bodied triclads and some pigmented microturbellarians. Other internal features, which may or may not be visible, are the sexual and egg-producing organs and, in the microturbellaria, the eggs themselves; these are not usually shown in the figures.

Triclads differ from microturbellarians in not producing free eggs; instead they make spherical cocoons, each containing several embryos. These are 1–4 mm in diameter and attached to a substratum, usually by a short stalk; at first they are pale in colour, often bright orange, later becoming dark. Asexual reproduction by budding is habitual in some microturbellarians, the buds remaining attached to the parent for some time, thus building up chains of individuals. Triclads are well known for their regenerative power when cut or otherwise damaged, and a number of species habitually reproduce by tearing themselves into two pieces which regenerate into two complete worms.

Flatworms are common amongst dead leaves or other plant debris, on the undersides of floating leaves such as water lilies, beneath stones, etc., and will appear in most collections made with a water net. They can also be attracted to bait: if a container with a small entrance is baited with raw meat or chopped earthworm and left in the water for a couple of hours, flatworms will enter it and be captured. This method can catch large numbers of flatworms, so take only one or two, and release the rest. This is also a useful method of sampling water too deep for a net, the trap being lowered on a cord.

Microturbellaria

Mostly small, flat or cylindrical worms, solitary or forming chains; usually 0.5–3 mm long with only a couple of species exceeding 6 mm; gut simple, broad and straight; mouth variously situated, often anterior; eyespots present or absent; colourless and translucent, or pigmented, sometimes green.

Identification of living specimens of this group is relatively easy in some cases, less so in others where internal detail may be obscured by pigmentation. Comparison of those features which *can* be seen – position of eyes, mouth (**m** in figures) and pharynx (**p**), etc. – will usually permit a diagnosis.

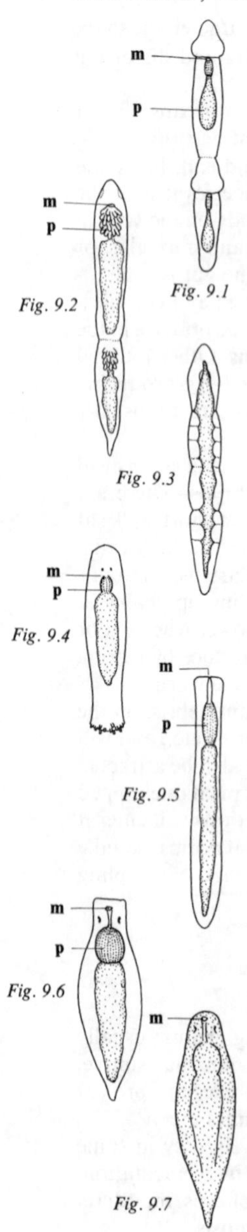

Fig. 9.1

Fig. 9.2

Fig. 9.3

Fig. 9.4

Fig. 9.5

Fig. 9.6

Fig. 9.7

Catenula *Fig. 9.1* Typically forms chains of individuals. each up to 1 mm long; mouth situated just behind a circular groove which divides 'snout' from body; pharynx small; eyespots absent; colourless or whitish. *C. lemnae* is probably the only species found in the area covered by this guide.

Stenostomum *Fig. 9.2* Forms chains, each individual being about 1.5 mm long; pharynx large and covered with papillae; head with two lateral, ciliated pits (inconspicuous): eyes usually absent; colourless or whitish; several species.

Microstomum *Fig. 9.3* Forms chains of 2 mm-long individuals, chains frequently exceed 10 mm; body nearly cylindrical with distinct cross-walls supporting gut; eyes absent or rudimentary; translucent brownish; several species, some of which feed on hydras and utilize their nematocysts (*p. 128*) for their own defence.

Macrostomum *Fig. 9.4* Solitary worms with adhesive papillae on the rear of the body; mouth sited close to small eyespots; several species, usually translucent whitish and up to 2 mm long.

Prorhynchus and ***Geocentrophora*** *Fig. 9.5* These two similar genera are very difficult to separate. They are solitary, flattened worms up to 6 mm long; eyeless; mouth at the anterior end of the body, and a broad gut occupying most of the body; translucent whitish; several species.

Castrella *Fig. 9.6* Solitary worms up to 1.5 mm long; mouth ventral with a large, muscular pharynx; eyes prominent and dumb-bell shaped; brownish, often strongly pigmented; several species.

Dalyellia *Pl. 120, Fig. 9.7* Solitary cylindrical worms up to 4 mm long, with eyespots; sole species is *D. viridis*, which is usually green due to abundant zoochlorellae, but these may be absent. It is common and often abundant where found; populations often contain non-green individuals which lack the symbiotic algae.

Other species which contain zoochlorellae are: *Typhloplana viridiata*, less than 1 mm long, with a centrally placed mouth and pharynx, and *Castrada viridis*, 1–2 mm long with central mouth and pharynx; both these species lack eyes.

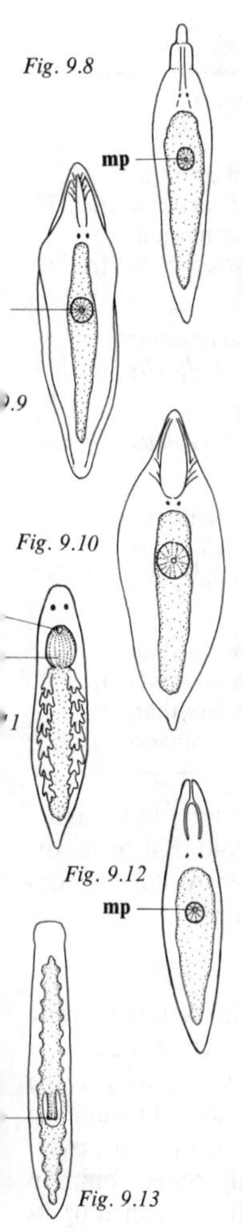

Rhynchomesostoma *Fig. 9.8* A flattish worm up to 3 mm long with a prominent retractile snout; mouth/pharynx in centre of body; translucent pinkish with reddish internal organs clearly visible. Sole species is *R. rostratum*.

Mesostoma *Pl. 121, Figs. 9.9, 10* Translucent, yellowish worms with a firmer texture than most other flatworms: internal organs clearly visible, snout long but not retractile; mouth/pharynx at centre of body; eyes present, preceded by a prominent nerve complex visible as an array of dark lines. Several common species: e.g., *M. tetragonum* which has flanged lateral edges to its body, up to 10 mm long; *M. ehrenbergi*, strongly flattened, up to 15 mm long.

Phaenocora *Fig. 9.11* Cylindrical or slightly flattened worms up to 3 mm long; mouth/pharynx just behind head; gut usually obscured by large and prominent, much-branched yolk-gland; yellowish or brown; eyes present or absent; several species.

Gyratrix *Fig. 9.12* Very contractile, colourless worms with a normally cylindrical body and a protrusible proboscis which can be withdrawn into a pouch; mouth/pharynx central. Sole species is *G. hermaphroditicus*, up to 2 mm long, which is remarkable for being found in fresh, brackish and fully marine habits.

Bothrioplana *Fig. 9.13* Solitary, eyeless, flattened worms up to 5 mm long; mouth and prominent tubular pharynx in posterior half of body; gut broad with slight lateral processes; whitish or colourless. *B. semperi* is probably the only species in the area covered by this guide.

References: Young, 1970. Brauer, 1909.

Tricladida

Flattened worms, always solitary, adults usually longer than 6 mm and up to 35 mm; gut three-branched with numerous secondary branches (*Fig. 9.15*): mouth in middle of ventral surface, with an evertible tubular pharynx; eyespots normally present; colour white, brown, grey or blackish, typically opaque, never colourless and transparent.

The genera are not difficult to distinguish by external characters but some of the species cannot be identified without detailed internal examination.

Key to the Genera of Triclads

1. a) Body white, or nearly so (gut may be dark due to its contents). . . . 2
   b) Body dark – grey, brown, blackish, etc. . . . 3
2. a) Space between eyes much smaller than their distance from anterior body margin *Phagocata* (*p. 136*)
   b) Space between eyes greater than their distance from anterior margin. *Dendrocoelum* (*p. 136*)
3. a) Eyes 2 (sometimes 3 or 4). . . . 4
   b) Eyes numerous (10 or more) arranged in a single row around anterior body margin. *Polycelis* (*p. 136*)
4. a) Head distinctly narrower than body, with a prominent ridge between the eyes. *Bdellocephala* (*p. 137*)
   b) Head more-or-less as wide as body, no ridge between eyes. . . . 5
5. a) Head truncated, with a short tentacle projecting from each anterior corner. *Crenobia* (*p. 137*)
   b) Head triangular, with small lateral projections. *Dugesia* (part) (*p. 137*)
   c) Head without tentacles or lateral projections. . . . 6
6. a) Head rounded or slightly pointed; distance between eyes greater than their distance from anterior margin. *Dugesia* (part) (*p. 137*)
   b) Head more-or-less truncated: distance between eyes much less than their distance from anterior margin. *Planaria* (*p. 137*)

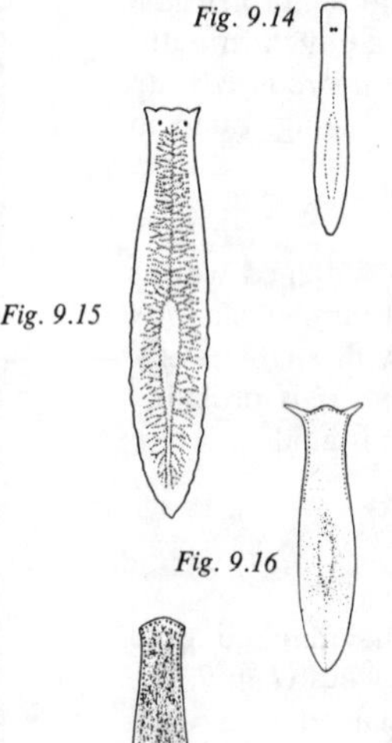
Fig. 9.14
Fig. 9.15
Fig. 9.16
Fig. 9.17 

Phagocata vitta *Fig. 9.14* Up to 15 mm long but usually less. A typical species of fast-running, cold, hill or moorland streams; in lowland areas sometimes occurs in subterranean streams and emergent cold springs; widespread.

Dendrocoelum lacteum *Pl. 122, Fig. 9.15* A large, white flatworm up to 25 mm long; body often with a reddish or greyish cast due to food in the gut, and very sticky to the touch. Very common in still or running waters throughout Europe, mostly in lowland areas. Probably does not overlap with habitats of *Phagocata*, the only other common white flatworm, from which it is easily distinguished by its larger size, stouter body, and eye position.

Polycelis *Pl. 123, Figs. 9.16, 17* Several species up to about 15 mm long. *P. felina*, with tentacles (*Fig. 9.16*), brown or blackish, is common in running waters or spring ponds. Other species *P. nigra* and *P. tenuis* lack tentacles and are not externally distinguishable; they are both brown or blackish and common in still or slow-flowing waters.

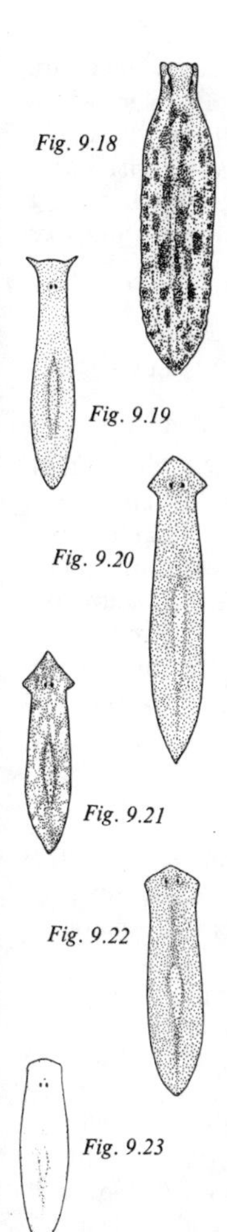
Fig. 9.18
Fig. 9.19
Fig. 9.20
Fig. 9.21
Fig. 9.22
Fig. 9.23

Bdellocephala punctata *Fig. 9.18* The largest European flatworm, up to 35 mm long; brownish, usually with darker spots and blotches. An uncommon species with a sparsely scattered distribution in north-west Europe; in lakes and canals, under stones and on mud.

Crenobia alpina *Fig. 9.19* Up to about 15 mm long; typically dark grey, sometimes brown or black. A characteristic species of cold upland streams, which also occurs sporadically in subterranean waters and emergent springs in lowland areas, often with *Phagocata*.

Dugesia *Pls. 124, 5, Figs. 9.20, 21, 22* A large genus with four main species in north-west Europe.
D. gonocephala Head triangular with lateral projections; uniform grey or brown with a pigmented pharynx; up to 30 mm long. Common in running waters in continental Europe but absent from Britain.
D. tigrina Head triangular with lateral projections; mottled grey or brown with an unpigmented pharynx: usually less than 15 mm. An introduced species from North America with a scattered distribution in Europe, but apparently spreading.
D. polychroa and *D. lugubris* both lack lateral projections on head and are difficult to separate. In *D. polychroa* the head is usually rounded, with a distinct 'neck', and the ventral surface is usually paler than the dorsal. *D. lugubris* has a more pointed head, little or no neck, and its ventral surface usually as dark as the dorsal. Both are brown or greyish, up to 20 mm long, and common in many habitats.

Planaria torva *Pl. 126, Fig. 9.23* Typically small, rarely exceeding about 12 mm; head usually square or very slightly pointed; brownish or dark grey. Widespread throughout north-west Europe but rather local and uncommon in Britain.

References: Reynoldson, 1978. Ball & Reynoldson, 1981. Brauer, 1909.

Class TREMATODA (Flukes)
Flukes are common parasites of vertebrates, mostly internal but with a number of genera living externally on aquatic hosts. Adult flukes resemble flatworms in their general form and texture, and are flattened or cylindrical in shape:

they always bear one or more distinct, usually rounded suckers. There are two subclasses: Monogenea, external parasites of fishes and amphibians with a simple life-cycle, the young flukes (often produced viviparously) usually developing on the same host species as the adult; and Digenea, internal parasites of aquatic and terrestrial vertebrates, with a complex life-cycle involving several intermediate stages, both parasitic (in invertebrate hosts) and free living, between egg and adult.

Subclass MONOGENEA
All monogeneans are aquatic, with many freshwater species; the two genera below are common in north-west Europe.

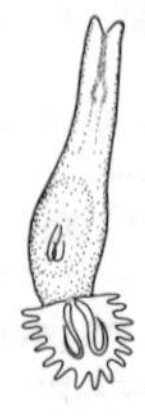
Fig. 9.24

Gyrodactylus and ***Dactylogyrus*** *Fig. 9.24* Tiny flukes. usually about 1 mm long but some species are larger: posterior end with a large, often fringed sucker, armed with a pair of grappling hooks; eyes present or absent. These two genera are difficult to distinguish externally: they are common on the skin, gills or fins of various fishes. especially sticklebacks, although difficult to find due to their small size and translucency. They move around on the host by 'looping'.

Subclass DIGENEA
Adult flukes of this subclass are mostly internal parasites and are not further described here. Other stages in the life cycle are usually aquatic, involving brief free-swimming stages and longer periods of parasitism in aquatic invertebrates, especially snails.

A typical life-cycle is as follows: Adult flukes release large numbers of eggs which pass out of the host with the faeces. These eggs hatch into microscopic ciliated, free-swimming miracidia larvae, which are unlikely to be found by the casual observer due to their very short lives. Immediately after hatching, the miracidia seek out a suitable host, usually a pulmonate snail, and bore into its tissues (the snail does not enjoy this but is powerless to resist) on which they feed and develop into the next stage, the sporocyst. Sporocysts are entirely parasitic, living in the digestive gland of the snail. When fully grown they start budding asexually, producing numbers of redia larvae, which also live parasitically within the snail. Mature rediae bud off great numbers of tiny larvae called cercariae, which are, effectively, baby flukes with tails (*Fig. 9.25*). The cercariae wait inside the snail until external conditions are suitable – warm sunny weather, preferably following rain – when they leave the snail in swarms and become free-swimming. These cercariae may enter their final host directly by attacking animals that come to drink; other species settle onto a substratum – water plants or overhanging grasses – where they form a cyst and wait for the substratum to be eaten by a final host (some herbivorous animal); yet others form cysts on, or become internal parasites of fishes and invertebrates. This final stage is known as a metacercaria; in all cases its aim is to be swallowed by the final host, inside which it completes its development into a mature fluke. Free-swimming cercariae sometimes attack humans, causing a skin irritation known as swimmer's itch.

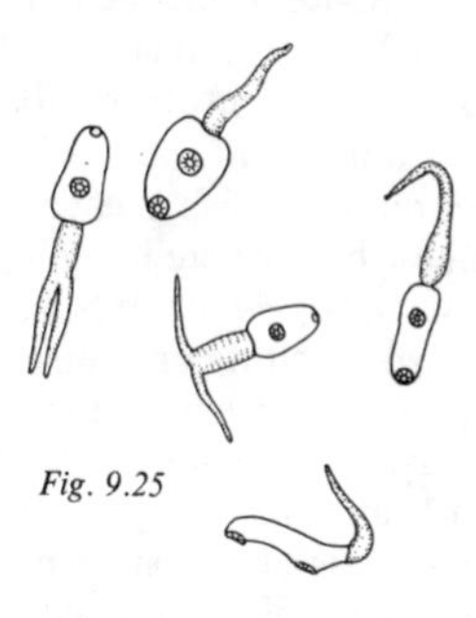

Fig. 9.25

Cercariae are sometimes released from snails taken in collections during the summer. If a selection of pulmonate snails (almost any species except *Lymnaea stagnalis*, which rarely seems to become infected) is placed in a small glass or plastic container and subjected to strong side-lighting a cloud of cercariae will often appear after two or three hours, sometimes much less. They are less than 1 mm long, translucent whitish, with single or forked tails, and swim with a frantic oscillating action; a cloud of cercariae appears to shimmer in certain lights. They may also settle on the bottom and move around by 'looping'. Encysted metacercariae are frequent on sticklebacks and leeches; these form circular or oval, translucent cysts about 0.5–1 mm in diameter; young flukes may be visible moving around inside them. The metacercariae of *Diplostomulum* live as parasites inside the eyes of fishes. They are externally visible as a greyish haze in the pupil of the eye; with a lens the tiny – 1 mm – active flukes can be seen crawling around inside the eye. They often cause blindness in the fish, which is thus less likely to evade capture by the fluke's ultimate host, a fish-eating bird. Rediae, and sometimes sporocysts, can be found by killing and dissecting a snail which is seen to be releasing cercariae. Both stages are found in the digestive gland, a dark mass near the top of the spine; they are amorphous, slow moving bodies which often contract to a sausage-shape when exposed.

Class CESTODA (Tapeworms)

Adult tapeworms are internal parasites of vertebrates. They consist of long chains of reproductive units (proglottids) budded off from a head or scolex which is attached to the wall of the host's intestine. The life-cycle is complex and the human broad tapeworm, *Diphyllobothrium latum*, which involves several aquatic stages, is described as an example.

Enormous quantities of eggs are produced in mature proglottids which pass out with the host's faeces. In water these hatch into microscopic, free-swimming coracidia larvae, which must be swallowed by a copepod crustacean – the primary intermediate host (*p. 220*). Inside the copepod they grow into worm-like procercoids. In turn the copepod must be eaten by a small fish if the worm is to develop further. Once inside the fish the procercoid grows and develops into a plerocercoid larvae, a relatively featureless worm which, when full-grown and by now several centimetres long, enters a resting phase, usually in the body cavity or muscles of the host. This stage can infect man if he eats the fish uncooked, but if the fish is eaten by a larger fish the plerocercoid can transfer to the new host where it again rests and waits until it is eaten by the final host. Such a life-cycle involves many 'ifs', all of which have to be realized if the worm is to reach its final host, but the numbers of eggs produced is astronomical and the efficacy of the life-cycle is proved by the success of tapeworms as parasites in a wide variety of hosts.

D. latum is an unpleasant beast which may attain a length of six metres or more in the human intestine. It is common in northern Europe and is an excellent reason for not eating uncooked freshwater fish!

Tapeworms and their larvae are common in freshwater hosts but with one exception are seldom evident by external observation:

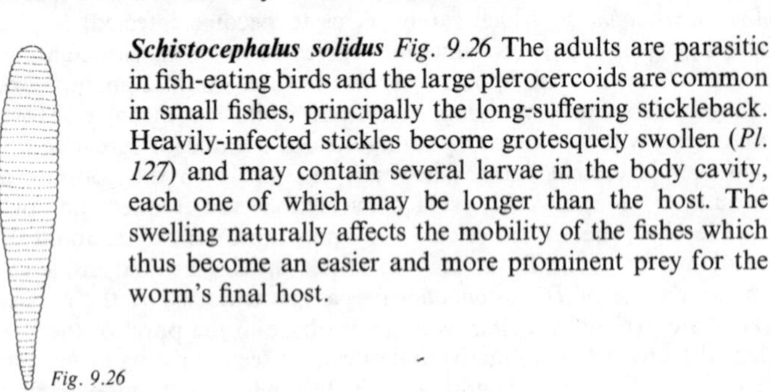

Schistocephalus solidus *Fig. 9.26* The adults are parasitic in fish-eating birds and the large plerocercoids are common in small fishes, principally the long-suffering stickleback. Heavily-infected stickles become grotesquely swollen (*Pl. 127*) and may contain several larvae in the body cavity, each one of which may be longer than the host. The swelling naturally affects the mobility of the fishes which thus become an easier and more prominent prey for the worm's final host.

Fig. 9.26

CHAPTER 10 Ribbon or Proboscis Worms
(Phylum Nemertea)

Recognition features Slender, very contractile worms with a long retractile proboscis (not visible when withdrawn); locomotion by ciliary gliding and stretching/contracting movements of the body.

This is probably the least well-known group of freshwater worms, specimens having been discovered only sporadically and in widely separated localities.

Ribbon worms resemble flatworms in their general texture and mode of locomotion, but are much more elongated when extended. Their major distinguishing feature is the proboscis, a long, tubular, mobile organ which at rest is kept withdrawn inside a long cavity in the body, the rhynchocoel. It can be protruded through a pore in the head, and when fully extended may exceed half the length of the body. The proboscis is used for sensing and capturing the prey, which consists of small annelid worms and other invertebrates.

Nemertines are common marine worms with numerous species, but in fresh water only a handful of genera are known, only one of which has been recorded in north-west Europe.

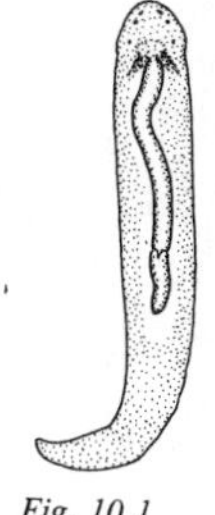

Fig. 10.1

Prostoma *Pl. 128, Fig. 10.1* Body slightly flattened, up to about 20 mm long and 1 mm wide when fully extended: 4–8, typically 6 eyespots; adults usually reddish-brown or orange, but juveniles are paler, pink or whitish. Several species of this genus have been found in the area covered by the guide, but they cannot be distinguished externally: *P. graecense* is the commonest. They occur in ponds, the shallow margins of lakes, or in slow-moving rivers, crawling on vegetation or amongst the general bottom debris.

References: Gibson & Moore, 1976. Gibson, 1982.

CHAPTER 11 Roundworms or Nematodes
(Phylum Nematoda)

Recognition features Slender, unsegmented, cylindrical worms, typically tapering to a point at the posterior end; microscopic or up to 2–3 mm long, rarely longer; unpigmented and translucent, usually appearing whitish; external cilia absent, locomotion is achieved by a highly characteristic, stiffly sinuous writhing or thrashing.

The Nematoda (*Pl. 129*) is an immensely successful and ubiquitous group of animals. It includes a vast number of free-living species which occur in probably all moist or aquatic habitats, and an even greater number of parasitic species which infest plants and animals. Most nematodes are microscopic or only a few millimetres long, but some of the parasitic species, such as *Ascaris lumbricoides* which infects man and pigs, may reach a length of 30 cm. In some environments the number of individuals present is almost unbelievable: more than 20,000,000 nematodes have been counted in 1 sq m (surface area) in some marine muds and sands! It is doubtful that they ever attain such density in freshwater habitats but they can still be extremely numerous and any collection of bottom sediments or plant debris is likely to contain plenty of specimens.

Despite the great variety of habitats in which nematodes live they exhibit

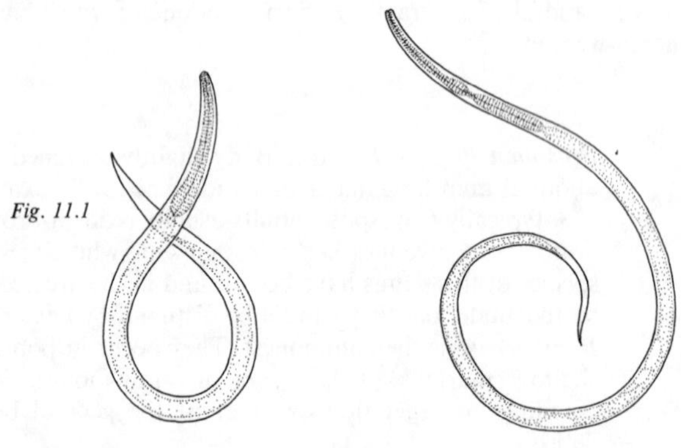
Fig. 11.1

a remarkable uniformity of external structure and appearance. The body is covered with a thick colourless cuticle which mechanically limits their movement to the strongly characteristic writhing or thrashing motion which, once seen, is subsequently easily recognized. The anterior end is tapered or truncated, more or less abruptly, and the posterior end tapers to a blunt or sharp point. This very uniformity of shape and movement which renders nematodes so easily recognizable is, however, a considerable obstacle to their further identification. This relies on microscopic study of the minutiae of internal detail and we will not attempt to distinguish the freshwater forms further.

Free-living nematodes, as a whole, feed on almost any organic matter, living or dead. Most species specialize on one type of food, e.g., diatoms or other types of algae, and their mouthparts have become adapted to dealing with specific food items.

The sexes are separate and the females produce eggs from which hatch fully formed young worms (there is no differentiated larval stage) or give birth to living young that have been brooded internally. In some species males are rare or unknown and the females reproduce by parthenogenesis; otherwise asexual reproduction is unknown.

Reference: Goodey, 1963.

CHAPTER 12 Horsehair, Hair or Gordian Worms
(Phylum Nematomorpha)

Recognition features Extremely elongated, unsegmented, cylindrical worms, often more than 20 cm long and typically less than 1 mm diameter; brown or reddish in colour.

Horsehair worms (*Pl. 130*) are poorly-known creatures bearing a superficial resemblance to nematodes, from which they are distinguished by their blunt-ended, pigmented bodies. The larval stages and developing worms are parasitic inside various insect hosts, and thus the only stage likely to be found is the free-living adult which enters fresh water to breed. Adults are found during spring and summer in shallow, weedy ponds and ditches, and are very erratic in occurrence, although sometimes abundant when they do occur. The name horsehair worm is said to derive from an ancient superstition that they arise by spontaneous animation of cast-off horsehairs.

The elongated body is covered by a thick layer of cuticle, reddish to dark brown in colour, the eyeless head usually being distinguished by a dark band just below the pale tip (*Fig. 12.1*). It is firm and wiry in texture and moves with a stiff, sinuous motion similar to a roundworm (Nematoda) but less energetic. Most specimens are about 10–20 cm long and 0.5 mm diameter but they can become much larger, continental specimens up to 1 m long having been recorded. The only obvious external structural variations involve the tail end.

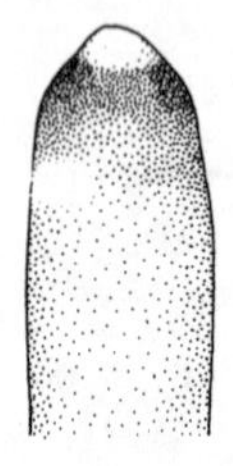

Fig. 12.1

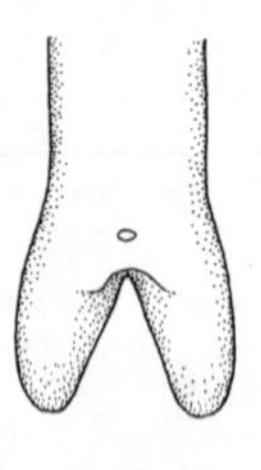

Fig. 12.3

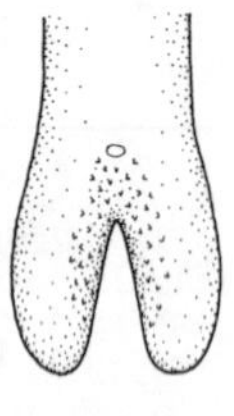

Fig. 12.2

This region may be simply rounded (females of most genera and some males) or divided into lobes by deep clefts. Males of most genera have two tail-lobes (*Fig. 12.2*); three-lobed specimens are females of *Paragordius*. Males of the best-known genus, *Gordius*, can be distinguished by a transverse crescentic fold in the cuticle just above the cleft (*Fig. 12.3*). The opposite sexes of these genera, and the remaining European genera *Parachordodes* and *Gordionus* cannot be distinguished without detailed microscopic study.

Adult females lay their eggs in strings of jelly attached to water plants. These hatch into microscopic larvae which, by some unknown mechanism, find their way into insect hosts; suitable hosts are not necessarily aquatic as adult worms have been seen to emerge from grasshoppers, cockroaches, beetles and other terrestrial insects. The worms grow to maturity in the body cavity of the host, eventually completely filling it, and then emerge to enter water where mating and egg-laying take place, after which the adult dies. The complete life-cycle of most species is unknown.

Reference: Goodey, 1963.

CHAPTER 13 Spiny-headed Worms

(Phylum Acanthocephala)

In the adult stage, these worms are all intestinal parasites of aquatic and terrestrial vertebrates. The larvae (cystacanths) of the aquatic species are parasitic in the body cavity of various arthropods, usually crustaceans, and are occasionally met with during the study of host species.

Both larvae and adults have relatively featureless, cylindrical or flattened, tapering bodies with no true segmentation, although sometimes there is irregular annulation. At the anterior end, there is a protrusible spiny probiscis (*Fig. 13.2*) with which the adults attach themselves to the host, embedding the proboscis in the intestinal wall. In cystacanths the proboscis is kept withdrawn inside the body until the larva is swallowed, together with its host, by the final host.

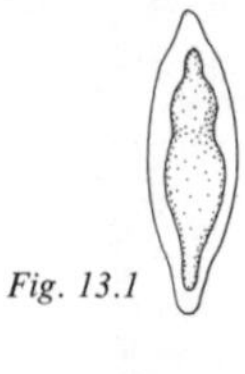

Fig. 13.1

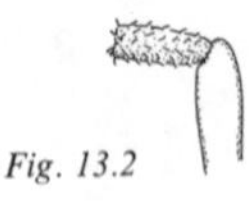

Fig. 13.2

Polymorphus *Pl. 131, Figs. 13.1, 2* Bright orange cystacanths found inside freshwater shrimps (*Gammarus*) are likely to belong to this genus. The conspicuous bright colour of the larvae no doubt increases the likelihood of the shrimp being spotted and eaten by a duck – the final host. Larvae are spindle-shaped, up to 5 mm long, and enclosed within a clear, yellowish cyst inside the shrimp. If a cyst containing a mature cystacanth is broken open in water, the larva will often protrude its proboscis (*Fig. 13.2*). Infected shrimps are local, but fairly common in streams and rivers.

Fig. 13.3

Echinorhynchus, Acanthocephalus, Pomphorhynchus, Neoechinorhynchus The cystacanths of these genera are all rather similar in appearance (*Fig. 13.3*), and mostly whitish in colour (*Pomphorhynchus* is pale yellow/orange). The first three are commonly found in *Gammarus* or *Asellus*, the last in ostracods. Due to their lack of bright colouration they are more difficult to discover than *Polymorphus*, although probably more common. Adults of these genera occur in fishes and birds.

CHAPTER 14 Rotifers, Wheel Animalcules

(Phylum Rotifera (Rotatoria))

Recognition features Microscopic or tiny worm-like animals up to 2 mm long; body soft and contractile, sometimes partly enclosed by a hardened lorica; cilia present only on head region, where they typically form encircling tracts; solitary or forming dense aggregations; free-swimming, epizooic, or fixed to a substratum, fixed forms often inhabiting tubes; some species move by looping (*Fig. 14.4*).

Rotifers are delightful little animals that have always charmed microscopists. Although small – their size range of about 25 μm–2 mm, averaging about 200–400 μm, is similar to the ciliated protozoans – rotifers are complex, multicellular animals with well organized bodies. The name of the group is derived from a distinctive ciliated region on the head, the wheel organ, which in some species imparts an illusion of rotating wheels. More than 1,000 species have been reported from north-west Europe, many of these being cosmopolitan in distribution, and the range of shapes and forms is very great. A few genera occur in the sea, but the majority are freshwater animals and are abundant in any wet or moist habitat, including such unlikely places as moss or lichens on roofs and in gutters. Many rotifers are free-swimming, being permanent or temporary members of the plankton: others are capable of temporary attachment (semi-sessile): some are truly sessile, permanently attached forms usually inhabiting tubes of various materials: some species clamber around amongst plants or creep on the mud surface, on stones, etc.; and some are epizooic on crustaceans or insects.

A rotifer has a worm-like, highly contractile body divided into three, rarely four, fairly distinct regions: head, (neck), trunk, and foot (*Fig. 14.1*). The outer skin, or cuticle, is flexible and transparent: in the trunk region of some species it is thickened to form a rigid shell, the lorica, which is often armed with spines. All rotifers are translucent and unpigmented, usually appearing yellowish or whitish.

The head can usually be retracted into the trunk. Its most conspicuous feature is the wheel organ (or corona), a series of ciliated tracts and tactile bristles, very variable in arrangement, which provide locomotory and feeding currents. Many species have one or two tiny eyespots inside the head.

Various organs are usually visible inside the trunk. The tubular gut leads from the inconspicuous mouth (usually near the ventral side within the corona) to the anus. It widens at the front of the trunk into the mastax, a bulbous. muscular crop which contains hardened, jaw-like elements (trophi). Although primarily for mastication of food the mastax in some species is protrusible and is used as a food-capturing device.

A little below the mastax the gut is again expanded to form the large sac-like stomach. Perhaps the most conspicuous organ, however, is the vitellarium or yolk-gland, a large mass containing eight prominent cells, lying to one side of the stomach. The ovary is usually hidden by the vitellarium, but mature eggs. which are often elaborately sculptured and very large for the size of the animal, are easily seen: some species carry one or more attached externally to the lorica (*Pl. 133*).

Most eggs develop without being fertilized, an asexual process known as parthenogenesis. Males are small, short-lived, and scarce; they never occur in bdelloids (see *p. 150*) and are known in only about ten per cent of the remaining species. Therefore nearly all the rotifers encountered are females, and it is to these that the descriptions apply.

Attached to the rear end or ventral surface of the trunk is the foot (absent in some planktonic forms). This is a very mobile, contractile organ usually terminating in two or more prehensile toes. The tip of the foot is adhesive and many species attach themselves temporarily to the substratum by this organ. Occasionally toes are absent or are replaced by other structures.

As stated above, rotifers are common in virtually all aquatic habitats. Perhaps the most productive environments are small weedy ponds, especially those containing an abundance of fine-leaved plants such as *Myriophyllum* or *Utricularia* (bladderwort); rotifer densities of 5,000 per litre are by no means uncommon in such places. Rotifers can, seasonally, be equally abundant in the plankton, and include many glassily transparent species adapted to this way of life. During these abundances they form an important food resource and many animals, especially fish larvae, avail themselves of this convenient 'meals-on-wheels' service.

Temporarily or permanently wet patches of bryophytes and lichens also harbour numerous rotifers, mostly bdelloids. If the environment dries up bdelloid rotifers can shrivel to a cyst-like form (*Fig. 14.2*) which can remain in a state of suspended animation and withstand years of desiccation or extremes of temperature (a state known as anabiosis). When re-wetted the rotifers revive quickly, usually in a matter of minutes, or at most, hours. These cysts are common in dried moss and are easily found by grinding the moss or lichen to a powder and mounting this on a slide in a drop of water. Cysts are usually pink or yellowish and about 50 μm long.

The sessile, tube-dwelling rotifers are quite common animals, typically occurring on the stems and leaves of fine-leaved plants or finely-branched. aquatic tree roots such as willow. They are all large enough to be seen with the naked eye, which aids their discovery, and have a great advantage as microscopical subjects: being fixed to a substratum they are easy to handle and do not wander out of the field of view. The exquisite coronas of such genera as *Floscularia* and *Stephanoceras* must be among the loveliest sights to be seen through a microscope. Like most tubicolous animals they withdraw rapidly into their tubes at the slightest disturbance, only emerging slowly and with great caution.

Precise identification of rotifers relies mainly on characters of the trophi and ciliation of the corona, a specialist's task. However, most of the common forms can be recognized with reasonable accuracy by various external features employed in the descriptions below. The drawings illustrate a typical example of each genus but other species, especially in the large genera *Lecane, Euchlanis, Cephalodella, Proales* and *Synchaeta,* may deviate considerably in overall shape from the example given.

The two classes of Rotifera are defined in the key below. The large class Monogononta is divided into three subclasses, two of which encompass such a widely-varying series of forms that a brief definition of each, based on external characters, is impossible to make. We have therefore split these into three artificial groups.

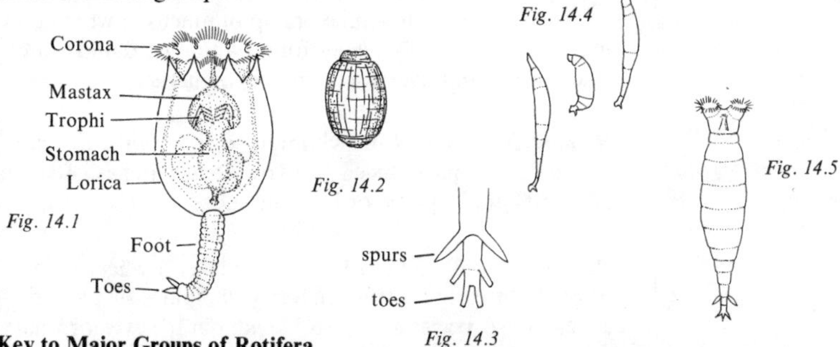

Fig. 14.1 *Fig. 14.2* *Fig. 14.3* *Fig. 14.4* *Fig. 14.5*

Key to Major Groups of Rotifera

1. a) Lorica absent, body very contractile, with the cuticle forming a series of apparently telescoping false-segments; corona typically consists of two wheels; foot usually with two spurs and two to four toes (*Fig. 14.3*); locomotion by ciliary swimming or looping (*Fig. 14.4*) on a substratum; all forms are rather similar in general appearance and easily recognized as a group (*Fig. 14.5*).

   Class BDELLOIDEA (*p. 150*)

   b) Lorica present or absent, cuticle not usually divided into segments except on foot; corona very varied in shape, rarely forming two wheels; foot present or absent, without spurs, not more than two toes: locomotion by ciliary swimming, never by looping, but the foot is often used as a lever for hopping: some forms are fixed to the substratum.

   Class MONOGONONTA . . . 2

2. a) Corona wide and cup-like, its rim often formed into lobes or arms: sessile forms lack mobile cilia but immobile, stiff bristles may be present: planktonic species (rare) bear a single ring of mobile cilia and inhabit a gelatinous case.

   Order COLLOTHECACEA (*p. 154*)

   b) Corona variously formed, always with mobile cilia: planktonic forms do not inhabit gelatinous cases. . . . 3

3. a) Semi-sessile or free-swimming rotifers with foot and toes.* **Group A** (*p. 151*)
   b) Exclusively free-swimming, planktonic rotifers lacking a foot. **Group B** (*p. 152*)
   c) Sessile species, mostly fixed, and lacking toes*; sometimes forming spherical, free-swimming aggregations. **Group C** (*p. 153*)

* *Testudinella* is free-swimming, possesses a foot but lacks toes; it belongs in **Group C**, *p. 153*.

Class BDELLOIDEA (DIGONONTA)

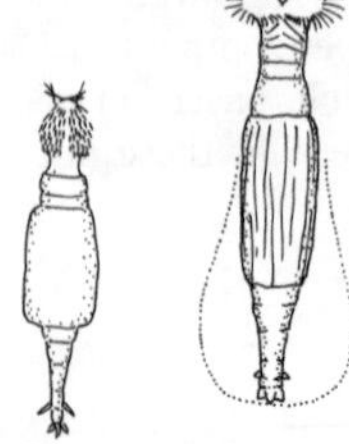

Fig. 14.7 *Fig. 14.6*

Habrotrocha *Fig. 14.6* A large genus which can only be distinguished from other bdelloids by the spongy, vacuolated appearance of the stomach, which lacks a true tubular cavity. Corona two-wheeled; three toes; up to 300 μm: many species secrete a tube or cup of mucus in which they live a sessile life. Typically found amongst damp waterside mosses and liverworts, rarely submerged.

Adineta *Fig. 14.7* Non-swimming rotifers with the non-retractile corona reduced to two lateral patches of small cilia; 500 μm: aquatic or in damp moss.

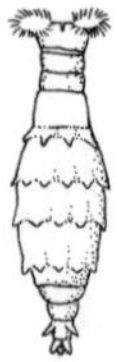

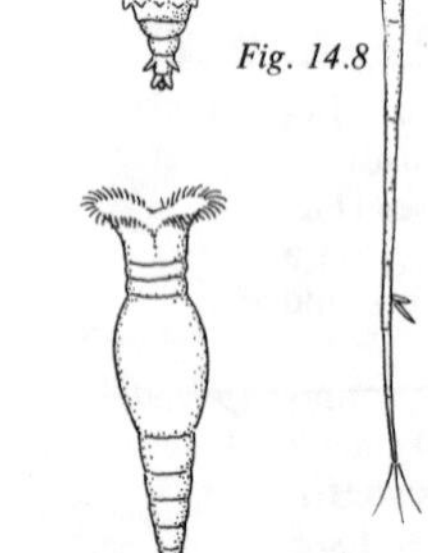

Fig. 14.9

Fig. 14.8

Fig. 14.10

Rotaria *Pl. 132, Figs. 14.5, 8* Corona two-wheeled; eyes present; three toes; trunk never with spines or protuberances; most species are up to 500 μm but the extraordinary *R. neptunia* (*Fig. 14.8*) may reach 1.5 mm; usually aquatic.

Macrotrachela *Fig. 14.9* Corona two-wheeled; eyes absent; three toes; trunk often with spines or various cuticular protuberances; 500 μm; usually amongst bryophytes.

Philodina *Fig. 14.10* Corona two-wheeled; four toes: cuticle smooth; foot usually shorter than rest of body, with small spurs; 500 μm; usually aquatic, sometimes epizoic on crustaceans.

Embata *Fig. 14.11* Corona two-wheeled; four toes; cuticle smooth; foot usually longer than rest of body, with long spurs; 600 μm; typically epizoic on crustaceans such as *Gammarus*.

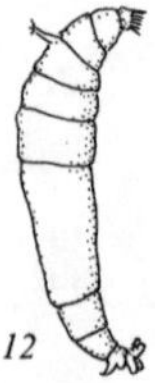

Philodinavus *Fig. 14.12* Corona reduced to a tiny, ciliated area on the tip of the body: 300 μm: movements rather maggot-like and unable to swim.

Fig. 14.12

Fig. 14.11

Class MONOGONONTA
Group A
Order PLOIMA (part)

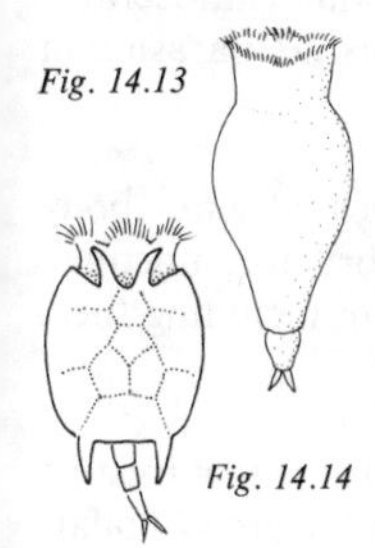

Fig. 14.13

Fig. 14.14

***Epiphanes* (*Hydatina*)** *Fig. 14.13* Lorica absent or very weak; corona wide and simple; foot short; overall shape variable, up to 500 μm long. A rather featureless rotifer that is often common in small temporary pools.

Brachionus *Pl. 133, Fig. 14.1* Lorica present, with up to six anterior spines and sometimes with posterior ones; 500 μm; foot usually long, annulated or segmented, very mobile and retractile; corona complex, with five lobes; many species, one is epizooic on water-fleas.

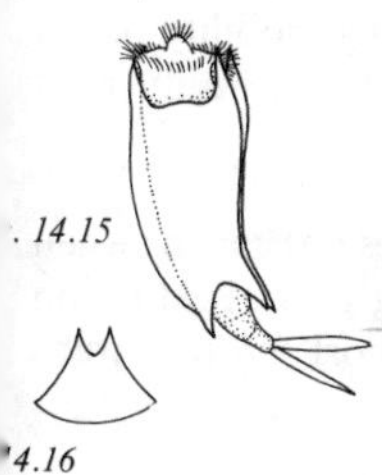

. 14.15

4.16

Platyias *Fig. 14.14* Resembles *Brachionus* but the segmented foot is not retractile; 350 μm; sole species is *P. quadricornis*.

Mytilina *Figs. 14.15, 16* Lorica present, triangular in cross-section with a deep dorsal groove (*Fig. 14.16*) and usually with spines at each end; 250 μm; toes usually long; benthic.

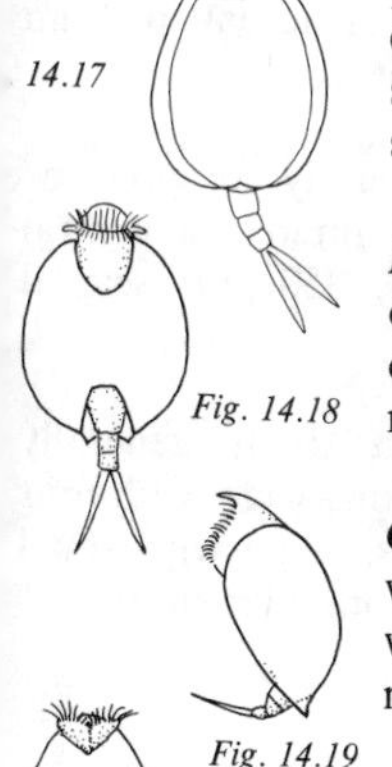

14.17

Fig. 14.18

Fig. 14.19

Fig. 14.20

Euchlanis *Fig. 14.17* Lorica thin, consisting of two unequal plates, a domed dorsal plate overlapping but not in direct contact with the smaller flat ventral plate; foot with two segments and long pointed toes; 600 μm; numerous species, planktonic or benthic.

Lepadella *Fig. 14.18* Lorica thin and flattened, consisting of two plates fused laterally with narrow head and foot openings; very transparent; 200 μm; foot with three segments and long pointed toes; mostly benthic.

Colurella *Fig. 14.19* Lorica laterally compressed, head with a separate protective plate but still retractile; foot with three segments, toes sometimes fused; small benthic rotifers up to 120 μm.

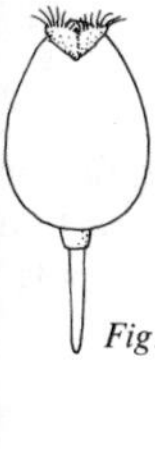

Lecane (inc. ***Monostyla***) *Fig. 14.20* Lorica weak, flexible, consisting of equal dorsal and ventral plates not united laterally; 300 μm; foot with two segments, toes long, may be fused together (*Monostyla*); numerous species, mostly benthic.

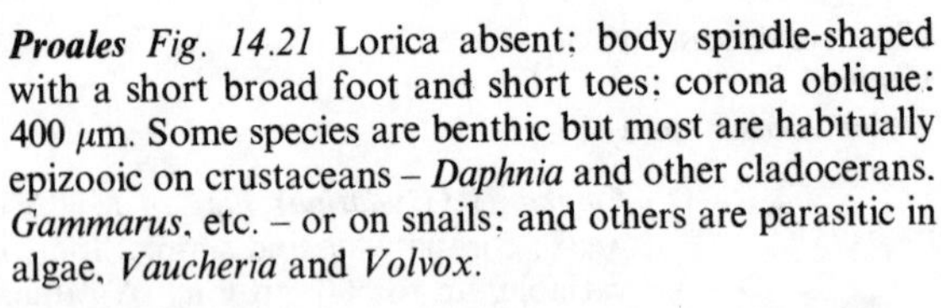

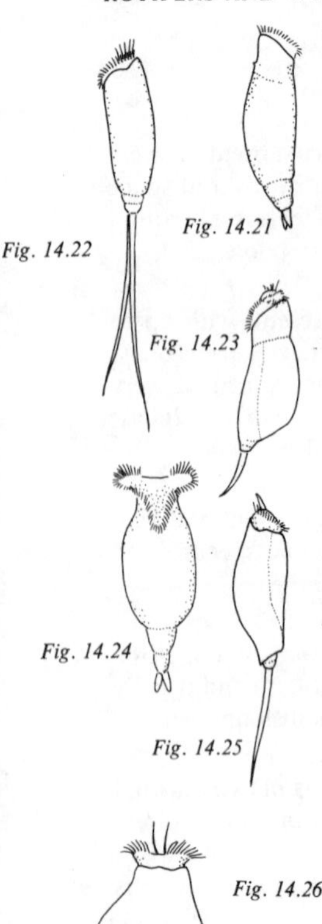

Fig. 14.21

Fig. 14.22

Fig. 14.23

Fig. 14.24

Fig. 14.25

Proales *Fig. 14.21* Lorica absent; body spindle-shaped with a short broad foot and short toes; corona oblique; 400 μm. Some species are benthic but most are habitually epizooic on crustaceans – *Daphnia* and other cladocerans, *Gammarus*, etc. – or on snails; and others are parasitic in algae, *Vaucheria* and *Volvox*.

Monommata *Pl. 134, Fig. 14.22* Lorica absent; body regions indistinct but foot obviously short; toes unequal, the longest longer than the body; 700 μm including toes; benthic or planktonic.

Cephalodella *Fig. 14.23* Lorica thin but distinct, consisting of three main plates, dorsal, ventral and a broad collar; ventral plate encloses the foot, restricting mobility of the toes; toes short to long, often curved; 300 μm; numerous species, mainly benthic.

Notommata *Fig. 14.24* Lorica weak or absent; corona when fully expanded (swimming) with a long ventral extension and two lateral lobes; foot and toes short, up to 1 mm; many species.

Trichocerca *Fig. 14.25* Lorica in one piece, asymmetrical, often twisted, often with neck spines; foot short, one toe is very long, the other being reduced or absent; 600 μm including toe; many species, benthic or planktonic.

Fig. 14.26

Gastropus *Fig. 14.26* Lorica thin, laterally compressed; foot variable in length, inserted near middle of ventral surface, toes small and often fused; 350 μm; several species, mostly planktonic.

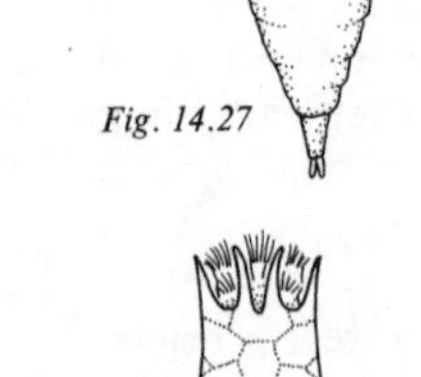

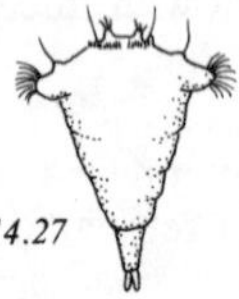

Fig. 14.27

Synchaeta *Fig. 14.27* Lorica absent; body generally conical or vase-shaped; foot tiny; corona wide, with four stout sensory bristles; several species, some with small lateral lobes on the corona, up to 600 μm, planktonic.

Fig. 14.28

Group B

Order PLOIMA (part)

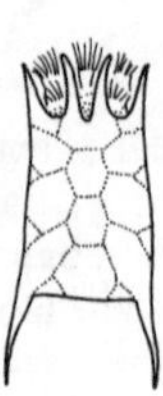

Keratella *Pl. 135, Fig. 14.28* Lorica strong, rectangular or triangular, with six anterior spines and usually one or two posterior ones, dorsal surface divided into distinct polygonal facets; numerous species, 300 μm.

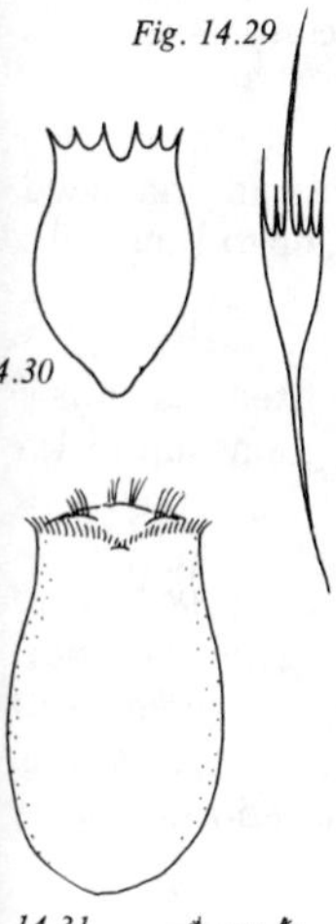
Fig. 14.29

14.30

Kellicottia *Fig. 14.29* Lorica strong, triangular, with about six long anterior spines, uneven in length and irregularly disposed, and one long posterior spine; 800 μm including spines.

Notholca *Fig. 14.30* Lorica strong, cup- or wine-glass-shaped, with about six short, usually regular anterior spines; rarely a posterior spine but this region is often drawn out; dorsal surface never faceted but may be striated; 200 μm; numerous species.

Asplanchna *Fig. 14.31* Lorica absent, body soft and bag-like, relatively featureless, up to 1 mm or more, very transparent.

14.31

Fig. 14.32

Polyarthra *Fig. 14.32* Lorica absent, body a short cylinder or squarish, up to 200 μm; twelve mobile cuticular appendages, blade or leaf-shaped, arise in four groups of three from the shoulders. By flicking these appendages the rotifer can dart through the water for a short distance, in addition to ciliary swimming.

Order FLOSCULARIACEA (part)

Hexarthra **(*Pedalia*)** *Fig. 14.33* Lorica absent, body conical with six unequal, muscular, leg-like appendages which terminate in bunches of bristles; 350 μm; several species. These very distinctive rotifers bear a strong resemblance to the *nauplius* larvae of certain crustaceans (*p. 220*) and like them, make short darting movements, using their 'legs'.

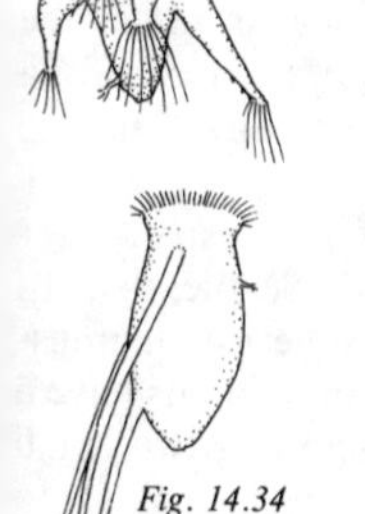
Fig. 14.33

Fig. 14.34

Filinia **(*Triarthra*)** *Fig. 14.34* Lorica absent; body cylindrical or tapering, with two mobile and often very long bristles arising from the shoulder, and one (rarely two) fixed caudal bristles; body up to 200 μm. The mobile bristles are used to make jumping movements.

Group C

Order FLOSCULARIACEA (part)

14.35

Testudinella *Fig. 14.35* Lorica well developed and very transparent, compressed, more or less circular in outline: 200 μm; foot inserted in ventral surface, long and very mobile, annulated, terminating in a very distinct ciliated

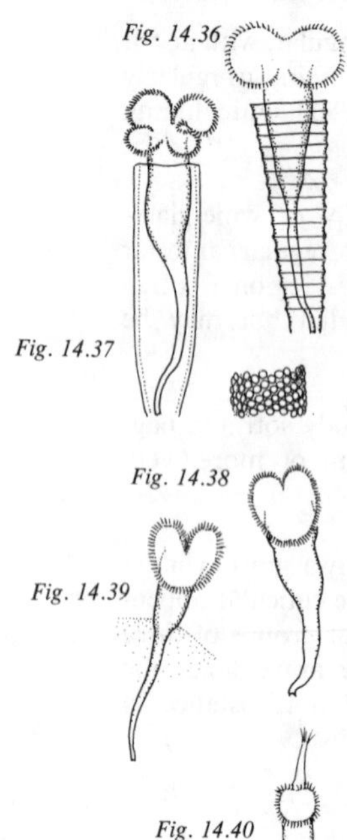

cup, toes absent; a free-swimming or semi-sessile genus with several species.

Limnias *Fig. 14.36* Sessile, fixed; inhabits an annulated chitinous tube, often covered with debris, up to 1 mm tall; corona consists of two, rounded lobes.

***Floscularia* (*Melicerta*)** *Fig. 14.37* Sessile, fixed; inhabits a tube made of tiny faecal pellets or sand grains, up to 1.5 mm tall; corona with four lobes.

***Sinantherina* (*Megalotrocha*)** *Pl. 136, Fig. 14.38* Sessile, detachable; tends to form spherical aggregations of individuals radiating from a common point; these aggregations sometimes detach and swim free; no tube or gelatinous matrix; corona heart-shaped, with a shallow dorsal notch; individuals up to 1.5 mm long.

Lacinularia *Fig. 14.39* Very similar to *Sinantherina* but secretes a gelatinous matrix in which the individuals become aggregated.

Beauchampia *Fig. 14.40* Sessile, fixed; inhabits a gelatinous tube encrusted with dirt and local debris, up to 1 mm tall; corona kidney-shaped; behind the corona is a prominent dorsal antenna.

Ptygura *Fig. 14.41* Sessile, fixed; lives in a gelatinous tube usually encrusted with debris, up to 750 μm tall; corona circular or oval, rarely with a shallow dorsal notch, dorsal antenna small or absent.

***Conochilus* (inc. *Conochiloides*)** *Fig. 14.42* Sessile, detachable; like *Sinantherina* it tends to form spherical aggregations which may detach and swim free; individuals have a distinctive horseshoe-shaped corona encircling two small antennae (in *Conochiloides* these are just ventral to corona); up to 750 μm.

Order COLLOTHECACEA

Collotheca (has been confused with *Floscularia* in nomenclature) *Figs. 14.43, 44* Sessile and fixed, or planktonic, inhabits a broad, gelatinous tube; corona varies in shape with species, usually circular, with or without marginal lobes, always with some stiff bristles typically arranged in

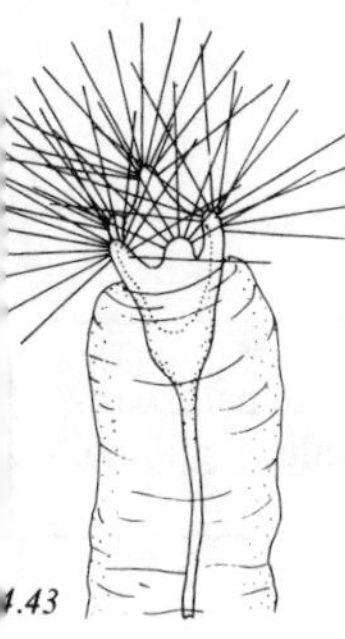

4.43

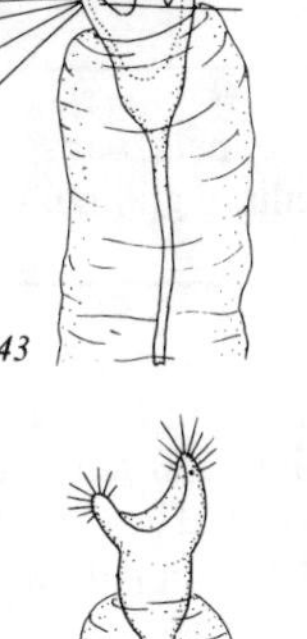

ig. 14.44

clusters, never in whorls on lobes; mobile cilia present only in planktonic forms; foot simple without toes; fixed species up to 1 mm tall, planktonic species to 500 μm.

Stephanoceras *Fig. 14.45* Sessile, fixed; inhabits a broad gelatinous tube; up to 1 mm tall, sometimes more; corona produced into five long arms bearing whorls of bristles.

Acyclus *Fig. 14.46* Sessile; corona wide, with a single elongated dorsal lobe, bristles and cilia absent; 1 mm tall: lives in aggregations of *Sinantherina*, feeding on their larvae.

***Cupelophagus* (*Apsilus*)** *Fig. 14.47* Sessile, attached by a small ventral sucker; about 700 μm long; body shape very unusual and unlike any other rotifer; corona forms a cavernous mouth with no bristles, which engulfs passing small animals; usually found attached to broad-leaved plants.

Atrochus *Fig. 14.48* Sessile but free; body very contractile: corona with a short dorsal lobe, no bristles or cilia: up to 1.5 mm long. This strange rotifer creeps about on the general substratum or plant leaves, engulfing the algae on which it feeds.

References: Donner, 1966. Pontin, 1978. Ruttner-Kolisko, 1974.

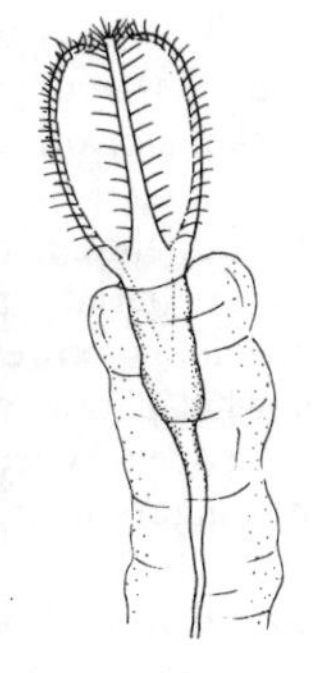

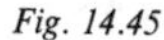

Fig. 14.45

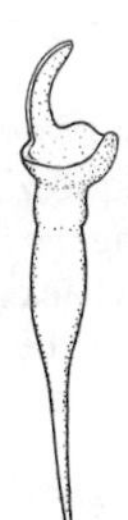

Fig. 14.46

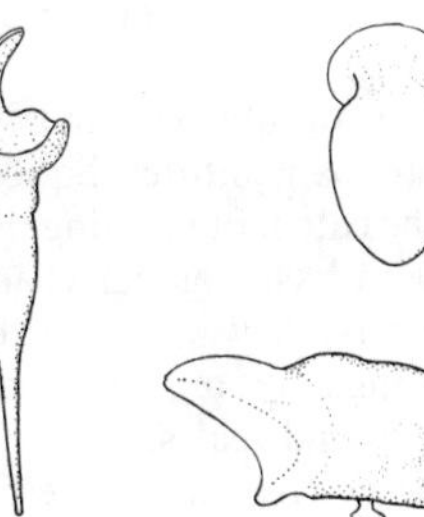

Fig. 14.47

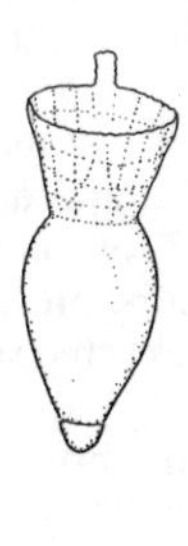

Fig. 14.48

CHAPTER 15 Hairybacks
(Phylum Gastrotricha)

Recognition features Microscopic, colourless, free-living metazoans with a characteristic body shape (*Fig. 15.1*), moving by ciliary gliding.

Gastrotrichs or 'hairybacks' are common in fresh water but are not often seen due to their secretive habits and small size – usually less than 250 μm and rarely exceeding 500 μm – smaller than many ciliated protozoans. They occur mostly amongst water plants, on the mud surface, or living between particles in sandy or gravelly sediments.

Gastrotrichs have a very characteristic body shape consisting of a rounded head bearing long sensory hairs, a slightly narrowed neck, and a streamlined body terminating posteriorly in a forked structure – the caudal furca. The flexible body has a distinct, flattened ventral surface covered with cilia; these provide the only means of locomotion – a smooth, flexuous gliding, rarely rising clear of the substratum. The dorsal surface is arched and its surface may be smooth, spiny or scaly. The straight tubular gut is formed into a strong muscular pharynx anteriorly, and can usually be seen through the translucent body wall. Gastrotrichs apparently feed on small algae and protozoans, or on organic debris.

In the freshwater species, male gastrotrichs are unknown and the large eggs develop by parthenogenesis. Developing eggs often cause a conspicuous bulge in the animal's body; eventually they are deposited on plants or other objects.

Hairybacks may be sought profitably in small, weedy ponds or ditches, and they often appear in old protozoan cultures. Squeezing out plant material is a good way of concentrating the catch, but sorting the results under a low-power microscope must be done with haste, as hairybacks are somewhat paranoid and quickly hide amongst debris; they do not like being out in the open. When spotted they appear like moving chips of glass, due to light reflecting from and refracting through their spines and scales.

About 100 species are known from north-west Europe, more than half of these belonging to the large genus *Chaetonotus*. Two other genera are reasonably common.

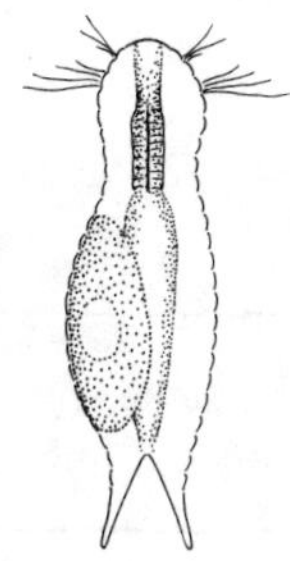

Fig. 15.1

Chaetonotus *Pl. 137, Figs. 15.1, 2* Dorsal surface of body covered with small scales, which are produced into spines varying in length according to species; in most species the spines are not difficult to make out, but where they are short, care must be taken to avoid confusion with *Lepidodermella*; usually up to about 250 μm but one species may exceed 500 μm.

Lepidodermella *Fig. 15.3* Dorsal surface of body covered with scales which are never produced into spines: up to 250 μm long.

Ichthyidium *Fig. 15.4* Dorsal surface smooth, without scales or spines; usually less than 100 μm long.

Reference: Edmondson, 1966.

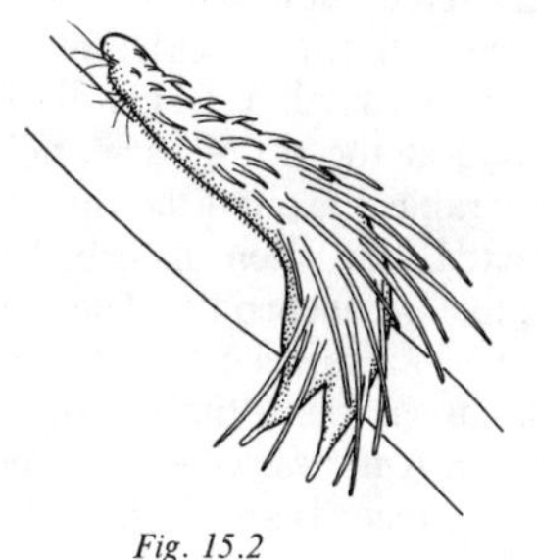

Fig. 15.2

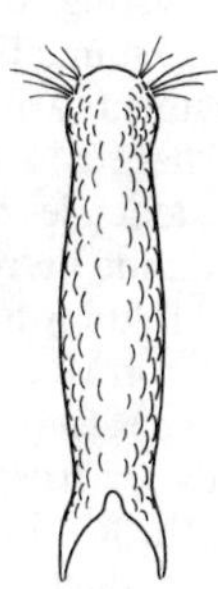

Fig. 15.3

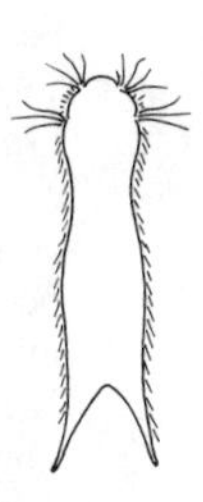

Fig. 15.4

CHAPTER 16 Entoprocts
(Phylum Entoprocta)

Recognition features Tiny, stalked animals with non-retractile tentacles, forming simple or branched attached colonies.

This is a very small phylum containing only a few marine genera and one rare freshwater one (absent from Britain).

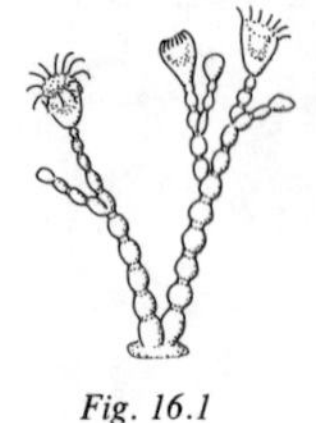

Fig. 16.1

Urnatella gracilis *Fig. 16.1* This animal forms small colonies consisting of clustered, sometimes branched stalks up to 2.5 mm tall. Each stalk comprises a number of bead-like segments arising from a small basal plate, with a single, cup-shaped individual at the tip of each stalk or branch. The tentacles are arranged around the rim of the cup; they cannot be retracted, but when disturbed they contract and fold inwards, like a clenched fist. The species occurs mostly in rivers but has also been found in large lakes. It attaches to rocks, mussel shells, timberwork, etc. Most records are from North America, where it is probably native; the few European records are possibly due to accidental introduction.

Reference: Edmondson, 1966.

CHAPTER 17 Segmented Worms and Leeches
(Phylum Annelida)

Recognition features Macroscopic, mostly free-living, elongated and contractile, cylindrical or more-or-less flattened worms, with the body divided into a number of segments (sometimes obscured by secondary annulation); most segments bear bundles of bristles or chaetae (not always conspicuous) or there is a rounded sucker at one or both ends of the body; sizes range from about 1 mm to upwards of 15 cm.

This phylum contains the familiar earthworms and their relatives; leeches: the marine bristleworms such as lugworms, ragworms and fanworms: and a host of smaller, less well-known aquatic forms.

Key to Major Groups of Segmented Worms

1. a) Body segments with lateral fleshy flaps (parapodia) bearing the chaetae: head region strongly differentiated, often with tentacles: some species inhabit tubes of hard calcareous material. A marine group with two brackish-water species occasionally found in fresh waters connected to the sea. Class POLYCHAETA *(p. 160)*
   b) Parapodia absent: head usually poorly differentiated, sometimes produced into a tentacle-like proboscis: never inhabiting calcareous tubes: numerous freshwater and terrestrial forms. Class CLITELLATA . . . 2
2. a) Chaetae always present but inconspicuous in earthworm-like forms: suckers absent. Subclass OLIGOCHAETA *(p. 160)*
   b) Chaetae absent: a sucker is present at the posterior, and usually at the anterior end. . . . 3
3. a) Small worms, less than 10 mm long, without secondary annulation: sucker at posterior end only: externally parasitic on crayfishes (uncommon). Aberrent oligochaetes, family BRANCHIOBDELLIDAE *(p. 166)*
   b) Mostly larger than 10 mm: body with secondary annulation and a sucker at each end: free-living or parasitic, but not on crayfishes. Leeches (common). Subclass HIRUDINEA *(p. 166)*

Class POLYCHAETA
(Bristleworms)

Polychaetes are almost exclusively marine but two species occasionally stray into fresh water from adjacent brackish waters. There is also a true freshwater genus *Manayunkia* (a small worm with a tentacled head, inhabiting a mucus tube) which is known from North America and Lake Baikal in the USSR, but has not been found in north-west Europe (see Pennak, 1953 for details).

Nereis diversicolor (Ragworm) A medium-sized worm up to 10 cm long, brownish or greenish, usually with a red centre-line. It walks using its parapodia and swims with strong lateral undulations of the body. A typical and very common estuarine species which occasionally wanders into adjacent freshwater regions of rivers.

Ficopomatus (Mercierella) enigmatica (Fanworm) *Pl. 138.* This worm inhabits a fixed calcareous tube about 1.2 mm diameter and up to 30 mm long; tubes typically become aggregated together to form massive colonies. The head of the worm bears a small fan of greenish and brown tentacles which are shyly protruded from the tube. Normally it occurs in the sea or brackish water but very rarely it has invaded freshwater lagoons and impounded lakes adjoining the sea.

Class CLITELLATA
(Earthworms, Leeches and others)

These are common freshwater and terrestrial worms characterized by the clitellum, a glandular, often collar-like thickening of the body wall covering several segments in the genital region. It usually occurs only in sexually mature specimens, and in some of the smaller aquatic forms it is completely absent, thus having little value as a recognition feature. Clitellates are hermaphrodite, each individual possessing both male and female organs, but the mating of two individuals is necessary for fertilization to occur. The eggs are deposited in cocoons secreted by the clitellum. Many of the aquatic species reproduce asexually by budding or fragmentation.

Subclass OLIGOCHAETA

The Oligochaeta includes a variety of worms of different shapes and sizes. At one extreme are the small, colourless aquatic forms with more-or-less prominent chaetae, at the other are the familiar large, pigmented earthworm types with inconspicuous chaetae. Many species have red blood and this generally influences the overall colouration which is typically some shade of pink, red or brown.

The number of body segments varies from less than a dozen up to several hundred. The first segment never bears chaetae and is divided into an anterior lobe, the prostomium, which precedes the ventrally-placed mouth, and is distinctively modified in certain genera, and the first segment proper. Each of the remaining segments typically bears four bundles of chaetae, two dorsolateral (dorsal) and two ventrolateral (ventral) but the dorsal bundles may be lacking in some or all segments. The form and arrangement of the chaetae is often the only external identification character available, but their study requires the use of a microscope. Ordinary chaetae are relatively short and stout, terminating abruptly at a notched or blunt-pointed tip, but some groups also possess hair chaetae, which are relatively long and slender, and taper to a fine point (*Fig. 17.1*). Hair chaetae occur only in some of the smaller, translucent worms and can usually be distinguished in living specimens that have been gently compressed on a glass slide (*p. 352*).

Most oligochaetes have a similar style of movement, alternately extending and contracting the body, using the chaetae to gain traction. Few species can swim and these thrash around in a very inelegant and inefficient manner. *Chaetogaster*, a very common worm, characteristically 'loops' with the anterior part of its body, shuffling and wriggling along with the remainder (*Fig. 17.2*). Only *Aeolosoma* employs cilia for locomotion.

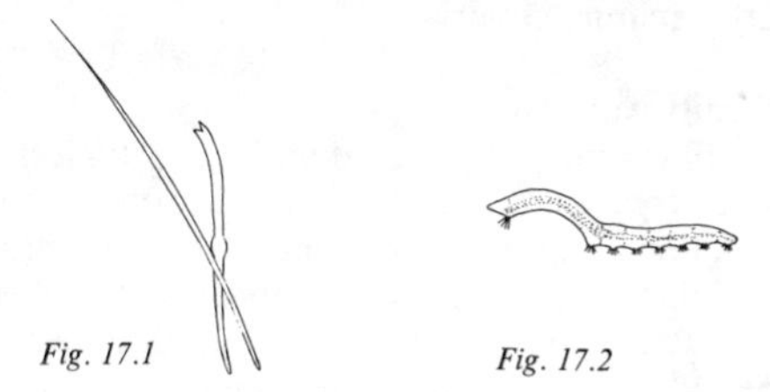

Fig. 17.1 *Fig. 17.2*

The majority of oligochaetes feed by swallowing large quantities of mud, soil or bottom debris, digesting the nutritious organic particles and discarding the remainder.

Identification below family level is notoriously difficult and even experienced biologists are reluctant to attempt it. Many otherwise detailed and painstakingly compiled species-lists include such lame items as 'unidentified oligochaete' or '?*Tubifex* sp.', in spite of the obvious importance of oligochaetes as a major faunal element in soft sediments!

The families of aquatic Oligochaeta are not difficult to recognize, but the construction of a satisfactory key (avoiding microscopic or internal characters) is not possible. The following notes should help to limit the search for the correct family for any given specimen. Small, usually translucent species will belong to Naididae, Tubificidae, Lumbriculidae (not always translucent), or Aeolosomatidae; large earthworm-like types are either Glossoscolecidae or Lumbricidae; *Haplotaxis* is extremely elongated and thread-like, and uncommon; enchytraeids are like miniature earthworms, usually white; and Branchiobdellidae are parasites of crayfishes.

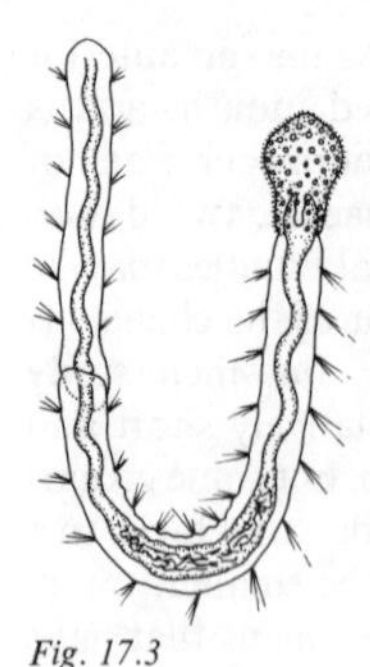

Fig. 17.3

Family Aeolosomatidae

Sole genus ***Aeolosoma*** *Pl. 139, Fig. 17.3* Small worms. typically 1–5 mm long but occasionally up to 10 mm: prostomium broader than rest of body, rounded in front. and ventrally covered with cilia; body transparent and colourless except for numerous, tiny oil droplets just under the skin – these may be red, blue, green, olive or colourless (these are only shown on the prostomium in *Fig. 17.3*). Hair chaetae are always present in dorsal and ventral bundles. Asexual reproduction by budding is habitual, two individuals joined nose-to-tail being a common sight. The body is contractile, but does not contribute to forward motion, this being achieved by the ciliated prostomium alone, the rest of the body trailing behind. The cilia are also used to gather together the particulate matter that constitutes the food.

Several European species are known, which are probably common but not often seen. They are found among bottom sediments and organic debris, commonly turning up in mature aquaria.

Family Naididae

Small worms rarely exceeding 20 mm; typically colourless and very transparent; hair chaetae often present in dorsal bundles if these are present. Asexual reproduction by budding from the tail is common, resulting in short chains of two or three individuals joined nose-to-tail. Some naidids are mud-burrowers, sometimes living in flimsy mud tubes with the head end protruding and waving in the water; others wander freely among plants or detritus. Several common genera are easily recognized:

Dorsal chaeta bundles absent:

Chaetogaster *Pl. 140, Figs. 17.2, 4* Anterior part of body very contractile, with a fan-like bunch of chaetae on each side near the mouth; locomotion by 'looping' with the anterior part of the body and wriggling, stretching, etc.. with the remainder. These are very common and easily-recognized worms with several European species. They are found in a variety of habitats and sometimes associated with other animals (e.g. *C. limnei* usually occurs in the mantle cavity of various pulmonate snails). *Chaetogaster* is one of the few carnivorous oligochaetes, feeding on any small animals it can catch and swallow, as well as algae and detritus.

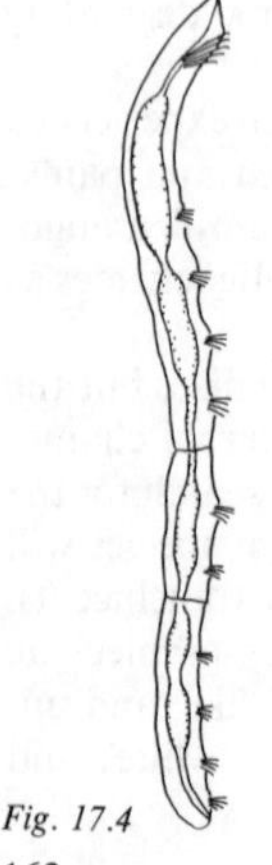

Fig. 17.4

Dorsal chaeta bundles present, hair chaetae absent:
Uncinais, Ophidonais, Paranais These are mud-dwelling worms with dorsal chaeta bundles on all but the first few segments; all rather similar in appearance. *Paranais* lacks eyes and is typically a brackish-water genus uncommon in fresh water. The other two genera have eyespots and some pigmentation on the first few segments: *Ophidonais* has only one chaeta in each dorsal bundle and is up to 30 mm long; *Uncinais* has two to five chaetae per dorsal bundle and is up to 18 mm.

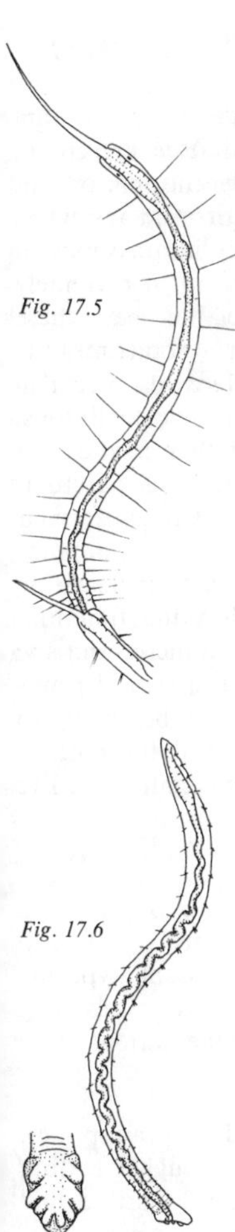
Fig. 17.5

Fig. 17.6

Dorsal chaeta bundles present, hair chaetae present:
Stylaria *Pl. 141, Fig. 17.5* Active worms with the prostomium drawn out to form a long, tentacle-like proboscis; eyespots present. *S. lacustris*, up to 15 mm long, is probably the sole species and is common in many habitats amongst filamentous algae and other plant-life, or in detritus; it swims with a stiff, wriggling action.

Dero *Fig. 17.6* Caudal segment truncated, with four to eight short, ciliated, papilla-like gills; up to 15 mm long. Several species occur in the area covered by this guide, living in mud or amongst plants, sometimes constructing flimsy mud tubes. A similar genus *Aulophorus* has caudal gills with two additional, slender unciliated appendates (*Fig. 17.7*); it is uncommon.

Pristina Dorsal chaeta bundles present on all segments. with long hair chaetae; prostomium produced as a slender proboscis in some species, eyespots absent; up to 15 mm long.

Nais *Fig. 17.8* Dorsal chaeta bundles absent from first three or more segments, hair chaetae present; eyespots present or absent; never with a proboscis or caudal gills. Many species of this large genus are common in a variety of habitats; up to 20 mm long.

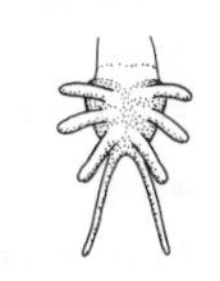
Fig. 17.7

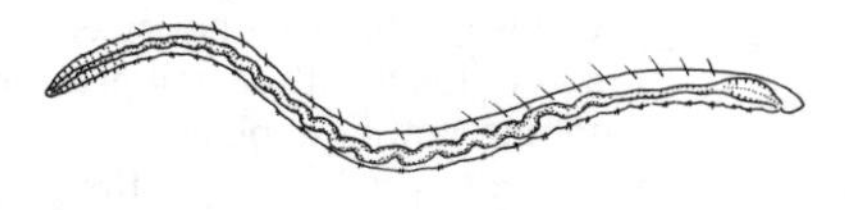
Fig. 17.8

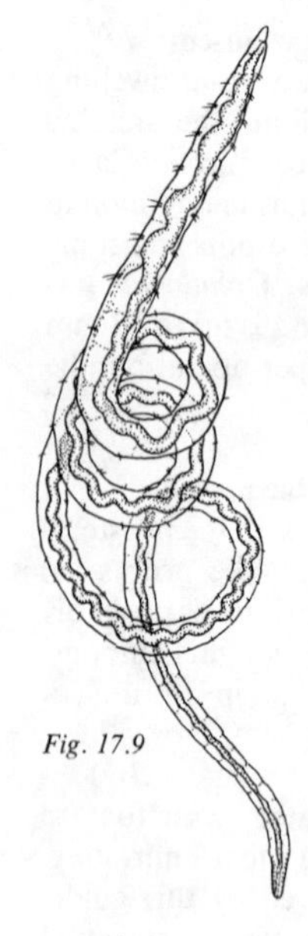

Fig. 17.9

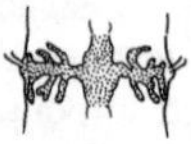

Fig. 17.10

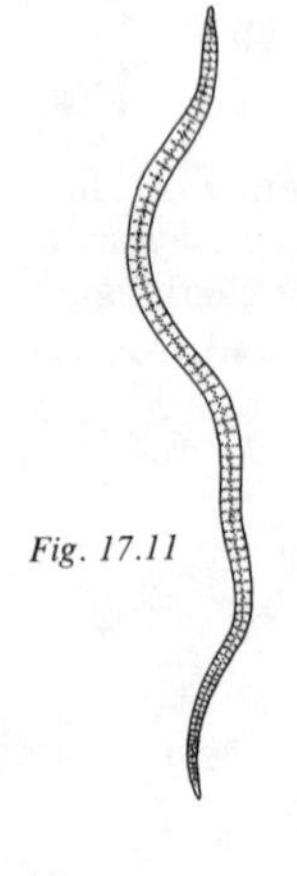

Fig. 17.11

Family Tubificidae (Sludge Worms, 'Tubifex') *Pl. 142, Fig. 17.9*

Small to medium-sized worms, typically 0.5–1.5 mm diameter and 20–70 mm long; hair chaetae present in dorsal bundles; most species are translucent pink or red, due to the colour of the blood. None can swim and when irritated they characteristically contract into a series of tight coils. These worms are common, often extremely abundant, in muddy sediments, especially where there are high levels of sewage contamination – sometimes the mud is turned red by their numbers. Many also occur in brackish water in estuaries, or in fully marine conditions.

Complete worms are distinctly larger than naidids but as asexual reproduction by fragmentation is habitual, small regenerating worms are common. Chains formed by budding never occur. Most species live buried in mud, often building soft mud tubes from which their posterior ends are protruded and waved about in the water, functioning as gills. Identification is extremely difficult and the only distinctive genera are rare. There is a typical genus *Tubifex*, with many species, but all worms belonging to this family are commonly referred to as 'tubifex worms'. They are collected and marketed commercially as a live food for aquarium fishes.

Family Lumbriculidae

Small or medium-sized worms up to 80 mm long and 1–2 mm diameter, resembling slender earthworms; only two chaetae in each bundle, hair chaetae are absent; typically deep red, due to plentiful, branched blood vessels (*Fig. 17.10*); asexual reproduction by fragmentation. Two genera are common:

Lumbriculus *Pl. 143, Fig. 17.11* The sole European species. *L. variegatus*, is an active worm with a tougher texture than naidids or tubificids, but nevertheless brittle and readily fragmenting; up to 80 mm long. It never forms tight coils like a tubifex worm and can swim with a powerful wriggling motion. The colour varies from deep red to brownish, usually with a green sheen. Normally it lives in mud or gravel with its tail protruding, but it is also found climbing amongst plants in the open water, apparently immune to predators – nothing seems to eat it.

Stylodrilus Generally smaller, up to 40 mm long, than *Lumbriculus*, and paler in colour; sexually mature worms are easily distinguished by a pair of pointed organs (penes) protruding ventrally from the tenth body segment; mostly found in fast-flowing rivers or clear upland lakes.

Family Enchytraeidae (White Worms)
Small worms resembling earthworms but seldom exceeding 20 mm in length; typically white, more rarely pink in colour, semi-translucent; more than two chaetae per bundle, hair chaetae absent; eyespots absent.

Only a few species of this family are truly aquatic, and these are not commonly seen as they hide amongst detritus or bury themselves completely in mud.

Family Haplotaxidae
This family contains one European species, *Haplotaxis gordioides*, which, as its name implies, is of similar proportions to a horsehair (gordian) worm, up to 30 cm long and 1 mm diameter. The body may consist of up to 500 segments, each with a single chaeta in each bundle, the dorsal ones smaller and disappearing posteriorly. It occurs in damp soil and marshy places, occasionally under stones in streams, but is probably not truly aquatic.

Families Glossoscolecidae and Lumbricidae
These are the large, earthworm-like oligochaetes, with only two small chaetae per bundle. There are only a few aquatic species but it is not uncommon to find terrestrial species that have crawled or fallen into the water.

Family Glossoscolecidae
Sparganophilus A distinctive, pinkish worm, with a bluish or greenish iridescence in some lights; rather stiff-bodied and firm to the touch; up to 20 cm long; clitellum located between segments 15 and 35. It occurs at the margins of lakes or rivers, under stones or in the mud amongst the roots of marginal plants; local in Britain, not known from continental Europe.

Criodrilus lacuum A large worm, up to 40 cm × 1 cm, with all but the most anterior segments square in section. The clitellum may be extensive, from segments 14 or 15 to about 45; behind the clitellum is a shallow dorsal groove. Found in similar places to *Sparganophilus*; widespread in Europe but absent from Britain.

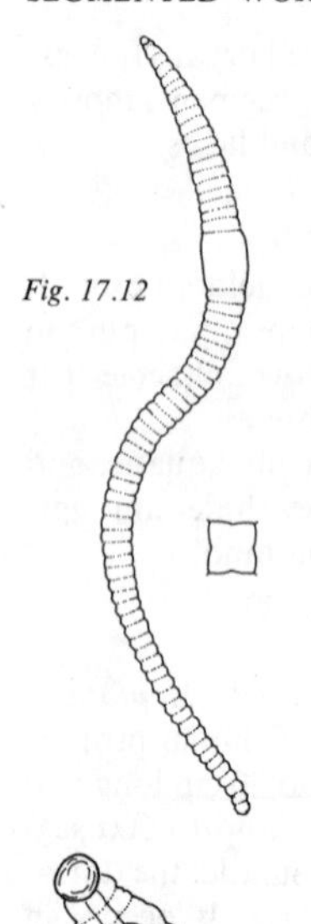

Fig. 17.12

Family Lumbricidae (Square-tailed worms)
Eiseniella tetraeda *Fig. 17.12* A medium-sized worm up to about 80 mm long, drab pink or greyish in colour; clitellum usually occupying segments 21–25 but sometimes further forward; posterior part of body markedly square in section. It occurs in mud amongst plant roots, under stones, etc.: often in fast-flowing rivers; widespread and fairly common.

Family Branchiobdellidae *Fig. 17.13*
Small, stout-bodied, eyeless worms up to 10 mm long but usually about 3–5 mm, lacking any chaetae and possessing a rounded sucker at the posterior end of the body. The only European genus is ***Branchiobdella***, with several species, which is parasitic on the freshwater crayfishes *Austropotamobius* and *Astacus*, usually being found on the gills but may also be on the legs or abdomen. Widespread in Europe but very rare in Britain.

References: Brinkhurst & Jamieson, 1971. Brinkhurst, 1971. Brauer, 1909.

Fig. 17.13

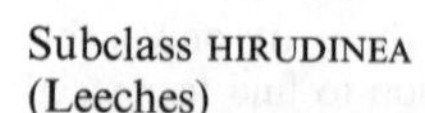

Subclass HIRUDINEA
(Leeches)

Leeches are easily recognized by their muscular, contractile bodies and the suckers, one at each extremity, with which they adhere to any firm surface. The division of the body into the 'true' segments characteristic of this phylum cannot be seen externally, but instead the surface is ringed, often minutely, by numerous narrow false-segments – annuli. The head is situated at the narrower end of the body and bears two to ten small dark eyespots. Beneath it is a sucker of variable form, sometimes only distinct when being used for attachment. The mouth either forms a small pore on the sucker, or occupies its whole concave centre. In the former type, a protrusible proboscis, inserted into the prey during feeding, is present: in the latter there may be formidable biting and piercing jaws, or jaws may be absent. The posterior sucker is always rounded and clearly marked off from the body.

Leeches are capable of great changes of shape, but they do spend long periods at rest, stuck to a substratum, and their resting outline is fairly constant and characteristic – this is the shape shown in the drawings. Measurements are also of resting leeches and these can be more than doubled during extension. Leeches typically gorge themselves when feeding and, naturally, a well-fed specimen is a good deal fatter than a starved one, as it may consume more than its own body weight at a single meal.

Movement over a solid substratum is by looping, often with sinuous, snake-like movements during the extension phase. Some species can swim by graceful, vertical undulations of the body.

Three of the four families produce egg-cocoons as is usual in this class, with a clitellum appearing only during the breeding season. The cocoons are usually stuck to a stone or plant stem. Glossiphoniids (see *p. 168*) are unusual in brooding their eggs – and later, the newly-hatched young – in a concavity formed by the ventral surface of their bodies. Asexual reproduction does not occur.

Leeches are very common in most freshwater habitats, usually being found clinging to a firm substratum out of the light – under stones or plant leaves – or buried in mud or gravel. Fish leeches lurk outstretched, camouflaged by their resemblance to plant stems, hoping to intercept a passing fish. Some species suck the blood of various vertebrates or invertebrates, becoming temporary parasites: others prey on various invertebrates – worms, molluscs, insect larvae – which they may swallow whole. Only two species, both very rare in Britain, are capable of feeding on human blood.

Key to the Families of Freshwater Leeches

1. a) Anterior sucker well-defined, with a complete rim (*Fig. 17.14*): cylindrical fish-leeches with four eyes.
   Family PISCICOLIDAE (*p. 168*)
   b) Anterior sucker fairly distinct but at least part of the rim merges with the body (*Fig. 17.15*), mouth a small pore in the sucker region: more-or-less flattened leeches with a pear-shaped resting outline: two to eight eyes.
   Family GLOSSIPHONIIDAE (*p. 168*)
   c) Anterior sucker poorly defined, mouth large and continuous with central concavity of sucker (*Fig. 17.16*): slightly flattened leeches with eight to ten eyes.
   . . . 2

2. a) Eight eyes: mostly small or medium-sized leeches not usually exceeding 40 × 5 mm at rest (but *Trocheta* can become very large – 100 mm).
   Family ERPOBDELLIDAE (*p. 171*)
   b) Ten eyes; large leeches up to 80 × 12 mm.
   Family HIRUDIDAE (*p. 170*)

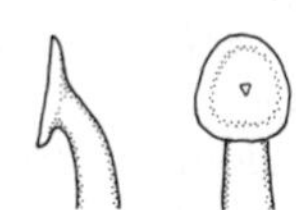
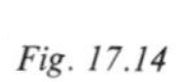

Fig. 17.14

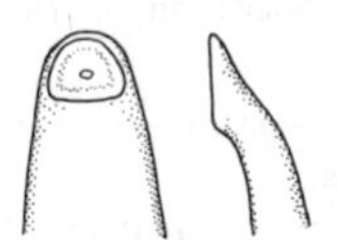

Fig. 17.15

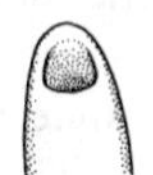

Fig. 17.16

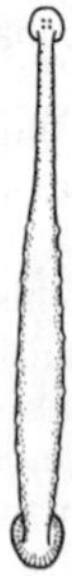

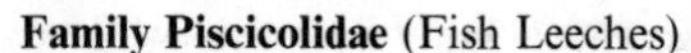

Family Piscicolidae (Fish Leeches)

Fig. 17.17

Piscicola geometra *Pl. 144, Fig. 17.17* A slender, wiry, scarcely contractile leech with prominent, well-defined suckers and four eyes; olive greenish, usually with about fifteen narrow pale bars; each flank with a row of inconspicuous, blister-like pulsatile vesicles, each one separated by about twelve annuli; up to 60 mm long. This species is a common external parasite of fishes, sucking their blood. It may be found on any part of the fish's body, including the gills or inside the mouth. When not attached to a fish, it lives free amongst water-plants, adopting a characteristic fishing posture. If taken in a net it becomes very active and mobile, and is unlikely to be overlooked. It is common and widespread throughout Britain and Europe.

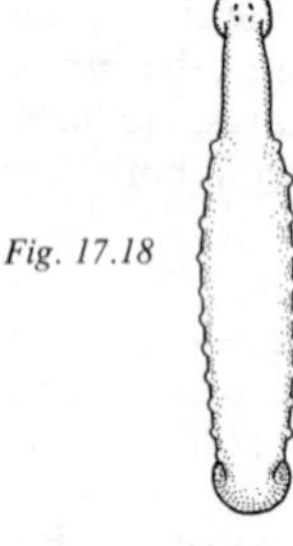

Fig. 17.18

Cystobranchus respirans *Fig. 17.18* A stouter leech than *Piscicola*, with prominent pulsatile vesicles separated by only two or three annuli; up to 70 mm long. This is probably the only species of the genus to occur in the area covered by this guide; it is absent from Britain.

Family Glossiphoniidae

Typically broad-bodied, markedly flattened leeches with a pear-shaped resting outline; the eggs and young are brooded beneath the body.

Key to the Genera of Glossiphoniidae

1. a)	Two eyes.	. . . 2
b)	Four eyes.	. . . 3
c)	Six eyes, sometimes partly or completely fused.	. . . 4
d)	Eight eyes.	*Theromyzon* (*p. 170*)
2. a)	With a hard scale in the skin of the dorsal surface (*Fig. 17.21*): small leeches up to 12 mm long: common.	*Helobdella* (*p. 169*)
b)	No hard dorsal scale: large leeches up to 70 mm long: very rare in Britain.	*Haementeria* (*p. 169*)
3. a)	Head and anterior sucker wider than anterior part of the body.	*Hemiclepsis* (*p. 169*)
b)	Head and anterior sucker not wider than anterior part of body.	*Batracobdella* (*p. 169*)
4. a)	Body firm in texture, slightly rough to the touch: dorsal papillae, if present, on every third annulus.	*Glossiphonia* (*p. 169*)
b)	Body very soft, with prominent dorsal papillae on two of every three adjacent annuli.	*Boreobdella* (*p. 170*)

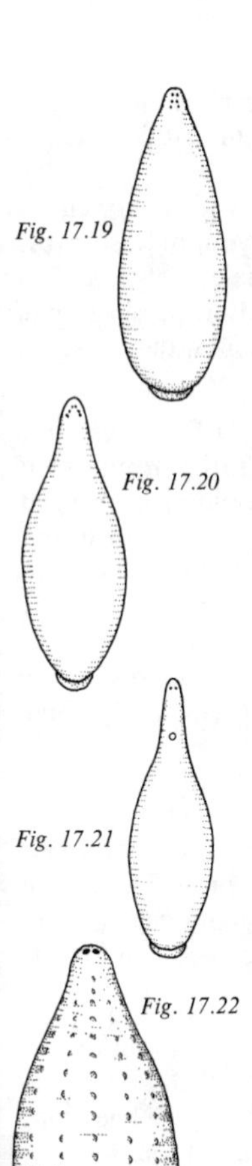
Fig. 17.19
Fig. 17.20
Fig. 17.21
Fig. 17.22
Fig. 17.23

Glossiphonia complanata *Fig. 17.19* Body of firm consistency, rubbery, and slightly rough to the touch, dorsal surface often with small papillae on every third annulus; pale green or brown with yellow spots, and two or more broken, dark longitudinal lines; six eyes, sometimes fused but the six elements can usually be distinguished; up to 30 mm. Very common in still or flowing waters, feeding on molluscs and other invertebrates; throughout Britain and Europe and in many other parts of the world.

Glossiphonia heteroclita *Fig. 17.20* Similar in form and texture to *G. complanata* but smaller, up to 15 mm long, lacking papillae, and pale translucent yellowish or cream in colour, usually unmarked. Common, mostly in ponds and ditches, in Britain, Europe and elsewhere.

Helobdella stagnalis *Fig. 17.21* A small, soft-bodied leech up to 12 mm long, with two eyes and a characteristic, small, hardened scale in the dorsal surface; translucent whitish or dull yellow, often with dark speckles. Common in many habitats, occurring throughout Britain and Europe.

Haementeria (Placobdella) costata *Fig. 17.22* Body broad and flat, up to 70 mm long and 25 mm wide; two eyes; dorsal surface with five or more longitudinal rows of papillae; yellowish or brownish with a dark median line. It feeds on the blood of reptiles, birds or mammals, including man. In Britain it is rare, having been discovered only recently, but is widespread in Europe except for the north.

Hemiclepsis marginata *Fig. 17.23* A small leech up to 20 mm long, with four eyes; body rather slender for a glossiphoniid, with both suckers prominent; attractively coloured, green or brown with rows of yellow spots on the dorsal surface. When feeding, it is parasitic on fishes or amphibians, but is commonly found resting free. A common species throughout Britain and Europe.

Batracobdella paludosa Body very soft with no dorsal papillae; four eyes; up to 10 mm long; green or brown, sometimes with dark longitudinal lines. It is found in many habitats, particularly small ponds, feeding on the blood and body fluids of snails and perhaps tadpoles. A widespread species in Europe, but rare in Britain.

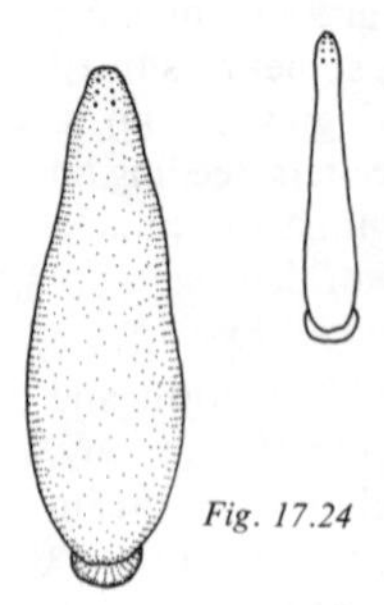

Boreobdella verrucata Body very soft, with conspicuous papillae on two of every three consecutive annuli; six eyes: up to 25 mm long; dark green with pale spots and two dark longitudinal lines. Widespread in Europe but rare in Britain, occurring in slow-flowing and still waters where it feeds on molluscs.

Fig. 17.24

Theromyzon tessulatum *Pl. 145, Fig. 17.24* Body very soft in texture, almost gelatinous in adults, rather irregular in shape but always broadest posteriorly; eight eyes arranged in two longitudinal rows; up to 30 mm long; translucent pale green, yellowish or grey, juveniles often with faint pale spots. This leech is a parasite of waterfowl, typically attaching itself inside the mouth, nostrils, or respiratory tracts, sucking the blood and often causing distress to the host. It is fairly common throughout Britain and Europe.

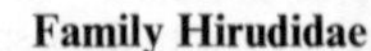

Family Hirudidae

Large leeches with more-or-less parallel-sided, slightly flattened bodies and ten eyes; the large mouth conceals strong jaws for biting and piercing the prey. They often leave the water and egg-cocoons are usually deposited in damp soil or under stones near the water's edge.

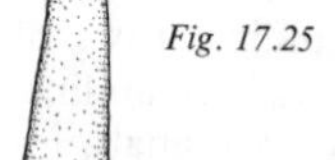

Fig. 17.25

Haemopis sanguisuga (Horse Leech) *Pl. 146, Fig. 17.25* Body up to 60 mm long at rest and about 6–10 mm wide: dorsal surface dark green or brown, mottled with black, and separated from the grey underside by a pale line along each flank. It preys on almost any invertebrate, swallowing whole those that are small enough, and sucking the blood of larger ones; but, despite its name, it does not attack horses or any other mammals. Widespread throughout the area covered by this guide, but rather local.

Fig. 17.26

Hirudo medicinalis (Medicinal Leech) *Fig. 17.26* Body a little stouter than *Haemopis*, up to 80 mm long at rest: dorsal surface dark olive-greenish, with several longitudinal lines of red or orange, each flank with a pale yellow line bordered with black, ventral surface greyish or olive. Generally more richly-coloured than *Haemopis*, with which it is often confused. It feeds on the blood of mammals, including man (*Haemopis* is unable to penetrate human flesh), amphibians or fishes.

Formerly widespread in Britain and Europe, this leech has become increasingly rare in most regions during the past century, due partly to destruction of its habitat – it prefers extensive marshland areas – and partly to collection on a large scale for medicinal bloodletting (a practice still common in some parts of eastern Europe), medical research, teaching (*Hirudo* is a standard biological student's dissection subject), and the pharmaceutical industry (it is the source of a powerful anticoagulant, *hirudin*). It still occurs in a few scattered localities in Britain and most continental countries but is definitely an endangered species: it is listed in the IUCN Red Data Book (Wells et al, 1983) and may soon receive official protection in western Europe.

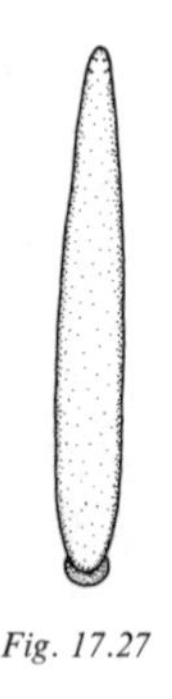

Fig. 17.27

Family Erpobdellidae *Fig. 17.27*
Slender, slightly flattened leeches with an indistinct anterior sucker, no jaws and eight eyes set roughly in two transverse rows. Some species are semi-terrestrial.

Erpobdella octoculata *Pl. 147* Body cylindrical to slightly flattened, with the annuli all of a similar width; up to 40 mm long; brown, with a variable amount of black speckling and usually with some yellow markings between the dark ones. It feeds on any small invertebrates that it can swallow. Very common in many habitats, especially fast-flowing streams, but not always conspicuous as it may burrow into soft substrata; throughout Britain and Europe.

Erpobdella testacea General form similar to *E. octoculata* but smaller, up to 30 mm long, and usually plain reddish-brown in colour, without any markings. Typically found in small weedy ponds but occasionally in other habitats. throughout the area covered by the guide.

Dina lineata A very slender leech up to 25 mm long. reddish-brown, usually with four dark longitudinal lines on the dorsal surface; every fifth annulus is wider than the others (cf. *Erpobdella*) but this is difficult to determine in life. Locally distributed in Europe and mainly confined to the north in Britain; usually in still waters.

Trocheta spp. Large cylindrical leeches which, at first glance, are easily mistaken for earthworms; up to 100 mm long at rest; green, grey or reddish-brown, sometimes with dark dorsal markings. These are amphibious leeches, often found in the soil where they feed on worms, slugs, etc., returning to water to breed in the summer. They occasionally turn up in drains, wells or even domestic waste pipes. Two species occur in Europe, widely distributed but rather local: *T. subviridis*, with a smooth dorsal surface, and *T. bykowskii*, with numerous fine dorsal papillae.

References: Elliot & Mann, 1979. Brauer, 1909.

CHAPTER 18 Moss Animals
(Phylum Bryozoa (Ectoprocta))

Recognition features Colonial, sessile animals, usually fixed to a substratum; colonies consist of fine-branching tubes up to 1 mm diameter, which form encrusting or erect structures, or else are gelatinous masses. Individual animals are small, with a retractile crown of ciliated tentacles which protrudes from the colony mass; colonies usually more than 10 mm across.

Bryozoans are fairly common animals in clean fresh waters, although due to their plant-like growth forms and cryptic habits, they are seldom noticed by the casual observer. With moderate magnification they are revealed as extremely beautiful organisms; the glassy delicacy and pleasing symmetry of their massed tentacles easily rival the attractions of any other inhabitants of fresh water.

The main colony mass is called the zooecium and the individual animals embedded in it are zooids. A zooid (*Fig. 18.1*) consists of two main parts: a body section containing the Y-shaped gut, and a circular or U-shaped structure surrounding the mouth, the lophophore, which bears a graceful crown of colourless ciliated tentacles. A whole zooid is about 0.5–4 mm long, depending on the species. When feeding, the lophophore and its tentacles are protruded clear of the zooecium but can be retracted into it when disturbed. Food particles – mostly unicellular algae and protozoans – are gathered and passed to the mouth on water currents produced by the ciliated tentacles.

Two very different types of zooecium occur: a mass of branching tubes made of cuticle-like material, or a gelatinous mass. Tubular zooecia are fixed to a substratum; they branch freely and often cover an area of tens of square centimetres. They may grow over the substratum, forming an encrusting network (*Fig. 18.3*), become erect branches (*Fig. 18.4*), or form aggregated masses of dense texture, often surrounding sticks or similar objects (*Fig. 18.7*). The zooids protrude from the ends of the branches. Gelatinous zooecia have a characteristic shape in each genus, with the zooids usually embedded in the jelly in a regular arrangement. They are not fixed to a substratum but adhere lightly to it; some are capable of slow, creeping movements.

During the summer and autumn circular or elliptical objects, about 0.5–1 mm across, may be seen within the zooecium. These are overwintering bodies called statoblasts, produced asexually by the zooids. Most freshwater bryozoans die off in the winter but the statoblasts remain, either anchored to the substratum or floating free, to hatch out in the spring and start new colonies.

Statoblasts, being very resistant to desiccation and extremes of temperature, have enabled bryozoans to distribute themselves far and wide; the majority of genera and many species are found worldwide. Statoblasts have a characteristic shape in each genus (*Fig. 18.2*) and in some genera are produced in great numbers. The presence of bryozoans in a water can usually be confirmed by searching the wind-blown debris on the shore for statoblasts.

Ripe statoblasts, which first need to serve an appropriate overwintering term, can easily be hatched out in aquaria. Young colonies, with one or two zooids, are ideal and fascinating subjects for microscopy; they can be fed on drops of algal cultures but seldom survive more than a week or two.

Sexual reproduction also occurs in the summer, resulting in the production of tiny, ciliated, free-swimming larvae which, after a short planktonic existence, settle on to a substratum and start budding off zooids to initiate a new colony.

Bryozoans are likely to occur on any firm substratum; most prefer shaded places but *Cristatella* seems to be indifferent to light. Colonies may be extensive but they are rarely conspicuous, and a keen eye is required to detect them. Sunken branches or submerged tree roots are favourite habitats, but submerged masonry or timberwork, stones or shells, and occasionally water plants (especially the undersides of floating leaves) may all be colonized. Colonies have been known to cause blockages in reservoir outlets and domestic water pipes. The tubular forms *Plumatella* and *Fredericella* provide a labyrinthine substratum amongst which many other creatures conceal themselves, and they are always worth searching for sponges, insect larvae, tube-dwelling rotifers and protozoans, and various worms.

The small class Phylactolaemata consists exclusively of freshwater genera. *Paludicella* is the only freshwater representative of an otherwise marine class, Gymnolaemata – this and other classes contain numerous marine genera. All the genera below, except *Pectinatella*, occur widely throughout the area covered by the guide, including Britain.

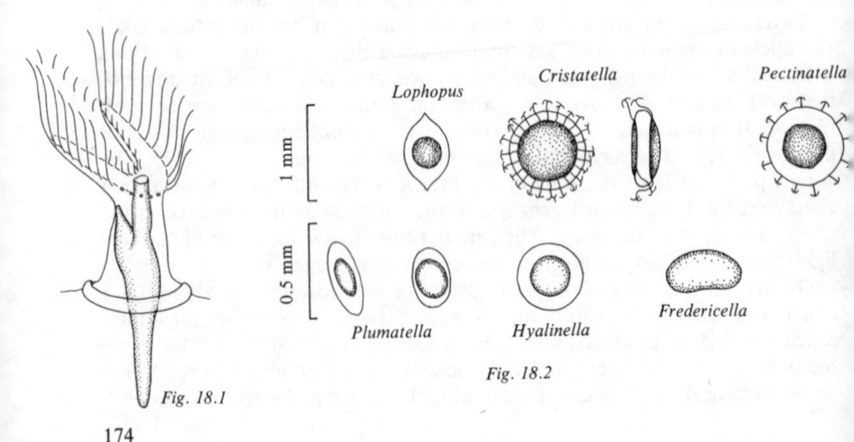

Fig. 18.1

Fig. 18.2

Key to the Genera of Freshwater Bryozoans

1. a) Lophophore circular: zooecium branching and tubular. . . . 2
   b) Lophophore U-shaped: zooecium branching and tubular, or gelatinous. . . . 3
2. a) Zooecium consisting of tiny, club-shaped joints (*Fig. 18.3*). *Paludicella* (*p. 175*)
   b) Zooecium consisting of plain, unjoined tubes (*Fig. 18.4*). *Fredericella* (*p. 175*)
3. a) Zooecium consisting of branching tubes. . . . 4
   b) Zooecium gelatinous. . . . 5
4. a) Tubes of zooecium without a gelatinous covering (ectocyst): statoblasts elliptical (common). *Plumatella* (*p. 176*)
   b) Tubes of zooecium with a gelatinous covering (ectocyst); statoblasts circular (rare). *Hyalinella* (*p. 176*)
5. a) Zooecium approximately triangular, with zooids facing in one direction. *Lophopus* (*p. 176*)
   b) Zooecium circular or elliptical when small (up to 15 mm), becoming elongated when larger (up to 200 × 8 mm): zooids facing in more than one direction. *Cristatella* (*p. 176*)
   c) Zooecium irregular in outline, situated on an irregular mass of firm jelly which may become very large. Zooids variously oriented (rare in Europe, absent from Britain). *Pectinatella* (*p. 177*)

Class GYMNOLAEMATA

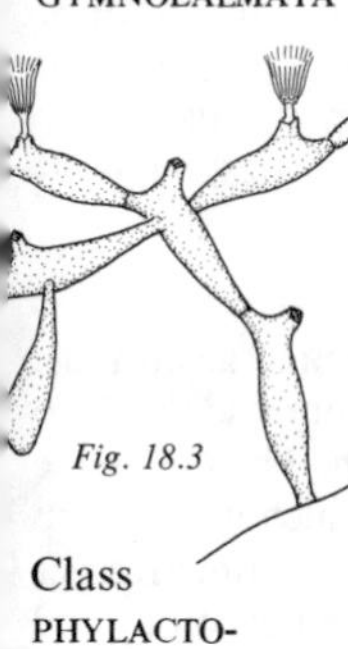

Fig. 18.3

Paludicella articulata *Pl. 148, Fig. 18.3* Zooecium consisting of tiny (1 mm long), horn-coloured, club-shaped joints, branching approximately at right angles, each with a short, lateral, square and tubular orifice from which the lophophore is extended; zooid with about twelve to twenty tentacles. Statoblasts are not produced but brownish nodules – hibernaculae – which have a similar function, are budded off externally in the autumn. This species forms delicate networks on various substrata in shaded places in rivers, canals and lakes; it is widespread and probably common but, being so small, is easily missed.

Class PHYLACTOLAEMATA

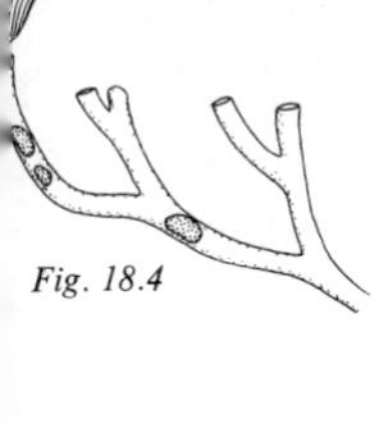

Fig. 18.4

Fredericella sultana *Fig. 18.4* Zooecial tubes slightly flattened, up to 0.7 mm wide, slightly translucent and sandy to dark brown in colour, rather soft in texture: zooids with up to thirty tentacles, shy, retracting at the slightest disturbance; statoblasts bean-shaped (in a related species, *F. australiensis*, not positively recorded in Europe, they are circular). Colonies form encrusting networks, often with a few erect branches, in flowing or still waters, often in company with *Paludicella*; common and widespread.

Fig. 18.5

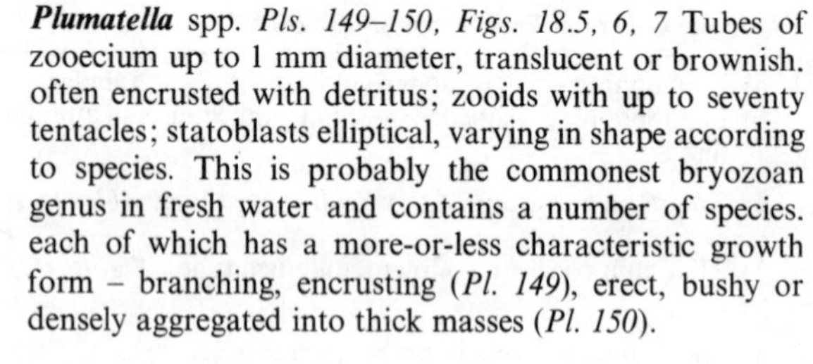

Plumatella spp. *Pls. 149–150, Figs. 18.5, 6, 7* Tubes of zooecium up to 1 mm diameter, translucent or brownish. often encrusted with detritus; zooids with up to seventy tentacles; statoblasts elliptical, varying in shape according to species. This is probably the commonest bryozoan genus in fresh water and contains a number of species. each of which has a more-or-less characteristic growth form – branching, encrusting (*Pl. 149*), erect, bushy or densely aggregated into thick masses (*Pl. 150*).

Fig. 18.6

Hyalinella punctata *Fig. 18.8* This genus differs from *Plumatella* only by the form of its zooecium, which is covered by a colourless layer of jelly, the ectocyst; statoblasts circular. It is widespread but uncommon, rare in Britain.

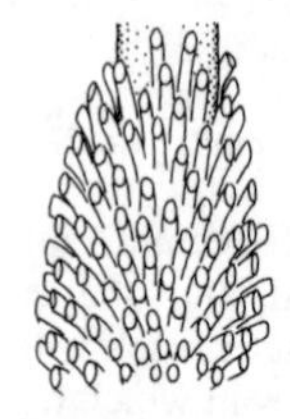

Fig. 18.7

Lophopus crystallinus *Fig. 18.9* Zooecium gelatinous with a fairly firm texture, translucent, colourless or yellowish; triangular in shape with the zooids all extending in the same direction; up to 25 mm across. A clear soft ectocyst covers the zooecium and may extend over adjacent colonies. Colonies are capable of creeping slowly over the substratum. Zooids with up to seventy tentacles, not very shy; statoblasts without hooks, oval with pointed ends. Usually found on water plants but also on stones, etc., in still or running waters; flourishing colonies of this species have been found in cold water in winter. Widespread but local in occurrence.

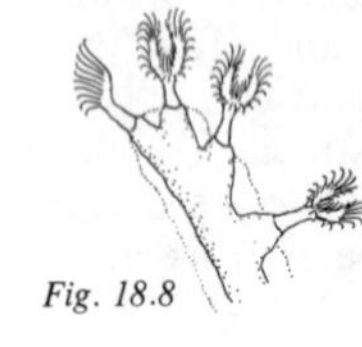

Fig. 18.8

Cristatella mucedo *Pl. 151, Fig. 18.1* Zooecium gelatinous, forming oval or elongated colonies 4–8 mm wide and up to 24 cm long, but usually about 2–10 cm. The underside of the colony forms a flattened 'sole' on which it can creep over the substratum at the rate of several centimetres per day. Zooids extend mostly from the sides of the colony and each has 80–100 tentacles; they are not very shy and expand readily; statoblasts are circular with a double row of hooks. The whole colony is translucent whitish, usually with a greenish tint due to the colour of the gut contents.

This lovely species may occur on any type of substratum. often in well-lit conditions, and usually in still waters. It is widespread but rather local although it may be very abundant where it occurs.

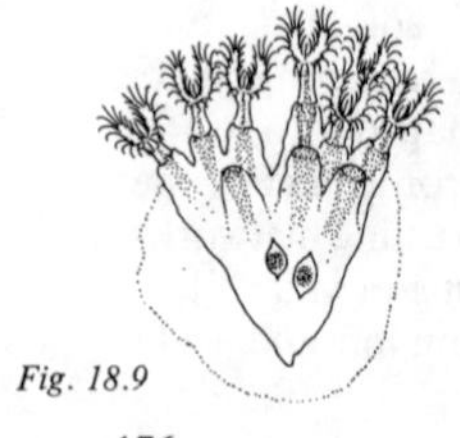

Fig. 18.9

Fig. 18.10

Pectinatella magnifica (The Jelly Ball) *Fig. 18.10* Zooecia forming flat, irregular, branching rosettes extending over the surface of a thick, clear, jelly-like mass which may exceed 1 m across but is usually smaller in European specimens. Statoblasts circular, with a single ring of hooks. An introduced North American species which is established in parts of Germany.

References: Lacourt, 1968. Mundy, 1980. Brauer, 1909 vol.

CHAPTER 19 Snails, Limpets, and Mussels
(Phylum Mollusca)

Recognition features Soft-bodied animals with hard shells. Snails are familiar creatures with coiled shells; limpets have small conical shells: and mussels possess two equal shells (valves) which fit closely together. Some freshwater molluscs are only 1–2 mm long but most are larger than this, up to 50 mm in snails and 150 mm in mussels.

This phylum contains several large classes: the squids, cuttles and octopuses (class Cephalopoda) are exclusively marine, as are the tusk shells (class Scaphopoda), chitons (class Polyplacophora), and solenogasters (class Aplacophora); only the Gastropoda (slugs and snails) and Bivalvia (mussels and clams) occur in fresh water.

Molluscs are relatively large invertebrates, common and often abundant in fresh water. Their distinctive shells make them unlikely to be confused with any other animals, although some small crustaceans, e.g., ostracods, are bivalved. The structure of the soft body is important in the classification of molluscs but the shell is the prime means of identification and all European freshwater species can be identified by shell characters alone. The shell is made of calcareous material covered by a thin layer of horny substance (periostracum). These materials are secreted and laid down in layers around the rim or aperture of the shell by a broad fold of the soft body wall called the mantle. Seasonal and other variations in the growth rate result in the formation of growth rings on the surface of the shell; these are well-developed and distinct on most mussels and some of the larger snails.

Because molluscs are such conspicuous and common members of freshwater communities we have provided descriptions of species as well as genera: only a few, strictly local and rare species in the guide's area have been omitted.

Two classes occur in fresh water:

1. a) Shell in one piece (sometimes with a separate lid which closes the aperture), coiled or conical.
   Class GASTROPODA, Snails and Limpets (below)

   b) Shell in two equal parts, typically oval or circular.
   Class BIVALVIA, Mussels (*p. 188*)

Class GASTROPODA (Snails and Limpets)

The characters of the gastropod shell are illustrated in *Fig. 19.1*. The small shells of freshwater limpets form low cones with a wide elliptical aperture and are not coiled. All remaining freshwater gastropods are snails, with tubular tapering shells coiled around a central axis or columella, which may be hollow with a small opening, the umbilicus, or enclosed by the coils. Each complete turn of the coiled shell is a whorl, and the sole opening, from which the soft body protrudes, is the aperture. The largest (outermost) whorl is the body whorl, the remaining whorls constituting the spire, which is obsolete in some species. If the shell is oriented spire-upwards, with the aperture facing the observer, then when the aperture is on the right, the shell is dextral, when it is on the left, sinistral (*Figs. 19.13, 19.17*). In the subclass Prosobranchia the aperture is closed by a flat lid, the operculum, when the body is retracted into the shell.

Freshwater snails, unlike many marine or terrestrial species, are usually dull in colour, typically some shade of brown, and rarely with any distinctive markings. Colourless or white-shelled specimens of any species occasionally turn up, but only one species is normally whitish.

The soft body can be divided into two regions: the coiled visceral hump, which contains most of the digestive organs and remains permanently hidden inside the shell; and the large muscular foot which is protruded from the aperture. The foot is retractile and its anterior end forms the head; if an operculum is present it is borne attached to the top of the foot behind the aperture. Locomotion is afforded by the flattened sole of the foot, which creeps over a substratum with complex rippling movements, clearly visible from below if the snail is caused to crawl over a sheet of glass. Its path is lubricated by copious amounts of secreted mucus – the familiar 'slime trail' – and some aquatic species have been observed to climb up and down ropes of mucus spanning open water. The head bears two sensory tentacles and two inconspicuous eyespots which in all truly aquatic species are located near the tentacle bases. In the Lymnaeidae and Ancylidae the tentacles are broad and flat, more or less triangular, but they are relatively long and thread-like in other families. The mantle, which secretes the shell, is a prominent fold of the body wall encircling the base of the visceral hump. It can be seen lining the shell rim. Prosobranchs breathe by means of a feathery or comb-like gill, usually concealed within the mantle but sometimes protruded in *Valvata* and other genera. Pulmonates (*p. 183*) breathe atmospheric air with an internal lung formed by a deep cavity in the mantle. This opens to the air, when the snail is at the surface, via a tubular opening at the edge of the aperture.

Freshwater snails are mostly herbivorous, feeding on water plants or, more commonly, on the film of algae which covers most submerged objects. Food is rasped away with the radula, a flexible, toothed, ribbon-like tongue.

Eggs are laid in firm masses of jelly on plants or other objects, but *Theodoxus* is exceptional in laying single eggs protected by small chitinous cocoons.

Water snails occur in all types of habitat, with the exception of very acidic waters, but are most common in clean, plant-rich lowland waters with a slight flow. In such localities it may well be possible to find as many as a score of species in the same place. There are also several genera that, although not aquatic, habitually occur near water or in wet places (see *p. 187*).

Snails are frequently the hapless hosts of the intermediate stages of various parasitic worms, particularly flukes (see *p. 138*). Their shells are often used as a substratum by many sessile micro-organisms, and a small worm, *Chaetogaster limnei*, habitually lives within the mantle fold of pulmonate snails.

Most snails are relatively easy to identify, provided adults are examined, but some species of Lymnaeidae and Planorbidae are differentiated by relatively slight characters that are often only apparent by direct comparison. For this reason it is desirable, if possible, to compare a series of specimens in order to determine the range of variation, maximum size, etc. Since most species are abundant where they occur this should present no difficulties, but do handle the snails carefully and return them to the same water afterwards. Always examine the smallest specimens carefully, as adults of species such as *Planorbis crista* may well be mixed in with juveniles of larger species. Juveniles do not always possess the key characters of the adults but their identity can be determined, with a series of specimens, by extrapolating backwards from the adult.

The large genera *Lymnaea* and *Planorbis* are commonly split into several smaller genera, especially by continental authors. These alternative names are given in brackets. The following key includes all common European genera except for the Planorbidae which are keyed separately. Brackish-water species are omitted. From time to time, exotic species are introduced, usually accidentally with aquatic plants, and in a few places permanent populations have become established. These are not included in the keys. Terrestrial species likely to be encountered near, or accidentally in, fresh water are described on *p. 187*.

Key to Genera of Freshwater Gastropoda

1. a) Shell coiled. (Snails) . . . 2
   b) Shell forming a low cone, not coiled. (Limpets) . . . 11

2. a) Operculum present. Subclass PROSOBRANCHIA . . . 3
   b) Operculum absent. Subclass PULMONATA . . . 8

3. a) Aperture large, much greater than half shell height or width. *Theodoxus* (*p. 181*)
   b) Aperture less than half shell height or width. . . . 4

4. a) Shell large, up to 40 mm high: usually banded, especially in young specimens. *Viviparus* (*p. 181*)
   b) Shell smaller, less than 20 mm high: not banded but may be blotched. . . . 5

5. a) Height of shell less than its width: shell sometimes a flat coil. *Valvata* (*p. 182*)
   b) Height of shell greater than its width. . . . 6

6. a) Operculum calcareous, with concentric growth lines: shell up to 15 mm tall, usually more than 6 mm. *Bithynia* (*p. 182*)
   b) Operculum horny, with spiral growth lines; shell less than 6 mm tall. . . . 7

7. a) Shell up to 5 mm tall, with about five whorls increasing evenly in size, not bulging conspicuously (common). *Potamopyrgus* (*p. 182*)
   b) Shell up to 3 mm tall, with about four whorls which bulge conspicuously (local and uncommon). *Bythinella* (*p. 182*)

8. a) Shell without a spire, forming a more-or-less flat coil, its orientation obscure.
   Family PLANORBIDAE, see separate key (*p. 185*)
   b) Shell with a spire, sinistral. . . . 9
   c) Shell with a spire, dextral. . . . 10

9. a) Aperture little more than half shell height. *Aplexa* (*p. 184*)
   b) Aperture nearly as tall as shell, at least three-quarters of its height. *Physa* (*p. 184*)

10. a) Shell very thin and delicate, mantle when expanded wraps around outside of shell. *Myxas* (*p. 183*)
    b) Shell of normal thickness: mantle never wraps around outside of shell. *Lymnaea* (*p. 183*)

11. a) Long axis of aperture less than twice its width: height of shell about equal to its width. *Ancylus* (*p. 187*)
    b) Long axis of aperture about equal to twice its width: height of shell much less than its width. *Acroloxus* (*p. 187*)

Subclass PROSOBRANCHIA

Family Neritidae

Fig. 19.2

Theodoxus fluviatilis (Nerite) *Pl. 152. Fig. 19.2* Shell thick and heavy, with a distinctive, smoothly streamlined shape and a broad ledge inside the aperture; up to 12 mm wide; brown marbled with white or yellow markings but often abraded or overgrown by algae. A common but inconspicuous species found on stones and other firm substrata; typically in rivers but also in lakes on wave-beaten shores; widespread throughout the area covered by this guide.

Family Viviparidae

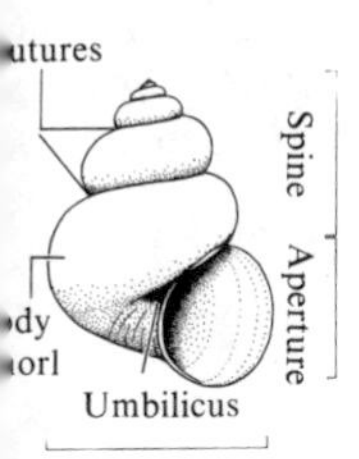

19.1

Viviparus spp. (River snails) *Pls. 153–154. Fig. 19.1* Shell with dark bands following the whorls, these bands more distinct on young specimens, which often have hairy shells (*Pl. 154*); up to 40 mm high. There are two species: *V. viviparus* has a non-glossy shell with an inconspicuous umbilicus, shallow sutures, and flatter whorls than the next species; widespread, mainly in rivers. *V. fasciatus* (*contectus*) has a glossy shell, broad umbilicus and deep sutures; less common than previous species, rather local in lakes and rivers.

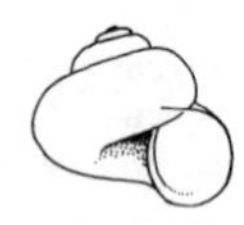
Fig. 19.3

Fig. 19.4

Family Valvatidae

Valvata spp. Small snails with broad low shells; the feather-like gill and an accessory 'mantle tentacle' can often be seen protruding from the aperture in this genus. Three species are widespread in the area covered by this guide. typically being found in still or slow-flowing waters amongst dense weed beds. *V. piscinalis* (*Pl. 155, Fig. 19.3*). Shell about equal in height and width, up to 7 mm wide; common. *V. cristata* (*Fig. 19.4*). Shell flat with no spire, easily mistaken for *Planorbis* but operculum is definitive: up to 4 mm wide; common. *V. macrostoma* Intermediate in shape between previous two species, up to 4 mm wide; local.

Family Hydrobiidae

Fig. 19.5

Potamopyrgus* (*Hydrobia*) *jenkinsi *Pl. 156, Fig. 19.5* Shell light to dark brown in colour, sometimes encrusted with a blackish deposit; rarely taller than 5 mm. Common, often very abundant, in still or, more usually, running waters throughout the area covered by the guide. Until late in the 19th century, this little snail was confined to brackish waters, but has since successfully invaded fresh water throughout much of Britain and Europe.

Fig. 19.6

***Bythinella* (*Marstoniopsis*)** *Fig. 19.6* Tiny snails distinguished from *Potamopyrgus* by the conspicuously blunt spire, inflated whorls with deep sutures, and the more nearly circular aperture; up to 3 mm tall. Several species have been reported in the area, of which *B. scholtzi* is the best known; it occurs very locally in north-west England and Scotland, and on the Continent.

Fig. 19.7

Bithynia spp. Small or medium-sized snails similar in shape to *Viviparus* but much smaller, never banded (but sometimes with dark blotches. Two species are widespread and common in still or slow-moving waters: *B. tentaculata* (*Pl. 157, Fig. 19.7*) up to 15 mm tall; shell with shallow sutures and indistinct umbilicus. *B. leachii* (*Fig. 19.8*) up to 8 mm tall; shell with deep sutures and distinct umbilicus.

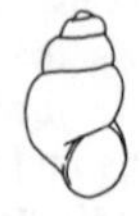
Fig. 19.8

Subclass
PULMONATA

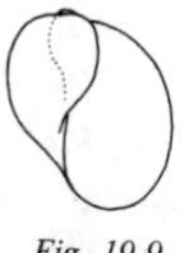
Fig. 19.9

Family Lymnaeidae

Myxas glutinosa *Fig. 19.9* Shell yellowish, translucent and glossy, very thin and fragile, up to 15 mm tall; a rare species, erratic in occurrence, with few recent records.

Lymnaea spp. (Pond snails) The native species of this common and well-known genus are defined in the following key:

Key to the Species of Lymnaea

1. a) Aperture distinctly less than half shell height. — *L. glabra* (*p. 183*)
   b) Aperture approximately equal to half shell height. ... 2
   c) Aperture large, at least three-quarters of shell height. ... 4

2. a) Outline of shell usually convex (*Fig. 19.11*); angle of coiling around umbilical axis obtuse: a small species up to 12 mm tall but usually only 5 or 6 mm. — *L. truncatula* (*p. 183*)
   b) Outline of shell straight or concave (*Figs. 19.13, 19.14*): angle of coiling around umbilical axis acute: larger species usually exceeding 12 mm tall, up to 50 mm. ... 3

3. a) Shell fairly thin and smooth: body whorl large in proportion to spire (*Fig. 19.14*). — *L. stagnalis* (*p. 184*)
   b) Shell fairly thick, with raised criss-cross lines on its surface: body whorl not disproportionately large. — *L. palustris* (*p. 184*)

4. a) Spire blunt: angle of aperture *a* (*Fig. 19.15*) less than 90°. (Species subject to great variation, sometimes the spire is completely enveloped by the body whorl.) — *L. peregra* (*p. 184*)
   b) Spire sharply pointed: angle *a* 90° or greater. — *L. auricularia* (*p. 184*)

Fig. 19.10

Lymnaea (Leptolymnaea) glabra *Fig. 19.10* Shell up to 25 mm tall, unlikely to be confused with other species. Widely but locally distributed in ponds and ditches, often in places which regularly dry up.

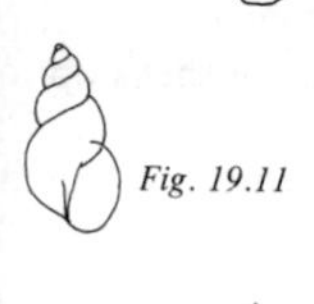
Fig. 19.11

Lymnaea (Galba) truncatula *Pl. 158, Fig. 19.11* Shell sometimes with very fine striations (cf. *L. palustris*); up to 12 mm tall but usually about half this; sometimes confused with young of *L. stagnalis* of comparable size – see *Fig. 19.12* for differences. A very widespread species that is typically found in shallow marginal water or on damp mud, in marshy places, or amongst grass roots in water meadows; often difficult to find even where common. This species has some economic importance as it is the intermediate host of *Fasciola hepatica*, the liver-fluke of cattle and sheep (see *p. 138* for details of the life-cycle).

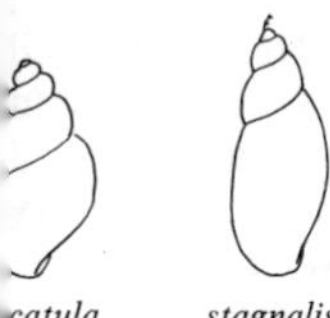

Fig. 19.12

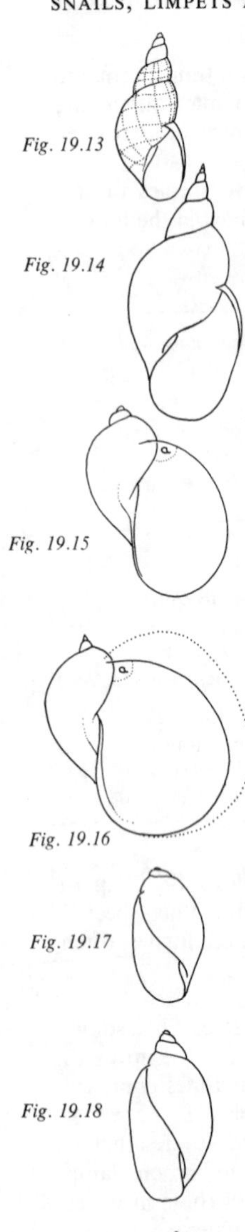

Fig. 19.13

Fig. 19.14

Fig. 19.15

Fig. 19.16

Fig. 19.17

Fig. 19.18

Fig. 19.19

Lymnaea* (*Galba*) *palustris (Marsh Snail) *Fig. 19.13* Shell rather thick and solid, often quite dark in colour; up to 30 mm tall. A widespread species, common in ponds, marshy places or temporary pools.

Lymnaea stagnalis (Great Pond Snail) *Pl. 159, Fig. 19.14* Shell usually pale or medium brown, slightly translucent, with a sharply pointed spire; up to 50 mm tall. Very common in still or slow-flowing waters.

Lymnaea* (*Radix*) *peregra (Wandering Snail) *Pl. 160, Fig. 19.15* Aperture of shell very variable in shape but its relative height is fairly constant; sometimes the spire is sunk into the body whorl; up to 30 mm tall but usually 15–20 mm. Probably the commonest water snail in Europe, occurring in a wide variety of habitats from lowland ponds to torrential mountain streams.

Lymnaea* (*Radix*) *auricularia (Ear Snail) *Pl. 161, Fig. 19.16* Aperture very wide, often curling backwards at the rim; spire very short but sharply pointed; up to 35 mm tall. Fairly common but, curiously, never abundant, in still and slow-flowing waters throughout the area covered by the guide.

Family Physidae

Physa spp. (Bladder Snails) *P. fontinalis* (*Pl. 162, Fig. 19.17*) Probably the only native European species. Shell glossy, thin, translucent and delicate, up to 10 mm tall; when expanded, the lobed edge of the mantle wraps around the outside of the shell. Common in many habitats throughout the area covered by this guide. Two other species, probably introduced, occur locally in the area, including Britain. They have thicker, opaque shells up to 16 mm tall: *P. acuta* (*Fig. 19.18*) has a relatively tall spire; *P. heterostropha* is plump, with a relatively short, blunt spire.

Aplexa hypnorum (Moss Bladder Snail) *Fig. 19.19* Shell very glossy, thin, translucent and delicate; up to 12 mm tall; body very dark, almost black. Widespread, but local in Britain and Europe, typically in small ponds that dry up in summer, sometimes found out of water.

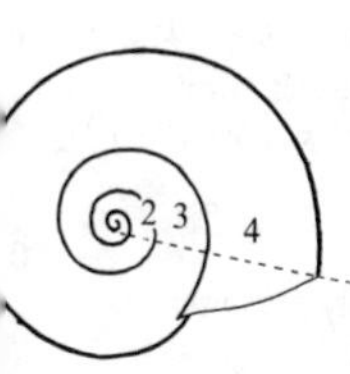

Family Planorbidae (Ramshorn Snails) *Pls. 163–166*

Many of the species in this very common family present difficulties in identification because their differences are slight, often somewhat subjective, and therefore liable to varied interpretation; so the key should be used with caution and discretion. As stated earlier, it is highly desirable to study a series of specimens rather than just one, especially in this group where typical sites are often inhabited by several species.

For the purposes of the terminology used here, planorbid shells are assumed to be dextral. The overall shape and proportions of the shell are determined by the number of whorls (a reasonably reliable character in adults only) and their degree of overlap; the shape of the whorls in cross-section also affects the overall appearance considerably. A few species possess a narrow, thickened ridge (keel) around their circumference. *Fig. 19.20* shows details of the features used in the key.

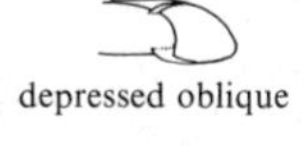

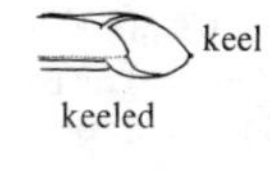

Fig. 19.20

Ramshorn snails are widely distributed throughout the area covered by the guide, and few species show any distinct habitat preference; most occur in small weedy ponds and ditches, they rarely live in fast-running waters. Individual species descriptions are not given, maximum sizes and other relevant details being included in the key.

Key to Native Species of Planorbidae

1. a) Shell with distinct raised transverse ridges, coiling sometimes not quite flat; up to 3 mm across; common (*Fig. 19.21*).
   Planorbis (*Armiger*) *crista*
   b) Shell without distinct transverse ridges, coiling flat, succeeding coils enclosing previous ones: adults more than 3 mm across. . . . 2

2. a) Not more than five whorls, which increase relatively rapidly in size. . . . 3
   b) Whorls five or more, increasing in size relatively gradually. (General aspect of whole shells in *b* is flatter and lower than in *a*.) . . . 9

3. a) Height of aperture greater than its width: whole shell up to 6 × 3 mm: (shape very distinctive with successive whorls greatly overlapping previous ones): common (*Fig. 19.22*)
   Planorbis (*Bathyomphalus*) *contortus*
   b) Aperture about equal in height and width, more-or-less rounded (*Fig. 19.20*); keel absent. . . . 4
   c) Height of aperture less than its width, aperture usually oval, depressed and oblique (*Fig. 19.20*); a keel may be present. . . . 6

4. a) A large, thick-shelled species, up to 35 × 12 mm: common.
   Planorbarius (*Coretus*) *vorneus* (*Pl. 163, Fig. 19.23*)
   b) Small species not exceeding 8 × 2.5 mm: shell not very thick. . . . 5

5. a) Shell normally whitish, with raised lines parallel to whorls: 8 × 2.5 mm: fairly common. *Planorbus (Gyraulus) albus (Fig. 19.24)*
   b) Shell normally brown, smooth, without raised lines: 6 × 2 mm: uncommon and local. *Planorbis (Gyraulus) laevis*

6. a) Width of outer whorl more than third of shell width: keel absent: shell typically translucent, smooth, and glossy. ... 10
   b) Width of outer whorl third or less of shell width: keel usually present: shell typically opaque, not usually glossy. ... 7

7. a) A small species up to 8 × 2.5 mm: keel weak but usually distinct: local and uncommon (Southern England and Scandinavia only). *Planorbis (Gyraulus) acronicus (Fig. 19.25)*
   b) Larger species up to 18 × 4 mm: keel always distinct. ... 8

8. a) Outer whorl about quarter of shell width: up to 18 × 4 mm: aperture less depressed, keel nearer to underside than in next species: common. *Planorbis (Tropidiscus) planorbis (Pl. 164, Fig. 19.26)*
   b) Outer whorl about third of shell width: up to 14 × 3 mm: aperture markedly depressed, keel nearer to mid-line than in previous species: common. *Planorbis (Tropidiscus) carinatus (Fig. 19.27)*

9. a) Aperture rounded, never with a keel (*Fig. 19.28*); up to 7 × 1.5 mm; common. *Planorbis (Anisus) leucostoma*
   b) Aperture oval; a slight keel sometimes present (*Fig. 19.29*); shell up to 10 × 2 mm; common. *Planorbis (Anisus) vortex (spirorbis) (Pl. 165)*
   c) Aperture oval: a slight keel sometimes present: shell up to 5 × 1 mm: local and uncommon. *Planorbis (Anisus) vorticulus*

10. a) Shell with internal transverse walls, visible by transparency: upper surface of shell strongly arched, underside almost flat: 5 × 2 mm: local and uncommon. *Segmentina (Hippeutis) nitida (Fig. 19.30)*
    b) Shell without internal transverse walls: upper and underside of shell about equally convex: 6 × 2 mm: fairly common. *Segmentina (Hippeutis) complanata (Pl. 166, Fig. 19.31)*

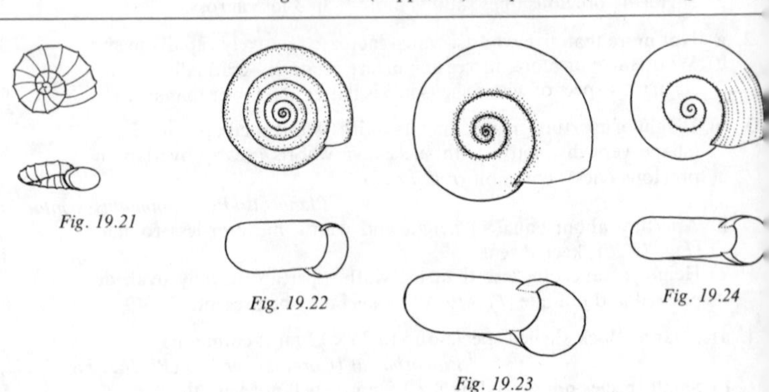

Fig. 19.21 *Fig. 19.22* *Fig. 19.23* *Fig. 19.24*

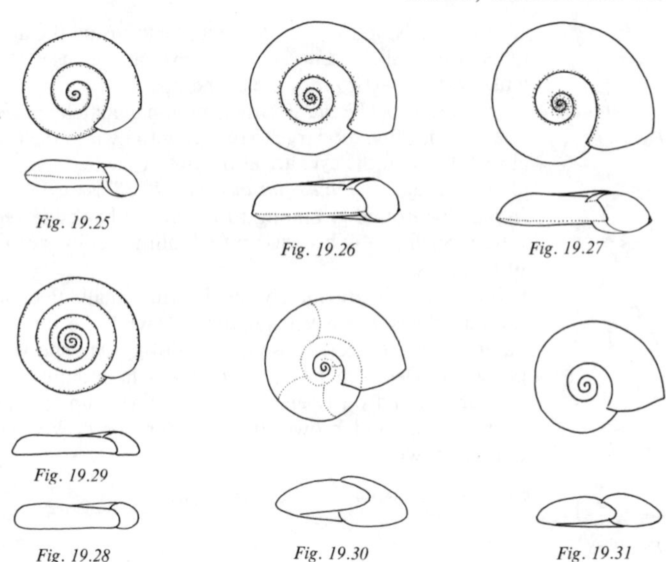

Fig. 19.25 Fig. 19.26 Fig. 19.27

Fig. 19.29

Fig. 19.28 Fig. 19.30 Fig. 19.31

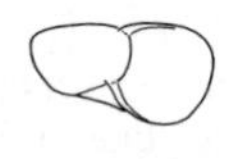

Fig. 19.32

Menetus dilatatus, an introduced species from North America, has become established in several localities in north-west England and Wales; it has a very distinctive shape, with about three whorls and markedly flared aperture, and is up to 4 × 1.5 mm (*Fig. 19.32*).

Family Ancylidae

Ancylus fluviatilis (River Limpet) *Pl. 167, Fig. 19.33* Shell relatively tall, wide and symmetrical; up to 8 mm long and 6 mm tall. Common in rivers and still waters, usually on hard substrata – stones, etc.

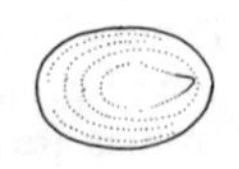

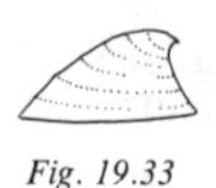

Fig. 19.33

Acroloxus lacustris (Lake Limpet) *Pl. 168, Fig. 19.34* Shell relatively low and narrow, its apex offset to one side: up to 7 mm long and 2 mm tall. Usually in still waters and may be found on plants or other substrata.

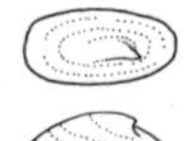

Fig. 19.34

In addition to the aquatic species described above there are several common terrestrial genera that live in damp marshy places or on overhanging waterside plants from which they may fall into the water or be brushed into the sweep net. *Carychium* and *Vertigo* spp. (*Figs. 19.35, 36*)

Fig. 19.35

Fig. 19.36

Fig. 19.37

Fig. 19.38

are tiny snails up to 2 mm tall with teeth inside the apertures of their shells. *Carychium* has eyes at the bases of its tentacles; in *Vertigo* they are at the tips.

Succinea spp. (*Fig. 19.37*) are common in marshy places. The body is often too large to retract entirely into the thin glossy shell and the eyes are at the tips of the tentacles; up to 20 mm tall. *Zonitoides nitidus* (*Fig. 19.38*) is also found in marshy ground or amongst the roots of bankside vegetation. Shell glossy brown, up to 8 mm wide; eyes at tip of tentacles.

Slugs, which are merely snails with small, uncoiled internal shells, are never aquatic but many are found in damp places, especially water meadows. One species in particular, *Deroceras* (*Agriolimax*) *laevis*, habitually occur in marshes and on river banks. It is small, up to about 25 mm long, and brownish, often dark or flecked with darker brown.

References: Macan, 1977. Ellis, 1969. Brauer, 1909.

Class BIVALVIA (PELECYPODA, LAMELLIBRANCHIA)
(Mussels and Cockles)

Freshwater mussels are common animals, often abundant in slow rivers and canals, but living specimens are inconspicuous, despite their size, due to their burrowing habits. Empty shells are a common feature of the bottom litter in many localities. Mussels range in size from little over 1 mm to more than 180 mm long.

The shell consists of two halves, valves, which are equal in size and shape and mirror images of each other (*Fig. 19.43*). They are hinged together at the dorsal edge by a tough flexible ligament; in some genera this hinge is reinforced by two or three interlocking teeth on the inside of the shell. Externally the valves show concentric growth rings centred around a low conical prominence, the umbo, which is anterior to the ligament. The line drawings are all of right valves – i.e., the front of the shell is to the right.

The inside of the shell is lined with mother-of-pearl (nacre), at one time the raw material for the pearl button industry. Pearls are frequently found in freshwater mussels but are usually very small, less than 1 or 2 mm, and irregular in shape; large, well-shaped pearls are very rare.

The soft body is very different from that of a snail. There are no head and eyes and most of the internal space of the shell is taken up by a complex, voluminous array of curtain-like gills, the digestive and other organs being relatively small. The mantle lines the valves and its edges may be visible when the mussel is open. The valves can be clamped shut by two powerful transverse adductor muscles.

When the animal is operational a pair of fleshy tubes, siphons, formed from the rim of the mantle, protrude posteriorly from between the valves. One of these is inhalant, the other exhalant. Water is drawn in through the inhalant siphon by ciliary action, filtered through the gills, and passed out through the exhalant siphon. The water currents are used for respiration and feeding, the suspended organic particles that are the sole food of bivalves being extracted by the gills and passed to the mouth on a conveyor belt of mucus, powered by cilia. Movement is provided by a fleshy, muscular, tongue-like foot that can be protruded between the valves, usually anteriorly. It is used for digging or for dragging the mussel along the surface of the substratum.

The life-cycles of freshwater mussels are varied and among their more interesting features. The unionid genera, *Margaritifera*, *Unio* and *Anodonta*, produce large numbers of distinctive glochidia larvae (*Fig. 19.39*) which are brooded internally during the winter and released in the spring. The larvae cannot swim but drift with the current, eventually sinking to the bottom. Each larva trails a sticky thread about 2 mm long, the function of which is to cause it to adhere to a fish, probably aided by turbulence caused by the fish swimming by. The larva then clamps on to its host by closing the valves and becomes encysted in the skin or on the fins. It lives parasitically, probably feeding on the blood and mucus of its host, which also provides transport. After a few weeks, the larva has developed into a tiny mussel which then drops off the host to settle into the substratum. Any species of fish may be used as a host but the luckless stickleback is possibly the most commonly infected.

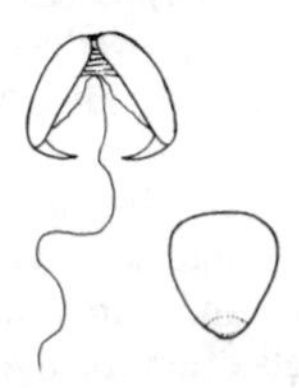

Fig. 19.39

Sphaeriids (*Sphaerium* and *Pisidium*) brood a small number of larvae in the shell cavity, eventually releasing them as fully formed young mussels. *Dreissena* produces tiny ciliated larvae called veligers which, after a short free-swimming life in the plankton, settle and attach to a substratum where they complete their development.

Most mussels live buried in sedimentary substrata, with just the posterior tip of the shell and the siphons protruding. In clear water the larger species are not difficult to discern. *Sphaerium* and *Pisidium* are sometimes found climbing among plant life. *Dreissena* has a different life-style, attaching itself to hard substrata by tough byssus threads; aggregations of this mussel may completely dominate large areas of substratum.

Key to Genera of Freshwater Mussels

1. a) Mussels living attached to firm substrata by a bunch of tough threads (but can climb free). *Dreissena* (*p. 190*
   b) Free-living mussels, burrowing in soft substrata or climbing on plants, etc. . . . 2
2. a) Larger mussels, adults longer than 25 mm: shells oval or elongated: typically dark – green, brown, or blackish – in colour. . . . 3
   b) Smaller mussels, rarely exceeding 25 mm, usually much less: shells circular or slightly oval: usually pale yellowish or whitish in colour. . . . 5
3. a) Umbos not very prominent, in side view level with dorsal edge of shell: widest part of shell about central; hinge teeth absent. *Anodonta* (*p. 191*
   b) Umbos prominent, in side view overlapping dorsal edge of shell; greatest width of shell just below umbos: internal hinge teeth present. . . . 4
4. a) Larger, adults exceeding 100 mm in length: ventral edge of shell straight or concave in mid-region: typically dark brown or blackish. *Margaritifera* (*p. 190*
   b) Smaller, less than 100 mm long: shell wedge-shaped with convex ventral edge: typically green or brown. *Unio* (*p. 191*
5. a) Siphon tubes long, separate: adult shells longer than 10 mm. *Sphaerium* (*p. 192*
   b) Siphon tubes short, fused together: usually less than 10 mm. *Pisidium* (*p. 192*

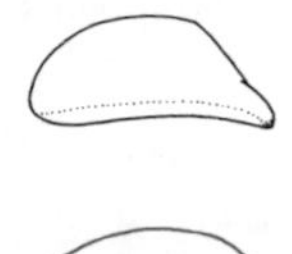

Fig. 19.40

Dreissena polymorpha (Zebra Mussel) *Pl. 169, Fig. 19.40*
Shell distinctively wedge-shaped at anterior end, up to 40 mm long; pale olive or yellowish, strikingly marked with dark zig-zag lines (but the surface may be worn away or overgrown by algae). It attaches itself to any firm substratum by tough, silk-like byssus threads, but can detach and crawl around using the slender foot. Large numbers of these mussels often form dense beds on submerged woodwork or masonry. During the past two centuries, this mussel has spread throughout much of western and northern Europe, including Britain, having apparently originated in eastern Europe, and is now a locally common animal in many rivers, lakes and canals.

Fig. 19.41

Margaritifera margaritifera (Pearl Mussel) *Fig. 19.41*
Shell elongated/oval, its ventral edge usually concave, up to 150 mm long. Typically found in soft-water districts, usually in large, fast-flowing rivers, buried in sand or gravel. Widespread in Europe and throughout the northern hemisphere; in Britain it is confined to the north and west. Since Roman times this species has been well known for

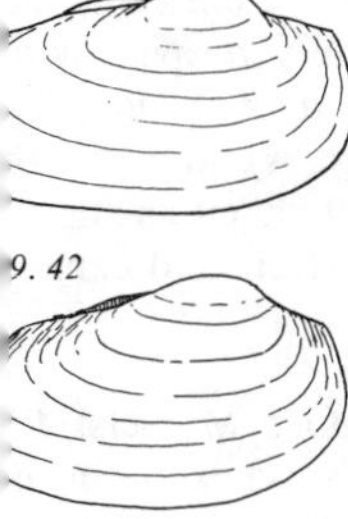

9. 42

its pearls, which are often of fine quality, and some local pearl fisheries still exist.

Unio spp. Shells generally wedge-shaped or elliptical with prominent umbos. There are three European species. usually found in hard calcareous waters.

Unio pictorum *Fig. 19.42* Shell with the dorsal and ventral edges more or less parallel, often more elongated than shown in the figure; up to 100 mm long; yellowish or brownish in colour. Widespread in Britain and Europe in still or slow-flowing waters, usually burrowing in mud.

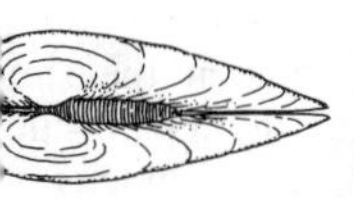

19.43

Unio tumidus *Fig. 19.43* Shell wedge-shaped, with dorsal and ventral edges converging posteriorly; up to 80 mm long; yellowish-brown, sometimes with a strong greenish tinge. Common in Britain and Europe, typically being found in rivers where it burrows in clean sand and gravel.

Unio crassus *Fig. 19.44* Generally deeper-bodied than the two species above, usually somewhat truncated at the posterior end; dorsal and ventral edges of shell approximately parallel but never as elongated as *U. pictorum*. Absent from Britain but frequent in rivers on the Continent.

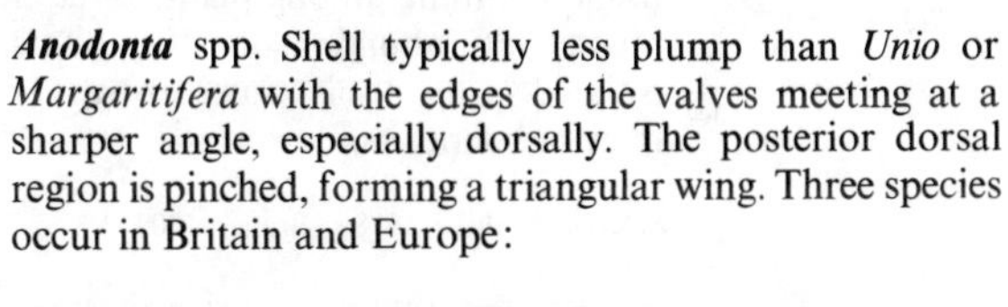

9.45

Anodonta spp. Shell typically less plump than *Unio* or *Margaritifera* with the edges of the valves meeting at a sharper angle, especially dorsally. The posterior dorsal region is pinched, forming a triangular wing. Three species occur in Britain and Europe:

Anodonta cygnea (Swan Mussel) *Fig. 19.45* The largest species, up to 150 mm long, exceptionally larger. Ventral edge of shell often concave; angle of wing about equidistant from umbos and posterior end of shell; colour varies from yellowish to green or brown. Common and widespread in still or slow-flowing waters, burrowing in mud.

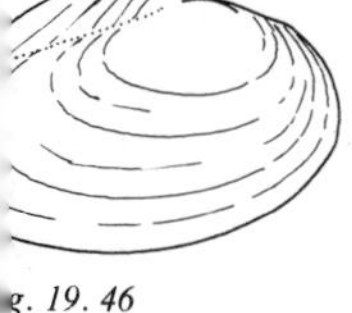

g. 19. 46

Anodonta anatina (Duck Mussel) *Pl. 170, Fig. 19.46* Shell deeper than *A. cygnea*; angle of wing usually nearer to umbos than posterior end; up to 100 mm long. Colour dark yellowish-brown or greenish. Typically found in rivers, burrowing in sand or gravel; common and widespread.

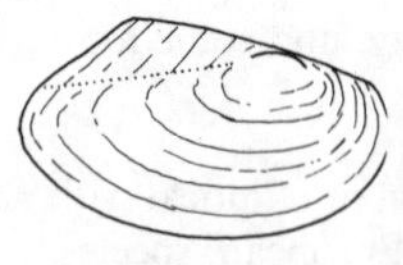
Fig. 19.47

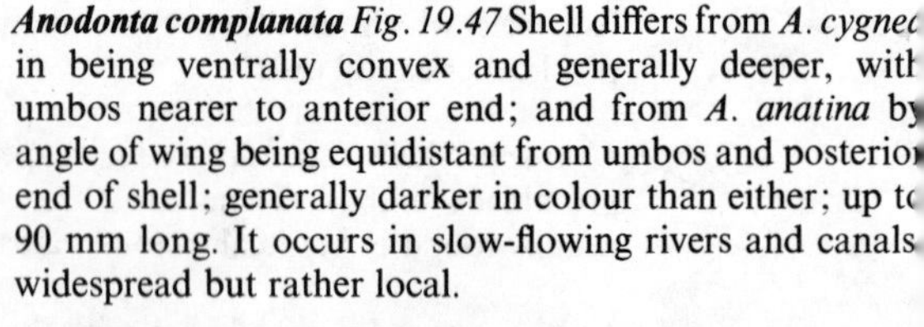
Anodonta complanata *Fig. 19.47* Shell differs from *A. cygnea* in being ventrally convex and generally deeper, with umbos nearer to anterior end; and from *A. anatina* by angle of wing being equidistant from umbos and posterior end of shell; generally darker in colour than either; up to 90 mm long. It occurs in slow-flowing rivers and canals; widespread but rather local.

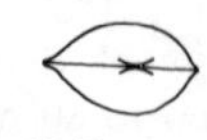
Fig. 19.48

Sphaerium spp. (Freshwater Cockles, Orb Mussels) *Pl. 171, Fig. 19.48* Several species of this genus occur in Europe, ranging in size from about 10–25 mm when adult. They are mostly whitish or yellowish in colour but the largest species, *S. rivicola*, more than 15 mm long, may be brown or greenish. These bivalves are common in many types of habitat, mostly burrowing in the sub-stratum, but young ones may clamber about on plant-life attaching themselves temporarily by threads of mucus.

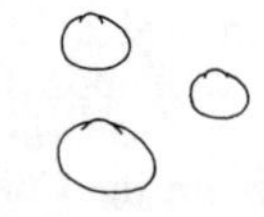
Fig. 19.49

Pisidium spp. (Pea Mussels) *Pl. 172, Fig. 19.49* The numerous species of this very common genus range in size from 2–7 mm long; only *P. amnicum* becomes larger than this, up to 12 mm long. Other species are extremely difficult to identify; they are all yellowish or buff in colour, but in certain conditions may become encrusted with a rust-like deposit. Like *Sphaerium* they may be found climbing among plants on a ladder of mucus. Pea mussels are abundant in all habitats, even occurring in isolated spring pools, water troughs or in interstitial water in gravel deposits.

References: Ellis, 1978. Brauer, 1909.

2▲

3▲

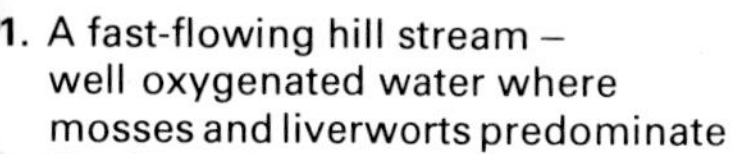

1. A fast-flowing hill stream – well oxygenated water where mosses and liverworts predominate
2. The River Tweed, a well known salmon river, beginning to emerge into the lowlands
3. A woodland pond, where frogspawn may be found, that dries up in summer
4. A eutrophic pond – shallow enough to maintain a fairly uniform temperature and for plants rooted on the bottom to emerge at the surface

4▶

5. An oligotrophic tarn with sedges and other emergent vegetation
6. Coniston Water – a deep, glacial hill lake
7. A flooded gravel pit – the most frequent form of new freshwater habitat
8. Lough Neagh – a lowland lake, and the largest freshwater body in Britain and Ireland

5▲

6▶

7▶

8▶

9▲

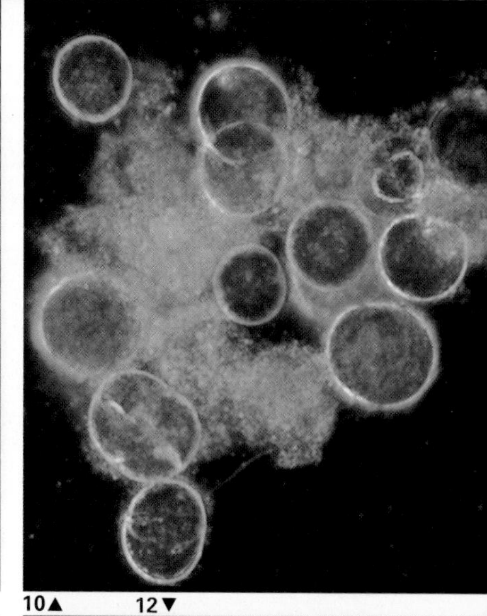

10▲ 12▼

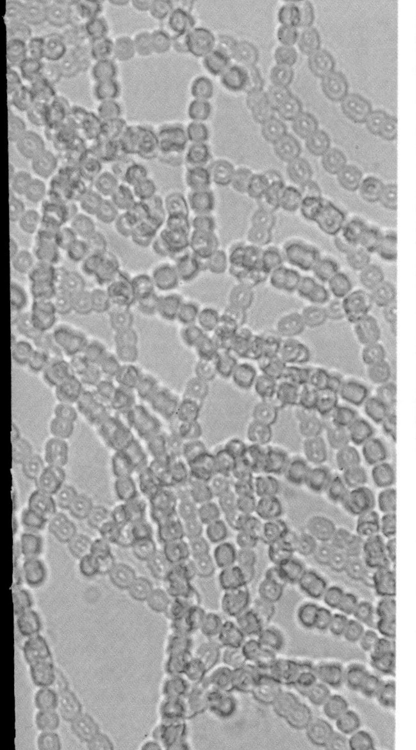

9. *Thiothrix*, a filamentous bacterium, ×6

10. *Chroococcus*, a spherical blue-green alga, ×1000 OSF

11. *Anabaena*, filamentous blue-green alga, ×350 OSF

12. *Gloeotrichia*, colonial blue-green algae, ×4

◄11

13▼

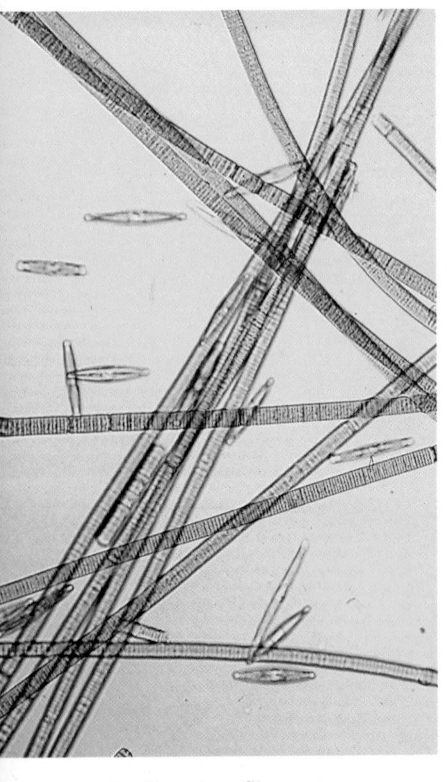

14▲

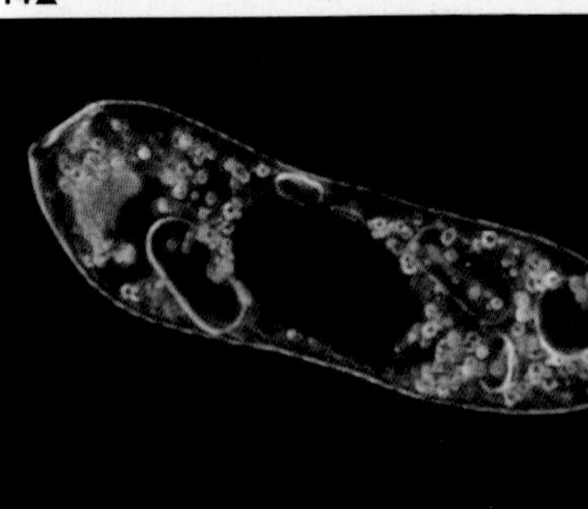

15▲

13. *Oscillatoria*, a filamentous blue-green alga, ×200 (with diatom *Nitzschia*)
14. *Saprolegnia*, fungus on dead stickleback, ×3
15. *Euglena*, green flagellated alga, ×1000 OSF
16. *Chlamydomonas*, green flagellated algae, ×750 OSF
17. *Volvox*, spherical green algal colonies with 'daughters', ×40 OSF
18. *Peridinium*, a dinoflagellate, ×700
19. *Pediastrum*, a non-motile algal colony, ×450
20. *Gonatozygon*, a cylindrical desmid, ×300 OSF
21. *Closterium*, a crescentic desmid, ×350 OSF
22. *Euastrum*, a desmid, ×1000 OSF

16▶

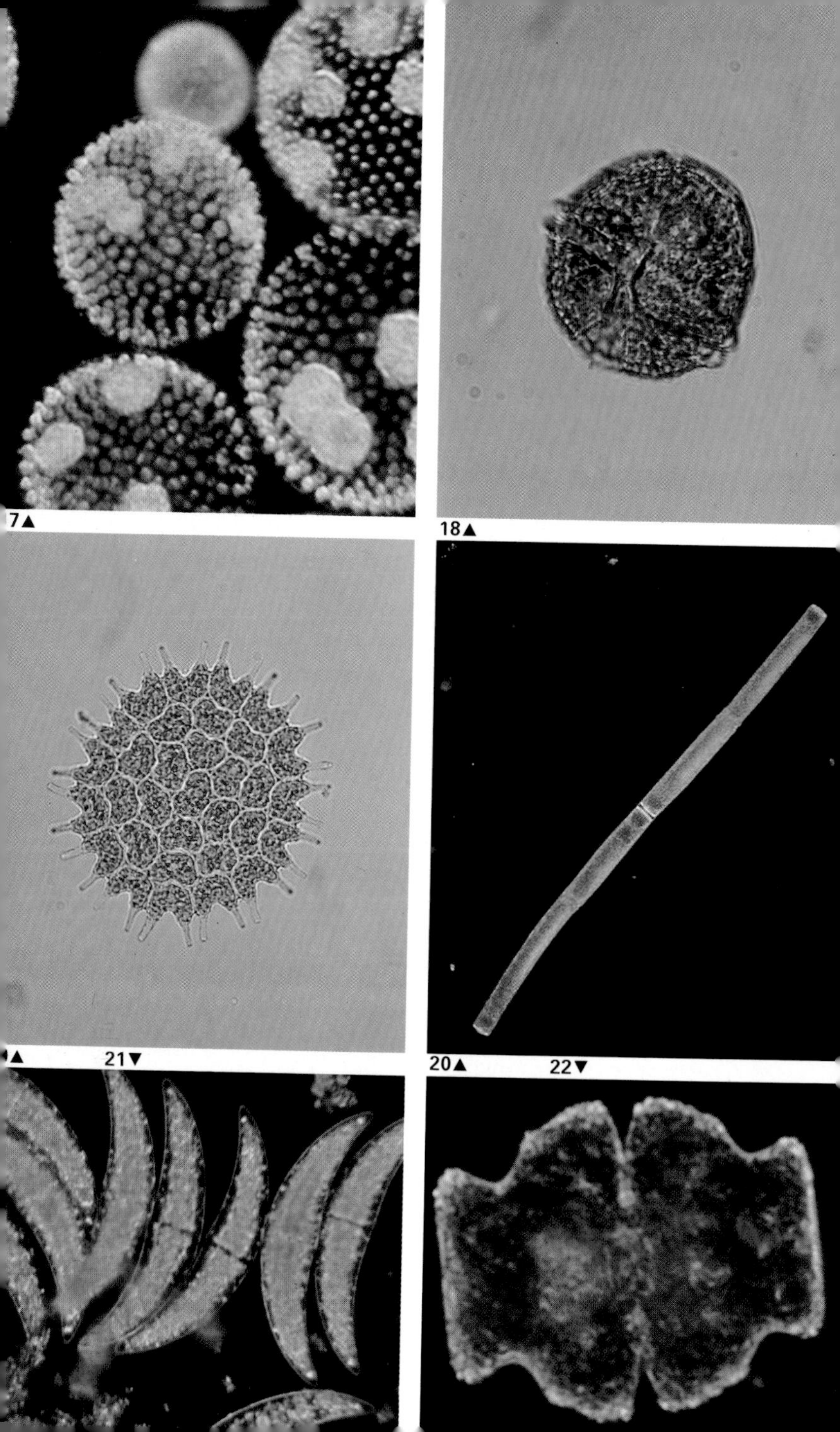

7▲
18▲
▲
21▼
20▲
22▼

◀ 23 24 ▼

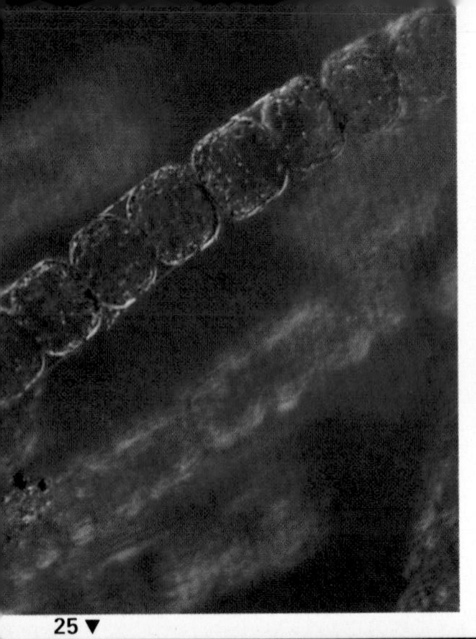

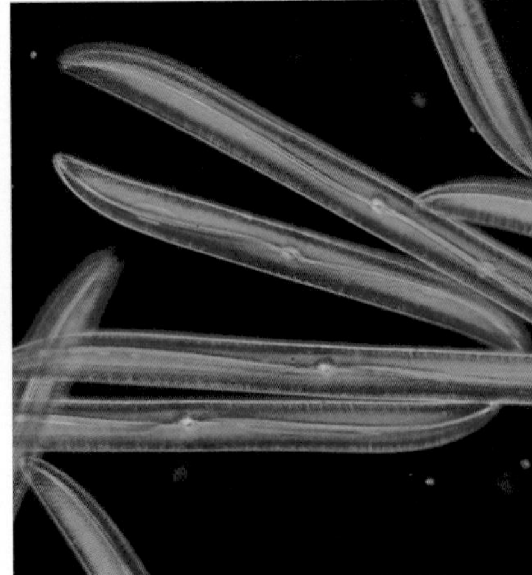

25 ▼

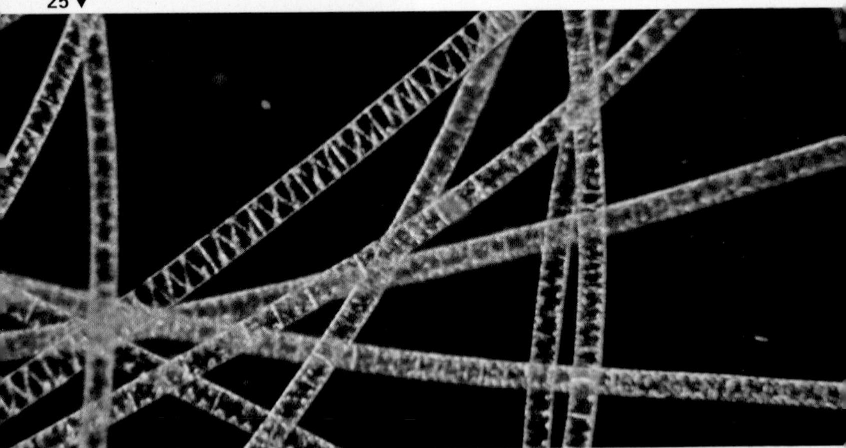

23. *Melosira*, a filamentous diatom, ×200 OSF

24. *Gyrosigma*, a solitary diatom, ×300 OSF

25. *Spirogyra*, a filamentous green alga, ×20 OSF

26. *Chara*, stonewort plant with antheridia, ×1.5

26 ▶

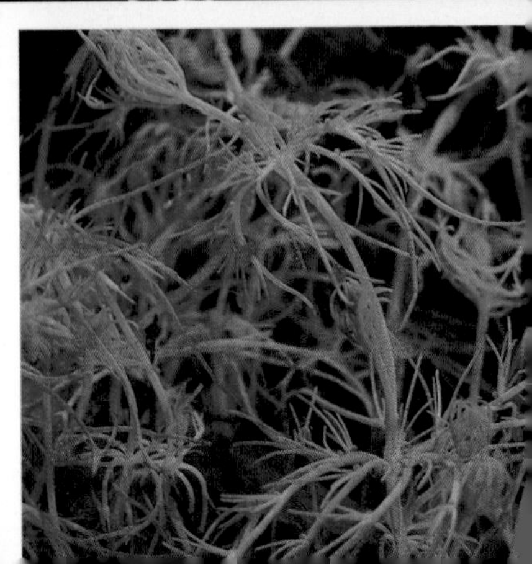

28▲

27. *Tolypella*, antheridia and oogonia, ×8 OSF
28. *Chaetophora*, several whole colonies, ×5
29. *Enteromorpha*, floating in pond
30. *Batrachospermum*, Frogspawn Alga, ×2
31. *Fissidens*, ×1.5

29▼

30▲ 31▼

32. *Fontinalis antipyretica* OSF
33. *Scapania undulata*, ×1

34. *Riccia fluitans*, aquatic liverwort OSF
35. *Equisetum fluviatile*, Water Horsetail OSF

36▲

37▲ 38▼

39▼

36. *Salvinia natans*, Floating Water Fern OSF
37. *Azolla filiculoides*, Water Fern OSF
38. *Azolla filiculoides*, Water Fern
39. *Salix fragilis*, Crack Willow

40. *Salix fragilis*, Crack Willow
41. *Salix caprea*, Goat Willow or Sallow, male flowers OSF
42. *Alnus glutinosa*, Alder OSF
43. *Alnus glutinosa*, Alder, fruits

40▲

41▲

42▼

43▼

4▲

45▲

6▲

44. *Polygonum amphibium*, Amphibious Bistort OSF

45. *Rumex hydrolapathum*, Great Water Dock OSF

46. *Nymphaea alba*, White Water-lily

47. *Nuphar lutea*, Yellow Water-lily

◀**47**

48▲

49▼

50▲ 51▼

48. *Ranunculus aquatilis*, Common Water Crowfoot OSF

49. *Ceratophyllum demersum*, Rigid Hornwort OSF

50. *Caltha palustris*, Marsh Marigold or Kingcup OSF

51. *Rorippa amphibia*, Great Yellowcress

2▲ 53▼

4▼

55▼

52. *Rorippa nasturtium-aquaticum*, Watercress OSF

53. *Crassula helmsii*, New Zealand Water Stonecrop

54. *Hypericum elodes*, Bog St. John's Wort OSF

55. *Lythrum salicaria*, Purple Loosestrife OSF

56▲

58▲ 60▼

57▲ 59▼

56. *Myriophyllum spicatum*, Spiked Water Milfoil OSF
57. *Myriophyllum verticillatum*, Whorled Water-Milfoil OSF
58. *Hippuris vulgaris*, Marestail OS
59. *Berula erecta*, Narrow-leaved Water-parsnip OSF
60. *Apium nodiflorum*, Fool's Watercress OSF

61▲

62▲

63▼

64▼

61. *Hottonia palustris*, Water Violet OSF

62. *Menyanthes trifoliata*, Bog Bean OSF

63. *Nymphoides peltata*, Fringed Water-lily OSF

64. *Myosotis secunda*, Creeping Water Forget-me-Not OSF

65▲

66▲

67▲ 68▼

65. *Callitriche stagnalis*, Common Water Starwort OSF
66. *Stachys palustris*, Marsh Woundwort OSF
67. *Mentha aquatica*, Water Mint OSF
68. *Mimulus guttatus*, Monkey-flower OSF
69. *Mimulus luteus*, Blood-drop Emlets OSF

69▼

71▼

72▲ 73▼

Veronica beccabunga, Brooklime OSF
Veronica anagallis-aquatica, Water Speedwell
Utricularia vulgaris, Great Bladderwort OSF
Alisma plantago-aquatica, Common Water Plantain OSF

74▲

76▲ 77▼ 75▲

74. *Sagittaria sagittifo*[…] Arrow-head OSF
75. *Butomus umbellat*[…] Flowering Rush OSF
76. *Hydrocharis morsus-ranae*, Frogbit OSF
77. *Stratiotes aloides*, Water Soldier OS[…]

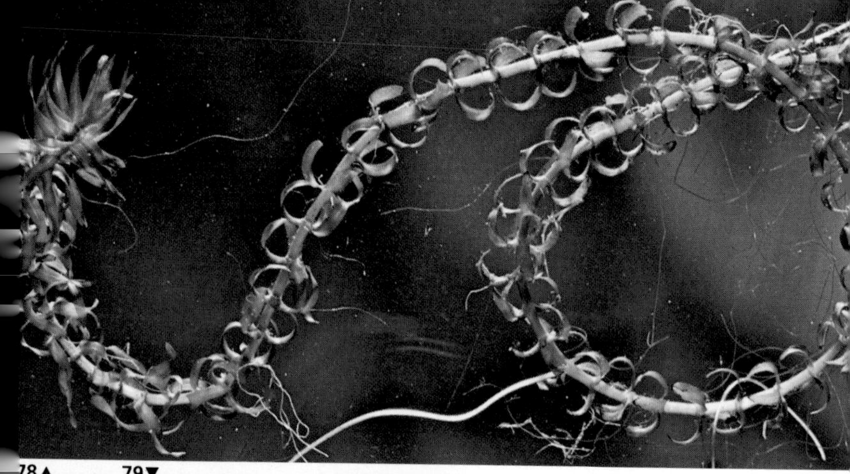

78▲ 79▼

78. *Elodea canadensis*, Canadian Waterweed OSF

79. *Potamogeton natans*, Broad-leaved Pondweed OSF

80. *Iris pseudacorus*, Yellow Flag OSF

81. *Juncus inflexus*, Hard Rush, in fruit OSF

80▼

81▼

83▼

82▲

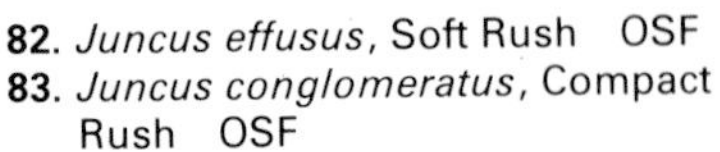

82. *Juncus effusus*, Soft Rush OSF
83. *Juncus conglomeratus*, Compact Rush OSF
84. *Phragmites australis*, Common Reed OSF
85. *Glyceria maxima*, Reed Sweet-grass OSF
86. *Calla palustris*, Bog Arum OSF

84▲ 85▼

86▼

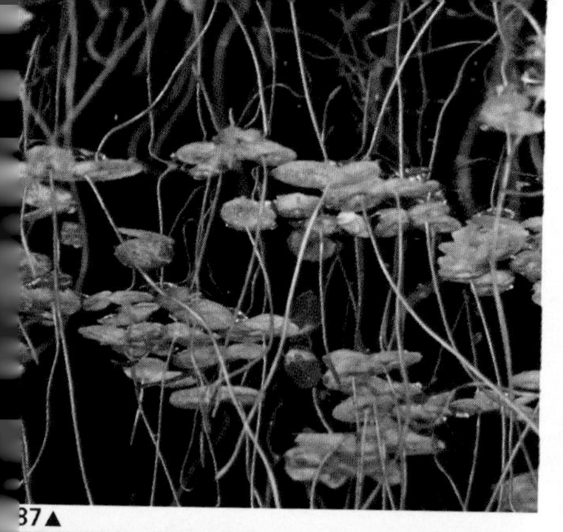

87▲

88▲

90▲

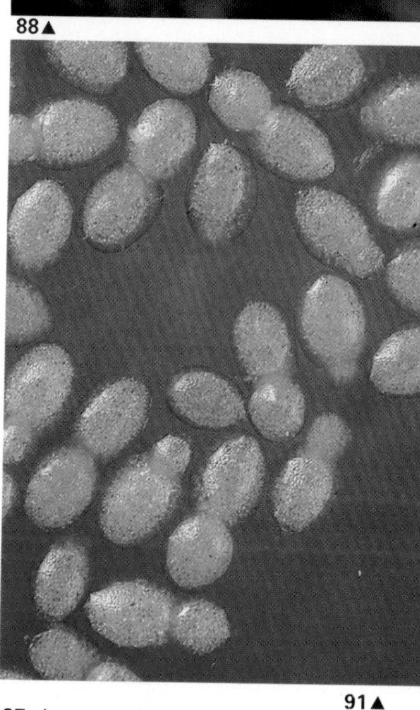

91▲

◀ 89

87. *Lemna minor*, Common Duckweed OSF

88. *Lemna trisulca*, Ivy-leaved Duckweed OSF

89. *Spirodela polyrhiza*, Great Duckweed OSF

90. *Lysichiton americanum*, Skunk Cabbage OSF

91. *Wolffia arrhiza*, Rootless Duckweed OSF

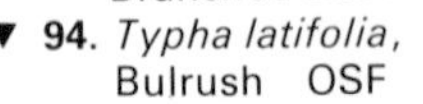

▲ **92**. *Sparganium erectum*, Branched Bur-reed OSF

▼ **94**. *Typha latifolia*, Bulrush OSF

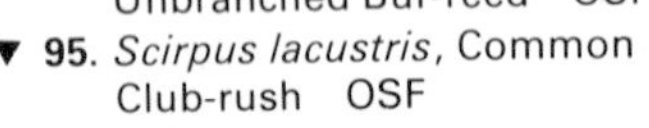

▲ **93**. *Sparganium emersum*, Unbranched Bur-reed OSF

▼ **95**. *Scirpus lacustris*, Common Club-rush OSF

97▲ 98▼

96. *Eleocharis palustris*, Common Spike-rush OSF
97. *Eriophorum angustifolium*, Common Cotton-grass OSF
98. *Carex acuta*, Slender Spiked Sedge OSF
99. *Carex riparia*, Great Pond Sedge OSF
100. *Carex acutiformis*, Lesser Pond Sedge OSF

99▼

100▼

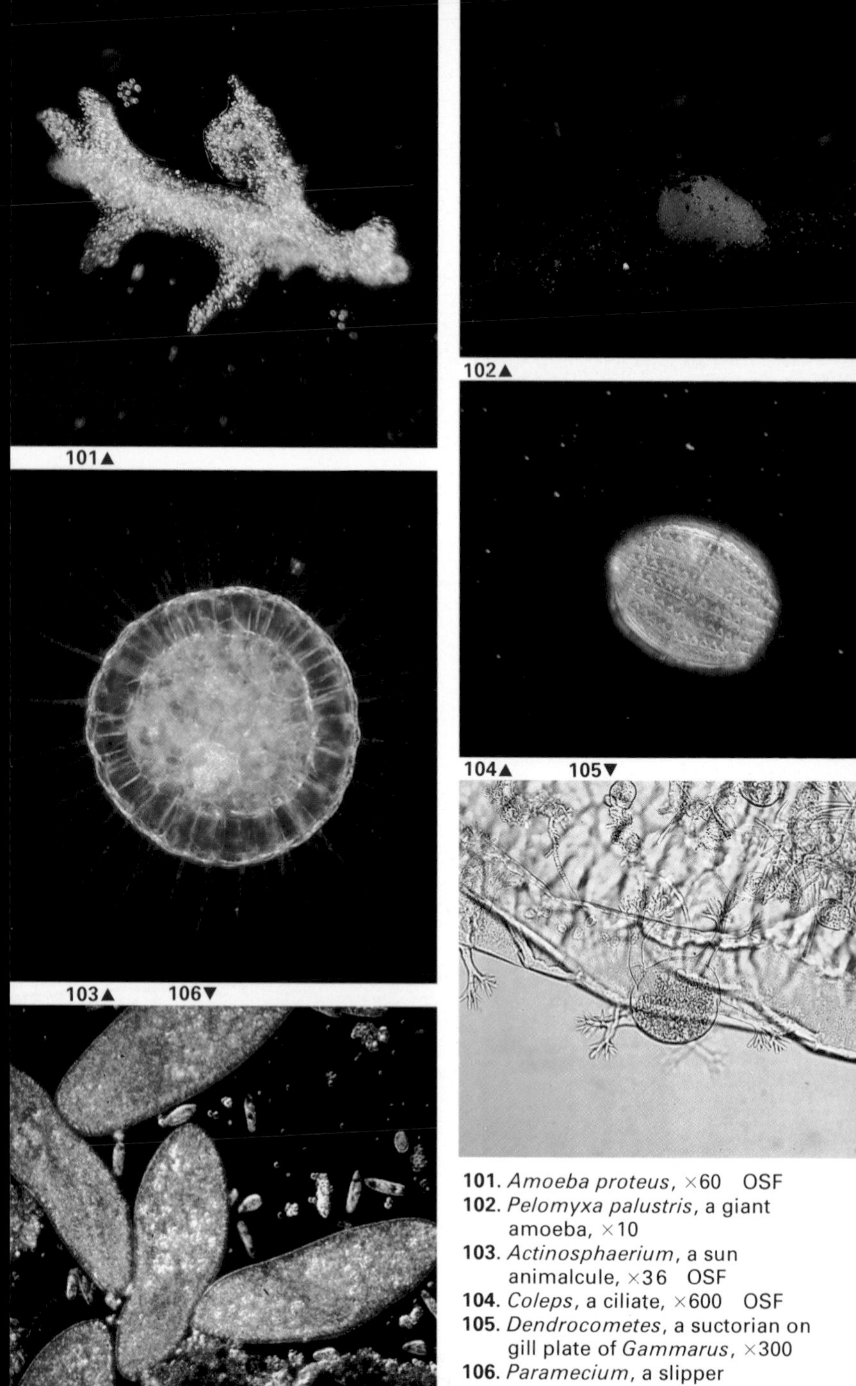

101. *Amoeba proteus*, ×60 OSF
102. *Pelomyxa palustris*, a giant amoeba, ×10
103. *Actinosphaerium*, a sun animalcule, ×36 OSF
104. *Coleps*, a ciliate, ×600 OSF
105. *Dendrocometes*, a suctorian on gill plate of *Gammarus*, ×300
106. *Paramecium*, a slipper animalcule, ×200 OSF

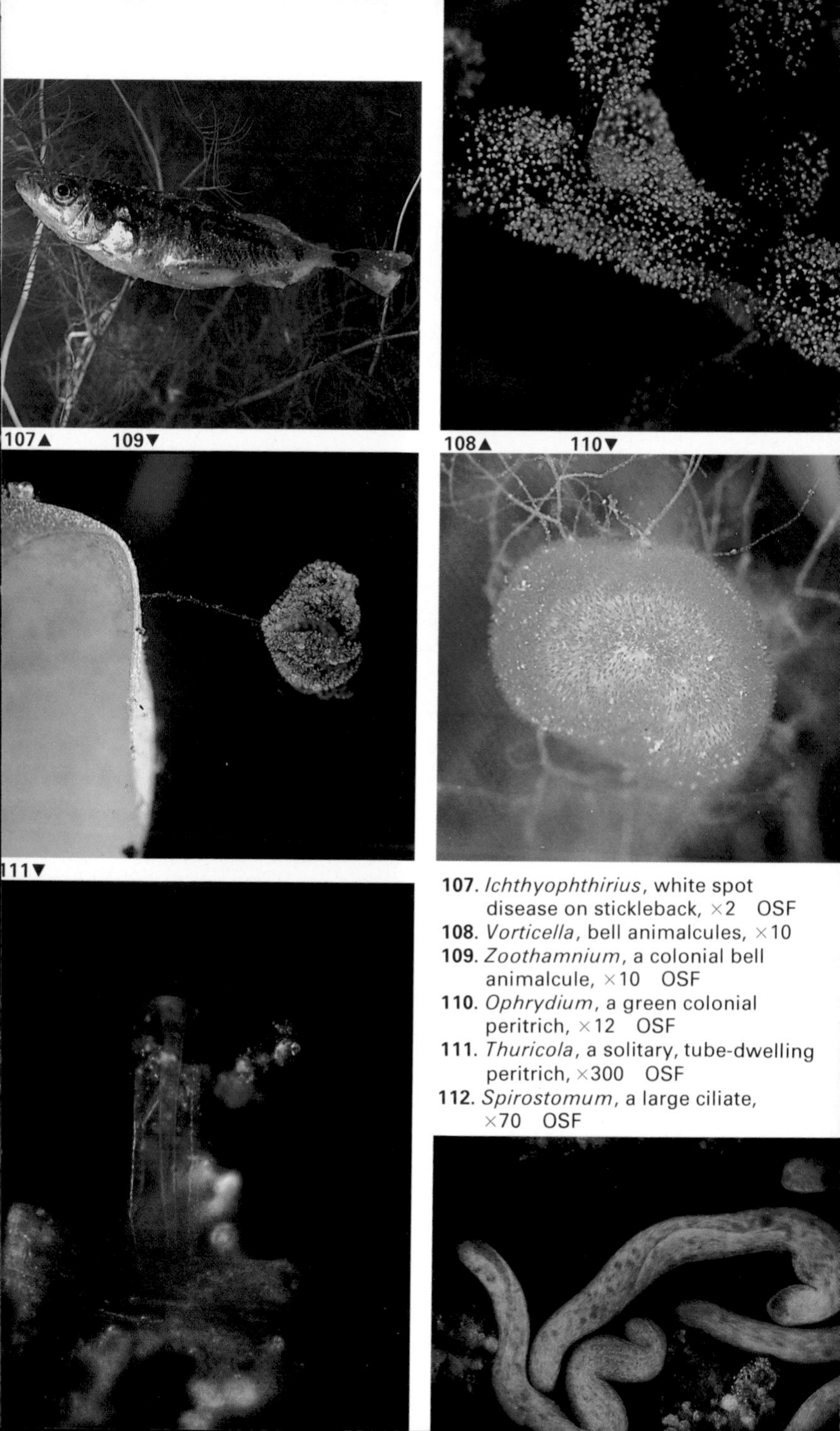

107. *Ichthyophthirius*, white spot disease on stickleback, ×2 OSF
108. *Vorticella*, bell animalcules, ×10
109. *Zoothamnium*, a colonial bell animalcule, ×10 OSF
110. *Ophrydium*, a green colonial peritrich, ×12 OSF
111. *Thuricola*, a solitary, tube-dwelling peritrich, ×300 OSF
112. *Spirostomum*, a large ciliate, ×70 OSF

113. *Stentor*, a group of green trumpet animalcules, ×8

114. *Stylonychia*, a ciliate, undergoing division, ×250 OSF

115. *Ephydatia fluviatilis*, gemmules of sponge, ×2.5

116. *Ephydatia fluviatilis*, 'green fingers' form, ×½

117. *Hydra viridissima*, green hydra, with buds, ×3

118. *Hydra oligactis*, brown hydra, with buds, ×1.5

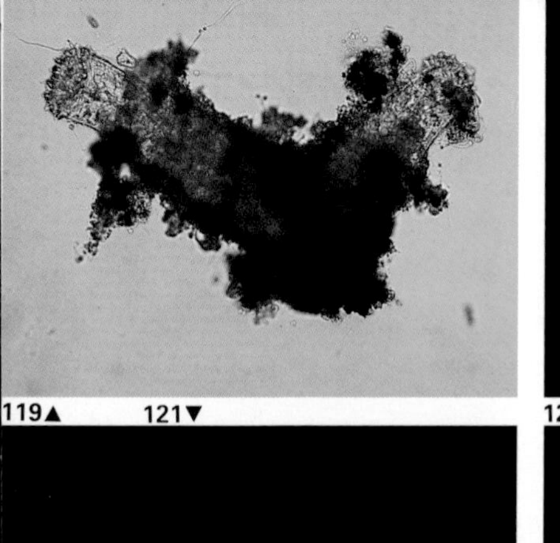

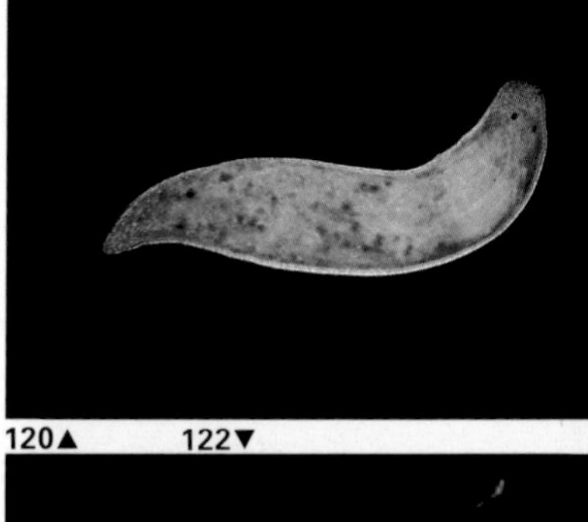

119▲ 121▼

120▲ 122▼

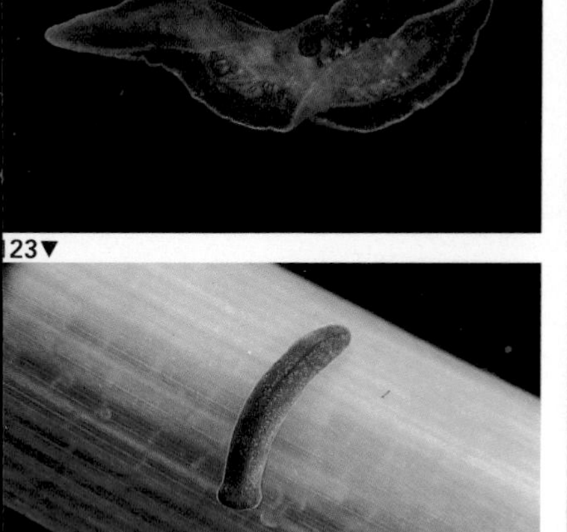

23▼

124▼

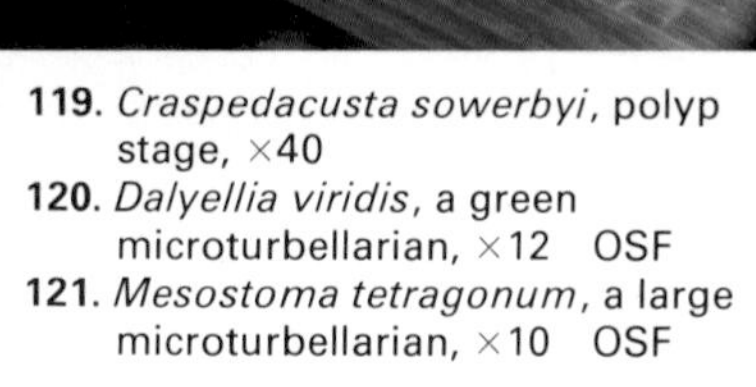

119. *Craspedacusta sowerbyi*, polyp stage, ×40

120. *Dalyellia viridis*, a green microturbellarian, ×12 OSF

121. *Mesostoma tetragonum*, a large microturbellarian, ×10 OSF

122. *Dendrocoelum lacteum*, flatworm feeding on *Asellus*, ×1.5

123. *Polycelis tenuis*, ×2 (whitish spots are *Urceolaria*)

124. *Dugesia tigrinum*, ×2

125. *Dugesia polychroa*, ×2

126. *Planaria torva*, ×2

25▼

126▼

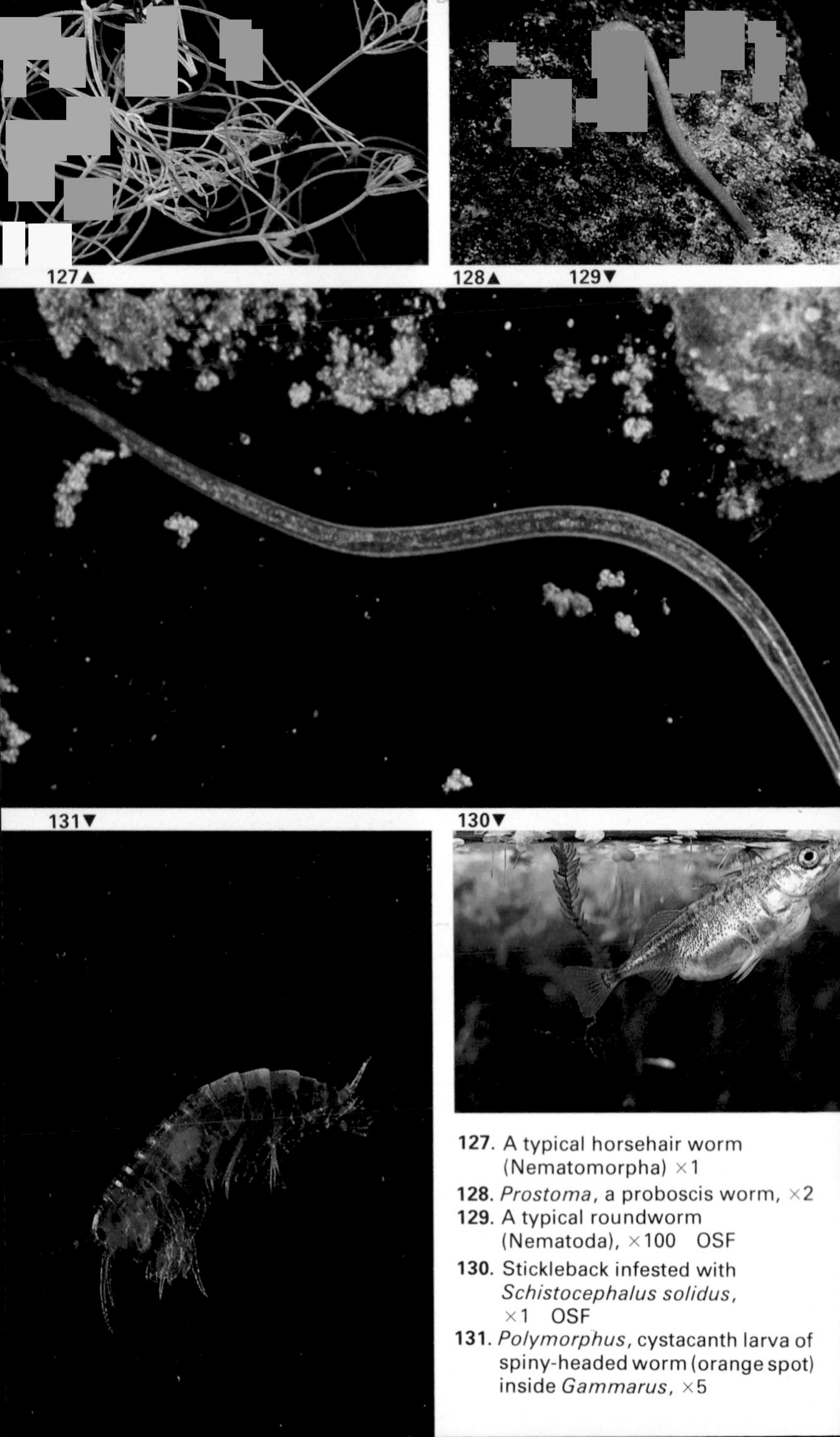

127▲

128▲

129▼

131▼

130▼

127. A typical horsehair worm (Nematomorpha) ×1

128. *Prostoma*, a proboscis worm, ×2

129. A typical roundworm (Nematoda), ×100 OSF

130. Stickleback infested with *Schistocephalus solidus*, ×1 OSF

131. *Polymorphus*, cystacanth larva of spiny-headed worm (orange spot) inside *Gammarus*, ×5

132▲

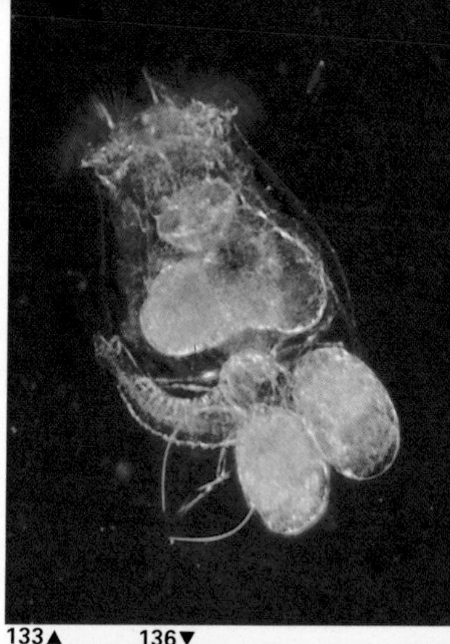

133▲

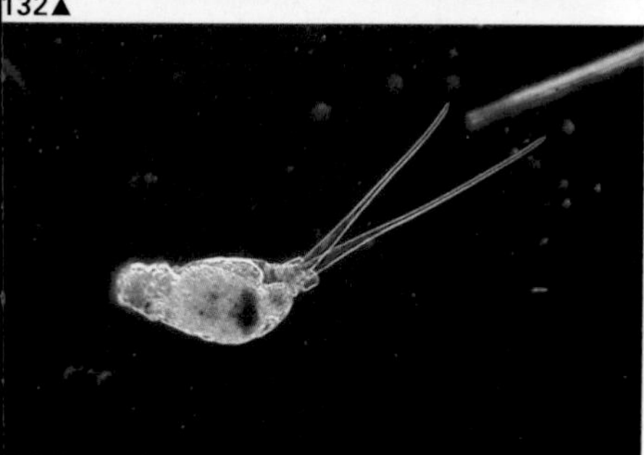

34▲

135▼

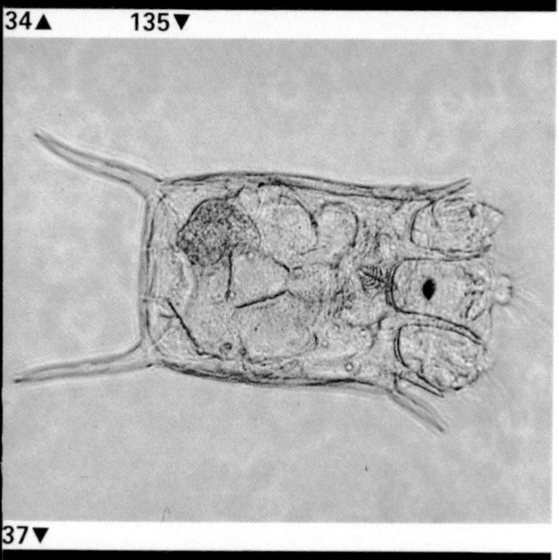

136▼

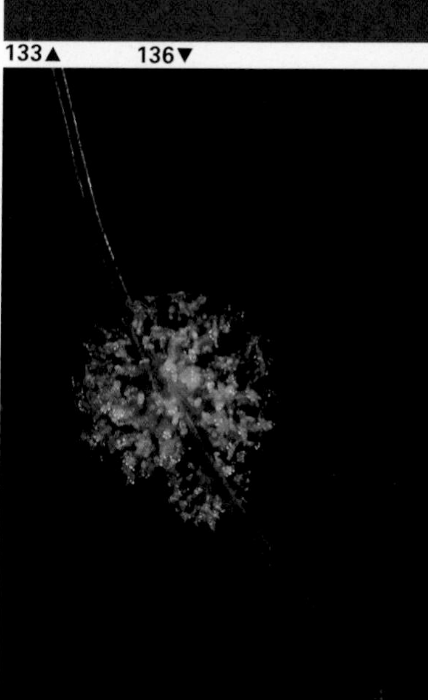

37▼

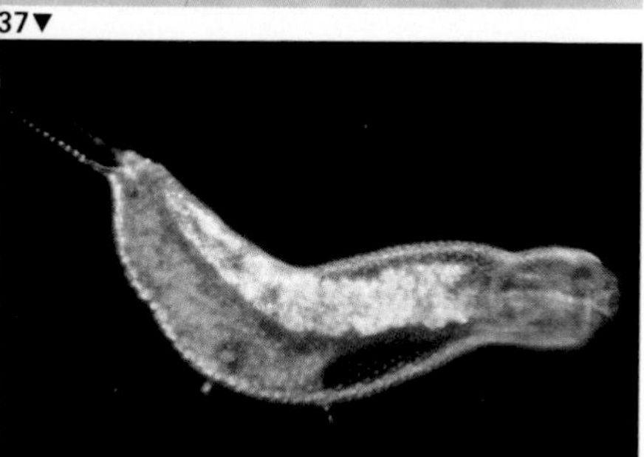

132. *Rotaria*, a bdelloid rotifer, ×100 OSF

133. *Brachionus*, ×150 OSF

134. *Monommata*, ×70 OSF

135. *Keratella*, ×150

136. *Sinantherina*, a colonial rotifer, ×10

137. *Chaetonotus*, a gastrotrich, ×150 OSF

138▲

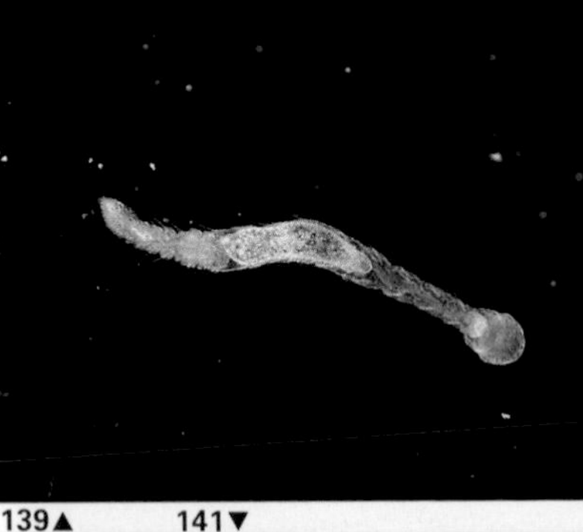

139▲

141▼

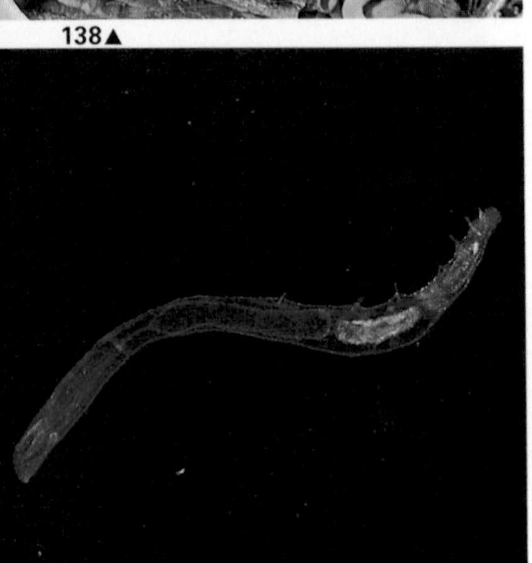

140▲

142▼

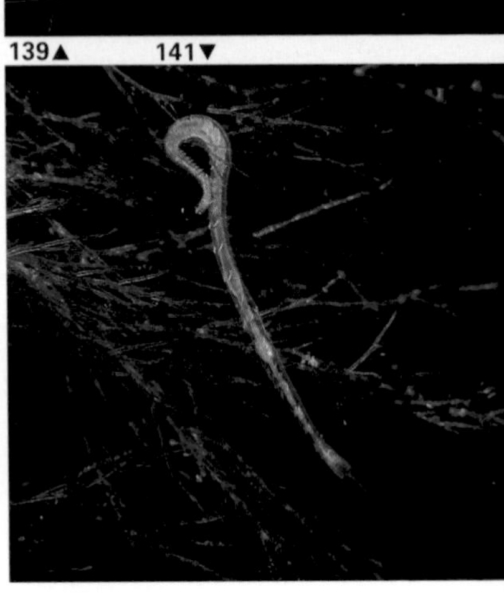

138. *Ficopomatus enigmaticus*, a tube worm, ×3
139. *Aelosoma*, ×10 OSF
140. *Chaetogaster*, ×10 OSF
141. *Stylaria lacustris*, ×4
142. *Tubifex*, ×12 OSF
143. *Lumbriculus variegatus*, ×3

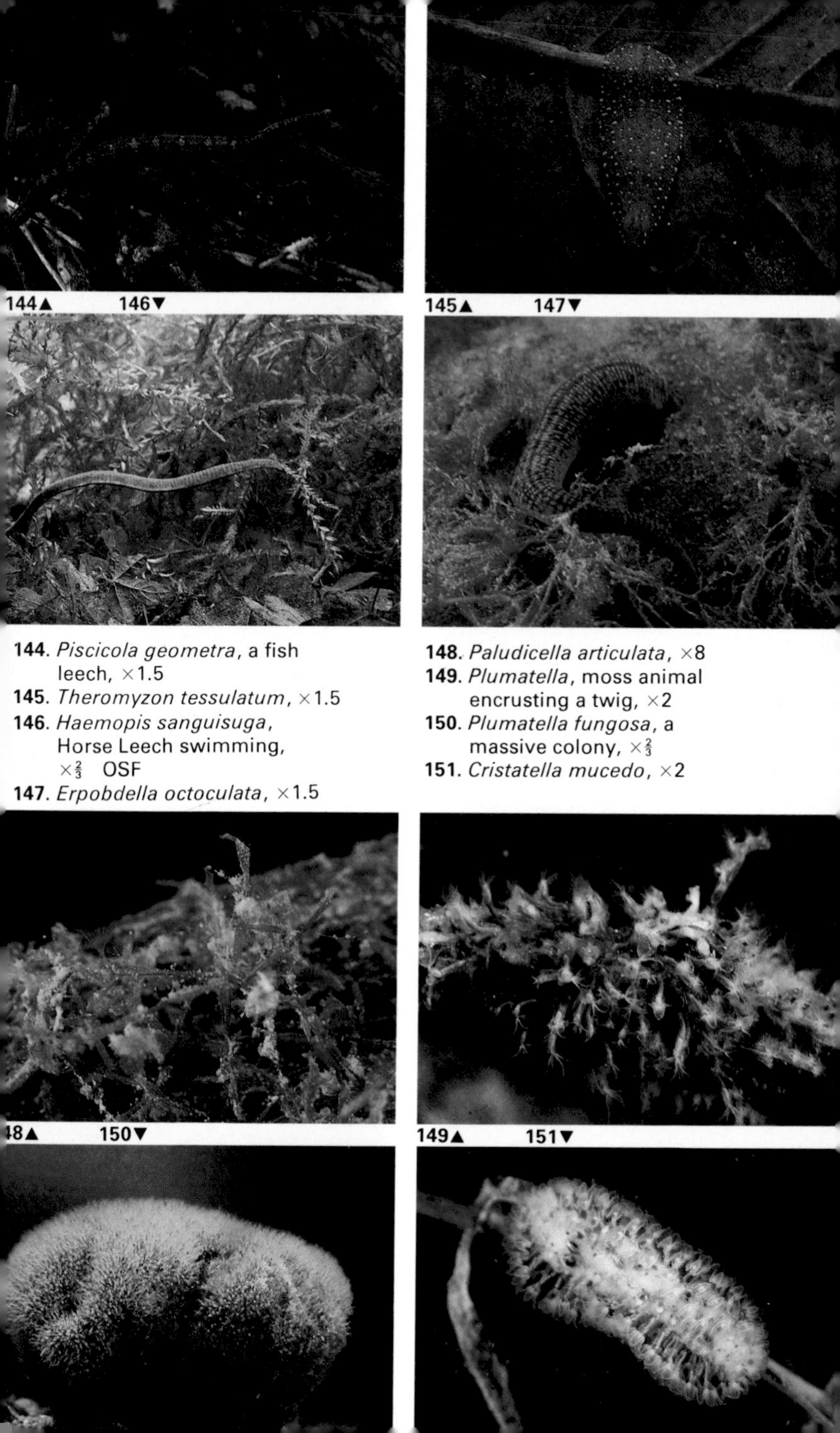

144. *Piscicola geometra*, a fish leech, ×1.5
145. *Theromyzon tessulatum*, ×1.5
146. *Haemopis sanguisuga*, Horse Leech swimming, ×⅔ OSF
147. *Erpobdella octoculata*, ×1.5

148. *Paludicella articulata*, ×8
149. *Plumatella*, moss animal encrusting a twig, ×2
150. *Plumatella fungosa*, a massive colony, ×⅔
151. *Cristatella mucedo*, ×2

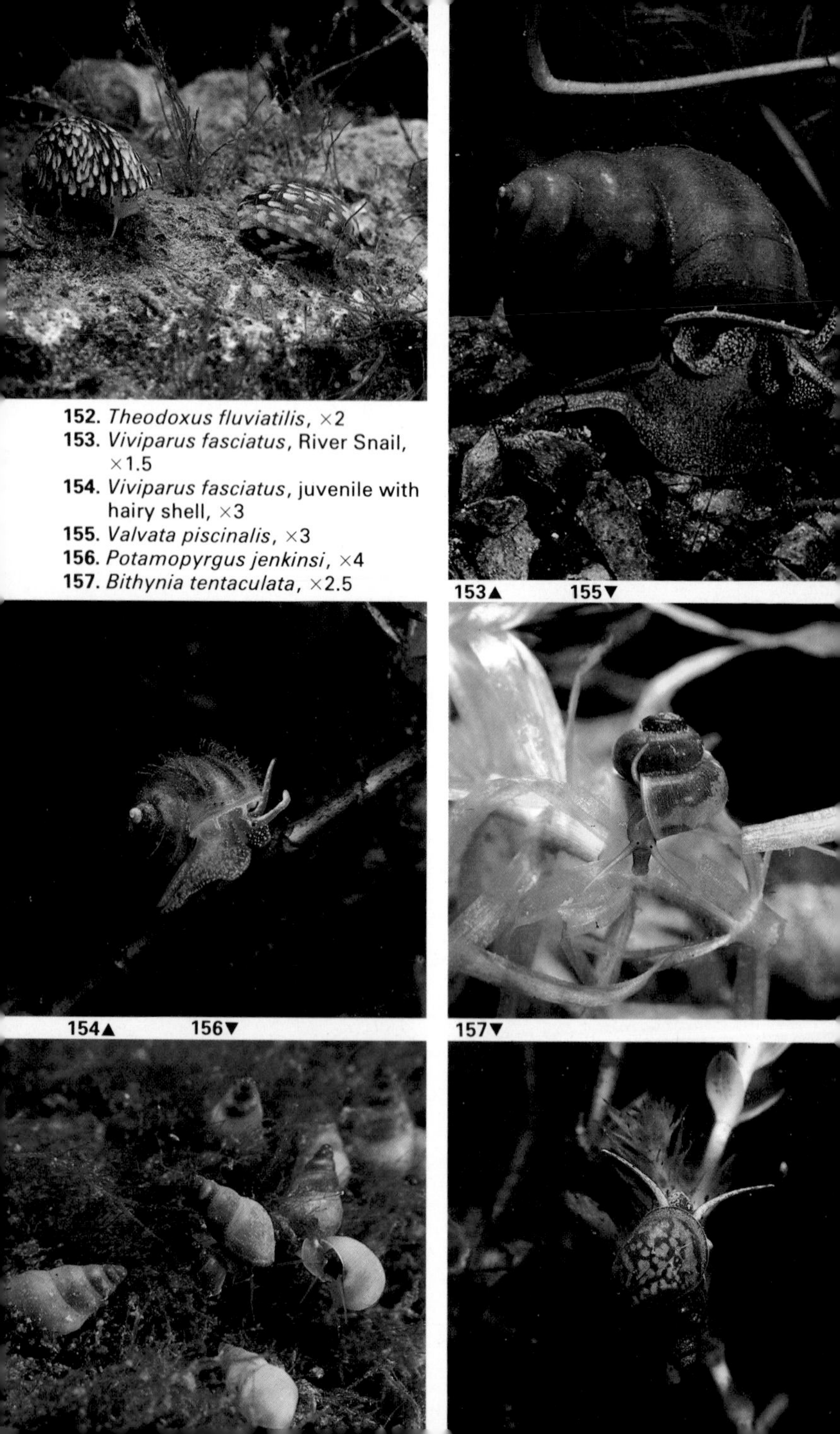

152. *Theodoxus fluviatilis*, ×2
153. *Viviparus fasciatus*, River Snail, ×1.5
154. *Viviparus fasciatus*, juvenile with hairy shell, ×3
155. *Valvata piscinalis*, ×3
156. *Potamopyrgus jenkinsi*, ×4
157. *Bithynia tentaculata*, ×2.5

153▲ 155▼

154▲ 156▼

157▼

158▲ 160▼

159▲ 161▼

158. *Lymnaea truncatula*, ×5
159. *Lymnaea stagnalis*, Great Pond Snail, ×1
160. *Lymnaea peregra*, Wandering Snail, ×2.5

162▼

161. *Lymnaea auricularia*, Ear Snail, ×1
162. *Physa fontinalis*, Bladder Snail, ×4 OSF
163. *Planorbarius corneus*, Great Ramshorn Snail, ×1.5

▼

164. *Planorbis planorbis*, ×1
165. *Planorbis vortex*, ×2
166. *Segmentina complanata*, on *Oedogonium*, ×4
167. *Ancylus fluviatilis*, River Limpet, on *Hildenbrandia*, ×2.5
168. *Acroloxus lacustris*, Lake Limpet, ×4
169. *Dreissena polymorpha*, Zebra Mussel, ×$\frac{2}{3}$

170. *Anodonta anatina*, Duck Mussel, ×1
171. *Sphaerium*, Orb Mussel, ×5
172. *Pisidium*, Pea Mussel, ×5
173. *Macrobiotus*, a water bear, ×100 OSF
174. *Hypsibius*, a water bear, ×100 OSF

175. *Argyroneta aquatica*, Water Spider with air bell, ×2 OSF

176. *Dolomedes fimbriatus*, Raft Spider, ×2.5 OSF

177. *Piona coccinea* ♀, Water Mite, ventral view, ×3

178. *Piona longipalpis*, ×3

179. Parasitic larvae of water mites on *Nepa cinerea*, ×4

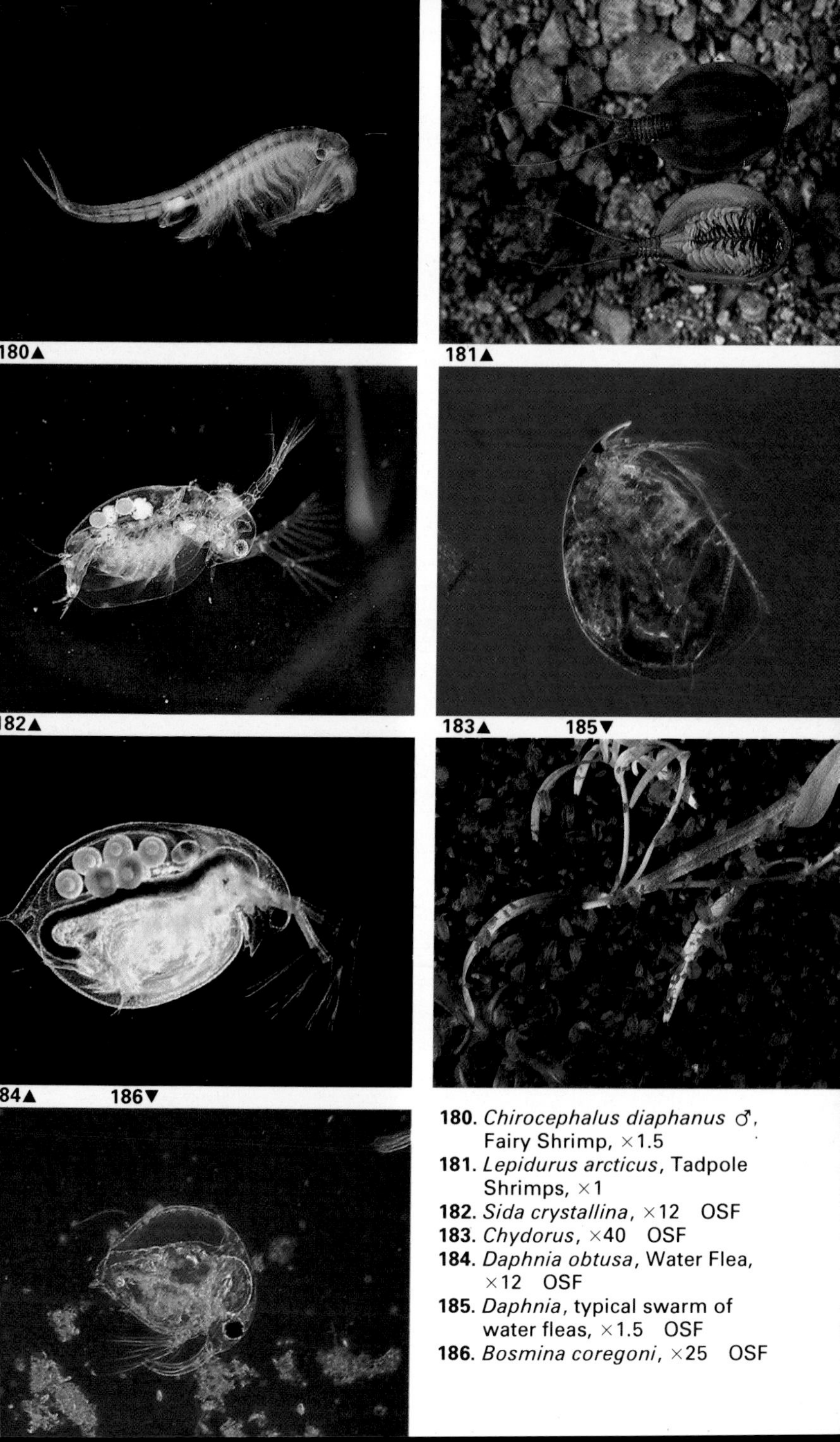

180. *Chirocephalus diaphanus* ♂, Fairy Shrimp, ×1.5
181. *Lepidurus arcticus*, Tadpole Shrimps, ×1
182. *Sida crystallina*, ×12 OSF
183. *Chydorus*, ×40 OSF
184. *Daphnia obtusa*, Water Flea, ×12 OSF
185. *Daphnia*, typical swarm of water fleas, ×1.5 OSF
186. *Bosmina coregoni*, ×25 OSF

187. Ostracods – seed shrimps, ×8 OSF
188. Harpacticoid copepod, ×50 OSF
189. Calanoid copepod, ×20 OSF
190. Cyclopoid copepod, ×20 OSF
191. *Argulus foliaceus*, Fish Louse on tail of stickleback, ×8 OSF

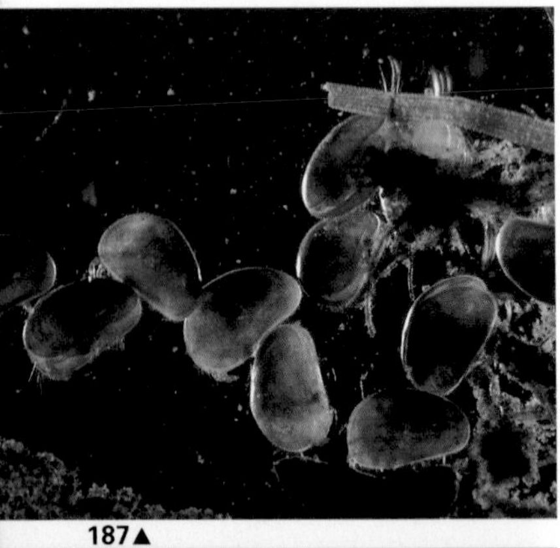

187▲

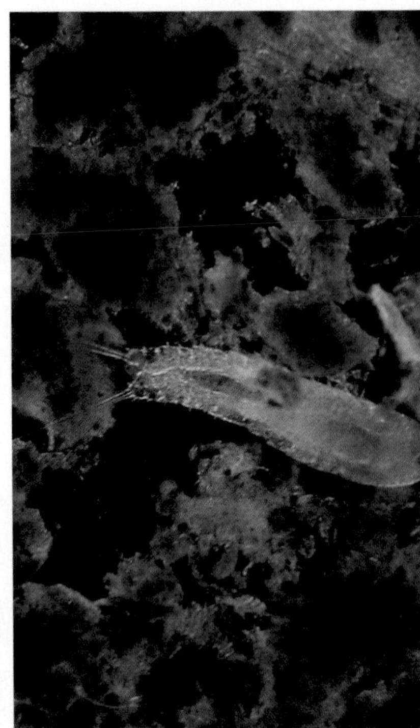

188▲

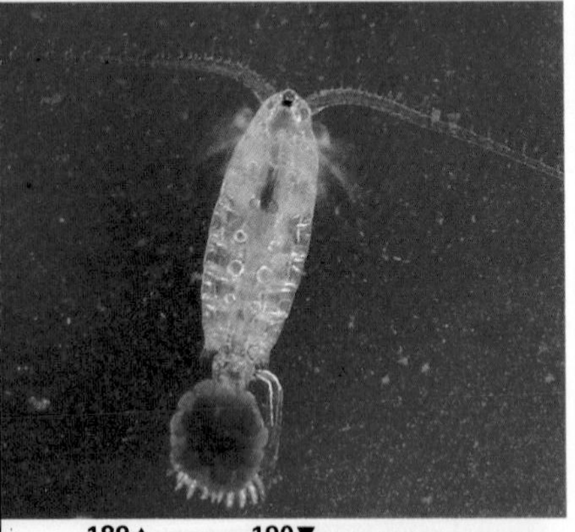

189▲

190▼

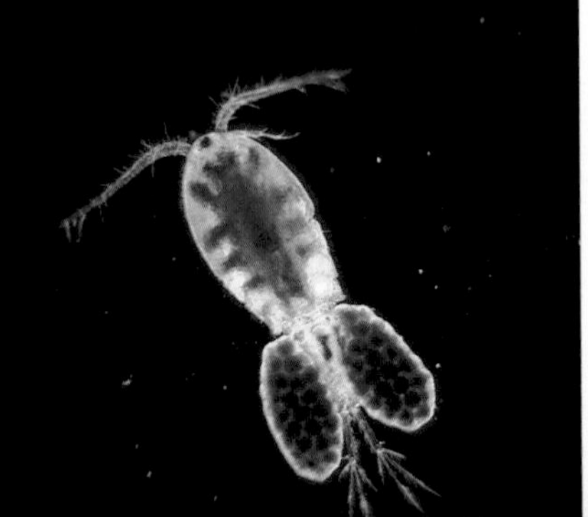

191▼

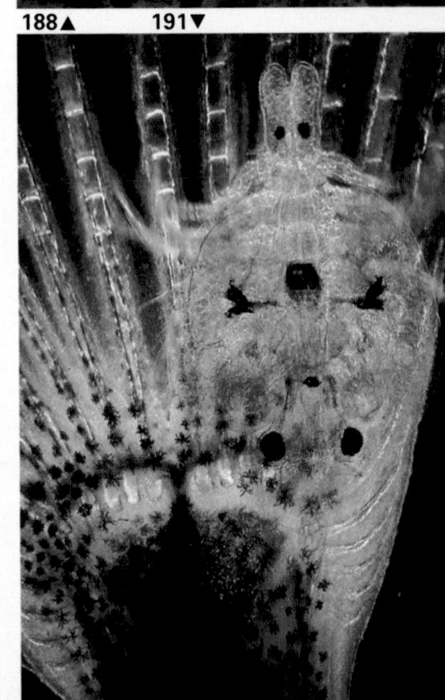

192▲ 194▼

193▲

192. *Orchestia cavimana*, Bank Hopper, ×2.5
193. *Gammarus pulex*, a freshwater shrimp, ×3
194. *Crangonyx pseudogracilis*, a freshwater shrimp, ×4
195. *Asellus aquaticus*, Water Slater, Hog Louse, ×3
196. *Austropotamobius pallipes*, Freshwater Crayfish, ×⅔

95▼

196▼

197▲ **199▼**

198▲ **200▼**

197. *Podura aquatica*, water springtails, $\times 12$ OSF

198. *Ephemera*, mayflies; dun (left), cast skin, spinner (right); $\times \frac{1}{2}$ OSF

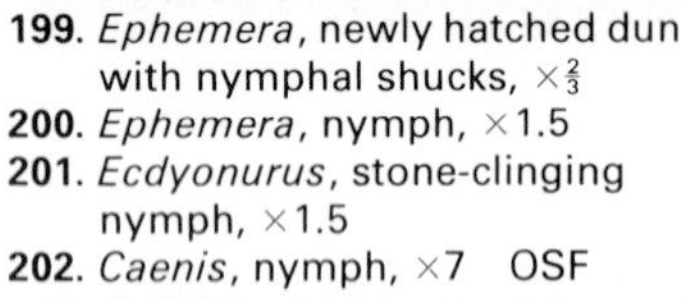

199. *Ephemera*, newly hatched dun with nymphal shucks, $\times \frac{2}{3}$

200. *Ephemera*, nymph, $\times 1.5$

201. *Ecdyonurus*, stone-clinging nymph, $\times 1.5$

202. *Caenis*, nymph, $\times 7$ OSF

201▼

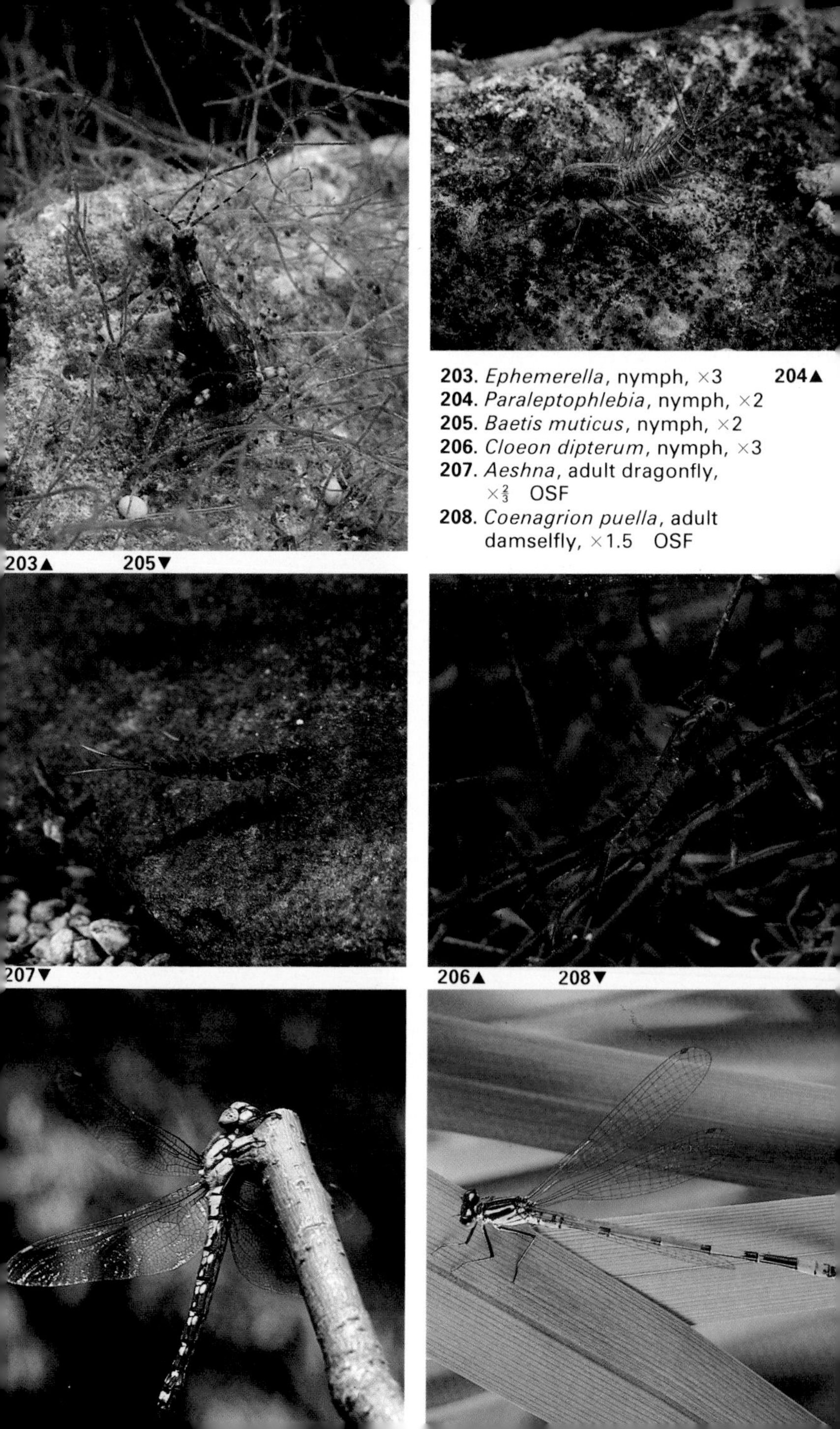

203▲ 205▼ 204▲ 207▼ 206▲ 208▼

203. *Ephemerella*, nymph, ×3
204. *Paraleptophlebia*, nymph, ×2
205. *Baetis muticus*, nymph, ×2
206. *Cloeon dipterum*, nymph, ×3
207. *Aeshna*, adult dragonfly, $\times\frac{2}{3}$ OSF
208. *Coenagrion puella*, adult damselfly, ×1.5 OSF

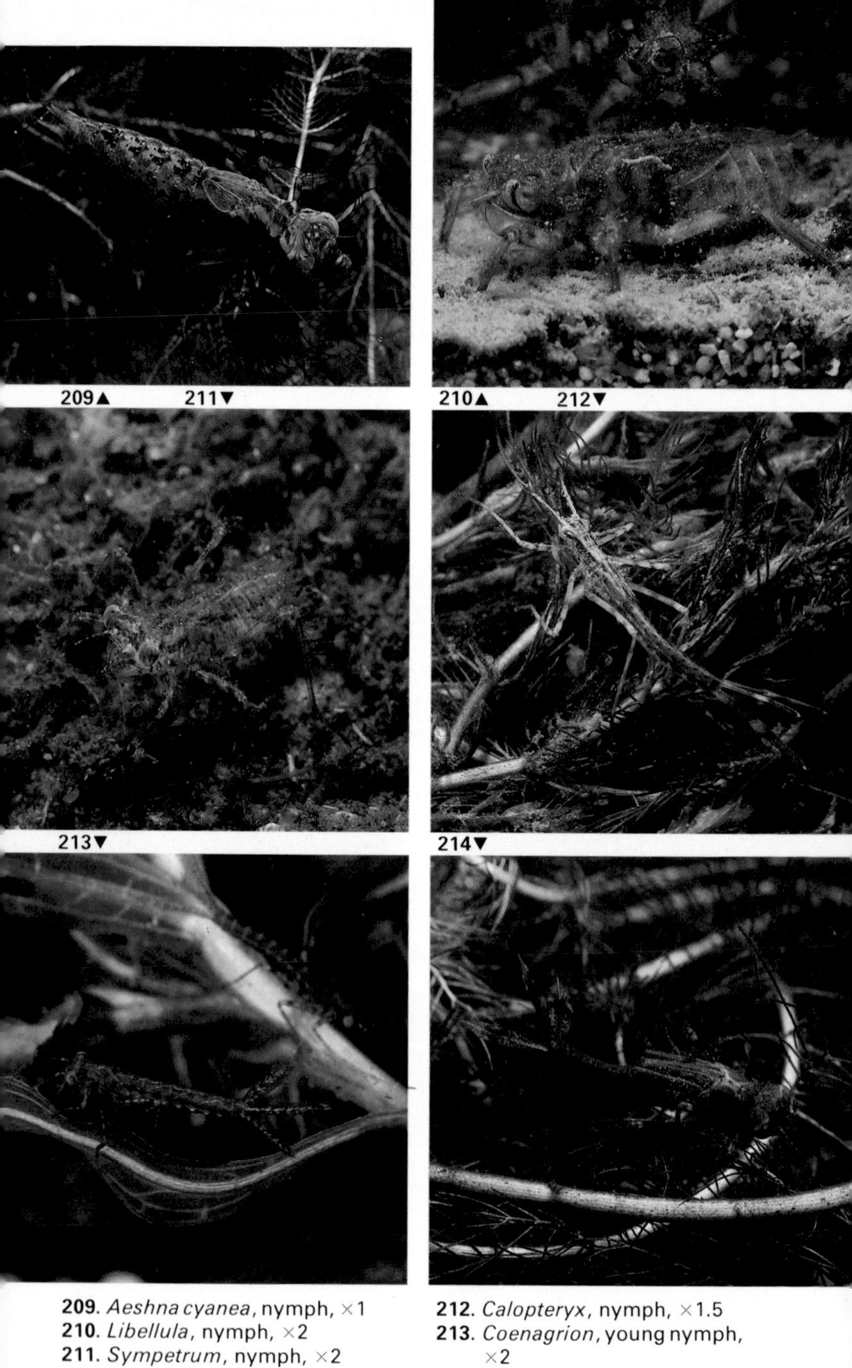

209. *Aeshna cyanea*, nymph, ×1
210. *Libellula*, nymph, ×2
211. *Sympetrum*, nymph, ×2

212. *Calopteryx*, nymph, ×1.5
213. *Coenagrion*, young nymph, ×2
214. *Pyrrhosoma*, nymph, ×1.5

15▲ 216▼ 217▶

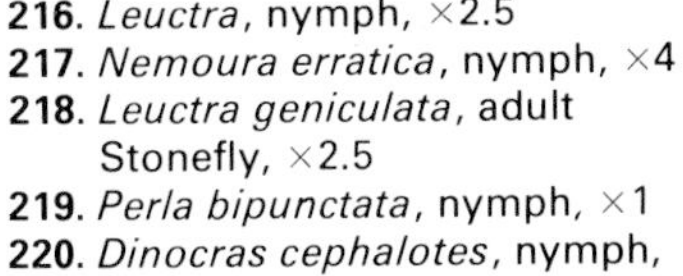

215. *Erythromma*, nymph, ×1.5
216. *Leuctra*, nymph, ×2.5
217. *Nemoura erratica*, nymph, ×4
218. *Leuctra geniculata*, adult Stonefly, ×2.5
219. *Perla bipunctata*, nymph, ×1
220. *Dinocras cephalotes*, nymph, ×1.5

218▼

219▼

▼

▲ **221.** *Hydrometra stagnorum*, Water Measurer, ×2.5 OSF
222. *Velia caprai*, nymph of Water Cricket, ×2
223. *Velia caprai*, adult Water ▼ Cricket, ×2

224. *Gerris*, adult Pond Skater, ×2 OSF
225. *Nepa cinerea*, Water Scorpion, ×1.5
226. *Ranatra linearis*, Long-bodied Water Scorpion, ×1

22

22

225▼

226▼

227▲

228▲

229▲ 231▼

230▲ 232▼

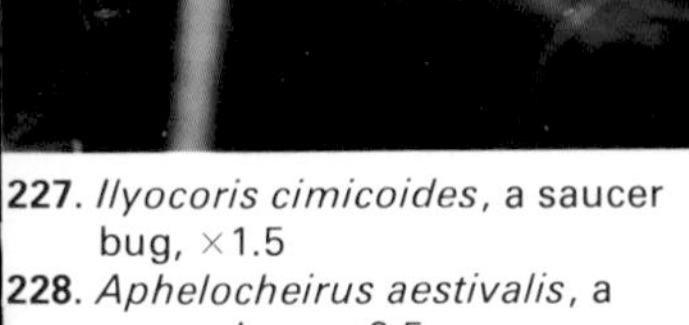

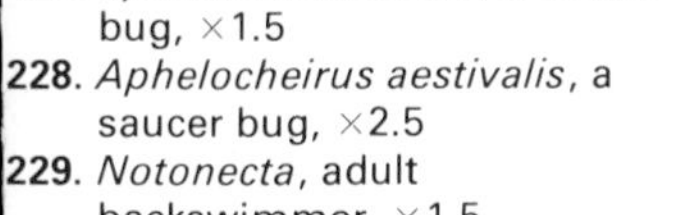

227. *Ilyocoris cimicoides*, a saucer bug, ×1.5
228. *Aphelocheirus aestivalis*, a saucer bug, ×2.5
229. *Notonecta*, adult backswimmer, ×1.5
230. *Notonecta*, nymph, ×2
231. *Plea leachii*, Lesser Backswimmer, adult, ×3
232. *Cymatia coleoptrata*, predatory Water Boatman, adult, ×4
233. *Corixa*, adult Water Boatman, ×2
234. *Corixa*, nymph, ×2

233▼

234▼

235▼

237▼

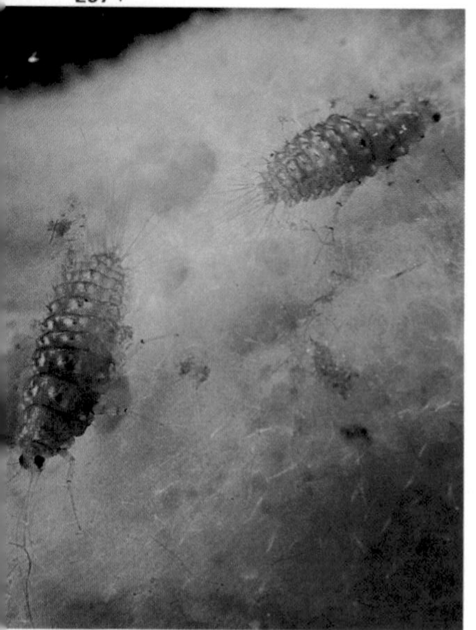

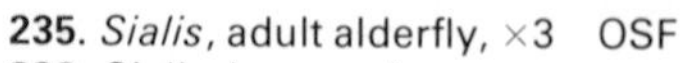

235. *Sialis*, adult alderfly, ×3 OSF
236. *Sialis*, larva, ×2
237. *Sisyra*, larva of spongefly, ×6 (on *Ephydatia*)
238. *Rhyacophila dorsalis*, free-living caddis larva, ×2.5
239. *Philopotamus montanus*, adult Sedge or Caddis Fly, ×2

236▲ 238▼

239▼

240. *Cyrnus flavidus*, polycentropid ▲ larva, ×2.5
241. *Hydropsyche*, larva, ×2.5
242. *Crunoecia*, lepidostomatid larva with square case, ×7.5
243. *Brachycentrus subnubilis*, larvae, ×2
244. *Agapetus fuscipes*, larva in stone case, ×7
245. *Agraylea*, larvae in secreted case, ×9

241▲

243▲ 245▼

242▲ 244▼

▲ **246**. *Phryganea grandis*, larva, ×1.5

247. *Limnephilus*, adult Sedge or Caddis Fly, ×1.5 OSF

248. *Limnephilus*, larva in case incorporating live snail!, ×1.5

249. *Stenophylax*, larva in stone case, ×2

250. *Leptocerus*, larvae in secreted cases, ×2

251. *Athripsodes*, pupa emerging from pupal case, ×1.5

252. *Athripsodes*, pupa at surface prior to 'hatching' into adult, ×1.5

253. Caddis pupa (unidentified) viewed through surface, ×1.5

24

248▲ 250▼

249▲ 251▼

252▼

253▼

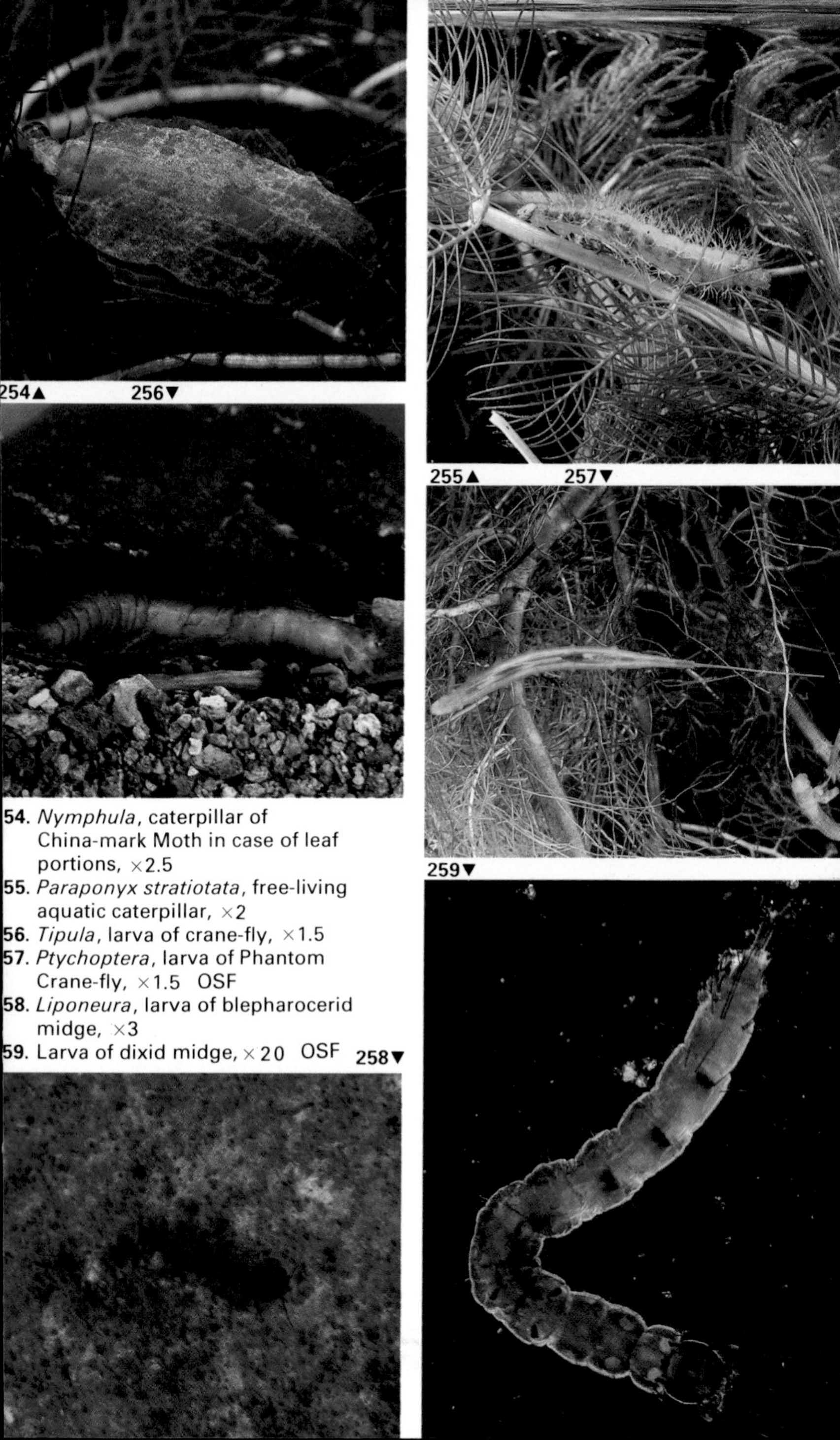

54. *Nymphula*, caterpillar of China-mark Moth in case of leaf portions, ×2.5
55. *Paraponyx stratiotata*, free-living aquatic caterpillar, ×2
56. *Tipula*, larva of crane-fly, ×1.5
57. *Ptychoptera*, larva of Phantom Crane-fly, ×1.5 OSF
58. *Liponeura*, larva of blepharocerid midge, ×3
59. Larva of dixid midge, ×20 OSF

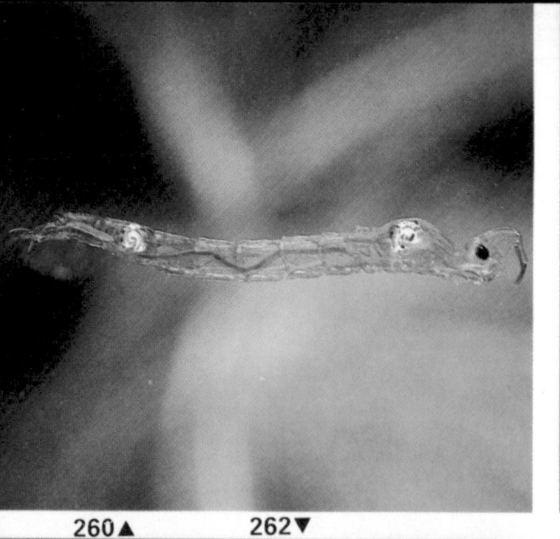

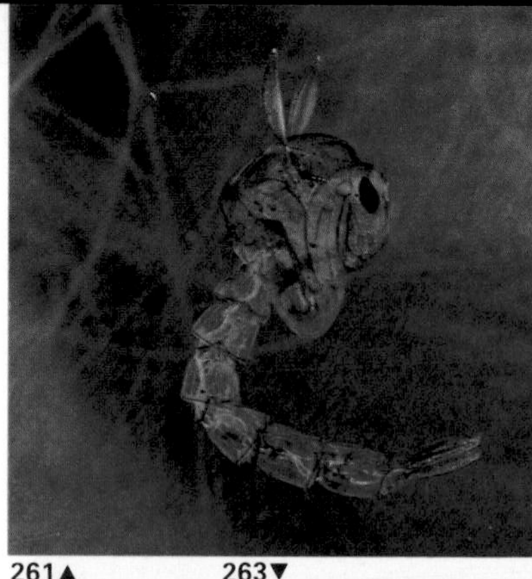

260▲ 262▼ 261▲ 263▼

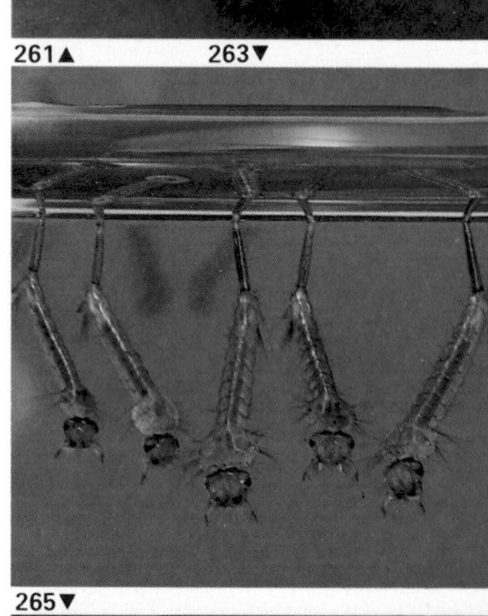

265▼

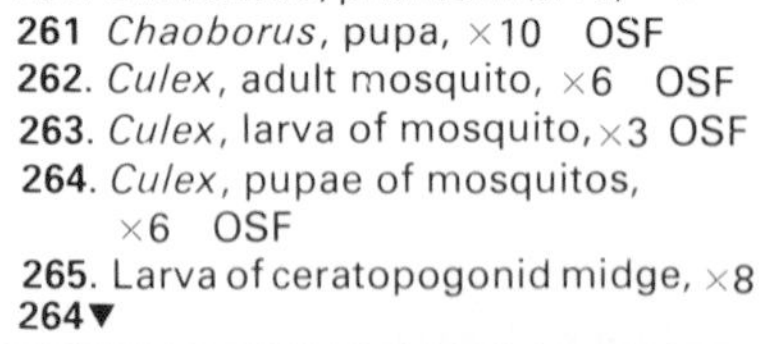

260. *Chaoborus*, phantom larva, ×4
261 *Chaoborus*, pupa, ×10 OSF
262. *Culex*, adult mosquito, ×6 OSF
263. *Culex*, larva of mosquito, ×3 OSF
264. *Culex*, pupae of mosquitos, ×6 OSF
265. Larva of ceratopogonid midge, ×8

264▼

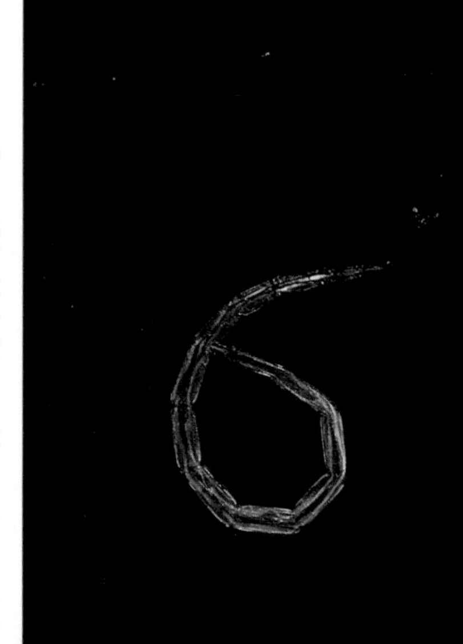

6. *Simulium*, larvae and pupae, ×2
7. Chironomid midge larva in mucus tube, ×3 OSF
8. Bloodworm larva of chironomid, ×3
9. Larva of chironomid midge, ×4
0. Pupa of chironomid midge, ×4 OSF
1. Larva of stratiomyid fly viewed through water surface, ×3

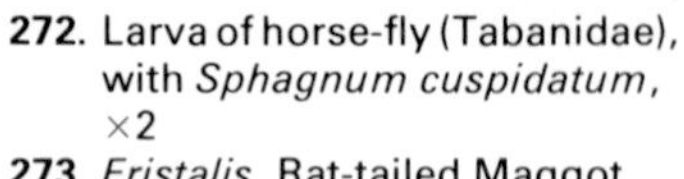

272. Larva of horse-fly (Tabanidae), with *Sphagnum cuspidatum*, ×2

273. *Eristalis*, Rat-tailed Maggot, ×2

274. *Gyrinus*, whirligig beetles on water surface, ×1.5 OSF

275. *Orectochilis villosus*, whirligig larva, ×2.5

276. *Haliplus*, adult beetle, ×4

277. *Haliplus*, larvae, ×4

279▲ 280▼

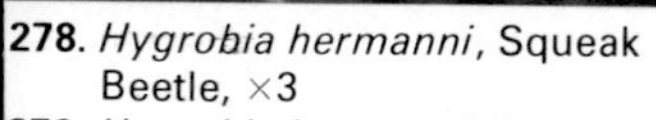

278. *Hygrobia hermanni*, Squeak Beetle, ×3

279. *Hygrobia hermanni*, larva, ×2.5

280. *Noterus capricornus*, adult ♂ ×4

281. *Laccophilus hyalinus*, adult, ×4

282. *Hyphydrus ovatus*, adult, ×4

281▼

282▼

283▲ 285▼

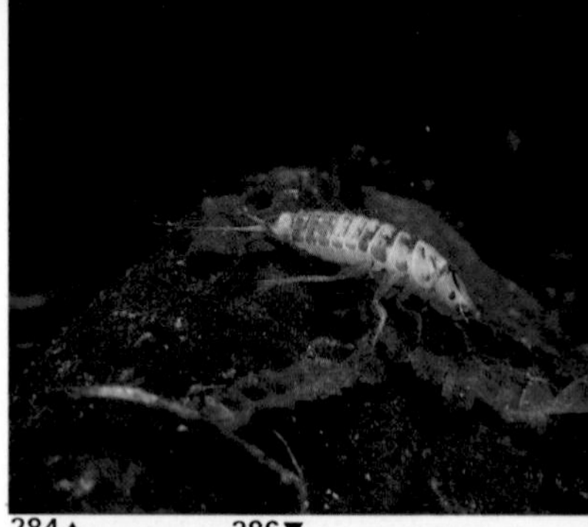

284▲ 286▼

287▼

288▼

283. *Hyphydrus ovatus*, larva, ×4 OSF

284. Hydroporine larva (unidentified), ×2.5

285. *Colymbetes fuscus*, adult, ×2

286. *Colymbetes fuscus*, larva, ×2

287. *Platambus maculatus*, adult, ×2

288. *Agabus bipustulatus*, adult, ×2

290▼

89▲ 291▼

289. *Ilybius fuliginosus*, adult, ×2
290. *Dytiscus*, adult ♂, ×2 OSF
291. *Dytiscus*, pupa, ×2 OSF
292. *Dytiscus*, larva, ×1.5
293. *Acilius sulcatus*, adult,
▼ ×2 OSF

2▼

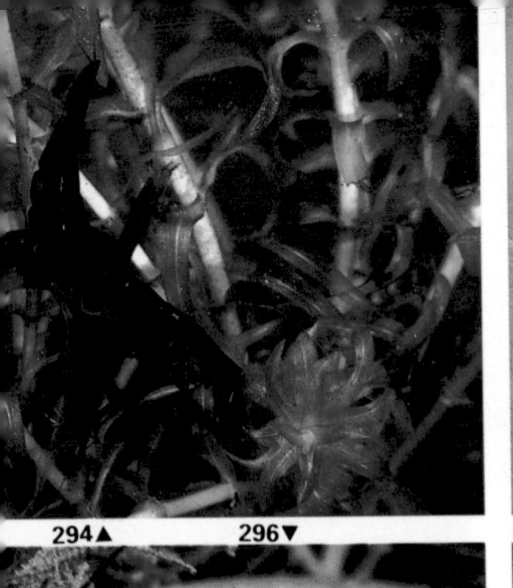

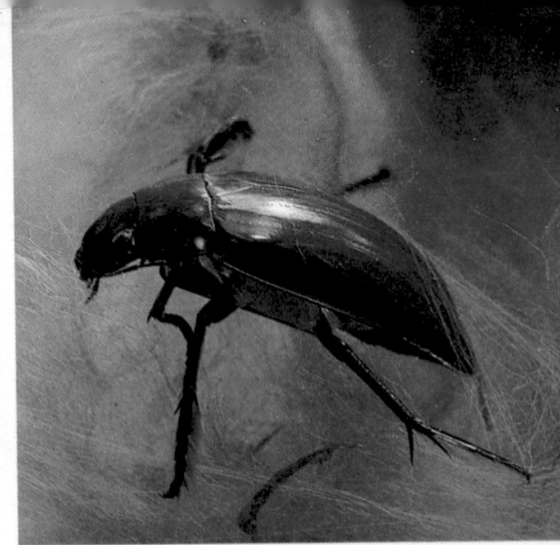

294▲ 296▼ 295▲

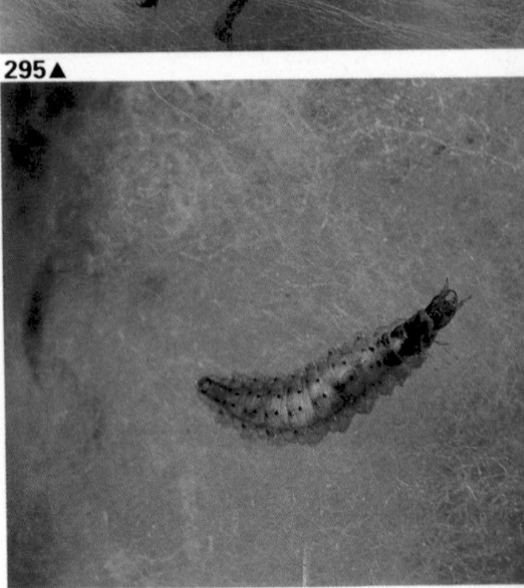

297

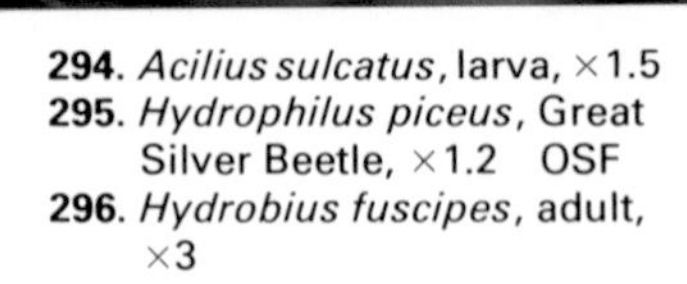

294. *Acilius sulcatus*, larva, ×1.5
295. *Hydrophilus piceus*, Great Silver Beetle, ×1.2 OSF
296. *Hydrobius fuscipes*, adult, ×3

297. *Hydrobius fuscipes*, larva, ×2.5
298. *Helophorus aquaticus*, adult, ×3
299. Larva of helodid beetle, ×3

298▼ 299

300▲ 302▼

301▲

300. *Dryops,* adult, ×3
301. *Elmis aenea*, adult, ×6
302. *Elmis aenea*, larvae, ×4
303. *Limnius volckmari*, larva, ×3
304. *Donacia*, adults, ×4 OSF
305 *Donacia*, larva, ×6 OSF

303▼

304▼

305▼

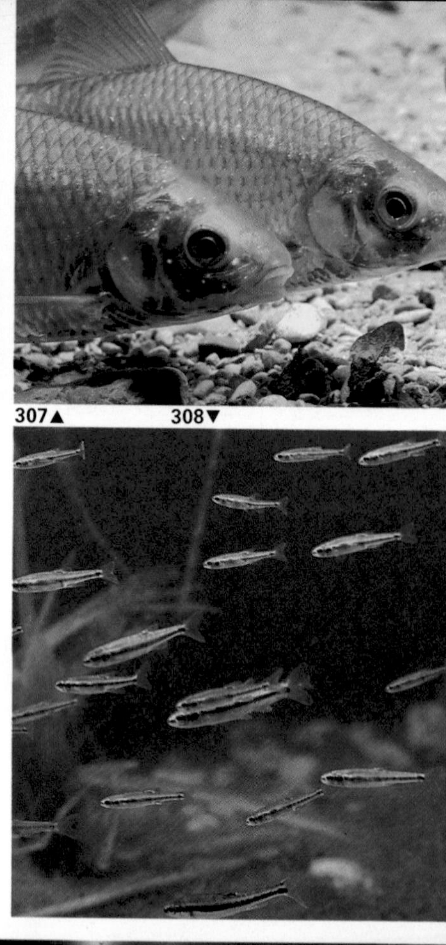

307▲ 308▼

▲ **306**. Common Carp *Cyprinus carpio* OSF
307. Roach *Rutilus rutilus* OSF
308. Minnow *Phoxinus phoxinus*, shoaling behaviour OSF
309. Tench *Tinca tinca*, juvenile OSF
▼

10▲ 312▼

311▲ 313▼

4▼

315▼

10. Perch *Perca fluviatilis* OSF
11. Three-spined Stickleback *Gasterosteus aculeatus* OSF
12. Ten-spined Stickleback *Pungitius pungitius* OSF
13. Common Newt *Triturus vulgaris*
14. Great Crested Newt *Triturus cristatus*
15. Great Crested Newt *Triturus cristatus*, larva OSF

316. Common Spadefoot *Pelobates fuscus*, burying itself in soil OSF

317. Common Toad *Bufo bufo* OSF

318. Common Toad *Bufo bufo*, spawn OSF

319. European Tree Frog *Hyla arborea* OSF

316 ▲

317 ▲

318 ▲ 319 ▼

320▲ 322▼

323▼ 321▲

320. Common Frog *Rana temporaria*

321. Common Frog *Rana temporaria*, tadpole with back legs developing OSF

322. Marsh Frog *Rana ridibunda*

323. Grass Snake *Natrix natrix* OSF

324 ▼

324. Dipper *Cinclus cinclus*, diving OSF

325. Little Grebe or Dabchick *Podiceps ruficollis* OSF

326. Kingfisher *Alcedo atthis* RSPB

325 ▼

326 ▼

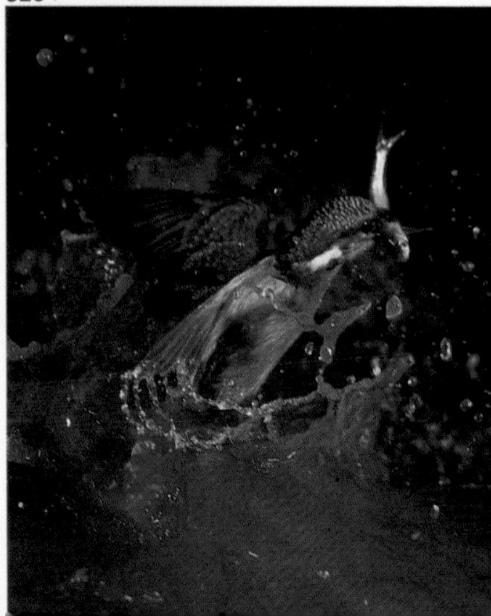

327▲ 328▼ 329▼

327. Heron *Ardea cinerea*, with lamprey RSPB
328. Mallard *Anas platyrynchos*, up-ending OSF
329. Shoveler *Anas clypeata*, feeding RSPB (Craig)

◀ **330.** Water Shrew *Neomys fodiens* OSF
331. European Beaver *Castor fiber* OSF
332. Water Vole *Arvicola terrestris* OSF

331 ▼

332 ▼

CHAPTER 20 Introduction to the Arthropods

Arthropods are animals that share the common characters of a markedly bilaterally-symmetrical segmented body; an external skeleton of more-or-less hardened (sclerotized) cuticle; and paired, jointed limbs. Apart from its regular segmentation (which is not always obvious), the body is frequently subdivided into two or three parts: head, thorax and abdomen (sometimes the first two are fused to form a cephalothorax). The cuticle consists of dead tissue which cannot grow with the body, so it is cast off (moulted) at intervals, to allow the body to grow, and is then replaced by a new layer that has formed beneath it. At first the new cuticle is soft, being stretched to a comfortable fit by expansions of the animal's body, before it finally hardens. Each stage between moults is called an instar.

By any criteria the arthropod design is immensely successful and these animals comprise a large majority of all known species. Until recently it was generally accepted that they constituted a single phylum (Arthropoda), a view still held by many authorities, but there is increasing evidence that the various groups within the arthropods are actually separate, unrelated phyla which have independently evolved a similar, highly successful body plan. Classification based on the latter arrangement is used here as it is more convenient in application and agrees better with our concept of phyla. 'Arthropod' is, however, still a useful general term, or common name, for animals of this type.

The four arthropod phyla occurring in fresh water are easily recognized:

Phylum TARDIGRADA – Water Bears (*p. 195*) 'Primitive', microscopic arthropods with indistinct segmentation and four pairs of short unjointed legs.

Phylum CHELICERATA – Spiders and Mites (*p. 198*). Body segmentation indistinct in all aquatic forms; body divided into two major parts (cephalothorax and abdomen) in spiders and some mites, or forming a single unit in other mites; four pairs of jointed, typically long legs.

Phylum CRUSTACEA – Water Fleas, Shrimps, Prawns, Crabs and Crayfish (*p. 205*). Body segmentation usually distinct; body often divided into three (not always distinct) major parts – head, thorax and abdomen; thorax with more than four pairs of jointed, often complex, legs, as well as other jointed appendages on head and (usually) abdomen. (In some groups the body and some or all limbs are hidden within an oval shell; one of these groups (Ostracoda) has only two pairs of legs.)

Phylum UNIRAMIA The only aquatic group is subphylum HEXAPODA – Insects (*p. 233*).

Adults: Body segmentation usually distinct, at least on underside; body

divided, usually clearly, into three major regions – head, thorax and abdomen; thorax with three pairs of jointed legs and (usually) with one or two pairs of membranous or partly hardened wings; other jointed appendages present on head and sometimes on last segment of abdomen.
Juveniles: Always wingless (but wing buds may be present). Some forms closely resemble their respective adults, apart from lacking wings. The others vary from soft-bodied, often legless grubs and maggots, to well-sclerotized, active larvae with well-developed legs. Body segmentation and division into head, thorax and abdomen usually distinct but thorax sometimes not differentiated from abdomen; thorax with three pairs of legs, sometimes rudimentary, or legless; abdomen sometimes bears unjointed prolegs.

CHAPTER 21 Water Bears
(Phylum Tardigrada)

Recognition features Microscopic animals with stout, indistinctly segmented bodies and four pairs of short legs terminating in hooked claws.

Tardigrades might have been specially designed for the delight and convenience of microscopists. They can be found almost anywhere, can be stored dry until required – just adding water will revive them – and they are without doubt the most engaging characters of the microscopic world.

The tardigrade body is a plump cylinder covered by a layer of flexible, translucent cuticle; this is sometimes hardened to form a number of semi-rigid plates on the dorsal surface. Four pairs of stumpy but mobile legs protrude from the body, the last pair pointing backwards. Each leg bears several hooked claws, the shape of which is important for identification (*Fig. 21.1*). The head is poorly differentiated, but the mouth and a pair of small eyespots can usually be made out; some species possess small sensory appendages. Internally a broad gut runs backwards from an oval muscular pharynx in the head region which can usually be seen by transparency. Ahead of the pharynx lies a complicated feeding apparatus consisting of a sucking tube and a pair of sharp spines (stylets) which can be protruded from the mouth to penetrate and suck out the contents of the plant cells on which most species feed.

Male tardigrades are rare and it seems likely that parthenogenetic reproduction is the rule. The large, often elaborately sculptured eggs (*Fig. 21.2*) are usually left in the cast skin after moulting.

Some species of tardigrade are truly aquatic and may be found when sorting plant material under the low-power microscope. But they are most common amongst mosses, lichens and non-aquatic algae, where they live in the water film that surrounds these plants for much of the time, and may be found wherever such plant-life occurs: on walls and tree-trunks, in the soil, in gutters, on roofs, and so on. Such habitats have an obvious snag – they periodically dry up – but tardigrades are well adapted to cope with this, they simply dry up themselves, shrivelling to form cyst-like structures, (tuns) (*Fig. 21.3*), which when re-wetted, return to life in an hour or two. This useful adaptation, known as anabiosis, has enabled tardigrades to exploit a very common and abundant habitat.

Tardigrades can be extracted from dry moss or lichens by crushing a small quantity (2–3 mm cube) between the fingers to break it up into a fine powder. This can first be searched for tuns under the microscope (but they are difficult to find) and then soaked in rain water. Any tardigrades present will soon wake up and betray their presence by their movements, along with various nematodes and rotifers which have similar habits.

Four genera of tardigrades are commonly found in fresh water and moss or lichen habitats.

Echiniscus *Macrobiotus*

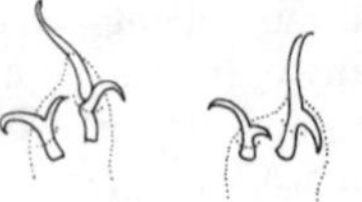

Hypsibius *Milnesium*

Tardigrade claws

Fig. 21.1

Ornamentation of tardigrade eggs

Fig. 21.2

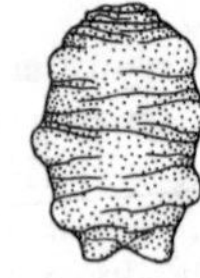

Fig. 21.3

Fig. 21.4

Echiniscus *Figs. 21.4* Cuticle of dorsal surface forming hardened plates often with a few spines or long filaments at their edges, usually bright chestnut-brown in colour. Each foot has four claws which are all similar in shape and orientation. Many species, 200–400 μm long.

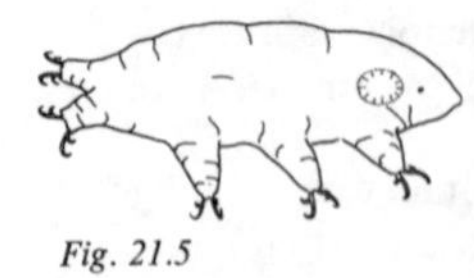

Fig. 21.5

Macrobiotus *Pl. 173, Fig. 21.5* Cuticle thin, not forming hardened plates or other ornamentation on the dorsal surface; colourless or tinted brown or pink. Each foot has two double claws which are similar in shape and size and are arranged symmetrically. This is a large genus, with numerous species, up to 1000 μm (1 mm) long; it includes several truly aquatic species and one or two that are carnivorous.

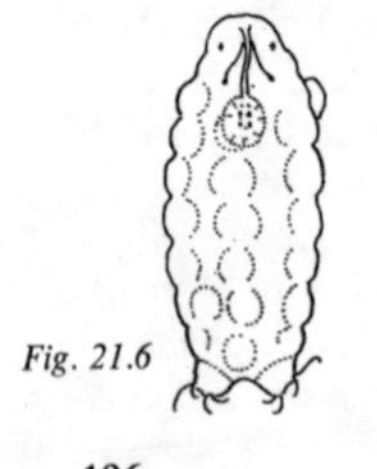

Fig. 21.6

Hypsibius *Pl. 174, Fig. 21.6* Cuticle of dorsal surface plain or ornamented with tubercles, papillae or spines, occasionally forming indistinct plates; colourless, brownish or pink, sometimes banded or striped. On each foot are two double claws which differ in size and shape and are not symmetrically arranged; sometimes one double claw is separated into two elements. Numerous species, 200–600 μm long.

Milnesium tardigradum *Fig. 21.7* The sole species of this genus is easily distinguished by the unique arrangement of the claws on its feet. The cuticle is thin and not formed into plates or tubercles. It is a large tardigrade, up to 1200 μm long, generally resembling *Macrobiotus* but the snout is long and bears sensory papillae at its tip. *Milnesium* is carnivorous, feeding on rotifers, nematodes and other tardigrades.

Reference: Morgan & King, 1976.

CHAPTER 22 Water Spiders and Mites
(Phylum Chelicerata)

Recognition features Body divided into cephalothorax and abdomen, or undivided (often globular); four pairs of long, jointed legs are present (three pairs in mite larvae). Head appendages are called palps and chelicerae: in spiders the palps are antenna-like, with the chelicerae forming curved fangs; in mites the palps are small and curved downwards beneath the rostrum, the chelicerae forming part of the feeding structure and being very difficult to see.

Chelicerates are mostly terrestrial animals, comprising such groups as the spiders, scorpions, false-scorpions, harvestmen, mites and ticks, in Europe, and several other tropical groups. In fresh water, only spiders and mites are represented. It is unlikely that anyone will have difficulty in recognizing a spider, and mites (*p. 199*) have a very characteristic form too, with most species being less than about 2 or 3 mm long.

Class ARACHNIDA
Order ARANEAE
(Spiders)

Fig. 22.1

Argyroneta aquatica Water Spider *Pl. 175, Fig. 22.1* This is the only truly aquatic spider, living submerged for virtually all its life, although making occasional trips to the surface for air. The body is dark greyish-brown and the abdomen is covered with fine hairs which give it a velvety texture. Unusually for spiders, the male, with a body length of up to 16 mm, is larger than the 11 mm female.

Water spiders build underwater nests by spinning a domed sheet of silk amongst water plants. This is filled from beneath with air carried down by the spider. When underwater, the spider's abdomen is surrounded by a glistening, silvery envelope of air trapped in the hairs which is brushed off to fill the nest. Each spider has its own nest but the eggs are laid in the female's; she guards them until they hatch and the young spiders disperse.

This species is common and widely distributed in Britain and Europe. It is mostly found in ponds and ditches where there is a dense growth of vegetation.

Dolomedes fimbriatus Swamp Spider, Raft Spider *Pl. 176*
Although not strictly aquatic, this impressive spider is always found in close association with water, which it frequently enters. It is one of Europe's largest spiders with a body up to 20 mm long, and long stout legs. The body is chocolate-brown, distinctively marked with a broad, yellowish stripe along each side.

Dolomedes is found mostly on bogs, swamps or weedy pools, where it lurks amongst surface vegetation. When alarmed it will dive into the water and remain submerged for some time. It may also enter the water to hunt small aquatic animals, including fish, for food. The female spins a spherical cocoon about 10 mm in diameter, in which the eggs are laid, and she carries it slung beneath her body until the eggs hatch. The bite of this spider is said to be mildly toxic to man.

This is a widely distributed species in Europe, though rather local.

Order ACARI
(Mites)

Recognition features Small arthropods, typically 0.5–2 mm long, occasionally up to 7 mm, with oval bodies which are often brightly coloured; adults with four pairs of six-jointed legs.

Key to Suborders of Aquatic Mites:

1. a) Eyes present; body not divided into regions.
   Suborder HYDRACARINA (below)
   b) Eyes absent; body divided into cephalothorax and abdomen.
   Suborder ORIBATEI (*p. 204*)

Suborder HYDRACARINA
(Water Mites)

Water mites are common inhabitants of fresh water and are found in a wide range of habitats. Despite their frequently bright colouration they are generally inconspicuous animals and have received less attention from biologists than their obvious ecological importance merits. More than 300 British and 900 European species are known.

The body is typically oval or circular in outline, more or less deep (varying from globular to flattened) and occasionally the posterior region is elongated. The texture of the dorsal surface (dorsum) varies from soft and 'squashy', to leathery but still flexible, or sclerotized and hard; some species possess chitinous plates (sclerites) on the dorsum. Two eyes are present and each consists of two elements (eyespots) set fairly close together; these 'double-eyes' may be well separated or close together near the mid-line of the body. The colouration of the dorsum, which may be red, green, yellow, blue, etc., is often enhanced by a pattern of light or dark markings, largely caused by the internal organs showing through the skin; these are somewhat variable and unreliable for identification.

The ventral surface of the body bears a number of chitinous elements (*Fig. 22.2*) whose shape and arrangement are of importance in classification. The mouthparts are combined to form a small structure, the capitulum, from which the jointed palps arise. The palps are usually held curved downwards, ready to seize the prey or grasp a convenient substratum. In some genera the capitulum projects forwards to form a rostrum, visible from above; otherwise the capitulum is only visible from below. Each of the eight legs arises from a single sclerite (epimeron). These epimera tend to join together, either forming four groups of two (*Fig. 22.2*), or fused to form a single large plate (*Fig. 22.5*); in either case, the first pair may be attached to the capitulum. Behind or between the epimera lies the genital region which usually bears two genital plates and a number of small chitinous discs (acetabulae). Although the structure of the genital region may vary between the sexes, its outline and position is constant for each species and is a useful identification feature. Sclerotized parts of the body are stippled in the drawings.

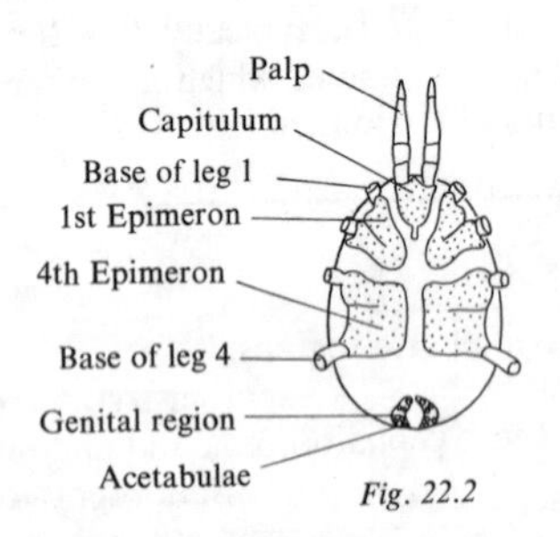

Fig. 22.2

Many mites can swim and in these the legs are equipped with swimming hairs, long hairs arranged in 'combs' or 'brushes'. Other mites have spines or hairs on their legs which are not specially arranged to aid swimming (even though their owners may be good swimmers!): these are omitted from the figures (except *Fig. 22.16*).

The orange or red eggs of water mites are enclosed within a matrix of firm jelly attached to any firm surface. Some species bore into plant stems and insert their eggs into the holes. The newly-hatched larvae, all less than 1 mm long, have only six legs and a greatly enlarged capitulum. These larvae seek out and attach themselves to various hosts, usually aquatic insects, on which they live parasitically. They are visible as small oval bodies, usually red.

yellow, or orange in colour, anchored to the host's cuticle by their capitulums. The larvae eventually enter a resting stage similar to an insect pupa, during which they moult, emerging as free-living nymphs which generally resemble the adult and have eight legs, but lack the external genitalia. Nymphs must undergo a further resting stage before they moult and become fully adult.

Adult water mites and nymphs are fierce predators, attacking small crustaceans, worms and various insect larvae, which they grasp with their palps and pierce with their capitula to suck out the body contents. Some species of one genus (*Pentatax*) live parasitically within freshwater mussels.

Most mites are so small that any attempt at identification requires the use of a good binocular microscope, and precise identification is a very complex procedure. The following selection of genera is grouped and identified by rather superficial characters and not arranged in families except for the rather odd family Limnohalacaridae (*p. 204*). Water mites are usually very active and must be preserved before they can adequately be studied.

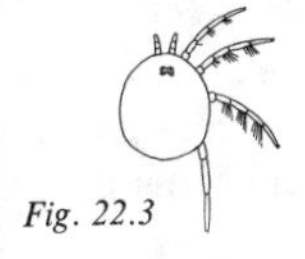
Fig. 22.3

A Eyes close together near midline of body

Eylais *Fig. 22.3* Body soft and globular, always bright red; swimming hairs on legs 1 (feeble), 2 and 3, absent from 4; numerous species from 3.5–7 mm long.

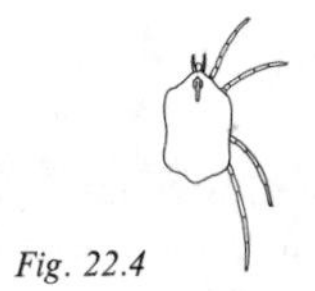
Fig. 22.4

Limnochares *Fig. 22.4* Body soft and irregularly rectangular, with the capitulum forming a small, protruding rostrum; legs without swimming hairs; sole species *L. aquaticus*, dull red and about 4 mm long, a sluggish bottom-dweller.

AA Eyes well separated, often near margin of body

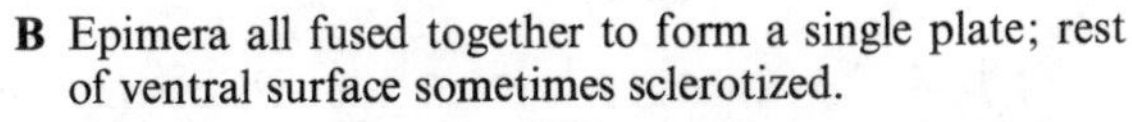

B Epimera all fused together to form a single plate; rest of ventral surface sometimes sclerotized.

Fig. 22.5

Lebertia *Fig. 22.5* Dorsum leathery; swimming hairs present but rather sparse in some species; about seventeen species ranging from 1–2.5 mm long, very varied in colour and often patterned; mostly in running water.

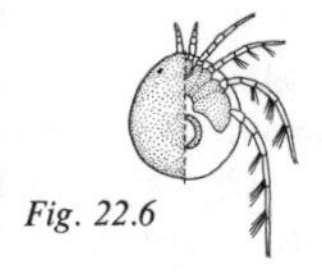
Fig. 22.6

Midea *Fig. 22.6* Dorsum strongly sclerotized and convex; swimming hairs present; sole species *M. orbiculata*, bluish or greenish and about 0.8 mm long.

Mideopsis Similar to *Midea* but body markedly flattened and disc-shaped; two species, both about 1 mm long, yellow or greenish with a dark pattern.

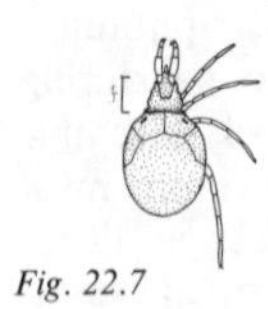
Fig. 22.7

Torrenticola *Fig. 22.7* Dorsal and ventral surfaces strongly sclerotized, dorsum consisting of five plates, anterior part of fused epimera projecting forward, forming a broad flange (f) on which the capitulum is mounted; legs without swimming hairs; many species all around 0.75–1 mm long, usually found in running waters.

BB Epimera forming four groups (1 & 2 and 3 & 4 joined on each side), first pair may be joined to capitulum.

C Dorsum hard, formed of many minute plates; posterior part of body of ♂ produced into a caudum.

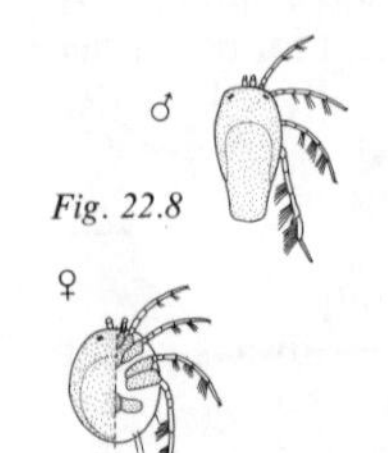

Fig. 22.8

Arrenurus *Fig. 22.8* A large and distinctive genus with marked sexual dimorphism: females are oval or circular in shape, with a circular dorsal groove forming a complete ring; males have the dorsum produced posteriorly, this part being circumscribed by the dorsal groove. Both sexes have swimming hairs. There are many species, mostly green or bluish in colour, sometimes red, ranging from 0.7–1.3 mm long.

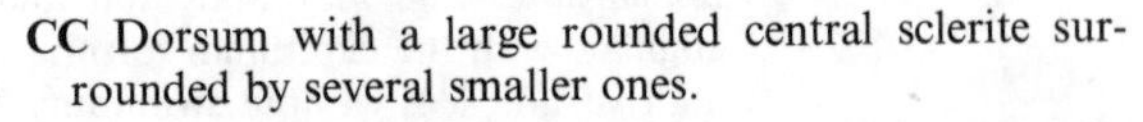

CC Dorsum with a large rounded central sclerite surrounded by several smaller ones.

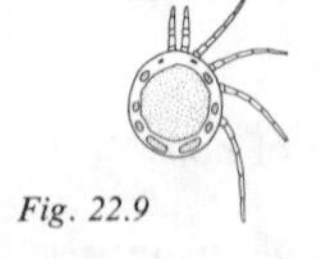
Fig. 22.9

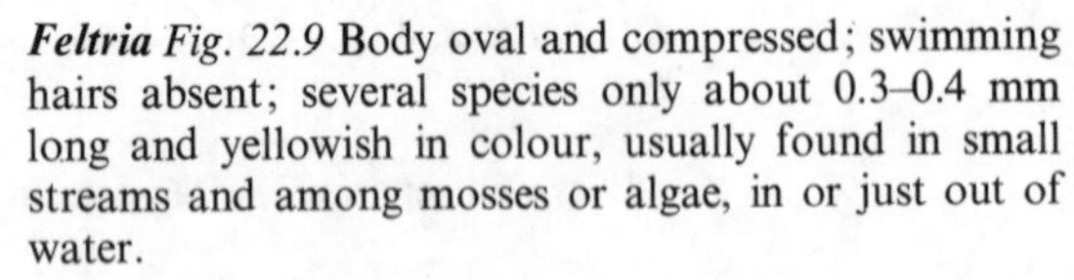

Feltria *Fig. 22.9* Body oval and compressed; swimming hairs absent; several species only about 0.3–0.4 mm long and yellowish in colour, usually found in small streams and among mosses or algae, in or just out of water.

CCC Dorsum soft, with at most a few small sclerites.

D Capitulum forming a rostrum visible from above.

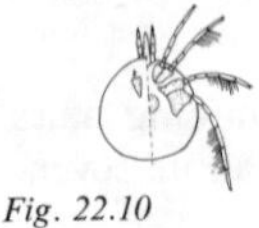
Fig. 22.10

Hydrachna *Fig. 22.10* Body globular with a small rostrum; swimming hairs strongly developed; about six species, 0.7–2 mm long, red or yellowish with dark markings.

Fig. 22.11

Sperchon *Fig. 22.11* Rostrum relatively large compared to *Hydrachna* (above); swimming hairs absent; about a dozen species, 1–2 mm long, yellow, brown or red; typically sluggish bottom-dwellers in fast streams and cold springs.

DD Capitulum small, not visible from above.

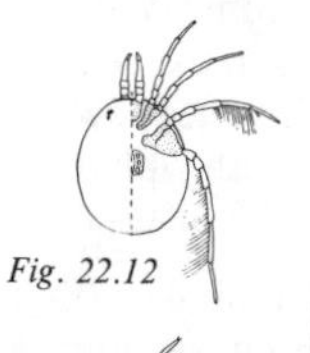
Fig. 22.12

Limnesia *Fig. 22.12* Body soft and globular, usually red or yellowish with a dark pattern; long swimming hairs present on legs 3 & 4, leg 4 lacks claws at tip; five species between 0.7 and 2 mm long.

Fig. 22.13

Hydraphantes *Fig. 22.13* Body rather compressed and always red; a single sclerite of moderate size is present between the eyes; swimming hairs well developed; six species all about 2 mm long.

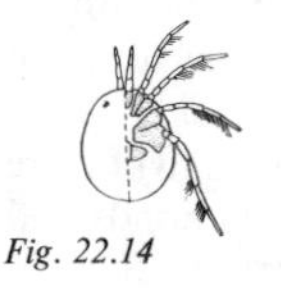
Fig. 22.14

Piona *Pls. 177–178, Fig. 22.14* A large genus of about twenty species which lacks any obvious identification characters. The concave hind border to the fourth epimeron is a constant feature but this also occurs in a few other genera; acetabulae are numerous, numbering from eight to about ninety on each side of the genital area, according to species; swimming hairs are present but often sparse. Most species are between 0.6 and 3 mm long and the range of coloration is wide – red, blue, green, brown, yellow, etc., often with dark markings.

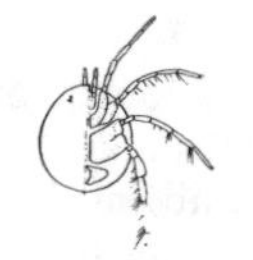
Fig. 22.15

Neumannia *Fig. 22.15* Body soft, oval; genital area with numerous (more than fifteen) acetabulae on each side; colouration and other features often resemble *Pentatax* (below) or *Piona* (above) from which this genus is distinguished by the position and shape of the structures on its ventral surface; several species up to 1.5 mm long.

***Pentatax* (*Unionicola*)** *Figs. 22.2, 16* Body oval and soft; legs longer than in most other water mites, first pair often thicker than the rest; swimming hairs very sparse, but the legs also bear many long spines and the mites can swim; several species, some of which live parasitically inside freshwater mussels (*Unio*, *Anodonta*), yellow or greenish with dark markings and 0.5–1.5 mm long.

Fig. 22.16

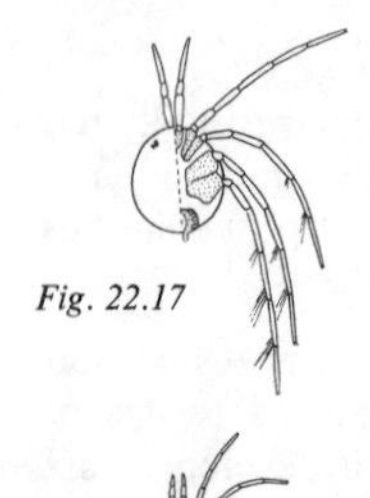

Fig. 22.17

Fig. 22.18

Hydrochoreutes *Fig. 22.17* Body oval and soft; legs very long; this genus is best distinguished from *Pentatax* (above) by its longer palps and differences in the ventral structures: males have a hook on the third leg; swimming hairs sparse; two species, both yellowish with dark markings, 1–1.7 mm long.

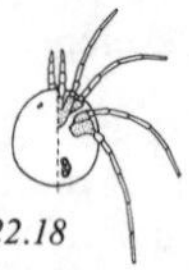

Hygrobates *Fig. 22.18* Body soft and globular; first epimera fused to capitulum; legs without swimming hairs; several species are known, mostly found in fast streams or cold springs, typically yellowish or greenish with dark markings, 1–2.5 mm long.

Family Limnohalacaridae *Fig. 22.19*
These are small mites, none exceeding about 0.75 mm in length, which are only distantly related to the other Hydracarina. They are distinguished from them by the prominent rostrum, the presence of four sclerites on the dorsum. and their weak legs which are attached to the side of the body, not underneath; none of them can swim. They are found crawling amongst algae or in detritus in ponds and streams; some occur in subterranean waters. Several genera are found in Europe but none is well known.

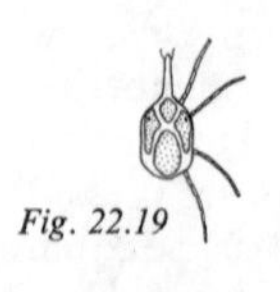

Fig. 22.19

Fig. 22.20

Suborder ORIBATEI
(Beetle Mites)

These mites (*Fig. 22.20*) have hard, often hairy, beetle-like bodies about 0.5–1 mm long, with a marked division of the body into cephalothorax and abdomen. The legs are strong with large claws for clinging; eyes are absent. Most beetle mites are terrestrial but a few occur in fresh water where they feed on surface plants such as duckweed (*Lemna*) or shallow-water species such as bog mosses (*Sphagnum*). *Hydrozetes* is the best-known genus.

References: Soar & Williamson, 1925, 1927, 1929. Hopkins, 1961. Michael, 1884, 1888.

CHAPTER 23 Crustaceans
(Phylum Crustacea)

Recognition features Arthropods with clearly-segmented bodies* which are usually also divided into head, thorax and abdomen; often partly or wholly enclosed by a curved or folded cuticular plate, the shell or carapace; head with two pairs of antennae; thorax with five or more pairs of variously-adapted legs (only two pairs in Ostracoda); abdomen with or without paired limbs; antennae and paired limbs often two-branched.

Crustaceans exhibit a diversity of form and numerical abundance of species and individuals rivalled only by the insects. Unlike insects, however, crustaceans are fundamentally aquatic animals (breathing by gills or direct exchange of gases through the body surface) with great numbers of marine and freshwater species, but relatively few on land. In fresh water they range in size from tiny 'entomostracans' – a general term for all non-malacostracan (*p. 223*) crustaceans – less than 0.5 mm long, to the large lobster-like crayfishes which may attain a length of 15 cm or more.

The crustacean body (see *Fig. 23.52*) consists of a variable number of segments that tend to form three regions – head, thorax and abdomen – although these are not always distinct. In some groups (e.g., Copepoda) there is a tendency for one or more of the thoracic segments to fuse with the head, forming a rigid cephalothorax. Most segments bear a pair of appendages which are frequently biramous (two-branched). In many groups the thorax is enfolded by a single large plate, the carapace or 'shell', originating from the head and usually attached to some or all of the thoracic segments. Sometimes the carapace encloses the whole body. The chitinous skin or cuticle is typically impregnated with calcium salts for strength and rigidity, particularly on the carapace. This is not an obvious feature in most freshwater crustaceans, due to their small size, but it is easily observed and felt in crayfishes and crabs.

The first two pairs of appendages on the head are antennae, denoted as either first and second antennae, or antennules and antennae, respectively. Antennae are basically sensory structures but in some groups are modified for swimming, or as copulatory claspers. The remaining head appendages are

* In some small forms – e.g., Cladocera and Ostracoda – the body is more-or-less entirely enclosed by the carapace and the segmentation may be difficult to discern (it is obsolete in the Ostracoda): but these forms are unmistakably crustacean, their jointed legs distinguishing them from small mussels. Some of the parasitic copepods have greatly modified bodies, with progressive loss of segmentation and limbs – see *Figs. 23.39, 40*.

mouthparts – mandibles, and first and second maxillae – and are generally inconspicuous. Most species have eyes on the head which are either set into the cuticle (sessile) or mounted on stalks. Eyes are usually paired, but in some of the smaller forms are single.

Each segment of the thorax bears a pair of appendages which are clearly different in structure from those of other regions, and often differ amongst themselves. They usually form walking or swimming legs and sometimes their tips are formed into 'pincers' (chelae) or other grasping structures. In many groups there are plate-like gills tucked away between the leg bases on the ventral surface of the thorax.

The form of the abdomen varies widely but it is typically a tapering cylinder, suffering various degrees of reduction in certain groups. Paired abdominal appendages may be present or absent; if present they are usually flattened and with hairs, beating continuously or intermittently to draw a current of water over the gills. They may also be used for swimming. The posterior part of the abdomen is often differentiated as a distinct caudal region (the urosome) with appendages called uropods. The urosome forms, for instance, the tail-fan of many malacostracans, which is used in their classic backward-darting escape reaction (and is thus responsible for the large muscles powering this movement – the main gastronomic attraction of crustaceans). The terminal abdominal segment is the telson, which may be forked (caudal furca), with fixed or movable branches (cercopods or cerci).

With a few exceptions the female carries her eggs in a brood-pouch or egg-sac, or simply glued to her abdominal limbs, until they hatch. Most newly-hatched crustaceans resemble their parents in shape but in some groups – Phyllopoda and Copepoda, for instance – the larva is a specialized type called a nauplius (*Fig. 23.36*), and in the Decapoda several different larval stages occur. Nauplii have a characteristic kite-shaped body and jerky swimming action. They start life with only three pairs of limbs, but more limbs and body segments are added at each moult until the adult form is reached.

Five classes, including about eighteen orders, of crustaceans are present in European fresh water.

A Simplified Classification of Freshwater Crustaceans

Phylum CRUSTACEA	
Class BRANCHIOPODA	
Order ANOSTRACA*	Fairy shrimps
Order NOTOSTRACA*	Tadpole shrimps
Order CONCHOSTRACA*	Clam shrimps
Order CLADOCERA	Water fleas, 'daphnia'
Class OSTRACODA	Seed shrimps

Class COPEPODA	'Copepods'. water fleas
Order HARPACTICOIDEA	
Order CALANOIDEA	
Order CYCLOPOIDEA	includes some parasitic forms
Order POECILOSTOMATOIDA } Order SIPHONOSTOMATOIDA }	wholly parasitic forms
Class BRANCHIURA	Fish lice – wholly parasitic
Class MALACOSTRACA	
Superorder SYNCARIDA*	
Order BATHYNELLACEA	Tiny subterranean crustaceans
Superorder PERACARIDA	
Order MYSIDACEA	Opossum shrimps
Order AMPHIPODA	Freshwater shrimps, hoppers
Order ISOPODA	Water slaters, hog-lice
Superorder EUCARIDA	
Order DECAPODA	
[Suborder NATANTIA – Swimming decapods]	
Infraorder CARIDEA	Prawns
[Suborder REPTANTIA – Walking decapods]	
Infraorder ASTACURA	Crayfishes
Infraorder BRACHYURA	Crabs

Representatives of nearly all crustacean groups occur in the sea, only the four groups marked * being found exclusively in fresh water: purely marine groups are omitted.

Key to the Groups of Freshwater Crustacea

1. a) External parasites of fishes (often greatly modified and may not be recognizable as crustaceans). ... 2
   b) Free-living crustaceans. ... 3

2. a) Body markedly flattened and disc-like, with a pair of circular suckers on its ventral surface (usually visible by transparency).
   Class BRANCHIURA (*p. 222*)
   b) Body variously formed: if flattened, not disc-like, and without ventral suckers.
   Class COPEPODA (parasitic forms) (*p. 220*)

3. a) Carapace absent. ... 4
   b) Carapace present, covering the thorax, sometimes also the head and/or abdomen. ... 8
   c) Carapace present but greatly reduced, not enclosing thorax, forming a small dorsal brood-pouch.
   Order CLADOCERA (aberrant families Polyphemidae and Leptodoridae) (*p. 217*)

4. a) Thorax with eleven (or more) pairs of limbs; eyes stalked.
Order ANOSTRACA (*p. 209*)
   b) Thorax with eight or less pairs of limbs; eyes sessile. ... 5

5. a) Thorax with four to six pairs of limbs; head and part of thorax fused to form a distinct cephalothorax; eye single. (Includes many free-swimming planktonic forms.)
Class COPEPODA (part) (*p. 218*)
   b) Thorax with seven to eight pairs of limbs; cephalothorax absent (or at least not obvious); eyes, if present, paired. (Does not include any habitually free-swimming planktonic forms.) ... 6

6. a) Thoracic limbs biramous; eyes absent (tiny, 1 mm or less, crustaceans confined to subterranean waters).
Order BATHYNELLACEA (*p. 223*)
   b) Thoracic limbs uniramous (unbranched); eyes present except in a few subterranean species; adults 3–25 mm long. ... 7

7. a) Body laterally compressed (or, if not, then second antennae conspicuously thickened).
Order AMPHIPODA (*p. 224*)
   b) Body depressed – woodlouse-like (second antennae long but not conspicuously thickened).
Order ISOPODA (*p. 228*)

8. a) Telson with two long filamentous cerci (tails).
Order NOTOSTRACA (*p. 211*)
   b) Telson without filamentous cerci. ... 9

9. a) Abdomen covered by carapace; eyes sessile. ... 10
   b) Abdomen not covered by carapace (folded beneath it in crabs); eyes stalked. ... 12

10. a) Carapace consisting of two separate halves (valves) united by a dorsal hinge, enclosing the head (head may protrude in one non-British species) and body. ... 11
    b) Carapace a single folded plate embracing thorax and abdomen but not enclosing the head, from which it originates; four to six pairs of thoracic limbs.
Order CLADOCERA (part) (*p. 213*)

11. a) 6–17 mm long when fully-grown; thorax with ten to twenty-eight pairs of limbs (not British).
Order CONCHOSTRACA (*p. 212*)
    b) Up to (exceptionally) 7 mm long but usually 0.5–3 mm; thorax with only two pairs of limbs (hidden by carapace).
Order OSTRACODA (*p. 217*)

12. a) Thoracic limb biramous, never forming pincers.
Order MYSIDACEA (*p. 223*)
    b) Thoracic limbs unbranched; one, two, or three anterior pairs chelate (forming pincers).
Order DECAPODA (*p. 228*)

Class BRANCHIOPODA
(Fairy Shrimps and Water Fleas)

Recognition features Small or medium-sized crustaceans with five to seventy pairs of flattened thoracic limbs, and limbless abdomens terminating in a caudal furca or a pair of cercopods. The body may or may not bear a carapace and is very varied in shape, each of the four orders being very distinctive. The first antennae are usually small and simple, the second antennae are often large, branched and modified for swimming or as copulatory claspers in the male.

Branchiopods are found mostly in fresh water with only a few species of Cladocera being marine. Their main characteristic is the form of the thoracic limbs which are flattened, usually equipped with accessory lobes and combs of bristles, and used for swimming and collecting the small particulate matter – algae, etc. – on which they feed. The first three orders, Anostraca, Notostraca and Conchostraca, are often collectively known as phyllopods and are mostly specialized for life in temporary water bodies. Most of the introductory remarks regarding the Anostraca also apply to the other phyllopods.

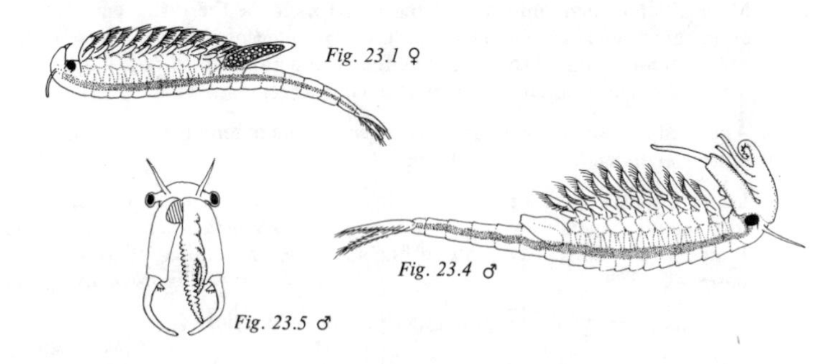

Fig. 23.1 ♀

Fig. 23.4 ♂

Fig. 23.5 ♂

Order ANOSTRACA
(Fairy Shrimps)

Fairy shrimps are delicate creatures which occur locally and sporadically in temporary bodies of water. Their general form is shown in *Fig. 23.1*; there is no carapace and the paired eyes are mounted on stalks. All the European species described below have eleven pairs of thoracic limbs, but there is a European Arctic species, *Polyartemia forcipata*, which has nineteen. Males are distinguished by their large, greatly modified second antennae which are used to grasp the female during copulation. Many species have a pair of fleshy pre-antennal appendages. Females have smaller, less extremely modified antennae, and their first two abdominal segments form a tubular or conical egg-sac. Fairy shrimps swim in a graceful, sedate manner, typically upside-down.

The eggs have a remarkable resistance to desiccation and extremes of temperature, remaining viable for long periods when the habitat is dry. The nauplius larvae hatch when their pond fills with rain or snow-melt water and develop rapidly; mature egg-carrying females may be present within one or two weeks. These resistant eggs are analogous to the 'winter eggs' of cladocerans (*p. 213*). Fast-hatching summer eggs are also produced which enable a rapid build-up of a population to take place during favourable conditions.

Temporary pools in field hollows and woodlands, which dry up periodically, are typical habitats, but fairy shrimps have also been found in such unlikely places as cart ruts and small puddles. They are defenceless animals and can only survive in the absence of such predators as fishes or carnivorous insect larvae, but are frequently found in the company of other branchiopods.

Key to the European Species of Anostraca

This is based on the structure of the male antenna and antennal appendage; females can often be identified by the form of the egg-sac, as in the descriptions.

1. a) Species restricted to inland saline waters. *Artemia salina* (*p. 210*)
   b) Species occurring in fresh water. . . . 2

2. a) Male antenna bent into a Z-shape and forked at the distal end; antennal appendage small (*Fig. 23.3*). *Streptocephalus torvicornis* (*p. 211*)
   b) Male antenna with a broad basal peduncle and an incurved tusk-like terminal portion; antennal appendage large (*Figs. 23.4–8*) . . . 3

3. a) Antennal appendage with lappets (fingers) on its outer edge. . . . 4
   b) Antennal appendage without lappets. . . . 5

4. a) Lappets consisting of a posterior triangular flap (not easily seen) and four lateral 'fingers' (*Fig. 23.4*). *Chirocephalus diaphanus* (*p. 211*)
   b) Lappets consisting of two lateral fingers (sometimes united at their bases – *Fig. 23.8*). *Tanymastix stagnalis* (*p. 211*)

5. a) Appendage slender, smooth, tapering and cylindrical (*Fig. 23.7*). *Branchipus schaefferi* (*p. 211*)
   b) Appendage broad and flattened, with a row of papillae along each edge (*Fig. 23.6*). *Siphonophanes grubei* (*p. 211*)

Fig. 23.2

Artemia salina (Brine Shrimp) *Fig. 23.2* This species is found in brine pools and salt lakes throughout the world; it has been recorded in southern England and in scattered localities throughout Europe. It varies in colour from pale pink to deep reddish and is up to 15 mm long. This is a well-known species, largely because of the widespread use of the newly-hatched nauplii as a convenient food for aquarium fishes and some laboratory animals. The eggs, which can be obtained from aquarists' suppliers, hatch readily in a salt solution and the larvae can be reared to adulthood by feeding with powdered yeast.

Fig. 23.3

Streptocephalus torvicornis *Fig. 23.3* The curiously-bent, forked antennae of the males are unmistakable; females have elongated egg-sacs which extend almost to the end of the abdomen. The maximum length is about 40 mm. This species occurs in southern and eastern parts of the guide's area, extending into Asia and southern Europe.

Chirocephalus diaphanus *Pl. 180, Figs. 23.1, 4, 5* This species attains a length of 35 mm; females have an egg-sac of moderate length, extending backwards over one to three segments. It is local and uncommon in Britain, mostly in the south, but widely distributed in Europe and North Africa.

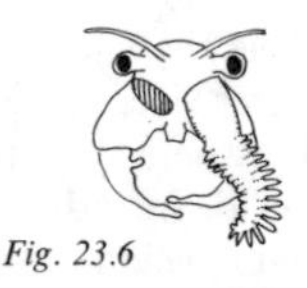
Fig. 23.6

Siphonophanes grubei *Fig. 23.6* Females have a short egg-sac which scarcely extends backwards; they reach a length of 35 mm; males are somewhat smaller. A common species in Europe northwards to Denmark; not known in Britain.

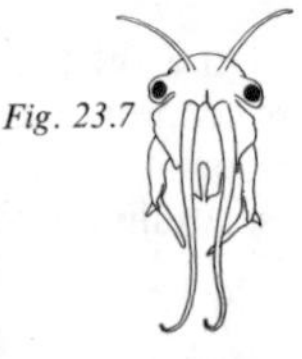
Fig. 23.7

Branchipus schaefferi *Fig. 23.7* The male antennae are often strongly chitinized and horn-coloured; the egg-sac of the female is short, not extending backwards; both sexes attain a length of about 20 mm. Widely distributed in Europe extending northwards to Denmark; one possible record from southern England.

Fig. 23.8

Tanymastix stagnalis *Fig. 23.8* The egg-sac of the female is very short and armed with two small ventral spines; both sexes reach about 20 mm in length. A widespread species found throughout Europe excepting northern Scandinavia; in Britain known only from western Ireland.

Order NOTOSTRACA
(Tadpole Shrimps)

Distinctive crustaceans with a broad, shield-like carapace, tapering cylindrical bodies bearing up to seventy pairs of limbs, and two long filamentous tails (cercopods). The eyes are sessile and set close together near the centre of the carapace. Two genera occur in Europe.

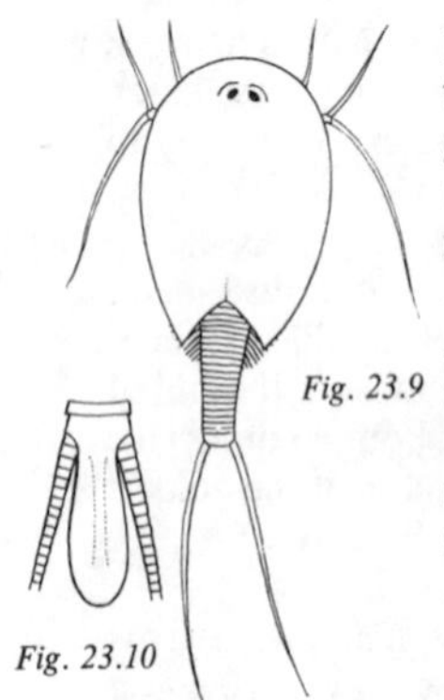

Fig. 23.9

Fig. 23.10

***Triops* (*Apus*)** *Fig. 23.9* This genus has no median lobe on the telson. The sole European species, *T. cancriformis*, is greenish or brown and up to 40 mm long. It occurs in temporary pools throughout Europe as far north as Denmark, but is very rare in Britain.

Lepidurus *Pl. 181* The distinguishing feature of this genus is the median lobe on the telson (*Fig. 23.10*). There are two European species: *L. arcticus* is a northern form with a circumpolar distribution, found in Europe in Iceland and northern Scandinavia, where it inhabits tundra pools and some large lakes; *L. apus* has a southerly distribution similar to *Triops*. but is not found in Britain.

Reference: Munro-Fox (1949).

Order CONCHOSTRACA
(Clam Shrimps)

These bivalved crustaceans occur in temporary waters and also in the shallow marginal waters of ponds and lakes. Superficially they resemble the Cladocera, from which they are distinguished by their generally greater size, more numerous thoracic limbs (ten to twenty-six pairs), and paired sessile eyes, set very close together. They are found in most parts of the world, including continental Europe, but are absent from Britain. The three main genera are:

Fig. 23.11

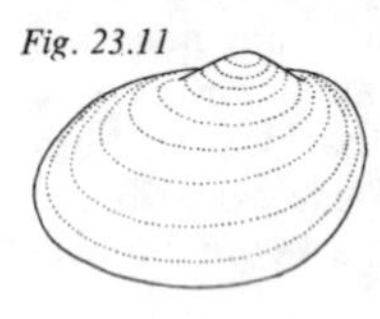

***Cyzicus* (*Estheria*)** *Fig. 23.11* Each valve of the shell is marked by concentric growth rings which are centred on a definite apex (umbo – see *p. 188*) near the dorsal hinge. *C. tetracerus* is the only European species, attaining a length of 12 mm.

Fig. 23.12

Limnadia *Fig. 23.12* The shell lacks umbos, but growth rings are present. A small frontal organ of unknown function protrudes from the front of the head (*Fig. 23.12*). The sole species in this genus, *L. lenticularis*, attains a length of 17 mm and is widely distributed throughout the northern hemisphere – Asia, Europe and North America.

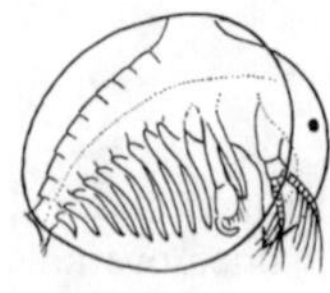

Fig. 23.13

***Lynceus* (*Limnetis*)** *Fig. 23.13* The head is very large, usually protruding from the shell which lacks umbos and growth rings. The main species is *L. brachyurus*, up to 6 mm long, distributed as *Limnadia*.

Order CLADOCERA
(Water Fleas)
Water fleas are small crustaceans ranging from 0.2–10 mm long, but more usually 0.5–4 mm with the thorax, which bears five or six pairs of limbs, and abdomen enclosed within a folded oval carapace. The head is separate from, but continuous with the carapace. In the aberrant families Leptodoridae and Polyphemidae the carapace is reduced to a small dorsal brood-pouch (*Figs. 23.29, 30*). The large eye is single and in some species is preceded by a smaller eyespot or ocellus. The first antennae are usually small, often vestigial, and unbranched; but the second antennae are large and prominent, branched, armed with long bristles, and used for swimming in most species. The abdomen is short and forms a broad foot terminating in a pair of parallel claws (the caudal furcae), which some of the bottom-dwelling species use to lever themselves along. Males, which are uncommon, are smaller than females and their first thoracic limbs are each armed with a hook and long bristle (*Fig. 23.14*). The descriptions below apply only to females, which are easily distinguished by the presence of large eggs in a dorsal brood-pouch within the carapace (*Fig. 23.15*).

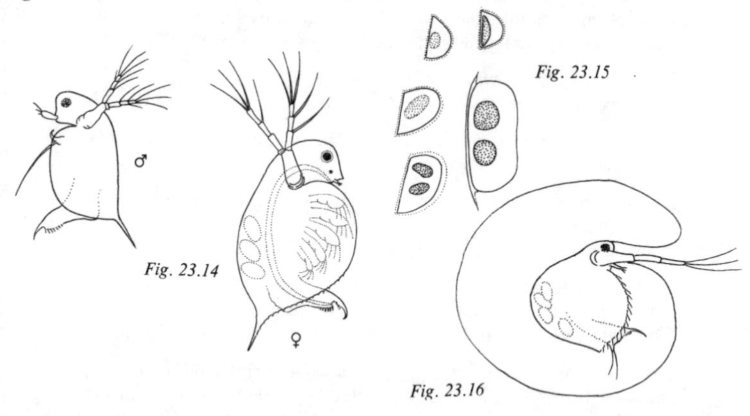

Fig. 23.14

Fig. 23.15

Fig. 23.16

Some species of water flea occur throughout the year but others are only present during the warmer months. The latter hatch from 'winter eggs' in the spring and at first only females are present. These reproduce rapidly, the 'summer eggs' developing without being fertilized by males (parthenogenesis). Later, towards the onset of winter, or other unfavourable conditions such as the pond drying up, males appear and fertilize the thick-shelled winter eggs produced by the females at this time. In the families Daphniidae, Chydoridae, Macrothricidae and Bosminidae, the winter eggs are carried by the female in a special saddle-shaped pouch (ephippium) which is shed at the following moult (*Fig. 23.16*). The ephippium persists and protects the eggs during the winter when the adults die off. Winter eggs are resistant to freezing and desiccation and, being easily carried by wind, flood or other agencies, have enabled many species to become widely dispersed. All the genera described below occur throughout the northern hemisphere – Asia, Europe and North America.

Cladocerans are very important in the economy of fresh water. Many species occur seasonally in vast numbers, providing a valuable food supply for many other creatures, and they in turn consume large quantities of planktonic algae. They are found in all types of standing waters – lakes, ponds, ditches, marshes and small temporary pools – as well as canals and slow-flowing rivers. Some species are planktonic, others clamber about on vegetation or amongst bottom debris. The jerky swimming motion of most species is highly characteristic and is responsible for the common name of water flea, but some swim more-or-less smoothly, or crawl rapidly along the substratum, in this respect resembling ostracods. Others are less mobile and push themselves along using the abdominal claws.

Key to the Families of Freshwater Cladocera (females only, see *p. 213*)

Some of the characters used in this key may require the use of a low-power microscope, but in practice, most families can easily be identified by simply glancing through the figures. Only a few typical genera in each family are described.

1. a) Carapace enclosing whole body. ... 2
   b) Carapace not enclosing body, reduced to a dorsal brood-pouch. ... 7

2. a) Living animal enclosed by a blob of jelly. Holopediidae (*p. 215*)
   b) Animal not living in a blob of jelly. ... 3

3. a) Branches of second antennae consisting of two and three joints respectively. Sididae (*p. 215*)
   b) Branches of second antennae consisting of three and four joints. ... 4

4. a) First antennae small and inconspicuous or, if large, not mounted on a pointed rostrum (*Fig. 23.26*) ... 5
   b) First antennae large and prominent, mounted on a pointed rostrum. ... 6

5. a) Carapace extending forward over the head, forming a pointed, hood-like structure; second antennae relatively small; mostly bottom-living forms. Chydoridae (*p. 215*)
   b) Carapace not forming a hood-like structure; second antennae relatively large; free-swimming forms. Daphniidae (*p. 215*)

6. a) First antennae fixed; carapace with a caudal spine; free-swimming forms. Bosminidae (*p. 216*)
   b) First antennae movable; carapace usually without a caudal spine (present in one rare genus only); bottom-living forms, poor swimmers. Macrothricidae (*p. 216*)

7. a) Not more than 3 mm long; eye very large, abdomen short (*Fig. 22.29*). Polyphemidae (*p. 217*)
   b) Up to 10 mm long; eye not very large; abdomen elongated (*Fig. 23.30*). Leptodoridae (*p. 217*)

3.17

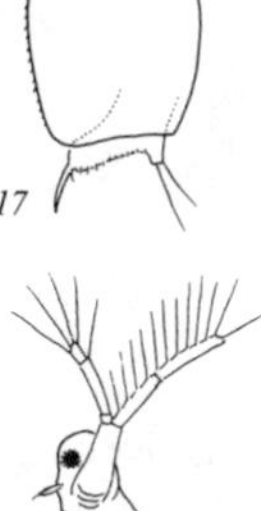

3.18

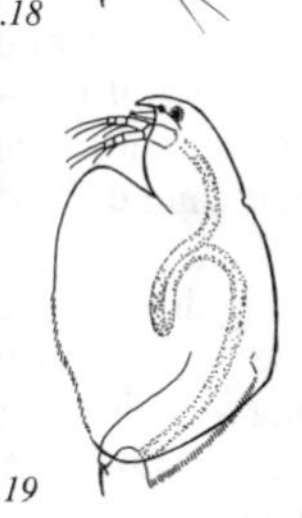

.19

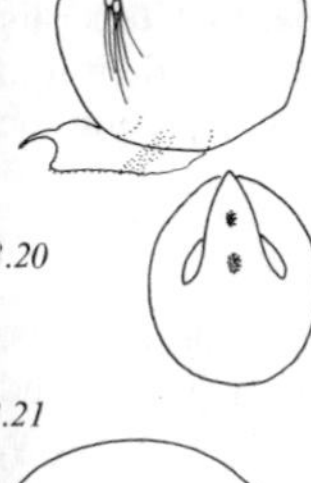

.20

.21

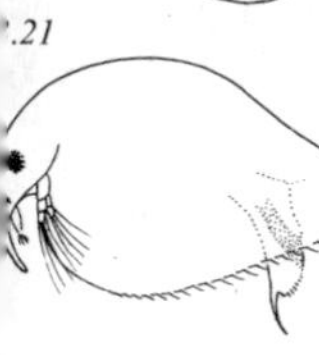

Family Holopediidae

Holopedium gibberum *Fig. 23.16* This species, about 2 mm long, with its hunched back and envelope of clear jelly, is unmistakeable. It is found in the plankton of large lakes.

Family Sididae

Sida crystallina *Pl. 182, Fig. 23.17* A transparent species, 3–4 mm long, with an adhesive organ behind the head which is used to attach the animal to water plants, etc. It is common in the weedy margins of clear lakes.

Diaphanosoma *Fig. 23.18* The several species in this genus are all small, about 1 mm long, and lack the adhesive organ that is characteristic of *Sida*. They are common in the marginal regions of lakes and ponds.

Family Chydoridae *Pl. 183, Figs. 23.19, 20, 21*

This family contains numerous species, many of which are very small; they range from 0.2–4 mm long but are usually about 0.5 mm. Although their faces immediately distinguish them from other claderocerans, to the unaided eye, chydorids are easily mistaken for ostracods because of their globular shape, small size, and smooth (not jerky) swimming motion. The most distinctive genus is *Eurycercus* (*Fig. 23.19*) which, at up to 4 mm long, is much larger than any other, and the dorsal edge of its foot is finely serrated along the whole length (partial in others). The other illustrated genera, *Chydorus* and *Rhynchotalona*, are typical of the group, but many other genera are equally common in a wide variety of habitats.

Family Daphniidae

Daphnia *Pls. 184–185, Figs. 23.14, 22* A large, well-known genus with many species, up to 5 mm long and mostly reddish or greenish in colour. The head has a distinct, pointed rostrum and the ventral edge of the carapace is strongly curved, leading to a prominent caudal spine. Some species develop tall crests on the head (*Fig. 23.22*). They are common in all types of still water, particularly small, organically rich ponds where they may be so abundant as to discolour the water. 'Daphnia' is also a general term for cladocerans and other small crustaceans such as copepods and ostracods which are collected *en masse* and used as a fish-food by aquarists.

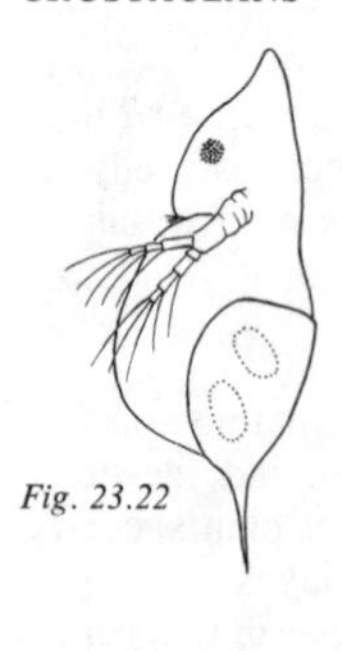
Fig. 23.22

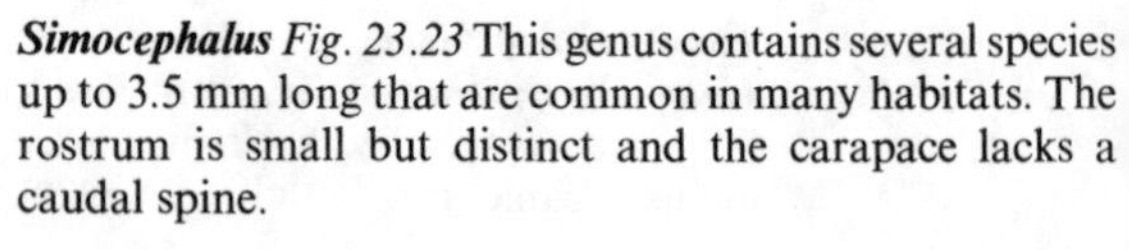
Simocephalus *Fig. 23.23* This genus contains several species up to 3.5 mm long that are common in many habitats. The rostrum is small but distinct and the carapace lacks a caudal spine.

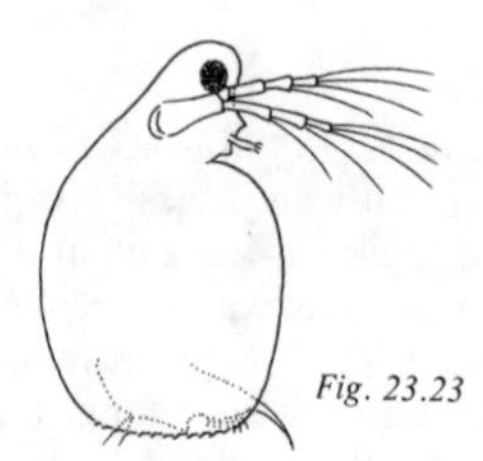
Fig. 23.23

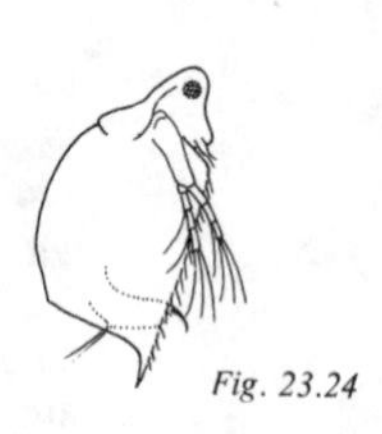
Fig. 23.24

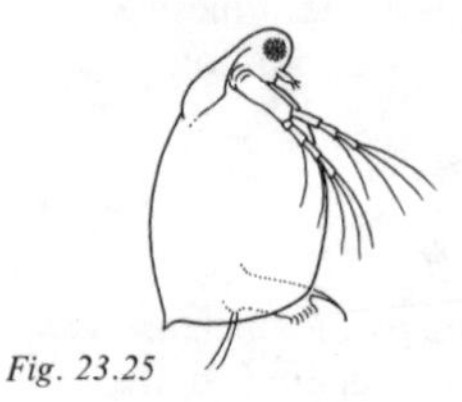
Fig. 23.25

Scapholeberis *Fig. 23.24* A small genus of distinctive cladocerans, up to 2 mm long, found mostly in smaller waters or the shallow parts of lakes, often hanging beneath the surface film. The head has a small rostrum and the ventral edge of the carapace is straight and in line with a caudal spine.

Ceriodaphnia *Fig. 23.25* A large genus with many common species, up to 2 mm long, that occur in a variety of habitats. The small head lacks a rostrum, the first antennae are tiny, and the carapace terminates in a small caudal spine or point.

Fig. 23.26

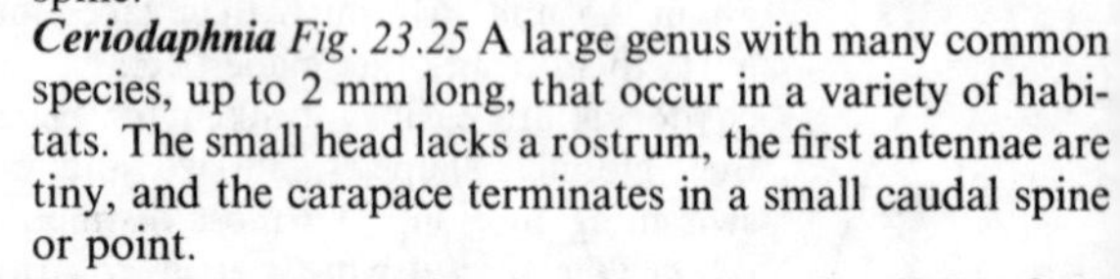
Moina *Fig. 23.26* This genus is easily distinguished from other daphniids by its large first antennae; it resembles certain macrothriciids but differs in lacking any sign of a rostrum. There are several species, about 1.5 mm long. usually found in small turbid pools or ditches.

Fig. 23.27

Family Bosminidae

Bosmina *Pl. 186, Fig. 23.27* This is the only European genus in the family; it is recognized by its fixed, parallel, tapering first antennae which bear small tufts of hair near their mid point. Several species, about 0.5–1 mm long, are common in the plankton of ponds and lakes.

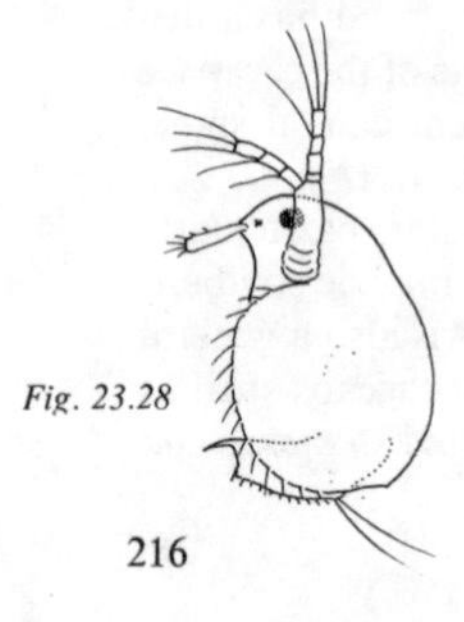
Fig. 23.28

Family Macrothricidae *Fig. 23.28*

These are characteristic bottom-dwelling cladocerans, generally found amongst dense vegetation or in the mud of ponds, marshes or lakes; some species can swim feebly, but rarely do so. Although not rare, they never occur in large numbers, and none of the genera is particularly common. The size ranges from about 0.5–1 mm and

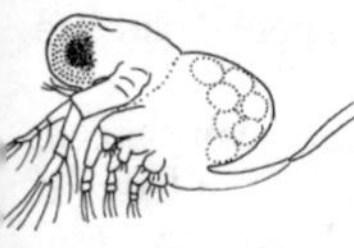

Fig. 23.29

colouration is commonly red but may be yellow, green or brown.

Family Polyphemidae *Fig. 23.29*
The appearance of these unusual cladocerans is unmistakable. There are two genera, both about 2 mm long: *Polyphemus*, with a caudal spine that is shorter than the body, is found in ponds and weedy lake margins; *Bythotrephes*, with a caudal spine much longer than the body, is a planktonic genus occurring in large lakes.

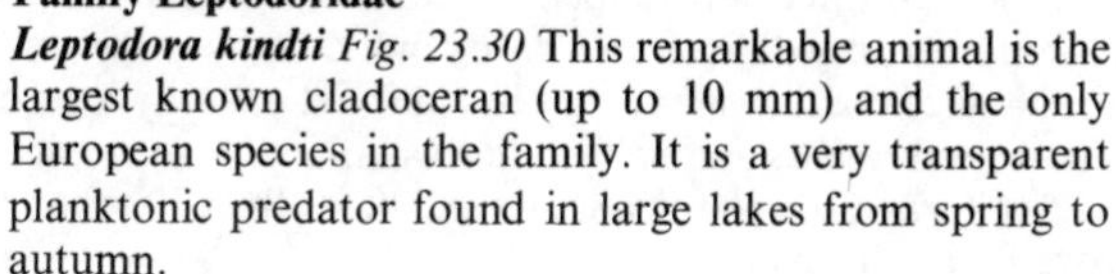

Fig. 23.30

Family Leptodoridae
Leptodora kindti *Fig. 23.30* This remarkable animal is the largest known cladoceran (up to 10 mm) and the only European species in the family. It is a very transparent planktonic predator found in large lakes from spring to autumn.

References: Scourfield & Harding, 1966. Brauer, 1909.

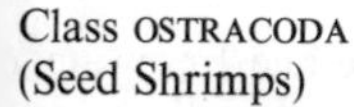

Class OSTRACODA
(Seed Shrimps)

Recognition features Tiny or small crustaceans, 0.2–7 mm long, with the entire body, including the head, enclosed by an oval, bivalved carapace. Only the tips of the appendages protrude from between the valves of the shell, even when the animal is active (*Pl. 187, Fig. 23.31*).

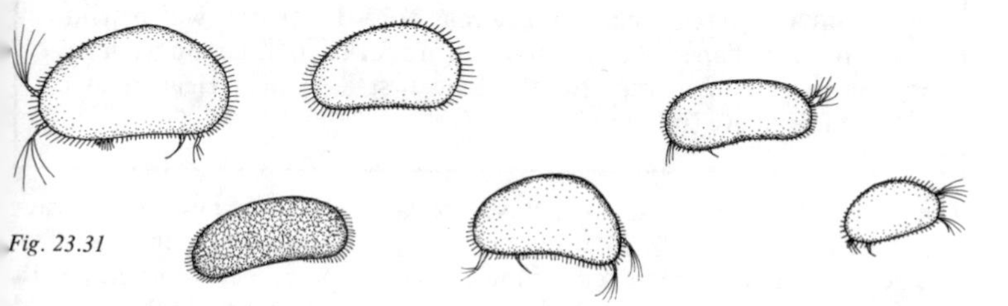

Fig. 23.31

Ostracods are common and often very abundant in marine and fresh waters. They are typically found scuttling about in vegetation or in the surface layers of bottom sediments, rarely venturing into open water but perpetually busy. They swim by beating their antennae, and crawl on the bottom using their legs and caudal furca. When threatened, they withdraw their appendages and clamp the two halves of the shell shut.

The carapace varies little in shape, being generally oval or bean-shaped, sometimes with fine surface sculpturing, and usually with a fringe of hairs. It is usually some shade of brown, sometimes patterned. Most European freshwater species fall into the size range 0.5–3 mm but some are larger. Internally the body forms a central mass with no discernible segmentation, from which arise seven pairs of appendages: first and second antennae which are usually large, branched, and terminate in a bunch of bristles; three pairs of complex mouthparts; and two pairs of short legs with long claws. At the hind end of the body is a leg-like caudal furca.

Their uniform external appearance renders ostracods easily recognizable as a group but very difficult to identify individually to genera, or even family. To the unaided eye they may easily be confused with chydorid cladocerans as their shape, movements and habits are so similar. There are many common genera (e.g., *Cypris*, *Candona*, *Eucypris*) and more than 300 species occur in the area covered by this guide.

References: Brauer, 1909.

Class COPEPODA
(Water Fleas)

Copepods are extremely abundant animals in both marine and freshwater habitats but have yet to acquire an exclusive common name, usually being lumped with other crustaceans as 'water fleas' or entomostraca; or all, erroneously, being called 'cyclops', the name of the most common genus. They can conveniently be divided into two groups: free-living and parasitic, which will be described separately.

Free-living Copepods

Recognition features Small crustaceans, 0.25–3 mm long, with a cylindrical or pear-shaped body terminating in a forked tail. The appendages are inconspicuous except for the long first antennae. Head with a single median eye (often red).

The three orders of free-living copepods are easily recognized by their characteristic shapes (*Figs. 23.32–35*) with the unaided eye, in spite of their small size. The head and first thoracic segment are fused to form a cephalothorax with long first antennae, shorter second antennae, inconspicuous mouthparts, and the first pair of thoracic limbs. The remaining four thoracic segments are stiffly articulated, and each bears a pair of short, flattened, branched limbs. The abdomen has no appendages other than the two branches of the caudal furca, which bear long bristles; its junction with the thorax is more flexible than other joints and is the main 'hinge' of the body. Females are more common than males and when mature, carry one or two prominent egg-sacs attached to the first abdominal segment. Males can be distinguished by their usually smaller size and slightly modified (crooked) first antennae which are used for grasping the female.

Key to the Orders of Free-living Copepods

1. a) Cephalothorax distinctly broader than abdomen, body pear-shaped; swimming or bottom-dwelling forms up to 3 mm long. . . . 2
   b) Cephalothorax not distinctly broader than abdomen, body more or less cylindrical; bottom-dwelling forms less than 1 mm long.
   Order HARPACTICOIDEA
2. a) First antennae very long, usually longer than whole body; egg-sac single.
   Order CALANOIDEA
   b) First antennae moderate, rarely reaching beyond hind end of cephalothorax; egg-sacs paired.
   Order CYCLOPOIDEA

Order HARPACTICOIDEA
Pl. 188, Fig. 23.32

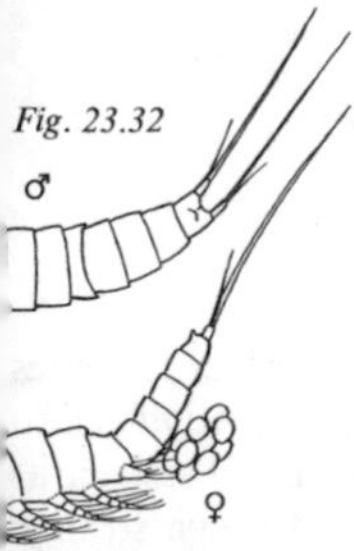

Fig. 23.32 ♂ ♀

These tiny, inconspicuous copepods are very common and widespread, and likely to be encountered in any aquatic habitat. They cannot swim but scuttle around on plants, stones or in bottom sediments, and are even found living interstitially between sand grains on lake shores. In collections of pond material they often become trapped in the surface film. Copulating pairs are commonly found travelling around in tandem, the male grasping the female's caudal bristles with his antennules. The females carry a single small egg-sac. About 200 species occur in Britain and northern Europe, the principal genera being *Canthocamptus, Laophonte, Bryocamptus* and *Atheyella*.

Order CALANOIDEA
Pl. 189, Fig. 23.33

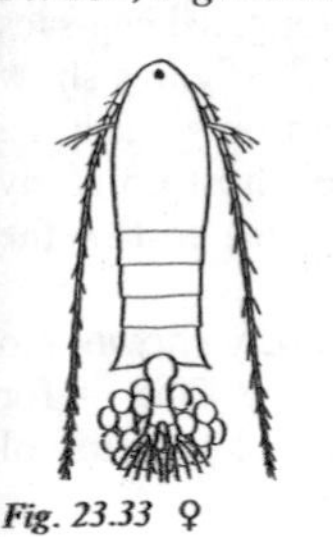

Fig. 23.33 ♀

These are planktonic copepods, 1–2.5 mm in length, colourless and translucent, or more rarely, reddish in colour. They swim with a flickering, darting motion almost too rapid for the eye to follow and stop so suddenly that they are easily lost. When stationary, the long first antennae are held in a characteristic horizontal attitude with the body hanging vertically from them.

Calanoids are common in most permanent still waters, either in the open water or, in smaller numbers but greater variety, in marginal areas amongst vegetation. There are several freshwater genera of which *Diaptomus* is the most common, the few other genera mostly occurring in or near brackish water.

Order
CYCLOPOIDEA
Pl. 190,
Figs. 23.34, 35, 36

This order contains the large and well-known genus *Cyclops* which is often subdivided into a number of smaller subgenera and encompasses about 100 species in the guide's area. They are 'general-purpose' copepods – equally at home swimming in the plankton or hunting amongst bottom debris. The characteristic twin egg-sacs of the female render the group unmistakable. Most species are drab brownish or greenish but some are brilliantly-coloured; they range in size from 0.5–3 mm long.

Cyclops are likely to be found in almost any permanent or semi-permanent body of water – even tree-holes – at any season. They feed on any small organisms, animal or vegetable, as well as organic debris.

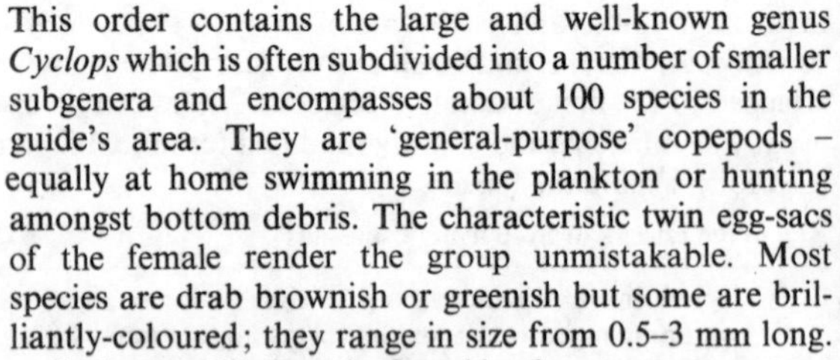

References: Brauer, 1909. Harding & Smith, 1974.

Fig. 23.34 ♀

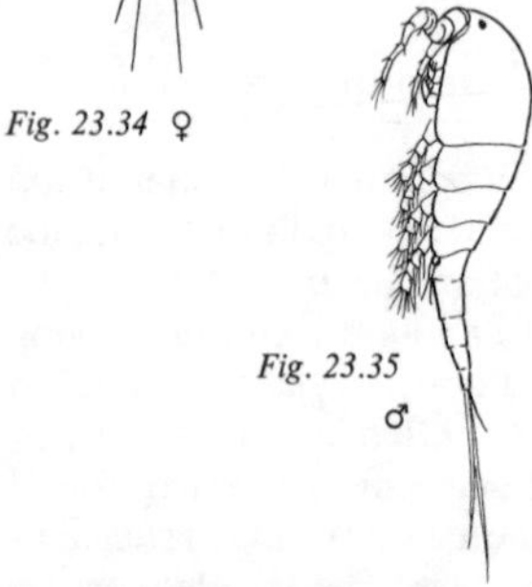

Fig. 23.35 ♂

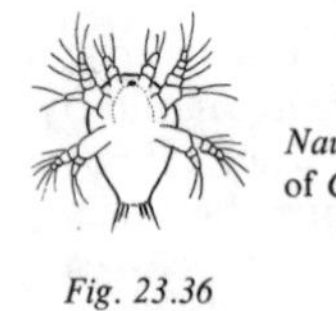

Nauplius larva of *Cyclops*

Fig. 23.36

Parasitic Copepods

A number of copepods have evolved into highly specialized external parasites of fishes. They are usually found temporarily attached or permanently fixed to the skin or fins, or inside the mouth or gill chambers. The form of these parasites ranges from the typical copepod (e.g., cyclopoid) shape to such strange and bizarre types as *Lernaea* and *Tracheliastes* (*Fig. 23.42*), which are scarcely recognizable, even as arthropods. Only the females show this extreme modification, the tiny, rarely-seen males usually retaining a recognizable copepod form. Mature females usually carry a pair of more-or-less elongated egg-sacs and, when present, these are a useful recognition feature. The figures show only females, as males, which are usually non-parasitic, can only easily be identified by context. Species which fix permanently to their host do so by means of an anterior anchor, which is embedded in the flesh of the host; in the figures these hidden parts are shown in black.

At present the distribution of many species in Europe is poorly known. No doubt the increasing trade in and transport by man of living fishes – for stocking fish farms, fisheries and ornamental ponds – will lead to many of these parasites becoming widely distributed.

Order
POECILO-
STOMATOIDA

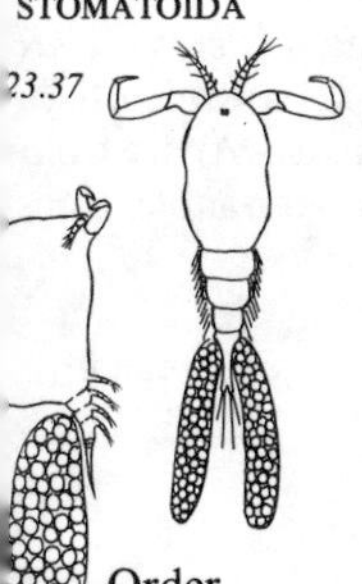
23.37

23.38

Family Ergasilidae

Ergasilus *Fig. 23.37* These retain the typical copepod form but the antennae are modified for grasping the gill bars of the host. Several species, about 1–2 mm long, occur in Europe; they are found on various members of the carp family, and eels.

Thersitina gasterostei *Fig. 23.38* A small (0.6–0.8 mm) parasite, similar to *Ergasilus* but with a very deep, almost globular, cephalothorax. It is found attached to the gills of sticklebacks (*Gasterosteus* and *Pungitius* spp.) and occurs mostly in the slightly brackish reaches of rivers.

Order
CYCLOPOIDEA

Family Lernaeidae

Lernaea cyprinacea (Anchor Worm) *Fig. 23.39* This species exhibits no sign of segmentation, and the limbs are tiny and not visible to the unaided eye. Anchor worms will attack any species of freshwater fish, fixing themselves to any part of the body. They are common and widespread in Europe and are being noticed with increasing regularity in Britain.

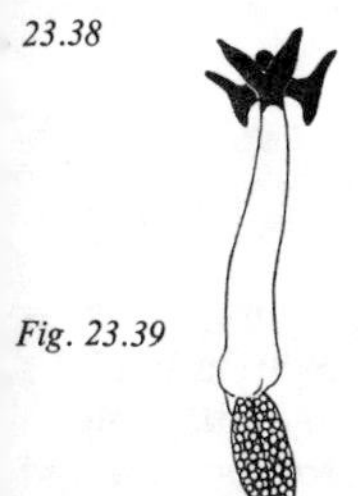
Fig. 23.39

Order
SIPHONO-
STOMATOIDA

Family Caligidae

Lepeophtheirus salmonis (Salmon Louse, Sea Louse) *Fig. 23.40* A parasite which infests the Atlantic salmon (*Salmo salar*) during its life in the sea, but which may survive for about a week in fresh water when the salmon enters a river to spawn. These large, up to 18 mm, and conspicuous parasites attach to the fish's skin, usually near the vent. The males are also parasitic but smaller (about 6 mm long) than the females. Sea lice are greeted with delight by the angler as they are a sign of a 'fresh run' fish – the best for eating.

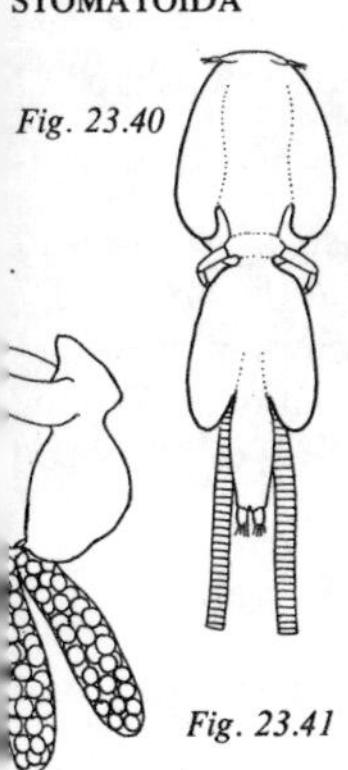
Fig. 23.40

Fig. 23.41

Family Lernaeopodidae

Salmincola (Gill Maggot) *Fig. 23.41* This genus contains several rather similar species which are parasitic on various salmonoid fishes: *S. salmoneus* on salmon (*Salmo salar*). *S. gordoni* on trout (*Salmo trutta*), *S. edwardsii* on charr

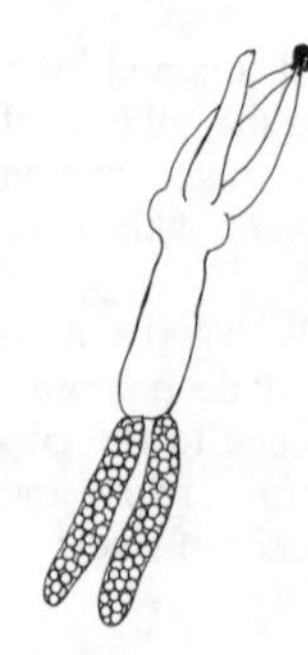
Fig. 23.42

(*Salvelinus alpinus*), and *S. thymalli* on grayling (*Thymallus thymallus*). They range in length from 3–8 mm and are usually fixed to the gills but sometimes to the body.

Achtheres percarum A parasite of the perch (*Perca fluviatilis*) and pike-perch (*Stizostedion lucioperca*) which fixes itself to the gills or inside the mouth cavity. Apart from its host preference it is almost indistinguishable from *Salmincola*.

Tracheliastes polycolpus *Fig. 23.42* This species is found on various cyprinid (carp) fishes, usually fixed to the fins. It is up to 10 mm long.

Reference: Fryer, 1982.

Class BRANCHIURA
(Fish Lice)

The sole European genus in this class is *Argulus*, the fish louse, which is sometimes included amongst the parasitic copepoda. Fish lice are capable of swimming freely in search of a host – any freshwater fish – and attaching to the fish's body, or inside the mouth cavity, where they pierce the skin and suck the host's blood. Their flattened, translucent bodies can be very difficult to discern and they are capable of shifting their position on the host by gliding over its surface. The general structure of the body is shown in *Fig. 23.43*, the two large suckers, which are diagnostic of the group, are prominent features. Unlike most other freshwater crustaceans *Argulus* does not carry its eggs until they hatch but sticks them to stones or plants. The young resemble the parents in form and seek out a host soon after hatching. There are two common species native to north-west Europe, distinguished by the shape of their abdominal lobes.

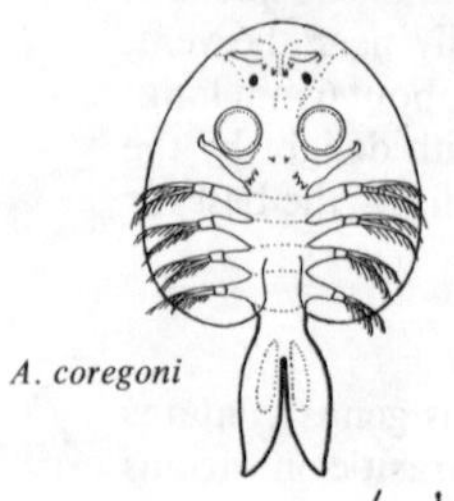

Fig. 23.43

Argulus foliaceus *Pl. 191, Fig. 23.43* This species is up to 10 mm long and is widespread throughout the area.

Argulus coregoni *Fig. 23.43* This is less common in Britain than *A. foliaceus* but widespread on the continent. Despite its name it has been found on a wide variety of hosts and reaches a length of 13 mm.

References: Fryer, 1982. Flossner, 1972.

Class MALACOSTRACA

Recognition features Small, medium or large crustaceans with five to eight pairs of thoracic limbs, typically modified for walking or grasping, and usually with a pair of limbs (pleopods, uropods) on each abdominal segment. A carapace is present in some genera.

This very large class includes the largest and most familiar crustaceans: all 'true' shrimps, prawns, crabs, crayfishes, etc. Most malacostracans are marine but five orders occur in European fresh water. The only common and widely distributed freshwater types are the freshwater shrimps (*Gammarus* and related genera, order Amphipoda), water slaters (order Isopoda) and crayfishes (Decapoda).

Reference to whole class: Gledhill, Sulcliffe & Williams, 1976.

Superorder SYNCARIDA
Order BATHYNELLACEA

Fig. 23.44

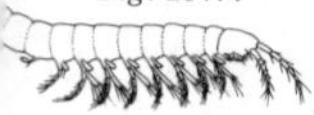

Bathynella spp. *Fig. 23.44* Tiny eyeless crustaceans, about 1 mm long, with eight pairs of biramous thoracic limbs and two pairs of abdominal limbs on segments 1 and 5. They are exclusively subterranean in habit, probably living in the interstitial water of sand or gravel beds. They have been found in caves, wells, spring outlets and pumped ground water. Several species occur in Britain and Europe.

Superorder PERACARIDA
Order MYSIDACEA

Fig. 23.45 ♂

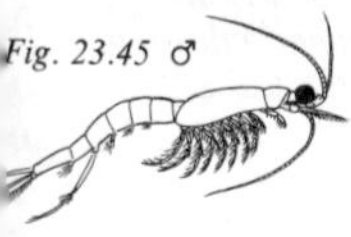

Mysis relicta (Opossum Shrimps) *Fig. 23.45* Shrimp-like crustaceans up to 18 mm long, with large stalked eyes and translucent, characteristically humped bodies. The thorax is covered by a carapace and bears eight pairs of branched limbs. Females carry their eggs in a ventral thoracic brood-pouch; males have enlarged fourth pleopods.

This is one of the relict species found in fresh waters in the northern hemisphere. It occurs in large lakes, more rarely in rivers, within the limits of glaciation of the last ice age. In Britain it is found in Ennerdale Water (Cumbria) and in many Irish loughs; also in Poland, Germany, Denmark and Scandinavia; outside Europe in North America and Asia.

Mysids are common marine animals and many species occur in brackish conditions in estuaries. One species,

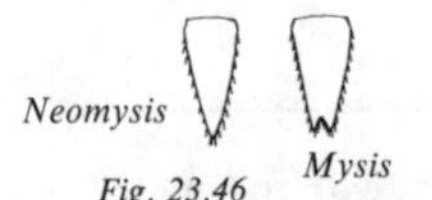

Fig. 23.46

Neomysis integer, is occasionally found in fresh waters near the sea; it is distinguished from *M. relicta* by the shape of its telson (*Fig. 23.46*).

Order AMPHIPODA

This group includes the familiar freshwater shrimps which are found in most aquatic habitats from fresh to fully saline waters. The typical, curved, laterally compressed shape (*Fig. 23.52*) is common to nearly all European genera which are diagnosed in the following key.

Key to Amphipoda

1. a) Eyes present. . . . 2
   b) Without eyes (subterranean forms). . . . 7

2. a) Body laterally compressed, antennae slender. . . . 3
   b) Body cylindrical or slightly depressed; antennae stout. *Corophium* (*p. 224*)

3. a) Antennules shorter than antennae. . . . 4
   b) Antennules longer than antennae. . . . 5

4. a) Antennules more than half length of antennae; aquatic (not British). *Pontoporeia* (*p. 225*)
   b) Antennules much less than half length of antennae; terrestrial (waterside). *Orchestia* (*p. 225*)

5. a) Abdominal segments 1 and 2 each with two dorso-lateral spines on posterior margin (*Fig. 23.50*); not British. *Pallasea* (*p. 225*)
   b) Posterior margins of most segments raised, forming a toothed dorsal outline (*Fig. 23.51*); not British. *Gammaracanthus* (*p. 225*)
   c) Body segments not spined or toothed as above (British and European). . . . 6

6. a) Tip of uropod 3 extending well beyond tips of uropods 1 and 2; dorsal edges of all three urosome segments with tufts of small hairs; often larger than 10 mm. *Gammarus* (*p. 226*)
   b) Tips of all three uropods approximately level; dorsal edges of urosome segments smooth, without tufts of hair; less than 10 mm long. *Crangonyx pseudogracilis* (*p. 227*)

7. a) Length of penultimate gnathopod segment (*Fig. 23.55*, arrowed) greater than its width. *Crangonyx subterraneus* (*p. 227*)
   b) Length of penultimate gnathopod segment about equal to its width. *Niphargus* and *Niphargellus* (*p. 227*)

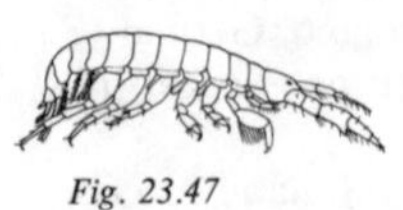
Fig. 23.47

Family Corophiidae

Corophium spp. *Fig. 23.47* This genus is easily recognized by its stout antennae (*Fig. 23.47* ♀, much longer in ♂♂) and cylindrical or slightly depressed body. The typical inland freshwater species is *C. curvispinum*, up to 6 mm long, but a number of others occur in brackish habitats and may wander into fresh water. *C. curvispinum* has spread across Europe from the Caspian area during this century and is well established in the Midland canal system of Britain and in scattered localities on the continent. It is found on plants, submerged brick or wood structures, or in silt, where it builds a flimsy tube of mucus and detritus in which it lives.

Family Haustoriidae

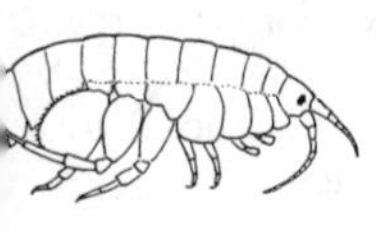

Fig. 23.48

Pontoporeia affinis *Fig. 23.48* A small amphipod up to 11 mm long; females have relatively short antennules and antennae but in males they may be longer than the body. This is a relict species with a similar distribution to *Mysis relicta* (see *p. 223*) although it is also found in estuaries of rivers flowing into the Baltic and Arctic seas. It is absent from Britain.

Family Talitridae

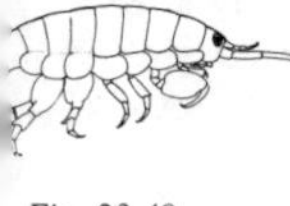

Fig. 23.49

Orchestia cavimana **(*O. bottae*)** (Bankhopper) *Pl.* 192, *Fig. 23.49* This is a semi-terrestrial relative of the 'sand-hoppers' of the sea-shore. It is the only British terrestrial crustacean apart from the woodlice (Isopoda) and is included here because of its close association with water. *O. cavimana* is dark grey-brown and walks with an upright gait, its legs splayed out sideways, and can spring away when disturbed. Males attain 22 mm in length, females 16 mm. It is a gregarious animal found amongst grass roots, under stones, in crevices, etc., on damp river and canal banks. This species was first discovered in Britain in 1942 and has since been recorded from many localities all over England (being particularly common in Cheshire) and northern Europe.

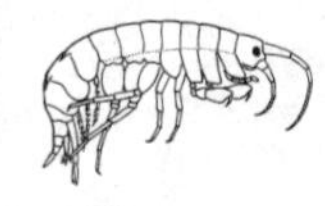

Fig. 23.50

Family Gammaridae (Freshwater Shrimps)

Pallasea quadrispinosa *Fig. 23.50* A relict species, up to 20 mm long, found in large lakes in northern Asia and Europe, from Poland, North Germany and Denmark northward into Scandinavia; not British.

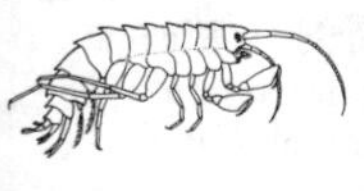

Fig. 23.51

Gammaracanthus lacustris *Fig. 23.51* A large gammarid, up to 30 mm long, occurring in deep lakes in Scandinavia and the east Baltic States; not British.

Gammarus spp. *Fig. 23.52* Freshwater shrimps of this genus are probably the most widely distributed and familiar of all freshwater crustaceans. Typically they occur in almost any clean running waters, large lakes and throughout the whole range of brackish conditions into the sea. They are found, often in great abundance, in or under any substratum that provides shelter from predators and a supply of organic debris which is their main food source – under stones, in gravel and other coarse substrata, amongst living or dead vegetation, etc. Many individuals carry a heavy load of parasitic or epizooic organisms (see *pp. 30, 118, 119, 121, 150*).

The three species most commonly found in British fresh water are identified by the following key.

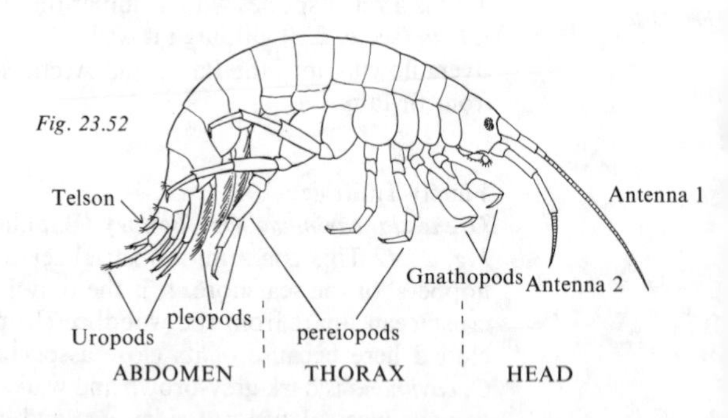

Fig. 23.52

---

1. a) Length : width ratio of eye less than 2 :1 – ; walking legs 3–5 with numerous small spines but few, if any, hairs; exclusively freshwater species. . . . 2
   b) Length : width ratio of eye 2 or more:1 – ; walking legs 3–5 with numerous spines and hairs; brackish-water species which may occur in fresh water. *Gammarus duebeni* and others

2. a) Posterior corner of epimeron 2 (*Fig. 23.53*) forming approximately a right angle. *Gammarus pulex*
   b) Posterior corner of epimeron 2 forming an acute angle and often extending backwards (*Fig. 23.54*) *Gammarus lacustris*

Fig. 23.53

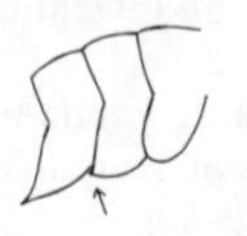

Gammarus pulex *Pl. 193, Fig. 23.52, 53* This species is very common in Britain, introduced (not native) to Ireland, and widespread throughout most of northern Europe. It is a typical species of running waters, including spring ponds and other 'still' waters with a regular flow, and is sometimes found in subterranean waters.

G. pulex attains a length of about 20 mm and varies in colour from gingery-brown to olive or greyish.

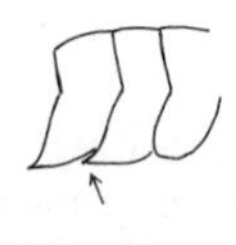
Fig. 23.54

Gammarus lacustris *Fig. 23.54* A species of large lakes, especially those of glacial origin, and sometimes in connected streams. It occurs throughout Britain and Ireland, except for central and southern England, and throughout most of Europe as well as North America. Only rarely does it co-exist with other *Gammarus* species. It reaches a length of about 25 mm and is coloured like *G. pulex*.

Gammarus duebeni This is a brackish-water species that in some areas has adapted completely to a freshwater existence. In British fresh waters it is found in many west coast localities and offshore islands where *G. pulex* is absent, and it is common in Ireland and many coastal areas of northern Europe.

Many other species of *Gammarus* occur throughout the European mainland in addition to those described above (see Karaman & Pinkster, 1977 Part 1 and Part 2).

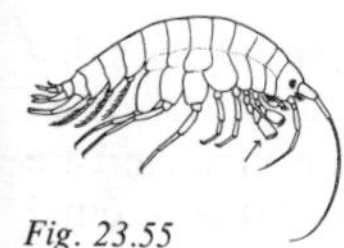
Fig. 23.55

Crangonyx pseudogracilis *Pl. 194, Fig. 23.55* This is an introduced species from North America which was first noticed in Britain in 1936 (London). It is now common in midland and southern England and has reached Scotland and Northern Ireland but not, apparently, continental Europe. *C. pseudogracilis* tends to occur in habitats where *Gammarus* is absent (being able to tolerate much lower levels of dissolved oxygen), typically in ponds and small lakes, ditches and canals, although the two genera do co-exist in some localities.

In life *C. pseudogracilis* is easily distinguished from *Gammarus* spp. by its smaller size, rarely exceeding 8 mm, its generally darker colouration which ranges from steely blue-grey to olive-greenish, the greater curvature of its thorax, and its more delicate appearance. It usually walks in an upright position whereas *Gammarus* shuffles along on its side; both may swim in any orientation.

Crangonyx subterraneus This is a small, 6 mm long, eyeless species found in underground waters in Britain and Europe.

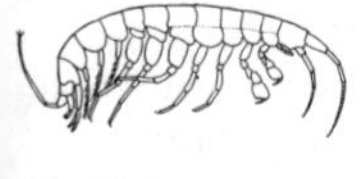
Fig. 23.56

Niphargus and ***Niphargellus*** *Fig. 23.56* These eyeless shrimps live in subterranean waters or interstitially amongst gravel within the water table. Occasionally they are washed out at the surface at spring outlets and may also be found in wells and pumped ground water.

Several species of *Niphargus*, up to about 15 mm long, occur in Britain and Europe. A related genus, *Niphargellus*

(3 mm long) in which the third uropod is not elongated as in *Niphargus*, has been found in ground water in Devon.

Order ISOPODA

Woodlouse-like crustaceans with depressed bodies consisting of a cephalothorax (head fused with first thoracic segment), seven free thoracic segments, each with a pair of limbs, the first modified for grasping the remainder for walking, and a reduced abdomen covered dorsally by a single plate and bearing a pair of branched uropods.

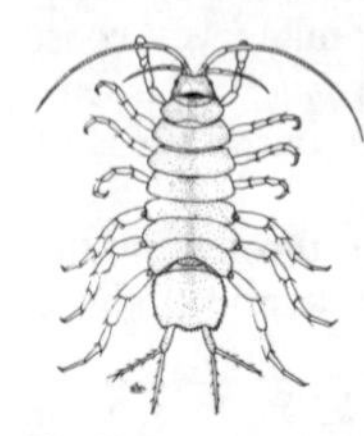

Fig. 23.57

Asellus (Water Slaters, Hog-Lice) *Pl. 195, Fig. 23.57* This is the only genus of freshwater isopods found in north-west Europe. Hog-lice are very common in any permanent, stagnant or slow-flowing waters where they crawl amongst bottom debris, especially dead leaves; they cannot swim.

The most common species is ***A. aquaticus*** which is up to 15 mm long and greyish-brown in colour. It occurs throughout Britain and Europe.

A. aquaticus

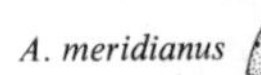

A. meridianus

Fig. 23.58

A. meridianus is a slightly smaller (12 mm) species distinguished from *A. aquaticus* by its head pattern (*Fig. 23.58*). It is fairly common in Britain, being characteristic of freshwater pools on offshore islands; and is also found in the lowlands of France and Belgium.

A colourless and eyeless species, *A. cavaticus*, occurs in subterranean waters.

Several species of brackish-water isopods of the genera *Sphaeroma* and *Jaera* have been recorded from fresh waters having direct connection with the sea. These are different in shape (*Fig. 23.59*) from *Asellus* and swim well. They are only transitory and do not form breeding populations in fresh water.

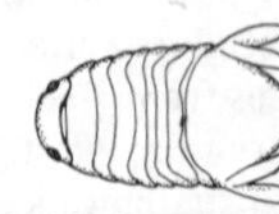

Fig. 23.59

Superorder EUCARIDA

Order DECAPODA

This order derives its name from the five pairs of walking legs possessed by most species. Only a few species occur in fresh waters in north-west Europe and these fall into three easily recognized groups (infraorders):

CARIDEA (Prawns)	Body cylindrical or laterally compressed, abdomen visible from above; thoracic limbs 1 and 2 chelate, both pairs small and delicate.
ASTACURA (Crayfishes)	Body more-or-less cylindrical, abdomen visible from above; thoracic limbs 1–3 chelate, the first pair forming large strong pincers.

BRACHYURA
(Crabs)

Body markedly depressed, carapace flat and more-or-less circular, abdomen folded beneath thorax, not visible from above; thoracic limbs 1 chelate (forming large strong claws).

Infraorder CARIDEA
(Prawns)

Family Palaemonidae

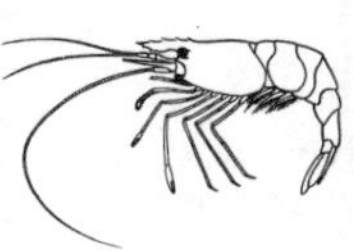

Fig. 23.60

Palaemonetes varians *Fig. 23.60* A very transparent prawn up to about 50 mm long. The rostrum usually has four to six teeth along its dorsal edge, and two ventrally. *P. varians* is a very common brackish-water species that frequently occurs in adjacent fresh waters, where it can survive indefinitely but cannot reproduce.

Palaemon longirostris A larger species than *P. varians*, up to 75 mm, usually with seven to eight dorsal and three to four ventral teeth on its rostrum. A brackish and marine species occasionally found in adjacent fresh water but local and uncommon.

Family Atyidae

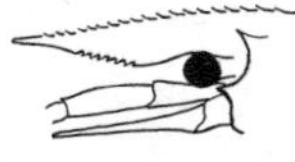

Fig. 23.61

Atyaephyra desmaresti (River Prawn) *Fig. 23.61* This small species, up to 30 mm long, is distinguished from the above species by the numerous – more than ten – dorsal teeth on its rostrum and a brush of hairs on each chela. It occurs in rivers flowing into the Mediterranean and west coasts of Europe as far north as Germany, including France, Belgium and Holland, but not Britain.

Infraorder ASTACURA
(Crayfishes)

Family Astacidae

Crayfishes are unmistakable, lobster-like crustaceans up to about 15 cm long. In life they are mainly greenish or brown but turn orange/pink when boiled! The first pair of chelae are greatly enlarged but, despite their fearsome appearance, crayfishes rarely use their pincers aggressively and can be handled with ease. Males can be distinguished from females by various modifications of their first two pairs of abdominal limbs (pleopods), which in females are normal (the same as the rest). The eggs are carried attached to the female's pleopods.

Crayfishes typically inhabit clean, flowing waters but also occur in some sluggish rivers and occasionally in lakes. They are mostly active at night, hiding by day under stones or in holes under the bank. The parasitic worm *Branchiobdella* (*p. 166*) is sometimes found on their legs and gills.

In many parts of Europe, crayfishes are a prized table delicacy and have been extensively fished and farmed. Unfortunately, during the last century, crayfish stocks in many regions were hit by a fungal disease (*Aphanomyces astaci*) which drastically reduced their numbers. Consequently, several species of American crayfishes (which are immune to the disease) have been introduced, of which *Orconectes limosus* and *Pacifastacus leniusculus* have become established in the area covered by this guide. These and the native species are defined in the key below.

Key to European Freshwater Crayfishes

1. a) Males with a hook (*Fig. 23.63*) near the base of leg 3 (sometimes also on others) and with the first pleopods cleft at the tip to form two branches. Females with first pleopods present and similar in size to the other four pairs. Carpopodite (*Fig. 23.64*) of first leg usually with one (sometimes two) distinct spurs.
   Subfamily Cambarinae, *Orconectes limosus*

   b) Males without hooks on the walking legs; tips of first pleopods rolled to form a loose tube (*Fig. 23.65*). First pair of pleopods in females lacking (leaving four pairs) or very small. Carpopodite of first leg without recognizable spurs but usually with several short, broad spines or rounded tubercles.
   Subfamily Astacinae . . . 2

2. a) Carapace with one pair of postorbital ridges (*Fig. 23.62*); median ridge smooth or absent. . . . 3

   b) Carapace with two pairs of postorbital ridges (*Fig. 23.66*) which may merge in older specimens; median ridge distinct, often toothed. . . . 4

3. a) Sides of rostrum in dorsal view (*Fig. 23.62*) converging towards tip; median ridge present but often small or obscure.
   Austropotamobius pallipes

   b) Sides of rostrum more or less parallel; median ridge absent.
   Austropotamobius torrentium

4. a) Chelae long and slender (*Fig. 23.67*); lateral edges of rostrum usually with five or six teeth. *Astacus leptodactylus*

   b) Chelae stout and strong (*Fig. 23.62*); lateral edges of rostrum smooth. . . . 5

5. a) Median ridge of rostrum toothed; posterior postorbital ridges distinct and equal in size to anterior ones. *Astacus astacus*

   b) Median ridge of rostrum smooth; posterior postorbital ridges smaller than anterior ones, often obsolete. *Pacifastacus leniusculus*

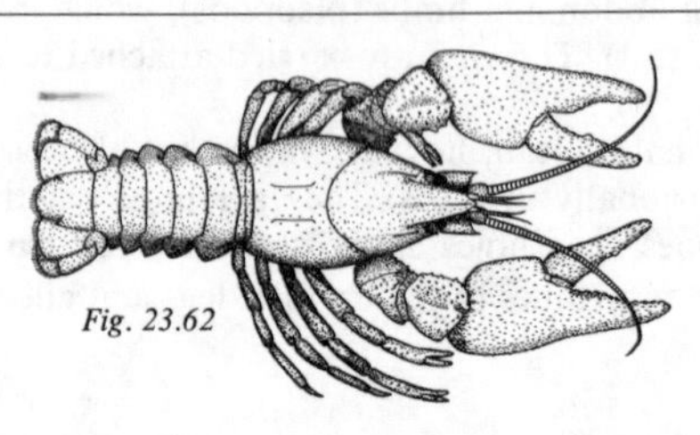

Fig. 23.62

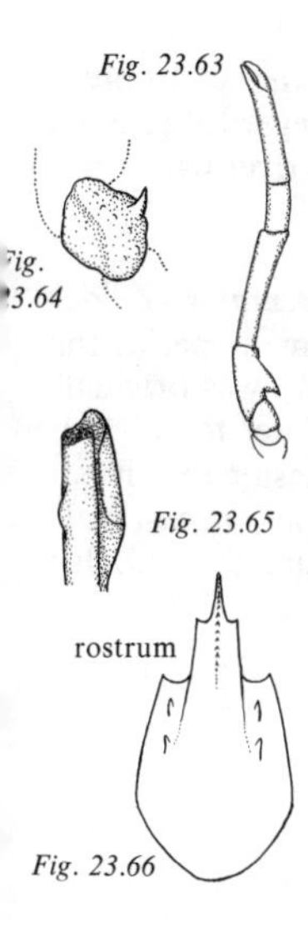

Fig. 23.63
Fig. 23.64
Fig. 23.65
Fig. 23.66
Fig. 23.67

Orconectes limosus *Fig. 23.63, 64* This introduced species is scattered throughout most of the area except for Britain and northern Scandinavia.

Austropotamobius pallipes (White-clawed Crayfish) *Pl. 196, Fig. 23.62* This is the only native species in Britain and is also common in lowland regions of France and Belgium. It attains a body length of about 12 cm.

Austropotamobius torrentium A smaller (8–9 cm) species that is found in mountain streams in southern and eastern parts of the area.

Astacus astacus (Noble Crayfish) *Figs. 23.65, 66* This is the main native species of continental Europe, ranging from southern Scandinavia to north-east France, Germany, Austria and eastern Europe. It reaches a length of 15 cm and is said to be more reddish in hue than the yellower *Aus. pallipes*; in France these two species are known as *écrevisses à pieds rouges* and *écrevisses à pieds jaunes.*

Astacus leptodactylus *Fig. 23.67* An Asian (Caspian) species that appears to be extending its range westwards into Poland, Germany and the Baltic States.

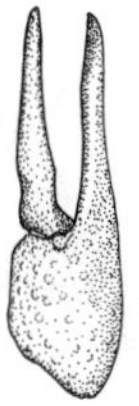

Pacifastacus lenuisculus (Signal Crayfish) An eastern American species recently introduced extensively into Sweden and Finland and scattered localities in central and southern Europe. It is very similar in size and overall proportions to *A. astacus*, but is generally brown in colour, with a whitish spot on the 'hinge' of the claw which is absent in *Astacus*.

Reference: Laurent & Forest, 1979.

Infraorder BRACHYURA
(Crabs)
No crabs are native to north-west European fresh waters but two species have been introduced.

Family Grapsidae

Eriocheir sinensis (Chinese Mitten Crab) *Fig. 23.68* This crab is easily identified by the dense fur covering the chelae; it attains a length of 60 mm. It is an Asian species introduced to Germany early this century which has spread to southern Scandinavia, Holland, Belgium and France, and has been found several times in the British rivers Humber and Thames. *E. sinensis* is a catadromus

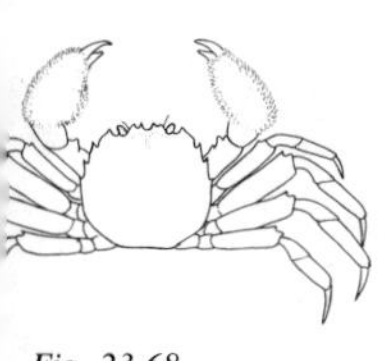
Fig. 23.68

species – it lives in rivers but returns to the sea to breed. In many rivers, such as the Rhine, it is regarded as a pest because its burrows cause erosion of the river banks.

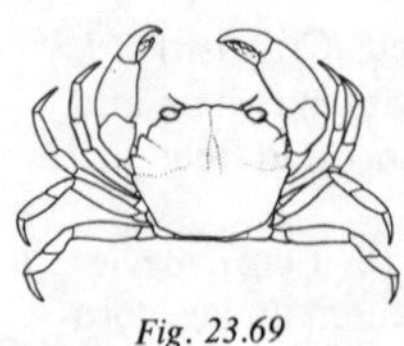

Fig. 23.69

Family Xanthidae

***Rhithropanopeus harrissi* (*Pilumnopeus tridentatus*)** (Zuiderzee Crab, Dwarf Crab) *Fig. 23.69* A small species that rarely exceeds 20 mm in length. This crab was originally discovered in Holland in 1874 but was later found to be identical with an American species; presumably it was introduced. It is now found in estuaries, occasionally wandering into fresh water, from the Seine to the Baltic, but not in Britain.

Reference: Ingle, 1980.

CHAPTER 24 Insects
(Phylum Uniramia, Subphylum Hexapoda)

Recognition features
Adults have segmented bodies secondarily divided into three regions: head, with one pair of antennae; thorax, with three pairs of jointed legs and usually two pairs of wings (sometimes one pair, sometimes reduced or absent); and abdomen, which lacks lateral appendages but may bear tails or other structures at the hind end.
Immature insects (nymphs and larvae), which never possess wings, vary enormously in their structure, appearance and habits. They range from relatively featureless, limbless maggots, through worm-like larvae and caterpillars, to active predatory forms with well-developed legs, large jaws, and often with lateral gills or other abdominal appendages. In spite of this variety they are rarely likely to be confused with other aquatic arthropods or worms. Nearly all species have a hardened (often shiny) head and most possess three pairs of jointed thoracic legs.

The insects form by far the largest group of invertebrate animals, probably outnumbering all others together. Although insects are primarily terrestrial (and aerial) creatures, some groups are aquatic, usually only during the immature stages, but sometimes also as adults. The adult forms of aquatic juvenile insects are usually closely associated with water, although not truly aquatic themselves. In this chapter we will concentrate on the aquatic stages: non-aquatic adults of aquatic juveniles are only briefly described.

Structure
A stonefly nymph (*Fig. 24.1*) is an ideal, generalized insect that illustrates most major features of insect structure. The head and three thoracic segments are constant features although often unequal, and one part may occlude another: e.g., the prothorax in beetles overlaps meso- and metathorax. The number of abdominal segments varies from group to group.

The skin or cuticle of each segment may be partly or wholly hardened (sclerotized) except for the soft 'hinges', each hard part (sclerite) forming an encircling ring, a single large plate or several small ones. In nearly all insects the head is sclerotized so that a protective capsule is formed. Frequently each thoracic segment bears a large sclerotized dorsal plate (notum); these are known respectively as pronotum, mesonotum and metanotum. Sclerotization of the abdomen is often less marked than elsewhere and it is sometimes completely unsclerotized and soft (e.g., caddis larvae). In two aquatic groups the forewings of the adult are partly (bugs) or wholly (beetles) sclerotized to form

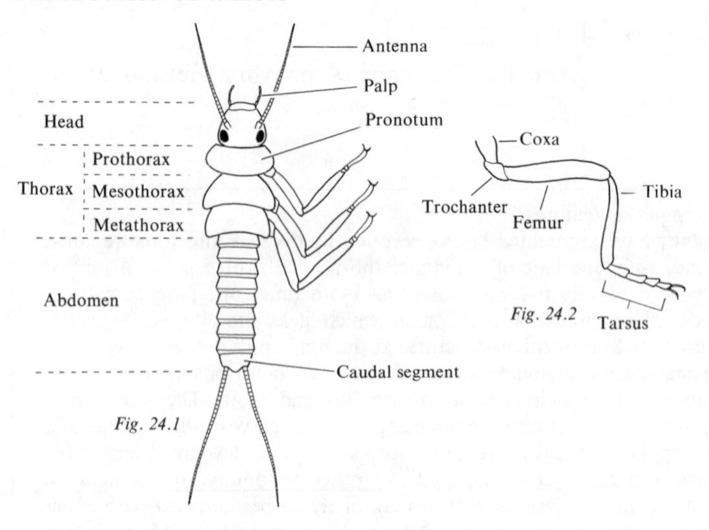

Fig. 24.1

Fig. 24.2

protective cases for the membranous hindwings which, in these groups, are solely responsible for flight.

Appendages: In most groups the only obvious appendages on the head are the segmented antennae which usually originate just above or anterior to the eye. Antennae occur in various shapes but mostly they are simple and filiform (thread-like) and composed of a number of cylindrical segments. In some insects the segmented maxillary and labial palps are prominent and may be confused with antennae, but palps always arise from the mouth region. Insect mouthparts, which are also paired appendages, are important in classification but are not used here except in very broad terms. They are very complex and their study requires special preparatory techniques.

All adult aquatic insects, all nymphs, and many larvae possess three pairs of legs, one pair on each thoracic segment, consisting of several (usually sclerotized) joints; see *Fig. 24.2* for their terminology. In addition to, or instead of these, some larvae possess prolegs – short, soft, unjointed, mobile projections usually ending in tiny spines or hooks. Prolegs may be single or paired and may occur on any segment except the head; they are easily observed on the abdomen of any caterpillar. The two pairs of wings always arise from the meso- and metathorax.

The abdomens of immature insects may bear various paired lateral appendages – gills, prolegs and other projections of various shapes which will be described where appropriate, as will 'tails' and other structures arising from the caudal segment.

Insects are fundamentally terrestrial animals which breathe atmospheric air through tiny openings (spiracles) on the surface of the body. Those aquatic forms which live more-or-less permanently submerged have evolved various methods of breathing whilst underwater. Some breathe atmospheric air at the surface via spiracles located at the tip of the abdomen, which may be elongated to form long breathing tubes. Many nymphs and larvae possess gills, usually on the abdomen but sometimes on the thorax, which are filamentous or plate-like appendages having a large surface area through which diffusion of dissolved gases takes place. Beetles and bugs breathe from a reservoir of air trapped beneath their wing cases or on hairs on their undersides. Most of them need to surface periodically to replenish this supply, but a few species can use their bubble as a kind of lung (plastron), through which sufficient diffusion occurs to enable them to remain permanently submerged.

Life History

Insects differ from many other animals in that most of their life-cycle is spent in the immature stages, adults in general being comparatively short-lived. Insects do not reach maturity until the final moult; hence adults do not (cannot!) grow and usually require little food, needing energy only for locomotion and egg production; some do not feed at all. Thus adult insects, although far more conspicuous and generally better known than the immature stages, play a relatively minor part in the life-cycle, their main function being simply to propagate the species. Their power of flight, which aids dispersal, can usually only work efficiently in warm conditions and so the majority of adults appear only during the spring, summer and autumn.

On the other hand, immature insects are present for most of the year – a typical aquatic insect's life-cycle occupies one year, but may be shorter or much longer. The major purpose of juvenile insects is to feed and grow. Like other arthropods they must moult their skin to allow for growth and each stage between moults is known as an instar. Most species have a fixed number of instars in their development, from as few as three to more than thirty.

Immature insects are called nymphs or larvae, depending on details of their development:

Nymphs are the young of the subclass Exopterygota ('external-winged'). Throughout their development they bear an increasing resemblance to the eventual adult form, and external wing buds appear on the thoraces of older instars. The final transformation into the perfect, winged insect is accomplished by a simple moult from the final nymphal instar.

Larvae, which are the young of the subclass Endopterygota ('internal-winged') never develop external wing buds and usually differ greatly in structure from their eventual adult stages (they may also be disproportionately large). When fully grown, larvae cease to feed and enter a comatose resting stage, either within the final larval skin, or inside a specially-constructed cocoon. This is the pupal or chrysalid stage, which lasts from several days to several months (some species overwinter in this condition). During this period, the fabric of the larval body is broken down and rebuilt into the adult form (metamorphosis). Two types of pupae are recognized: a hard immobile, featureless cocoon or chrysalis such as those produced by fly-maggots; or a more flexible object, best described as a mummified insect with distinct appendages that may be bonded to the body or free and mobile; such pupae may be capable of swimming or crawling, but most remain stationary.

The adult insect that emerges from the pupa or nymph has wings that are soft and crumpled. Before it can fly they must be expanded, by pumping blood into their veins, and allowed to harden. Some aquatic insects choose to emerge on the water surface, a hazardous place accessible to predators both above and below water, so speed in wing expansion is vital to the insect's survival and many accomplish it in a matter of seconds.

A Simplified Classification of Aquatic Insects

Subphylum HEXAPODA (INSECTA)		
Class APTERYGOTA	Order COLLEMBOLA	Springtails
Class PTERYGOTA		
Subclass EXOPTERYGOTA	Order EPHEMEROPTERA	Mayflies
	Order ODONATA	Dragonflies
	Order PLECOPTERA	Stoneflies
	Order HEMIPTERA	Bugs
Subclass ENDOPTERYGOTA	Order MEGALOPTERA	Alderflies
	Order NEUROPTERA	Spongeflies
	Order TRICHOPTERA	Caddisflies
	Order LEPIDOPTERA	Moths
	Order DIPTERA	True-flies
	Order COLEOPTERA	Beetles
	Order HYMENOPTERA	Wasps

The following key should enable the user to assign any aquatic insect to its correct order. Only the pupal stages and some tiny, obscure, parasitic Hymenoptera (*p. 295*) have been omitted from the key. A good hand-lens may be necessary to determine certain features but, as far as possible, characters visible to the unaided eye are used. Very young nymphs or larvae may not work out in this key; identification of these is very difficult, often impossible.

Key to Adult and Immature Stages of Aquatic Insects

[NB This key only applies to aquatic members of the various orders, not to all insect orders in general.]

1. a) Insects without wings (larvae, nymphs, wingless adults). . . . 2
   b) Insects with wings (forewings may form wing cases). . . . 13

2. a) Insects living *on* the water surface. . . . 3
   b) Insects living *in* the water. . . . 4

3. a) Tiny (less than 2.5 mm), short-legged insects usually capable of springing into the air; mouthparts inconspicuous.
   Order COLLEMBOLA (*p. 239*)
   b) Larger (mostly more than 2.5 mm) often long-legged insects incapable of springing; mouthparts forming an elongated piercing proboscis.
   Order HEMIPTERA (part) (*p. 255*)

4. a) Active or relatively inactive, soft-bodied larvae, with short stumpy thoracic legs, or legless, and often incomplete differentiation of head, thorax, and abdomen. ... 5
   b) Relatively active larvae or nymphs with three pairs of well-developed, jointed thoracic legs, and bodies clearly divided into head, thorax, and abdomen. ... 7

5. a) Larvae with three pairs of stumpy, jointed, thoracic legs and a well-developed head.
   Order COLEOPTERA (part) (*p. 281*)
   b) Caterpillars: as a) but additionally with five pairs of prolegs on abdomen.
   Order LEPIDOPTERA (*p. 270*)
   c) Maggot or worm-like larvae: thorax without jointed legs, prolegs often present; head distinct or not. ... 6

6. a) Larvae living inside (mining) plant tissues. Head relatively large, not retractile; body typically more-or-less permanently curved, rather fat, not very mobile; prolegs absent.
   Order COLEOPTERA (part) (*p. 281*)
   b) Larvae probably never living inside plant tissues. Form very varied: head often retractile, from small and indistinct to quite large and prominent; body not permanently curved, often very mobile; prolegs present in many groups.
   Order DIPTERA (*p. 272*)

7. a) Caudal segment with a pair of prolegs, each bearing a sclerotized claw; abdomen soft and fleshy; larvae often inhabiting mobile tubular cases manufactured from various materials, but some are free-living (caddises).
   Order TRICHOPTERA (*p. 262*)
   b) Caudal segment without prolegs, abdomen usually sclerotized; free-living larvae and nymphs. ... 8

8. a) Nymphs with a hinged, protrusible grasping organ (mask) folded beneath the head (*Fig. 24.24*); often quite large, 15–60 mm long when mature.
   Order ODONATA (*p. 246*)
   b) No hinged protrusible mask beneath the head; often smaller than 15 mm when mature but some attain 60 mm ... 9

9. a) Small hairy larvae (up to 5 mm) living on or in freshwater sponges.
   Order NEUROPTERA (*p. 261*)
   b) Nymphs with two or three long segmented tail filaments ('tails'). ... 10
   c) Nymphs or larvae not as a) or b). ... 11

10. a) Abdomen with paired lateral gills and three tails.
Order EPHEMEROPTERA (*p. 240*)
b) Abdomen without paired lateral gills; two tails.
Order PLECOPTERA (*p. 251*)

11. a) Larvae with pincer-like biting jaws; a wide variety of forms often possessing lateral abdominal gills. . . . 12
b) Nymphs with mouthparts forming a tubular or triangular rostrum ('snout'), never pincer-like; never with lateral abdominal gills.
Order HEMIPTERA (part) (*p. 255*)

12. a) Abdomen with seven pairs of lateral filamentous gills and a single, long flexible caudal filament.
Order MEGALOPTERA (*p. 260*)
b) Abdomen without or with lateral gills in various numbers, caudal filaments paired or absent (caudal segment sometimes produced into a stiff rod).
Order COLEOPTERA (part) (*p. 281*)

13. Winged insects.
a) Insects living more or less permanently in the water or on the water surface; with forewings hardened to form wing cases. . . . 14
b) Insects living in the vicinity of water (with aquatic juvenile stages) which may enter or alight on water to deposit eggs. (NB Of course *any* insect may turn up near water and allowance must be made for non-aquatic species.) . . . 15

14. a) Wing cases meet along centre-line of body; mouthparts form biting jaws.
Beetles, order COLEOPTERA (adults) (*p. 281*)
b) Wing cases oblique and overlapping; mouthparts form a tubular or triangular rostrum.
Bugs, order HEMIPTERA (adults) (*p. 255*)

15. a) Insects with two pairs of wings. . . . 16
b) Insects with one pair of wings. . . . 21

16. a) Insects with jointed filamentous tails, usually long but very short in some Plecoptera. . . . 17
b) Insects without jointed filamentous tails. . . . 18

17. a) Forewings markedly larger than hindwings; two or three tails.
Mayflies, order EPHEMEROPTERA (part) (*p. 240*)
b) Wings about equal; two tails only.
Stoneflies, order PLECOPTERA (*p. 251*)

18. a) Abdomen usually slender and elongated; all species longer than 30 mm.
Dragonflies and Damselflies, order ODONATA (*p. 246*)
b) Abdomen not markedly elongated; all species less than 30 mm long. . . . 19

19. a) Mouthparts forming a long slender proboscis normally kept rolled up beneath the head; wings covered with minute scales.
Moths, order LEPIDOPTERA (*p. 270*)

b) Mouthparts inconspicuous, not as in a); wings not covered with scales. ... 20

20. a) Wings covered with fine hairs and hence rather opaque, their veins inconspicuous and with few cross-veins.
Caddis or sedge flies, order TRICHOPTERA *(p. 262)*

b) Wings not covered (but may be fringed) with fine hairs and generally transparent; veins conspicuous and with many cross-veins – netlike.
Alderflies and Spongeflies, orders MEGALOPTERA and NEUROPTERA *(p. 260)*

21. a) Wings held erect at rest; long filamentous tails present.
Mayflies, order EPHEMEROPTERA (part) *(p. 240)*

b) Wings not held erect at rest, typically laid flat over body; long filamentous tails absent.
True-flies, order DIPTERA *(p. 272)*

References to whole group: Harris, 1952. Miall, 1895. Rousseau, 1921. Wesenberg-Lund, 1943. Chinery, 1986.

Springtails
(Class Apterygota, Order Collembola)

Recognition features Tiny wingless insects, 0.5–2.5 mm long, crawling on the water surface and springing into the air when disturbed.

The springtails are common insects of the soil and litter layers. A few species are found on water, where they occur all year round. They have soft cylindrical or globular bodies covered with a fine velvety fur, fairly short legs and well-developed antennae. Folded beneath the rear of the abdomen is a forked, lever-like appendage, which can be suddenly flicked open through 180° to project the insect into the air, faster than the eye can follow. Otherwise slow crawling is the only means of locomotion.

Being so small, aquatic springtails are easily overlooked but close scrutiny of the surface around the margins and amongst emergent vegetation will usually reveal numbers of these tiny creatures. The four main genera containing aquatic species are easily recognized.

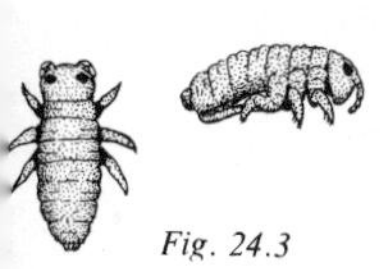
Fig. 24.3

Family Poduridae

Podura aquatica *Pl. 197, Fig. 24.3* Body stumpy, cylindrical, rather wrinkled, dark blue-grey (young may be pink or reddish), up to 1.2 mm long. Very common and widely distributed.

Fig. 24.4

Family Sminthuridae
Sminthurides *Fig. 24.4* Body globular and hunched, reminiscent of a crouching rabbit; external signs of body segmentation lacking. Several species are commonly found on water or in damp situations; usually yellowish or greenish; up to 1 mm long.

Family Isotomidae
Isotomurus palustris *Fig. 24.5* Body elongated, up to 2.5 mm long; yellowish or green, often mottled with brown or purplish. The last few abdominal segments bear stiff sensory bristles.

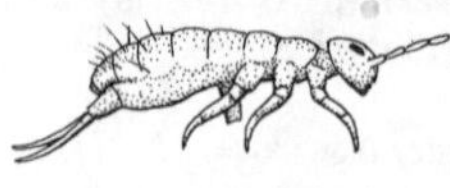
Fig. 24.5

Isotomus This genus resembles *Isotomurus* but lacks the sensory bristles. It includes both terrestrial and aquatic species, usually less than 1 mm long.

Reference: Gisin, 1960.

Mayflies
(Class Pterygota, Subclass Exopterygota,
Order Ephemeroptera)

Recognition features
Winged forms: Body more-or-less cylindrical, 3–25 mm long; forewings large, hindwings small or absent; at rest, wings held vertically together, never folded; antennae usually short, legs long, tails usually longer than the body, two or three; silhouette (*Fig. 24.6*) very characteristic and unmistakable.
Nymphs: Body cylindrical or flattened, with plate-like or filamentous gills arranged laterally or dorso-laterally along the abdomen, and three tails (even when the winged form has only two); 3–30 mm long.

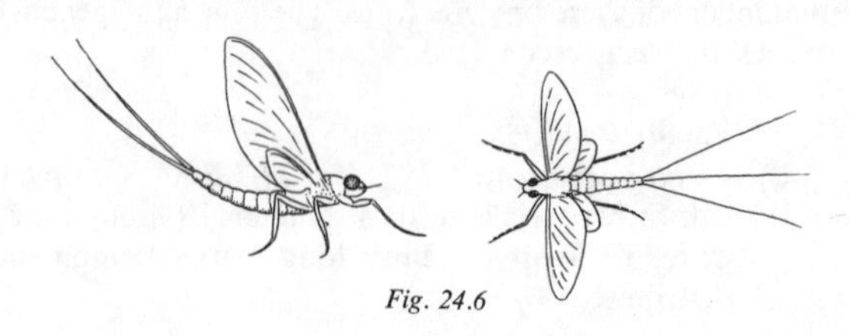
Fig. 24.6

The mayflies, also known as dayflies, upwing flies, olives and numerous more specific or local names, form one of the most characteristic groups of aquatic insects. The winged forms are rarely found far from water and the nymphal stages are always aquatic. Despite their name 'mayfly', adults of one species or another can be found during all but the coldest months of the year.

Unlike all other insects except dragon- and damselflies (Odonata), mayflies cannot fold their wings back over the body, which accounts for their characteristic resting attitude. Although rarely very colourful, winged mayflies are attractive insects with delicate bodies and dusky or transparent wings; nymphs are nearly always drab olive or brownish. Nymphs are a favourite food of fishes and larger invertebrates and consequently spend much of their lives hidden under stones, amongst weeds or detritus, or buried in the sediment. They are rarely found free-swimming except when they rise to the surface to moult into the winged forms.

In some species the eggs hatch soon after they are laid, in others the egg is an overwintering stage. Growth of the nymphs from first to final instars takes from a couple of months to two years, according to species, and may involve as many as twenty-seven instars. Sometimes there are two generations in one year. Mature nymphs, recognizable by their darkened and enlarged wing-buds, swim or crawl to the water surface where the nymphal skin or 'shuck' splits open and the winged insect emerges. Wing expansion is rapid, often occupying only a few seconds, and the insect then flies off to hide in bankside vegetation. This newly emerged winged form (sub-imago) is known to anglers as the 'dun' because of its drab colouration and opaque wings. Large numbers of duns of the same species usually emerge together within a short space of time – an event known to fishermen as the 'hatch'.

Mayflies are unique amongst insects because the sub-imago, within minutes or hours of emerging, undergoes a further moult into a second winged stage, the sexually mature imago. The imago is usually brighter in colour than its dun, with sparkling translucent wings and longer tails; to the angler it is a 'spinner'.

The winged stages do not feed, and live for only a few days at most – hence the name Ephemeroptera – 'day-winged'. Courting males often dance in swarms over the water, usually in the evening, in the hope of attracting females. Mating takes place on the wing, although copulating pairs often fall to the ground or water surface. Later the female deposits her eggs in the water, either dropping them through or brushing them off on the water surface (which may involve tricky low-flying manoeuvres), sitting on the surface and pushing them through, or crawling down some emergent plant or stone. She dies soon afterwards, but will already have outlived her mate.

The appearance of winged mayflies is seasonal and an experienced waterman can often predict the occurrence of a hatch to within a few minutes. So can the birds and fishes which enjoy the orgy of feasting provided by these dainty flies. Not surprisingly, mayflies have long been of importance to fishermen, especially those who pursue trout with artificial flies (mayflies are the model for many of these). At one time the study of these insects by anglers assumed cult status, with the publication of countless articles and books discussing in infinite detail the anatomy, biology and behaviour of these flies, and methods of manufacturing copies of nymphs, duns and spinners – see Harris, 1952.

Apart from the seasons of emergence, which will indicate when mature nymphs are likely to be found (these times are only approximate and will vary according to climate and geography), we do not describe the winged forms in further detail. Many of them are beautifully illustrated in Chinery (1982). The common names given are mainly anglers' terms for various duns and spinners and have little relevance to the nymphs.

Ephemeropteran nymphs can be divided into four groups, according to their form and life-styles.

Group 1 Stout cylindrical nymphs with tails densely fringed with hairs; mostly burrowing forms more than 12 mm long when full grown. (*p. 242*)

Group 2 Nymphs with markedly flattened bodies, up to 15 mm long; typically found clinging to stones in fast-running waters, sometimes in lakes; tails with short sparse hairs. (*p. 243*)

Group 3 Cylindrical or slightly-flattened nymphs up to about 12 mm long; typically slow-moving, and poor swimmers living amongst detritus; tails with short sparse hairs. (*p. 243*)

Group 4 Streamlined, fast-swimming nymphs up to 18 mm long; tails densely fringed with hairs. (*p. 244*)

Group 1 Medium to large nymphs with cylindrical bodies: in burrowing forms, the legs are short and stout, adapted for digging, the appendages on the head are large and prominent and the gills are held close to the body, curved up over the back; gills in all genera consisting of two branches, usually fringed with numerous fine filaments.

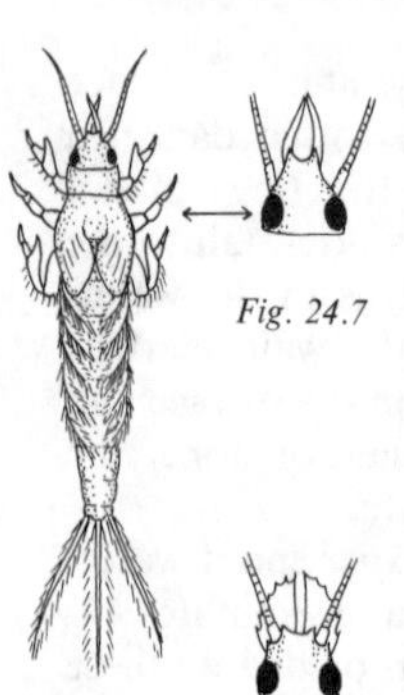

Fig. 24.7

Fig. 24.8

Family Ephemeridae

Ephemera (Mayflies, Greendrakes) *Pls. 198–200, Fig. 24.7* Large nymphs up to 30 mm long; cream-coloured with dark markings; head appendages as in *Fig. 24.7*; gills fringed with numerous fine dark filaments, held over the back in life; burrowing nymphs living in mud or gravel in running waters and lakes. Several species occur in Britain and Europe; duns (three tails) emerge May–July.

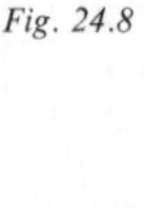

Family Palingeniidae

Palingenia *Fig. 24.8* Nymphs similar to *Ephemera*, differing in the structure of the head appendages and having a middle tail shorter than the laterals. *P. longicauda*, the sole species, is the largest European mayfly, common in large rivers on the continent but absent from Britain; duns (two very long tails) emerge June–July.

Family Polymitarcidae

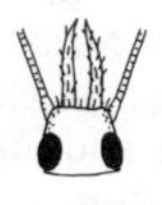
Fig. 24.9

Ephoron *Fig. 24.9* Nymphs similar to *Ephemera* but smaller, not more than about 15 mm long, with different head appendages and sparsely-fringed gills. They live in burrows in river banks. *E. virgo* is a common species throughout Europe but is absent from Britain; duns (two tails) emerge August–September in darkness.

Family Potamanthidae

Potamanthus Nymphs similar to *Ephemera*, up to 18 mm long, but with inconspicuous head appendages and gills that spread out laterally from the body. They live amongst stones or detritus in the slacker parts of large rivers. The sole European species, *P. luteus*, is very rare, possibly extinct, in Britain, but widespread on the continent; duns (three tails) emerge May–June in darkness.

Group 2 Nymphs with broad flattened bodies up to 15 mm long; gills plate-like with a tuft of filaments at the base. They cling to, and can run rapidly across, the surface of stones, and swim strongly if necessary; typically occurring in fast-running waters. Duns have two tails and emerge throughout most of the year, February–October, being more seasonal locally.

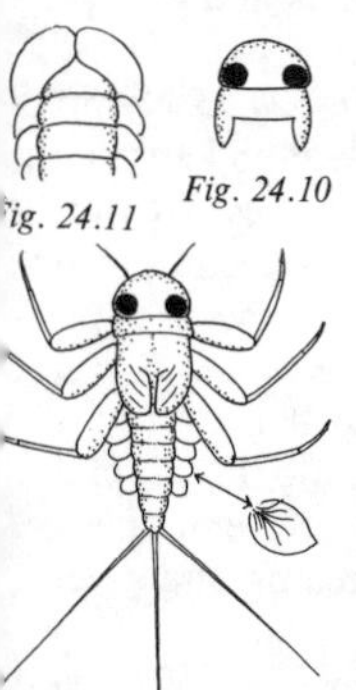
Fig. 24.10 *Fig. 24.11* *24.12*

Family Ecdyonuridae

Ecdyonurus *Pl. 201, Fig. 24.10* Nymphs distinguished from others in this group by the shape of the pronotum (*Fig. 24.10*). Several species occur in north-west Europe, in running waters and near exposed shores in large lakes.

Rhithrogena (March Brown) Nymphs with pronotum shaped as in *Fig. 24.12*; first pair of gills enlarged to meet beneath the body (*Fig. 24.11*); two species occurring only in running waters.

Heptagenia *Fig. 24.12* Nymphs with pronotum as in *Fig. 24.12*; first pair of gills normal (not as in *Fig. 24.11*). Several species are found in running water or large lakes.

Group 3 Nymphs with slightly flattened or cylindrical bodies, up to 12 mm long; gills plate-like or filamentous, varying widely in form; tails not strongly fringed with swimming hairs. These nymphs are generally sluggish movers, swimming poorly with a laboured action, but some are capable of a short burst of speed if pursued. They occur mostly in running waters and live amongst dense weed growth, aquatic mosses, in detritus, etc.

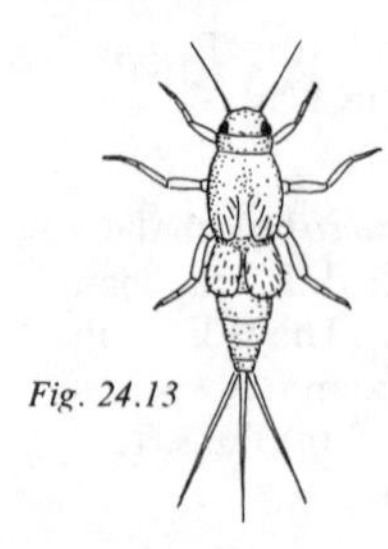
Fig. 24.13

Family Caenidae

Caenis (Broadwings, Angler's Curse) *Pl. 202, Fig. 24.13* Small nymphs less than 9 mm long, often much less; legs and bodies fairly short and stout, first pair of gills small and inconspicuous, second pair enlarged to form a flap which covers the remaining gills; body often covered with silt or detritus. Many species occur in Britain and Europe, in still or flowing waters. Nymphs are common in mud, detritus, under stones, etc.; duns (tiny, one pair of wings, three tails) emerge June–Sept, often in dense swarms.

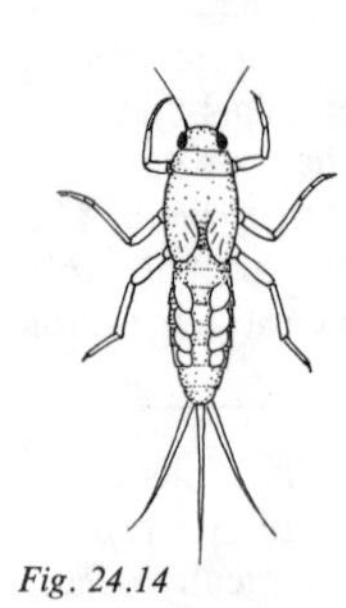
Fig. 24.14

Family Ephemerellidae

Ephemerella (Blue-winged Olives) *Pl. 203, Fig. 24.14* Slightly flattened nymphs up to 9 mm long, often strikingly marked with dark and light bands; gills plate-like, small, attached to dorsal edges of abdomen, not laterally, only four pairs visible. Two species, of which *E. ignita* is the more common, occur in the guide's area, mostly in running waters. Nymphs are present only during spring and summer, the overwintering egg stage being exceptionally long-lasting; duns (three tails) emerge May–September.

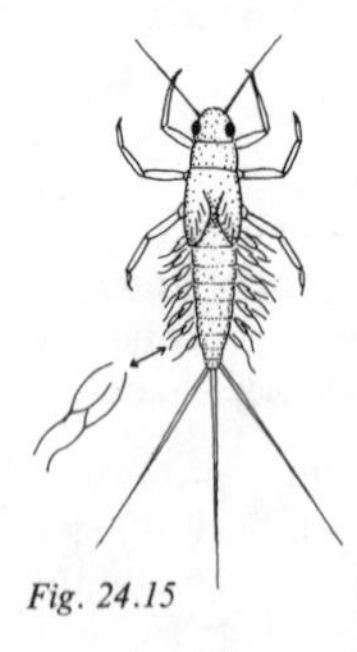
Fig. 24.15

Fig. 24.16

Fig. 24.17

Family Leptophlebiidae

Nymphs up to 12 mm long, with long tails and gills produced into laterally spreading filaments.

Leptophlebia (Claret and Sepia Duns) *Fig. 24.15* Nymphs with gills consisting of a small basal plate and two long filaments; dark reddish-brown. Two species are fairly common in still or slow-flowing waters, typically in upland, acid-water areas; duns (three tails) emerge April–August.

Paraleptophlebia (Turkey Brown) *Pl. 204, Fig. 24.16* Gills each consisting of two filaments with no basal plate; nymphs reddish-brown; 8–11 mm long. Several species, none very common, occur in small streams and rivers; duns (three tails) emerge May–August.

Habrophlebia *Fig. 24.17* Gills each consisting of several filaments arising from a flat plate; 6–8 mm long. *H. fusca*, the sole species, is found in sluggish streams amongst detritus and dead leaves, rather local; duns (three tails) emerge May–September.

Group 4 Nymphs with plate-like, never filamentous gills, and tails strongly fringed with swimming hairs. Typically streamlined, 'agile darters' – swimming well and running rapidly over the substratum.

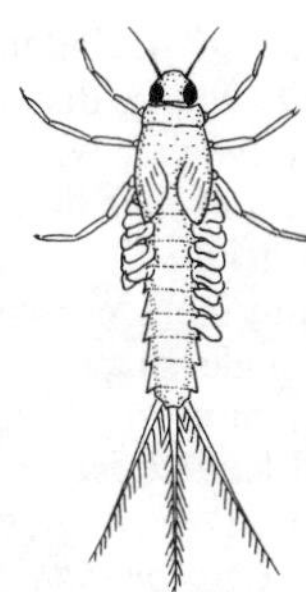
Fig. 24.18

Family Siphlonuridae

Siphlonurus (Large Summer Dun) *Fig. 24.18* Large, robust, olive-brown nymphs up to 18 mm long; abdominal segments produced into short lateral spines (*Fig. 24.18*, several gills removed to show this); gills large, often consisting of double plates. Several European species are known, of which *S. lacustris* is the commonest, typically occurring in large lakes or in slow-flowing rivers; duns (two tails) emerge June–September.

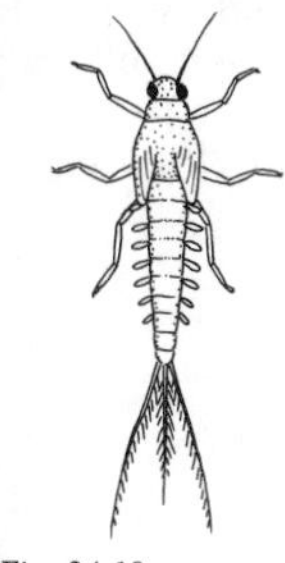
Fig. 24.19

Family Baetidae

A large family with many common species. Identification of nymphs is often difficult, but the four commonest genera can be distinguished.

Baetis (Various Olive, Iron Blue, Pale Watery Duns) *Pl. 205, Fig. 24.19* Middle tail shorter than laterals (equal in other genera); body various shades of brown, from sandy to dark olive, tails usually plain, lacking the fine dark rings found in the genera below but sometimes with dark patch; gills consisting of single plates rounded at the tip; up to 12 mm long. These nymphs occur exclusively in running waters and soon die in small containers unless aerated. There are numerous common and widespread species, with duns (two tails) emerging throughout most of the year (February–November).

Cloeon, Procloeon and ***Centroptilum*** Three very similar genera distinguished from *Baetis* by their more-or-less equal tails, which are marked with fine, dark rings in all species, and best separated by the shape of the middle gills (nos. 3–5, see *Fig. 24.21*) and habitat considerations.

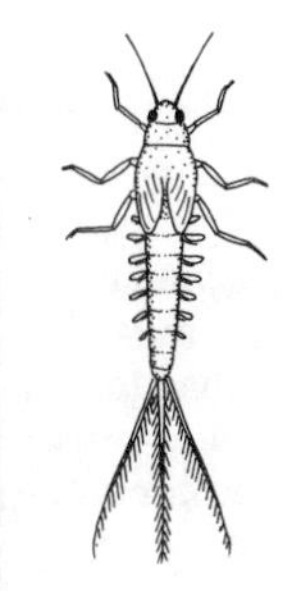
Fig. 24.20

Cloeon (Pond and Lake Olives) *Pl. 206, Fig. 24.20* Gills consisting of two plates except for last pair (single); body olive-brownish, tails with a dark patch on the distal half; up to 10 mm. Two common species are widespread in the area: *C. dipterum* (Pond Olive) with mid-gills as in *Fig. 24.21*; and *C. simile* (Lake Olive), mid-gills as in *Fig. 24.21*, the secondary plate much smaller than in *C. dipterum*. Both are characteristic of still waters, although sometimes occur in slow rivers. *C. dipterum* is typical of small ponds, and *C. simile* tends to favour larger ponds and lakes, but they are found together in some localities; these are often the only mayfly nymphs present in lowland still waters. Duns (one pair of wings, two tails) emerge from April–September.

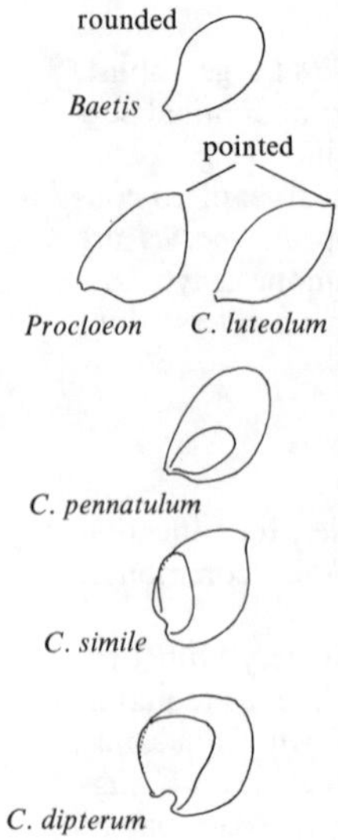

rounded

Baetis

pointed

Procloeon *C. luteolum*

C. pennatulum

C. simile

C. dipterum

Fig. 24.21 Gill shapes of Baetidae (bases at lower left)

Procloeon (Pale Evening Dun) Gills single, shaped as in *Fig. 24.21*; body olive, often mottled, tails with a distal dark patch; up to 8 mm. Sole species *P. bifidum* is widespread in slower-flowing regions of rivers. Duns (two tails) emerge May–October, in late evening or darkness.

Centroptilum (Pale Watery Duns) This genus includes two rather different species: *C. luteolum* Gills single, shaped as in *Fig. 24.21*; body pale buff with darker mottling, tails finely-ringed, with blackish but with no dark patch (sometimes dark at tips); up to 8 mm; common in running waters and occasionally in upland lakes. *C. pennatulum* Gills double (*Fig. 24.21*) except for last; colour as in *luteolum* but dark patch present on tails; up to 10 mm; typically in running waters, rather local. Duns of both species (two tails) emerge April–October.

References: Elliot & Humpesch, 1983. Macan, 1979.

Dragonflies and Damselflies

(Order Odonata)

Recognition features

Suborder ANISOPTERA (dragonflies)

Adults: Large insects with elongated, cylindrical, or slightly flattened bodies up to 80 mm long, often brilliantly-coloured; eyes prominent, very large; wings equal in length, hind-wings slightly broader than fore-wings, they cannot be folded and are held open horizontally at rest; legs rather weak, used for perching, not walking, and for the capture of prey in flight; strong and skilful fliers capable of hovering in one place (*Fig. 24.22*).

Nymphs: Large, stout nymphs with tapered, cylindrical or slightly flattened bodies up to 55 mm long; eyes large or very large, but not always prominent; a hinged, protrusible 'mask', used for prey capture (see *Fig. 24.24*) lies beneath the head and thorax; external gills are absent. They crawl slowly or swim in rapid spurts by jet propulsion.

Suborder ZYGOPTERA (damselflies)
Adults: General structure as for dragonflies, but bodies very slender, cylindrical, up to about 40 mm long; wings equal in size and shape, held together over the back at rest; relatively weak fliers, not capable of hovering.
Nymphs: General structure as in Anisoptera, but bodies relatively slender with three, leaf-like caudal lamellae at the tail end; up to about 40 mm long; swimming by lateral undulations of the body, not jet propulsion.

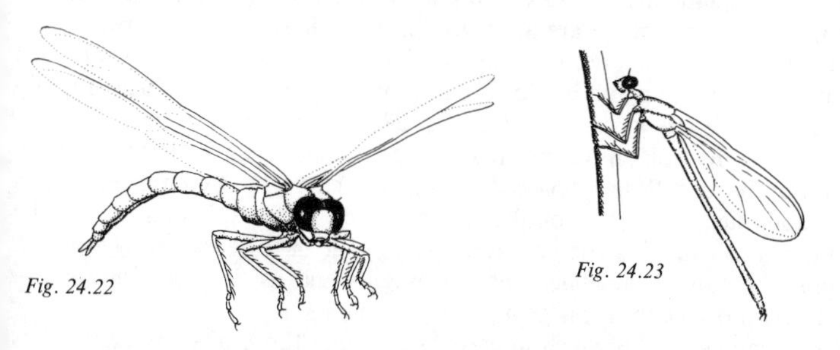

Fig. 24.22

Fig. 24.23

Dragonflies are amongst the largest and most conspicuous of British insects, and are familiar creatures to all who spend time at the waterside. They are instantly recognized by their fast, whirring flight and often vivid colouration – red, yellow or metallic blue or green. Although in the past they have been attributed with various sinister abilities they are harmless to man – they cannot sting – and their diet, which includes many noxious insects such as mosquitoes and biting midges, must be positively beneficial.

The smaller damselflies are equally well known and at least as colourful. They are easily distinguished from dragonflies by their slender bodies and wing position when resting (*Fig. 24.23*).

These insects hunt entirely by sight, using the enormous eyes whose efficiency is increased by the mobility of the head swivelling on the slender neck. The prey is captured in flight with the legs, which are held together to form a 'scoop-net', and then transferred to the powerful jaws.

During courtship the male grasps the female by the head (Anisoptera) or thorax (Zygoptera) with a pair of claspers located at the end of his abdomen. The pair may fly in tandem like this for some time before mating takes place, usually on bankside vegetation. The eggs are deposited soon afterwards, often with the male in attendance, or even still grasping the female. Damselflies and some dragonflies (Aeshnidae) reach through the surface or climb down into the water, penetrate the tissues of water plants with the tips of their abdomens, and deposit their eggs inside. Others drop their eggs, singly or in gelatinous masses, into the water, usually in the safety of weed beds. If laid in summer, the eggs hatch in two or three weeks, but later ones may overwinter. Nymphal life lasts from one to five years, depending on species and climate.

Emergence, the final moult from nymph to winged adult, takes place out of water and usually during the hours of darkness or at first light. The newly-emerged (teneral) insects are dull in colour and do not attain their full glorious colouration for two or three weeks, at which time they also become sexually mature.

If the adult dragonflies, patrolling their beats along the reedy margins, are the bandits of the waterside, then their ferocious nymphs, of grotesque appearance and menacing demeanour, are surely the ogres of the underwater world. Few aquatic invertebrates are safe in their presence, and young fishes, tadpoles, and newt larvae are all devoured in their season. The prey is captured with a unique and fearsome apparatus, the labium or 'mask' – a hinged structure, terminating in a pair of broad pincers, which can be thrust rapidly forward to seize any passing animal (*Fig. 24.24*).

Anisoptera nymphs have internal gills contained in an abdominal cavity opening at the rear. Water is passed through this cavity by rhythmic pulsations of the abdomen wall; jet propulsion is obtained by forcibly contracting the abdomen and squirting the contained water out backwards. In zygopteran nymphs, the caudal lamellae function as gills and as fins for swimming – jetting does not occur in this group.

Most nymphs are brownish in colour, although those of some damselflies are greenish and can even change their colour slightly to harmonize with their surroundings. The large dragonfly nymphs are usually found amongst dense vegetation, the roots of marginal plants, or buried in the sediment. Damselfly nymphs are almost invariably found amongst vegetation, climbing in the foliage where they are well camouflaged.

The identification of nymphs is not easy, and certain species are virtually impossible to separate, the only reliable method being to rear them to maturity and identify the adult. The shape of the mask, viewed from below, is helpful (see *Figs. 24.25, 26*).

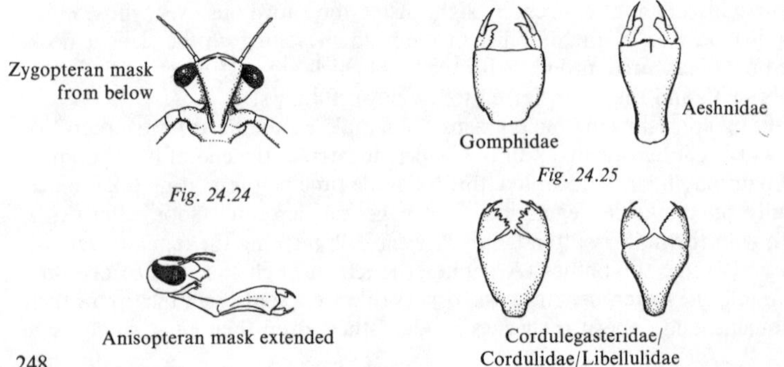

Fig. 24.24

Fig. 24.25

Suborder ANISOPTERA (Dragonflies)
Stout, powerful nymphs without caudal lamellae.

Fig. 24.27

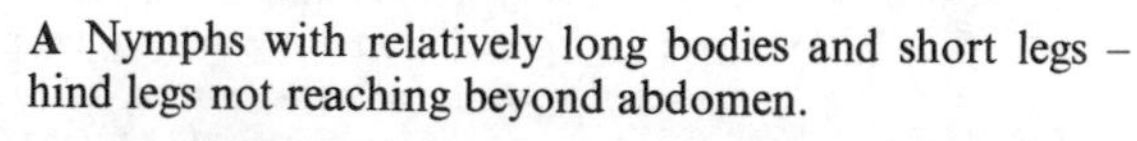

A Nymphs with relatively long bodies and short legs – hind legs not reaching beyond abdomen.

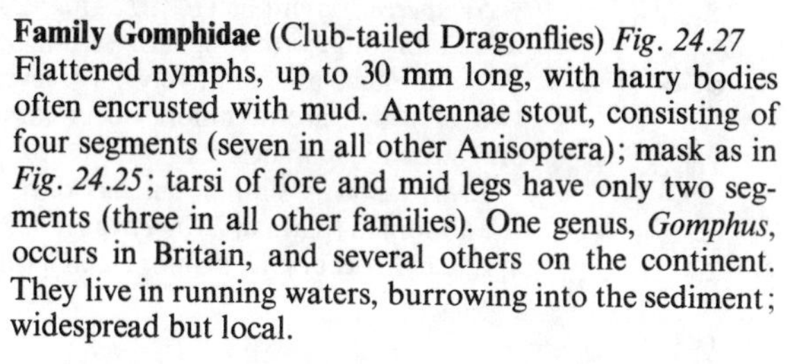

Family Gomphidae (Club-tailed Dragonflies) *Fig. 24.27*
Flattened nymphs, up to 30 mm long, with hairy bodies often encrusted with mud. Antennae stout, consisting of four segments (seven in all other Anisoptera); mask as in *Fig. 24.25*; tarsi of fore and mid legs have only two segments (three in all other families). One genus, *Gomphus*, occurs in Britain, and several others on the continent. They live in running waters, burrowing into the sediment; widespread but local.

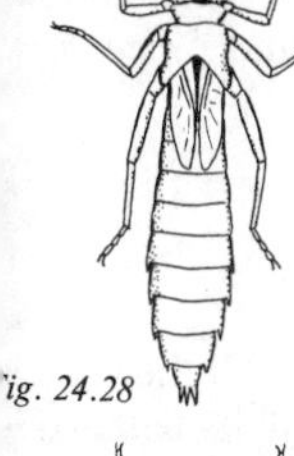

Fig. 24.28

Family Aeshnidae (Hawkers) *Pls. 207, 209, Fig. 24.28*
Cylindrical, smooth-bodied nymphs up to 55 mm long; mask as in *Fig. 24.25*. Three common genera: *Aeshna* and *Anax* have very large eyes, and hunt amongst weeds in still waters; *Brachytron* (up to 40 mm) has relatively small eyes and lurks amongst reed beds, aquatic roots or submerged twigs, clinging to the substratum.

Fig. 24.29

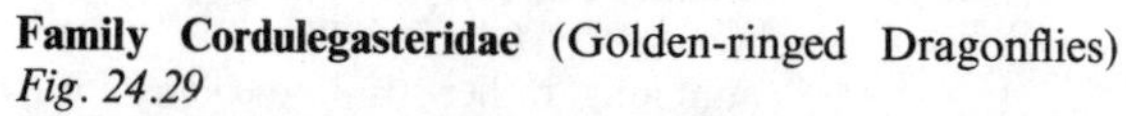

Family Cordulegasteridae (Golden-ringed Dragonflies) *Fig. 24.29*
Slightly flattened nymphs up to 42 mm long, with hairy bodies often encrusted with mud; mask as in *Fig. 24.25*; eyes small, projecting slightly upwards. The sole genus, *Cordulegaster*, is widespread in running waters, living buried in the sediment.

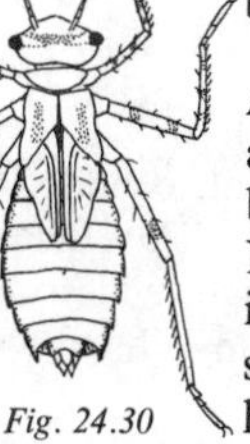

Fig. 24.30

AA Nymphs with long legs, hind legs reaching beyond abdomen; labium as in *Fig. 24.25*, markedly convex; bodies generally short, squat, often somewhat flattened. No reliable, easily-determined characters for distinguishing the two following families are known. All are fairly similar in appearance although found in a variety of habitats.

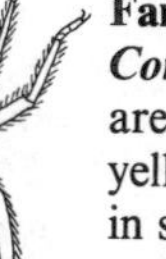

24.31

Family Corduliidae (Emeralds) *Fig. 24.30*
Cordulia is the only common genus in this family. Nymphs are up to 25 mm long, with relatively long antennae and yellowish bodies, boldly marked with dark brown. Found in small, still waters living on, but not buried in, the sediment; rather local.

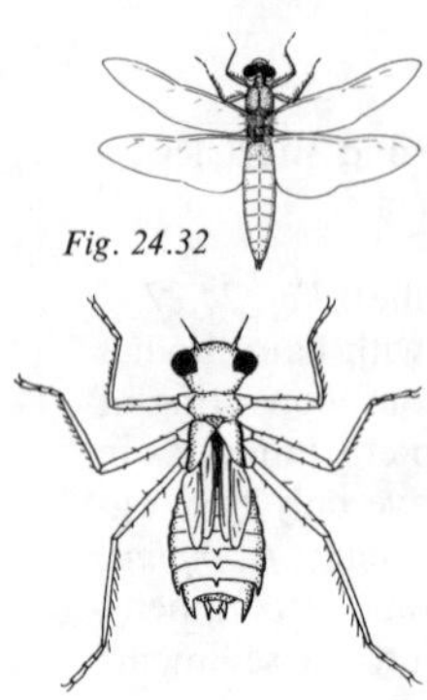
Fig. 24.32

Family Libellulidae (Chasers, Skimmers and Darters) *Pl. 210, Fig. 24.31, 32*

Libellula *Fig. 24.31*) and ***Orthetrum*** are flattened, hairy nymphs, up to 25 mm long, which live partly buried in the sediment in still waters; common and widespread.

Sympetrum nymphs (*Pl. 211, Fig. 24.33*) are small, up to 18 mm long, with long legs, strong spines on the end of the abdomen, and relatively large eyes. Several species, two of which are common, occur in small weedy ponds and streams, marshes, etc.

Leucorhinia is similar but has dark markings beneath the abdomen (plain in *Sympetrum*); it is rather uncommon and local, occurring in moorland ponds and *Sphagnum* pools.

Fig. 24.33

Suborder ZYGOPTERA (Damselflies)

Slender nymphs with three prominent, leaf-like caudal lamellae.

B First segment of the antenna longer than all the other together – very distinctive; lateral caudal lamellae triangular in section.

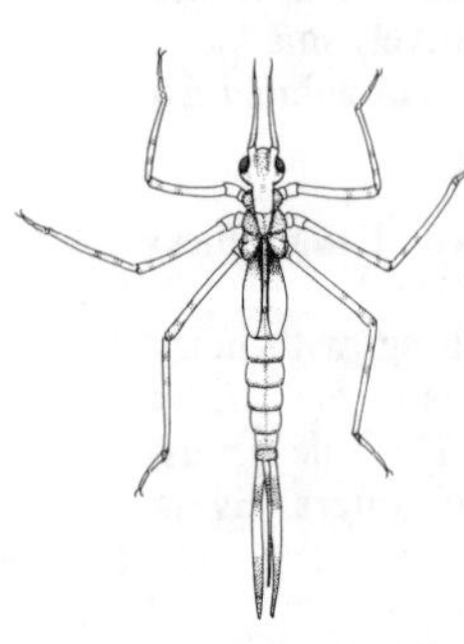

Family Calopterygidae (Agriidae) (Demoiselles)

Calopteryx **(*Agrion*)** *Pl. 212, Fig. 24.34* Nymphs up to 45 mm long, rather stiff-bodied and stick-like compared with other zygopterans; two species are common in running waters.

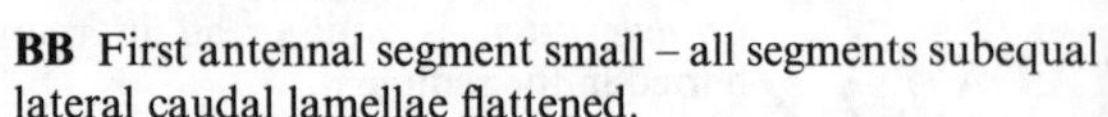
BB First antennal segment small – all segments subequal lateral caudal lamellae flattened.

Fig. 24.34

Family Lestidae (Emerald Damselflies)

Nymphs up to 35 mm long, with blunt, rounded caudal lamellae and distinctive labium (*Fig. 24.26*). One genus *Lestes*, which is rather local, though widespread, in weedy ponds and ditches with clean, unpolluted water.

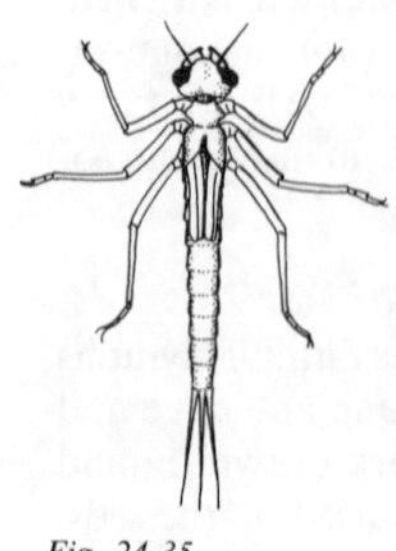
Fig. 24.35

Family Coenagrionidae *Pls. 208, 213–5, Figs. 24.35, 36*

Mature nymphs 15–30 mm long; labium as in *Fig. 24.26* caudal lamellae usually, but not invariably, pointed. This family contains several common genera which are difficult to distinguish as nymphs: *Coenagrion*, *Enallagma* (*Fig. 24.36*), *Ischnura*, *Pyrrhosoma*. All live amongst weeds in slow-flowing or still waters.

Family Platycnemidae
One genus, *Platycnemis*, similar to Coenagriidae but the caudal lamellae are produced into a slender point; labium as in *Fig. 24.26*; lives in unpolluted streams and rivulets, rather local.

References: Lucas, 1930. Hammond, 1983. Corbet, Longfield & Moore, 1960. d'Aguilar, Dommanget, Préchac, 1986.

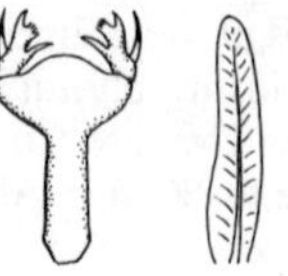

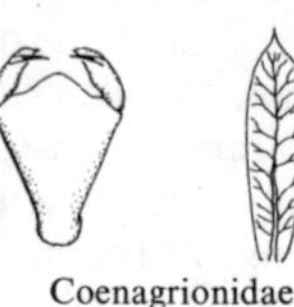

Fig. 24.26 Masks of Zygoptera

Stoneflies
(Order Plecoptera)

Recognition features
Winged adults: *Pl. 218, Fig. 23.37*. Body rather soft, cylindrical or slightly flattened; two pairs of large, equal wings (reduced in some males) which at rest, are folded flat along the back or sometimes rolled around the abdomen; long antennae and two tails (sometimes absent).
Nymphs: Body as in adult, with two well-separated pairs of wing buds in mature nymphs; long antennae and two tails; never with pairs of gills on the sides of the abdomen (cf. Ephemeroptera); up to about 30 mm long.

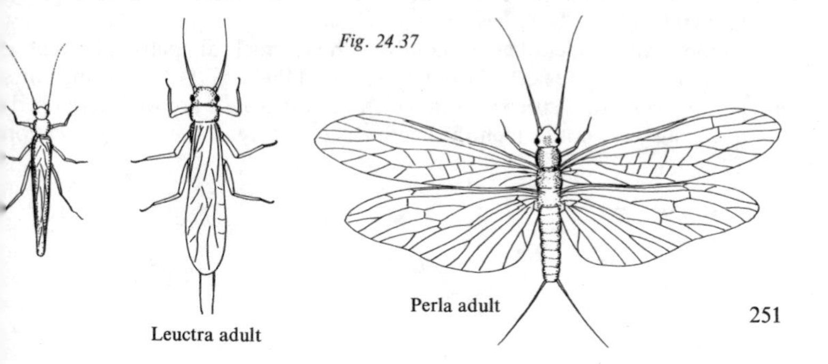

Stoneflies are common waterside insects, but are not generally well known due to their secretive habits, and a tendency for most species to live in out-of-the-way places. The adult flies spend most of their short lives, of one to three weeks, hiding amongst waterside stones or plants; only a few species feed in this state. When they fly, which is neither often nor far, they exhibit a characteristic lack of skill, travelling mostly in a straight line until they land or collide with something. In flight their large wings make them appear much bigger than they really are. Most species are dull brownish or greyish in colour, but a few are brighter, greenish or yellow. Although seasonal in their emergence, winged adults of one or more species can be found at most times of the year.

The nymphs, also known locally as creepers, are invariably aquatic and vary in appearance from small, slender, inconspicuous creatures to fairly large, predatory beasts bearing a superficial resemblance to earwigs. Although usually a dingy brown in colour, some species (e.g., *Perla bipunctata, Pl.* 219) are strikingly patterned. They have strong legs for clinging to stones in fast currents, and are often found in the company of the flattened, stone-clinging mayfly nymphs such as *Ecdyonurus*. The presence of only two tails distinguishes them immediately from mayfly nymphs, which have three.

Stonefly nymphs typically inhabit fast-running, stony streams and rivers in upland areas where they live beneath stones, in aquatic mosses or in drifts of accumulated debris, dead leaves, etc. Some species are also found on wave-beaten stony lake shores, and others, mostly the smaller, inconspicuous types, occur in lowland rivers or lakes. They are often overlooked in the latter habitats because of the greater abundance and prominence of other aquatic insects such as mayflies (Ephemeroptera), sedges (Trichoptera) and dragonflies (Odonata).

Adult flies mate on the ground or amongst vegetation, and the fertilized females deposit their eggs in the water by flying or running over the surface and dropping or brushing off their egg masses. Up to thirty or so nymphal instars may occur before the mature nymphs crawl out of water to undergo the final moult to adulthood. The length of the whole life-cycle is usually one year but it may take two or three in some larger species.

Adults and nymphs are eaten eagerly by fishes and, in areas where they are abundant, they are of some importance to anglers. The nymphs are used as bait (creepers) and the adults are imitated by artificial flies; some, confusingly, are known locally as 'mayflies'.

Nymphs can be identified to family without much difficulty, although a good hand lens may be needed to study details of the tarsi. The following notes apply only to mature specimens in which the wing-buds are well developed; younger stages cannot be identified with any certainty. A useful identification

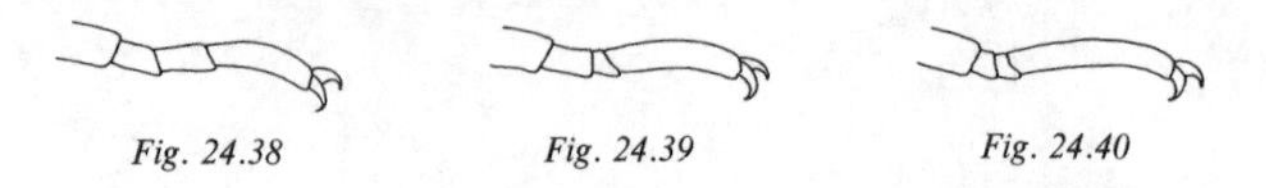

Fig. 24.38 *Fig. 24.39* *Fig. 24.40*

feature is the relative lengths of the three tarsal (foot) segments: *Fig. 24.38* – segments subequal (middle segment fairly long); *Fig. 24.39* – middle segment much the shortest; *Fig. 24.40* – segments 1 and 2 very short.

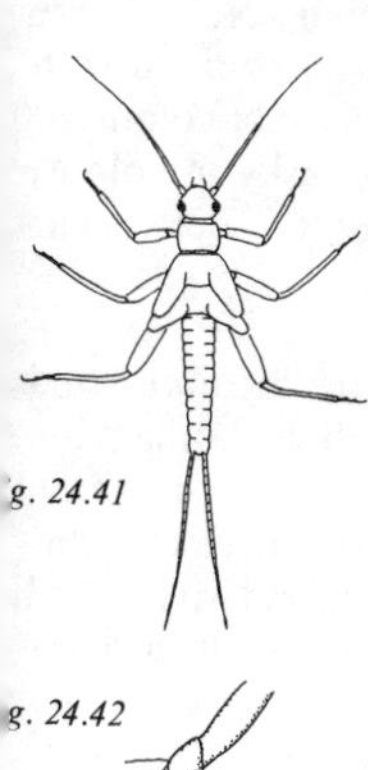

g. 24.41

Family Taeniopterygidae *Fig. 24.41*
Medium-sized nymphs, 8–12 mm long, with prominent, obliquely-set wing buds which project strongly over the sides of the body. Tarsi as in *Fig. 24.38*; legs long and strong; antennae and tails long. The principal genus *Taeniopteryx* is distinguished by the presence of a peculiar, jointed gill filament at the base of each leg (*Fig. 24.42*). This genus is unusual as the nymphs inhabit slow-flowing, muddy rivers, usually hiding in vegetation. Other genera, e.g., *Brachyptera*, do not possess leg gills and are found in stony upland streams.

g. 24.42

Family Nemouridae *Pl. 217, Fig. 24.43* Small or medium-sized nymphs, 5–10 mm long, generally resembling Taeniopterygidae, but tarsi are as in *Fig. 24.39*. Legs very long – hind legs reaching well beyond abdomen. Several common genera which occur in a wide range of habitats: *Nemoura* and *Nemurella* lack gills, *Amphinemura* and *Protonemura* possess tiny bunches of filamentous gills beneath the 'neck'.

. 24.43

Family Leuctridae (Needle or Willow Flies) *Pls. 216, 218, Figs. 24.37 (left), 44* Small, slender, cylindrical nymphs usually less than 10 mm long. Legs relatively short (not reaching tip of abdomen); tarsi as in *Fig. 24.39*; wing buds more-or-less in line with body; last few abdominal segments entire (cf. Capniidae). Principal genus *Leuctra*, which has many species occurring mostly in stony streams, but one, *L. geniculata*, is found in lowland rivers; common and widespread.

24.44

Family Capniidae *Fig. 24.45* Small nymphs, up to 9 mm long, resembling Leuctridae but the last few abdominal segments are divided into dorsal and ventral plates (*Fig. 24.45*). The only genus, *Capnia*, is found in small, stony streams or upland lakes, mostly in the north and west.

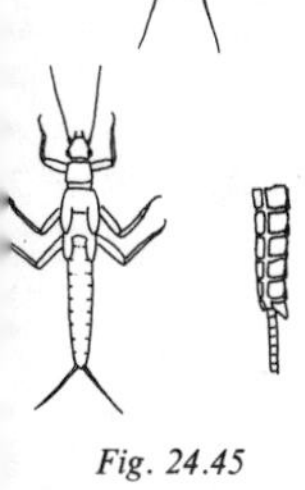

Fig. 24.45

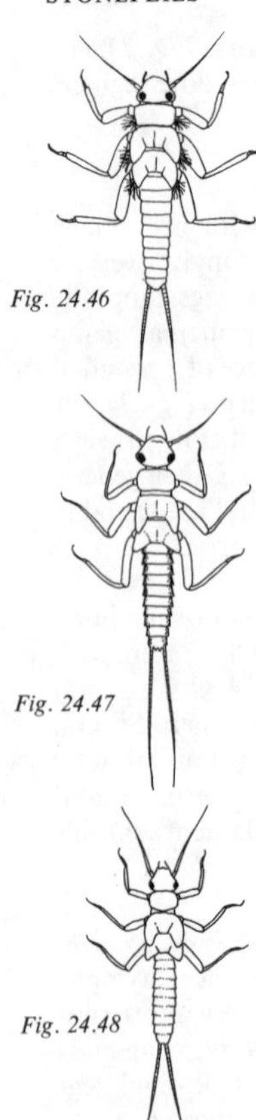

Fig. 24.46

Fig. 24.47

Fig. 24.48

Family Perlidae (Large Stoneflies) *Pls. 219–220, Figs. 24,37, 46* Large, stout-bodied nymphs up to 30 mm or more long. Each thoracic segment bears lateral bunches of fine, white filamentous gills; tarsi as in *Fig. 24.40*. There are two common genera, *Perla* (*Pl. 219*) patterned with black and yellow, caudal segment distinctly bi-coloured; and *Dinocras* (*Pl. 220*), generally more dusky in colour, with a uniformly dark caudal segment. Both genera occur in fast-running upland rivers.

Family Perlodidae *Fig. 24.47* Large or medium-sized nymphs, 10–28 mm long, resembling Perlidae but lacking thoracic gills; tarsi as in *Fig. 24.40*. Several common genera: *Isoperla*, the adult of which is the distinctive 'yellow sally', 8–14 mm long, is common in lowland rivers and some lakes; other genera, *Diura*, *Perlodes*, occur in rivers in upland areas.

Family Chloroperlidae *Fig. 24.48* Small nymphs resembling rather slender Perlodidae but not exceeding 10 mm in length. *Chloroperla* is a common genus with upland and lowland species, including the 'small yellow sally' (5–8 mm).

Reference: Hynes, 1977.

Water Bugs: Pond Skaters, Water Scorpions, Water Boatmen
(Order Hemiptera)

Recognition features Small or medium-sized insects (one genus is very elongated) living on the water surface or submerged; mouthparts forming a more-or-less prominent, pointed rostrum (*Fig. 24.49*) usually tucked under the head (the rostrum is blunt in Corixidae); forewings (if present) partly sclerotized to form overlapping wing cases; nymphs and adults essentially similar in form, but all nymphs and some adults lack wings.

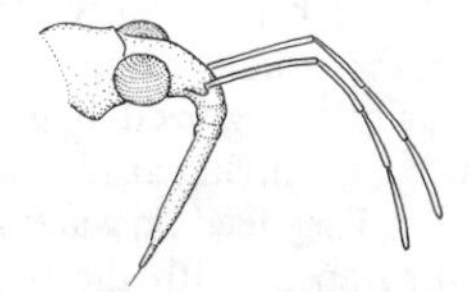

Fig. 24.49

The Hemiptera is a large order of mostly terrestrial insects known as bugs; unfortunately, the term 'bug' is often used to denote almost any small creepy-crawly animal. Water bugs belong to the suborder Heteroptera, in which the forewings are sclerotized to form wing cases that fold flat and overlap on the back, giving a characteristic X- or Y-shaped pattern of fold-lines (*Fig. 24.67*). They are distinctive creatures, unlikely to be mistaken for any other aquatic insects, although some bear a superficial resemblance to beetles (immediately distinguished by the form of the mouthparts and wing cases).

Water bugs are aquatic in both the nymphal and adult stages. Some are exclusively surface-dwelling, walking or skating on the surface film, others live submerged. Nymphs, at least in their older instars, are generally similar to their respective adults, except that they lack wings (*Fig. 24.68*). Adults of many surface-dwelling forms are wingless or have reduced wings; even members of the same species may occur as winged or wingless populations. This may cause confusion between nymphs and adults but does not affect identification. Although most winged adults can fly, in some, the wing muscles are atrophied and individuals are not capable of flight.

Most species are predatory, feeding on any small creatures they can subdue – even tadpoles and young fishes. The prey is pierced with the rostrum and a toxic saliva injected, which partially digests the soft parts; the resultant soup is then sucked up through the rostrum. Because of these endearing habits, the larger, more active forms such as *Notonecta* and the saucer bugs should be handled with care – their bites can be extremely painful to man.

Eggs are deposited on or in plant stems or leaves and the nymphs pass through about five instars before becoming adult. The adults are long-lived and many species are capable of surviving through the winter by hibernating.

Water bugs fall into two easily defined groups:

1. Bugs living on the water surface. (Below)
2. Bugs living submerged. (*P. 257*)

Group 1 Bugs living on the water surface. These have long and prominent antennae and winglessness is common. This group includes the familiar pond skaters and their relatives, most of which prey on insects which fall on to and become trapped in the surface film.

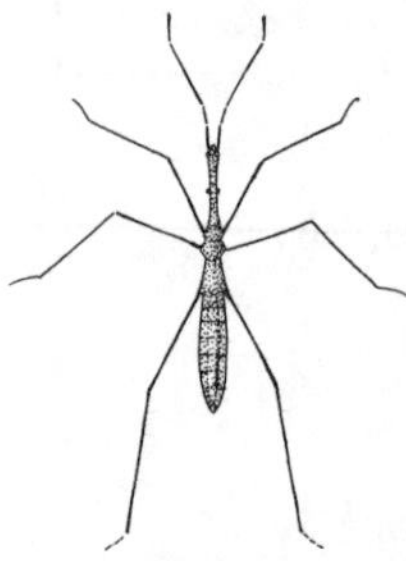

Fig. 24.50

Family Hydrometridae (Water Measurers)
Hydrometra *Pl. 221, Fig. 24.50* Very slender bugs up to 12 mm long; head greatly elongated, about quarter of total length; usually wingless; dark brown or blackish. These easily recognized bugs are common and widely distributed, occurring at the margins of still or slow-flowing waters. They feed on water fleas, mosquito larvae, etc., which they spear with the rostrum through the surface film.

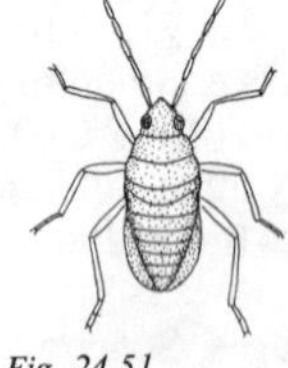

Fig. 24.51

Family Hebridae
Hebrus *Fig. 24.51* Tiny, brownish bugs, not more than 2 mm long, with five-segmented antennae; winged or wingless. They are semi-aquatic, being found amongst *Sphagnum* or other mosses in bogs or at stream margins; widely distributed.

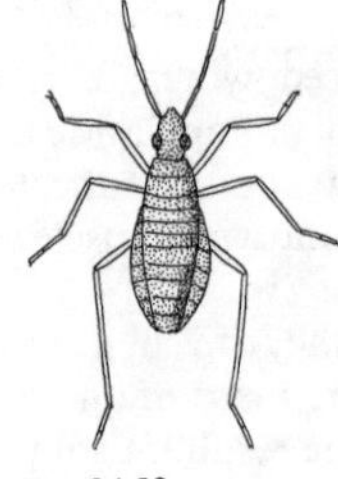

Fig. 24.52

Family Mesoveliidae
Mesovelia *Fig. 24.52* Small, greenish bugs with dark markings, up to 3.5 mm long, with four-segmented antennae; usually wingless. The middle and hind legs are attached to the body near its ventral mid-line – in all other bugs of this group they are attached at or near the edge (*Fig. 24.53*). These bugs are found on floating leaves of pondweeds in still waters, but their distribution is poorly known.

Fig. 24.53

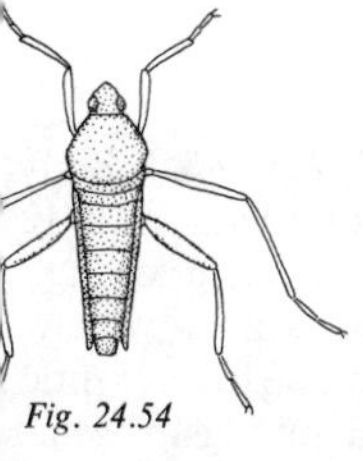

Fig. 24.54

Family Veliidae (Water Crickets)

Velia *Pls. 222–223, Fig. 24.54* Up to 8 mm long, with four-segmented antennae, winged or wingless. The commonest species, *V. caprai*, is brown with two crimson stripes. Common and widely distributed, typically occurring on slacks in flowing waters but occasionally in ponds.

Microvelia *Fig. 24.55* Tiny, brownish bugs up to 2.5 mm long, with four-segmented antennae, winged or wingless. Common, usually found hiding amongst emergent vegetation.

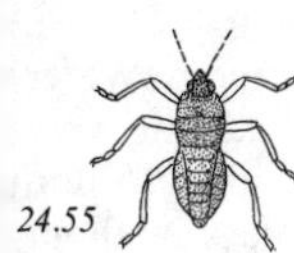

24.55

Family Gerridae (Pond Skaters, Wherrymen)

Gerris *Pl. 224, Figs. 24.49, 56* Slender bugs up to 17 mm long, distinguished from all other surface-dwellers by the spacing between front and middle legs, which is much greater than between middle and hind legs. There are many common species, usually blackish-brown in colour, winged and capable of flight, found in all sorts of habitats including brackish water.

Fig. 24.56

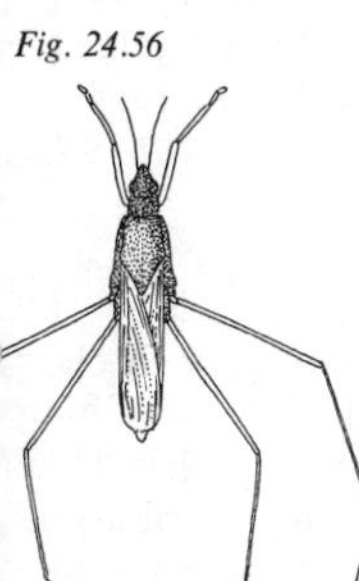

Group 2 Bugs that live in the water. The antennae are tiny and hidden beneath the head. Except for *Aphelocheirus*, all genera in this group are winged, but not necessarily capable of flight, and breathe at the surface via the tip of the abdomen. Four types can be recognized, defined in the following key.

---

1. a) Non-swimming bugs with a long caudal breathing tube.
   (Water scorpions) Family NEPIDAE *(p. 258)*
   b) Swimming forms with hair-fringed hind legs; caudal breathing tube absent. . . . 2
2. a) Rostrum blunt and flattened (*Fig. 24.64*).
   (Water boatmen) Family CORIXIDAE *(p. 259)*
   b) Rostrum sharply pointed, usually needle-like (*Fig. 24.60*). . . . 3
3. a) Body flattened, oval or nearly circular.
   (Saucer bugs) Families NAUCORIDAE and APHELOCHEIRIDAE *(p. 258)*
   b) Body deep, boat-shaped; swimming upside down.
   (Backswimmers) Families NOTONECTIDAE and PLEIDAE *(pp. 258, 9)*

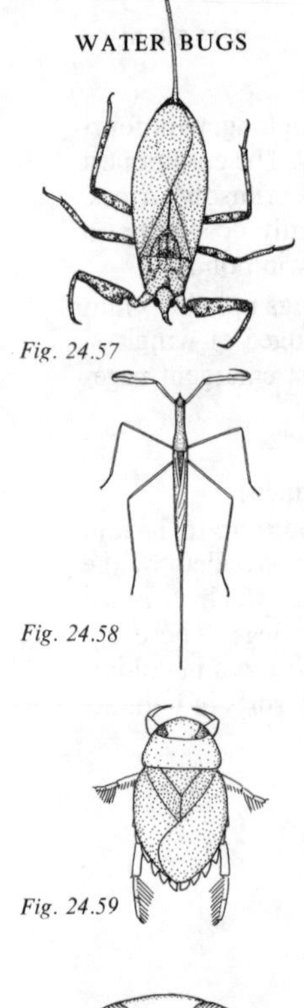

Fig. 24.57

Fig. 24.58

Fig. 24.59

Family Nepidae

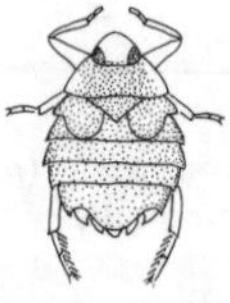

Nepa cinerea (Water Scorpion) *Pl. 225, Fig. 24.57* Body flattened and leaf-like, up to 23 mm long with a 10–12 mm breathing tube; forelegs raptorial (prey-grasping), hind legs without swimming hairs; brown. A widely-distributed species, but easily overlooked due to its cryptic appearance and slow movements; found amongst vegetation in still or running waters.

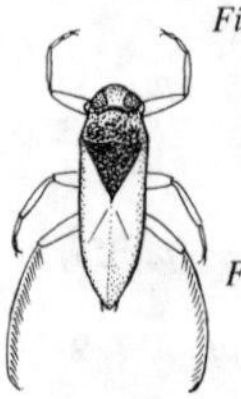

Ranatra linearis (Long-bodied Water Scorpion, Water Stick Insect) *Pl. 226, Fig. 24.58* Body cylindrical, very slender and elongated, up to 35 mm long with a 15 mm breathing tube; forelegs raptorial, hind legs without swimming hairs; brown. This species has a mainly southerly distribution in the area covered by the guide. It lurks in dense vegetation, motionless and mantis-like, waiting to seize its prey.

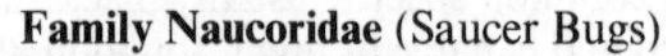

Family Naucoridae (Saucer Bugs)

Ilyocoris (Naucoris) cimicoides *Pl. 227, Figs. 24.59, 60* Broad-bodied, flattened bugs up to 16 mm long; forelegs broad and raptorial, hind legs fringed with swimming hairs; rostrum not reaching behind base of forelegs (*Fig. 24.60*); dark brown. Mostly southerly in distribution, found in weed beds in still waters; a fierce predator with a bite that is painful to man.

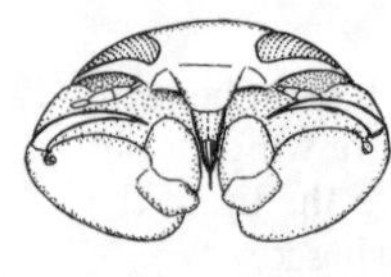

Fig. 24.60

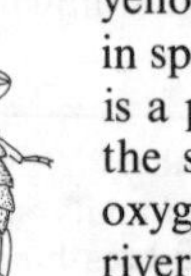

Fig. 24.61

Family Aphelocheiridae (Saucer Bugs)

Aphelocheirus aestivalis *Pl. 228, Fig. 24.61* Body almost circular, markedly flattened, up to 12 mm long; rostrum long, reaching to base of middle legs; dark brown above, yellowish below. Wings are usually absent but may occur in specimens from the extreme south of the area. This bug is a plastron breather (*p. 235*) and hence it does not visit the surface to breathe; but it does require clean, well-oxygenated water, and is usually found in fast-running rivers, where it lives buried in the gravel or clinging to stones. Like *Ilyocoris* it can inflict a painful bite.

Fig. 24.62

Family Notonectidae (Backswimmers, Greater Water Boatmen)

Notonecta *Pls. 229–230, Fig. 24.62* Streamlined, deep-bodied bugs up to 16 mm long, which swim vigorously on their backs using the long, hair-fringed, hind legs as paddles; green, brown and yellowish. Several species are common and very widespread, mostly in still waters.

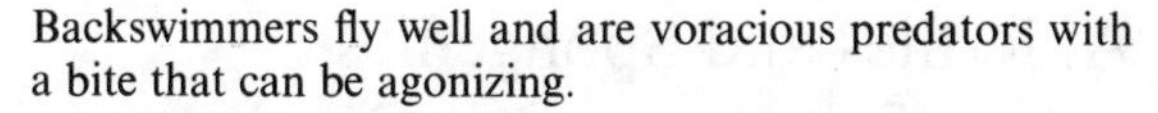

Backswimmers fly well and are voracious predators with a bite that can be agonizing.

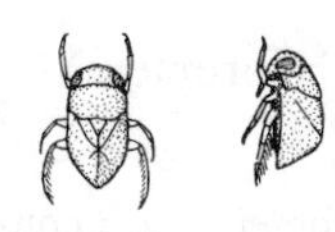

Fig. 24.63

Family Pleidae (Lesser Backswimmers)
Plea *Pl. 231, Fig. 24.63* Tiny, deep-bodied bugs up to 3 mm long which swim on their backs – but their swimming is less powerful and jerky than other swimming bugs; yellowish-brown. Widespread and common, found in dense weed beds, usually clinging to the stems.

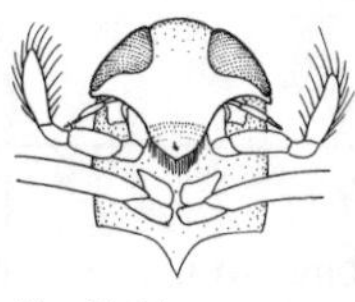

Fig. 24.64

Family Corixidae (Lesser Waterboatmen, 'Corixas') This is the largest family of water bugs and differs from all others by the short, blunt rostrum and short forelegs which terminate in single-jointed, usually flattened tarsi (*Fig. 24.64*). The elongated-oval, slightly flattened body is very characteristic and easily recognized, and the oar-like hind legs are thickly fringed with long swimming hairs (but this feature is also found in *Notonecta* and saucer bugs). Most species feed on algae and plant debris but members of the genus *Cymatia* are predacious. There are two subfamilies:

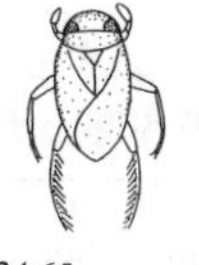

Fig. 24.65

Subfamily Micronectinae (Scutellum visible, *see p. 282*)
Micronecta *Fig. 24.65* Tiny, oval-bodied bugs less than 3 mm long; fairly common in lakes and rivers with a clean sand or gravel substratum.

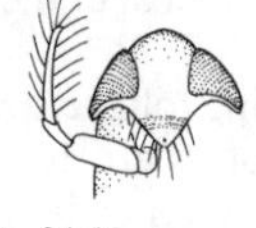

Fig. 24.66

Subfamily Corixinae (Scutellum not visible) *Fig. 24.67*
Cymatia *Pl. 232, Fig. 24.66* Pronotum plain brown, without cross-striations; fore-tarsus narrow and cylindrical; rostrum with a sparse fringe of hairs around its apex; 3–6 mm long. Two species of this carnivorous genus are common in still or slow-flowing waters, typically amongst dense weed-beds.

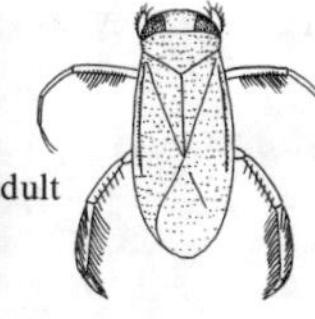

Fig. 24.67

Corixa, Sigara, Hesperocorixa, Callicorixa *Pls. 233–234, Figs. 24.64, 67, 68* Pronotum (and wing cases in many species) handsomely cross-striated with black or brown; fore-tarsus broad and more-or-less flattened; apex of rostrum with a dense brush of hairs; 3–15 mm long.

These are the commonest of the herbivorous corixids, but individual genera are not easily distinguished, although any specimen longer than about 11 mm is almost certainly *Corixa* itself. 'Corixas' are very widely distributed and few habitats and localities in fresh water do not harbour at least one species.

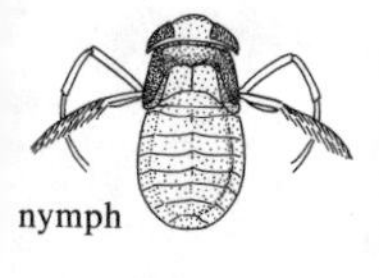

Fig. 24.68

Reference: Macan, 1965.

Alderflies and Spongeflies

(Subclass Endopterygota, Orders Megaloptera and Neuroptera)

These are two small orders, sometimes united under Neuroptera, each containing a single family with aquatic larvae, and a further neuropteran family with semi-aquatic larvae.

Order MEGALOPTERA
Family Sialidae (Alderflies)
Sialis *Pls. 235–236, Figs. 24.70, 71*

> **Recognition features**
> **Adults:** Blackish, soft-bodied flies about 10–16 mm long, with conspicuously broad, blunt heads, long antennae, and four glossy wings with prominent dark veins, folded roof-like over the body when at rest.
> **Larvae:** Body elongated, with strongly sclerotized head and thorax and well-developed pincer-like jaws; abdomen relatively soft, with seven pairs of lateral, filamentous, jointed gills and a single long caudal filament (all fringed with fine hairs); dark brown; up to 26 mm long.

Alderfly larvae are common inhabitants of still or flowing waters, where they live buried in silt or decaying vegetation, or under stones. Adults are typical waterside insects of early summer; they are poor fliers and spend most of their time hiding amongst vegetation.

The life-cycle occupies two years. Eggs are laid on overhanging foliage, and after hatching, the larvae drop into the water. The earliest instars are active swimmers so that the brood can disperse before settling to a life on or in the substratum. Mature (tenth instar) larvae leave the water in the spring and pupate in damp soil or leaf litter near the water; adults emerge a few weeks later, usually in May and June.

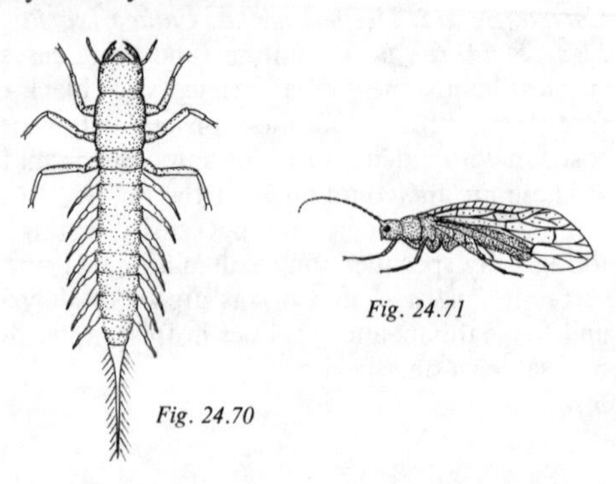

Fig. 24.71

Fig. 24.70

Order NEUROPTERA
Family Sisyridae (Spongeflies)
Sisyra *Pl. 237, Figs. 24.72, 73*

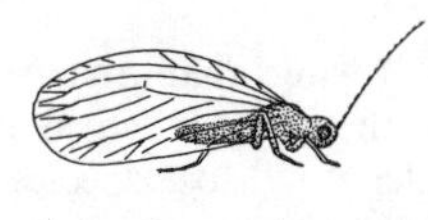

Fig. 24.73

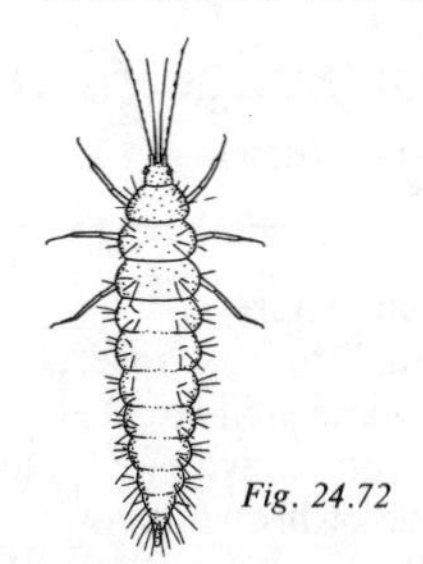

Fig. 24.72

Recognition features
Adults: Small, brownish flies about 4 mm long, with four large, delicate wings with prominent veins; resembling the related terrestrial lacewings.
Larvae: Small, soft-bodied, bristly larvae about 5 mm long, living on or in freshwater sponges on which they feed and which they resemble in colour, usually green or yellowish-brown. The mouthparts form two long slender sucking tubes and there are seven pairs of inconspicuous gills on the underside of the abdomen.

The larvae of spongeflies crawl out of the water in the autumn and spin a cocoon in some sheltered crevice – in bridge-work, tree bark, etc. – where they remain throughout the winter. In the spring they pupate, and the adults emerge around May or June. Eggs are laid in crevices overhanging the water and the newly-hatched larvae drop in and seek out a sponge on which to live. There are only three larval instars. Occasionally, in exceptionally favourable conditions, two generations may occur in one year.

Family Osmylidae
Osmylus

Recognition features
Adult: Resembles *Sisyra* but is much larger, about 12 mm long, and has spotted wings.
Larva: Similar to that of *Sisyra* but less hairy and up to 15 mm long.

The larvae are usually found in damp moss at the edges of small streams, particularly in woodlands, but they occasionally enter the water to feed on chironomid and other aquatic larvae. The life-cycle is broadly similar to *Sisyra*.

Reference: Elliot, 1977.

Caddis or Sedge Flies
(Order Trichoptera)

Recognition features
Adults: Small to large insects with two pairs of wings that are covered with fine hairs and are held roof-like over the body when at rest; antennae very long, often exceeding body length. Most species are some shade of brown, plain or variegated, or black.
Larvae: Body soft and cylindrical, with head, pronotum and often other parts of thorax sclerotized; legs strong and well-developed; abdomen often with filamentous gills, variously arranged, and always with a pair of short or moderate anal prolegs, each armed with a sclerotized hook, on the caudal segment; length ranges from about 3 mm to more than 40 mm. Larvae live in tubular cases made of various natural materials, or free, usually associated with fixed nets, tubes or tunnels made of silk and debris.

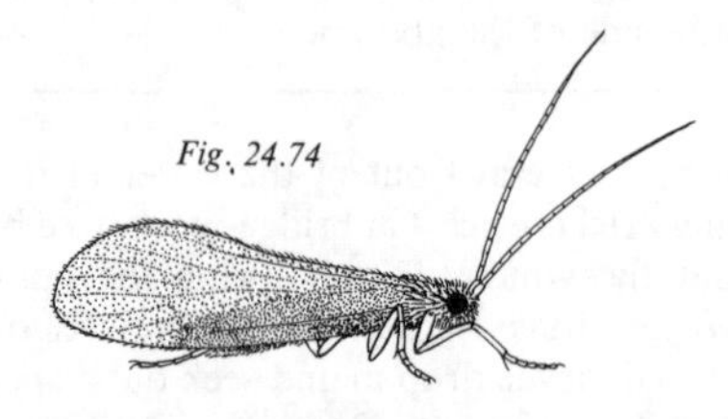
Fig. 24.74

Caddis worms are some of the most common and familiar of aquatic insect larvae (there is one rare terrestrial species in Britain), many being relatively large and conspicuous. The adult flies are active mostly at dusk, hiding amongst vegetation during the day, but a few species fly in daylight. They bear a close resemblance to moths, from which they are distinguished by their simple mouthparts which do not form a long tube (see *p. 270*), and hairy, not scaly, wings.

The eggs are laid in the water or on overhanging plants. On contact with water, the egg masses swell up to become globular or doughnut-shaped blobs of soft jelly, with the eggs arranged neatly within. The larvae of case-building species manufacture their houses from a wide variety of materials – sand grains, small stones, twigs, leaves or pieces of vegetation cut to size, small shells, etc., – which are bound together with a silk-like secretion. In a few species the case is made entirely of this secreted material which, when hardened, acquires a translucent, chitinous texture. The structure of the case is often characteristic of a species but much local variation occurs and it is not unknown for vacant cases of other species to be occupied. When collected, caddis larvae sometimes quit their homes which can lead to confusion of house ownership. Caddis larvae are mostly omnivorous, but some of the larger species are fierce carnivores and should be segregated from their smaller relatives when collected.

When ready to pupate, the case is anchored with silk to some firm substratum and the entrance sealed. The larva then spins a silken cocoon inside the case in which pupation takes place. The legs, wings and antennae of the pupa are not fused to the body, as in most insects, so that when it chews its way out of the pupal case (*Pl. 251*) it can swim or crawl to the bank or water surface (*Fig. 24.76*). Some species 'hatch' at the surface, others crawl up emergent vegetation into the air.

The free-living (non-case-building) forms usually construct nets or tubes of silk, often incorporating local debris, attached to stones or plantlife, in which they live and ensnare their food. The shape of the net varies according to species and local conditions – water current, topography, etc. They pupate within a cocoon of silk, sometimes with a few small stones stuck to it, attached to some nearby solid substratum.

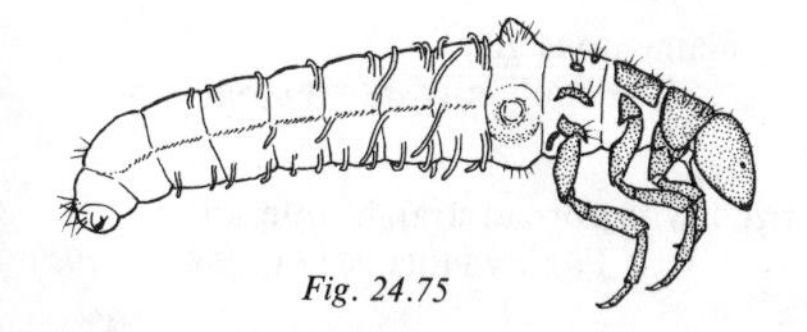

Fig. 24.75

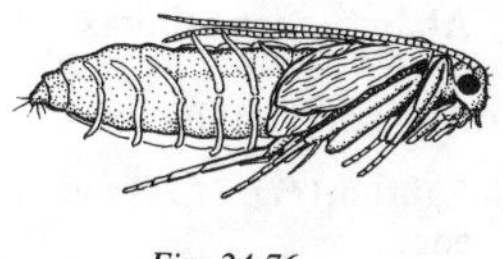

Fig. 24.76

Identification of larvae, at least to family, is reasonably straightforward, but most characters can only be satisfactorily observed if the larva is first preserved. Living larvae are usually far too lively to permit a close examination. If required, however, they can often be evicted from their cases by encouraging them with a pin introduced at the closed end of the case. In a few species or genera the form of the case is sufficient for identification.

The characteristic features require some elaboration, see *Figs. 24.75, 91, 93*. The head and pronotum are always sclerotized; meso- and metanota may be membranous (unsclerotized), membranous with small separate sclerites, or fully sclerotized. Do not confuse the sclerotized leg bases (coxae) with notal sclerites. A few families possess a prosternal horn, a sclerotized spike on the underside of the prothorax between the front legs. The first abdominal segment in case-bearers is often produced into fleshy tubercles, one dorsal and two lateral. Many species have fleshy filamentous gills on the abdomen, either single or in bunches, and a lateral line of tiny hairs or scales is sometimes present. The abdomen terminates in a pair of anal prolegs, which are typically longer and more prehensile in the caseless species and less conspicuous in cased forms. The larvae of certain species are poorly known or even unknown. Readers who wish to rear larvae through to adulthood for precise identification, or simply to identify adults, will find keys and descriptions in Mosely (1939).

Caddises are found in most clean aquatic habitats, from cold, fast-flowing mountain streams to lowland ponds and ditches. They crawl around on the bottom or climb amongst vegetation and a few species can swim, having very light-weight cases. Because of their relatively large size and frequent abundance, caddis grubs are an important item in the diet of fishes, amphibians and many invertebrates. To the angler they are one of the most important groups of aquatic insects, as the pupae and adults are taken avidly by trout, which often pursue the newly-hatched flies along the surface in exciting and spectacular fashion.

Key to the Families of Caddis Larvae

1. a) Larvae lacking a case; free-living, typically associated with a fixed net, tube, or tunnel of silk and debris. ... 2
   b) Larvae living in mobile (rarely fixed) tubular cases. ... 7

2. a) Mesonotum and metanotum membranous (unsclerotized). ... 3
   b) Mesonotum and metanotum sclerotized. ... 6

3. a) Abdomen (and thorax) with tufted, filamentous gills. Family RHYACOPHILIDAE (*p. 265*)
   b) Abdomen without gills. ... 4

4. a) Labrum (*Fig. 24.79*) soft, unsclerotized, with a broad straight anterior edge. Family PHILOPOTAMIDAE (*p. 265*)
   b) Labrum sclerotized (*Fig. 24.79*). ... 5

5. a) Anal prolegs long, sclerotized for distal half of their length. Family POLYCENTROPIDAE (*p. 265*)
   b) Anal prolegs short, sclerotized along their whole length. Family PSYCHOMYIDAE (*p. 266*)

6. a) Abdomen with tufted gills on its ventral surface. Family HYDROPSYCHIDAE (*p. 266*)
   b) Abdomen without gills. Family ECNOMIDAE (*p. 266*)

7. a) First abdominal segment without tubercles. ... 8
   b) First abdominal segment with two lateral tubercles only; (case often square in section). Family LEPIDOSTOMATIDAE (*p. 266*)
   c) First abdominal segment with two lateral and one dorsal tubercle. ... 9

8. a) Case tubular, made of chitin-like, brownish-translucent material, often attached to substratum. Family BRACHYCENTRIDAE (*p. 266*)
   b) Case free, made of small stones, convex with a flat base (*Fig. 24.85*). Family GLOSSOSSOMATIDAE (*p. 267*)
   c) Case free, small (8 mm or less), flattened – seed or flask-shaped, made of chitin-like material, sometimes covered with fine sand or algal filaments. Family HYDROPTILIDAE (*p. 267*)

9. a) Case made of sand grains, surrounded by a broad flat flange (*Fig. 24.89*). Family MOLANNIDAE (*p. 267*)

b) Case made of sand or fine gravel, with a few larger stones arranged laterally (*Fig. 24.90*). Family GOERIDAE (*p. 267*)

c) Case made of various materials, usually more-or-less cylindrical; if made of sand not constructed as a) or b) above. ... 10

10. a) Mesonotum and metanotum membranous. Family PHRYGANEIDAE (*p. 268*)
    b) Mesonotum at least slightly sclerotized; metanotum with a few distinct, separate sclerites. ... 11
    c) Mesonotum at least slightly sclerotized, metanotum membranous without clearly-defined sclerites. ... 12

11. a) Prosternal horn present; metanotal sclerites generally as in *Fig. 24.93*. Family LIMNEPHILIDAE (*p. 268*)
    b) Prosternal horn absent; metanotal sclerites as in *Fig. 24.95*. Family ODONTOCERIDAE (*p. 269*)

12. a) Hind legs not conspicuously elongated as in b) below. ... 13
    b) Hind legs relatively long, extending forward beyond other legs and generally more than twice their length. ... 14

13. a) Pronotum with small lateral projections, case not more than 10 mm long. Family BERAEIDAE (part) (*p. 269*)
    b) Pronotum without lateral projections, case up to 20 mm long. Family SERICOSTOMATIDAE (*p. 269*)

14. a) Hind claw longer than tarsal segment; hind leg with sparse hairs. Family BERAEIDAE (part) (*p. 269*)
    b) Hind claw shorter than tarsal segment; hind legs often fringed with Family LEPTOCERIDAE (*p. 269*)

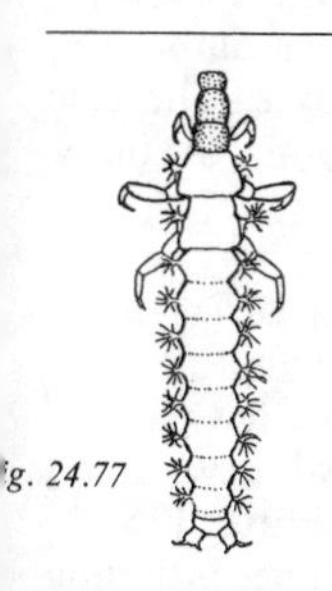

'g. 24.77

Family Rhyacophilidae *Pl. 238, Fig. 24.77* Distinctive larvae up to 20 mm long; mesonotum, metanotum and abdomen green or yellowish, each of these segments with small lateral bunches of filamentous gills. These larvae are active, free-living predators living under stones in running waters, never building nets or tubes of silk. They pupate in roughly-constructed cases of silk and small stones attached to a rock. The sole genus is *Rhyacophila*, with many common species.

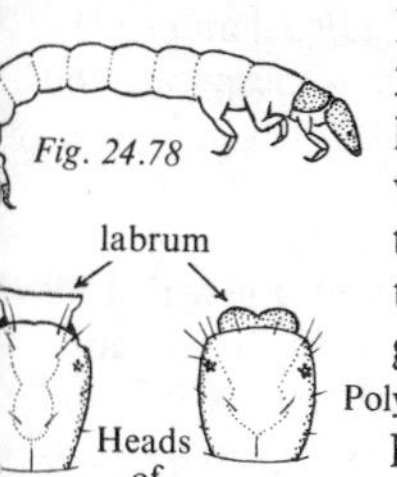

Fig. 24.78

Fig. 24.7

Family Philopotamidae *Pl. 239, Fig. 24.78, 79* Larvae up to 22 mm long, soft parts usually white or yellowish. The soft labrum which distinguishes this family is not always obvious as it may be partly withdrawn. The larvae construct tubular nets up to 40 mm long, closed at one end, attached to rocks in fast-running water. There are three common genera: *Philopotamus, Wormaldia* and *Chimarra.*

Family Polycentropidae *Pl. 240, Fig. 24.79, 80* Body of larvae slightly flattened, up to 25 mm long, soft parts

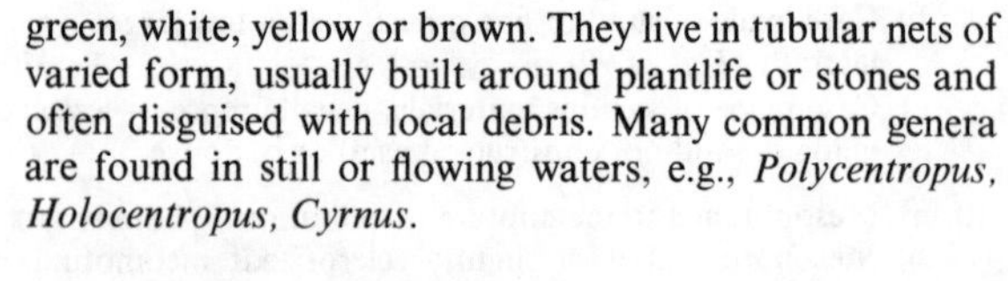

green, white, yellow or brown. They live in tubular nets of varied form, usually built around plantlife or stones and often disguised with local debris. Many common genera are found in still or flowing waters, e.g., *Polycentropus, Holocentropus, Cyrnus.*

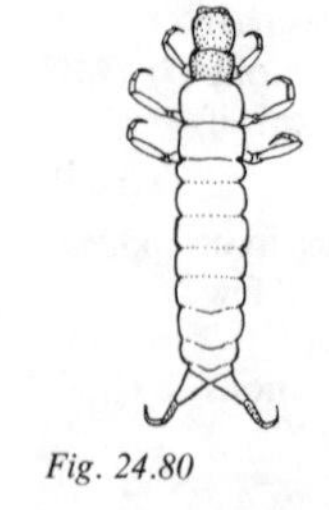

Fig. 24.80

Family Psychomyidae *Fig. 24.81* Larvae small, not more than 11 mm long, soft parts usually brown or green. These larvae live in narrow tubular galleries of silk, often camouflaged with debris, attached to boulders and stones, sometimes on permanently-wet rocks out of water; species of *Lype* scrape out grooves in submerged timber, roofing them over with silk and debris. *Psychomyia, Tinodes* and *Lype* are the commonest genera, found in still or flowing waters.

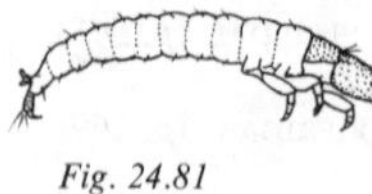

Fig. 24.81

Family Hydropsychidae *Pl. 241, Fig. 24.82* Larvae up to 18 mm long; abdomen usually brownish, typically arched, and covered with short dark hairs. They build flimsy, trumpet-shaped nets amongst plantlife or on stones. *Hydropsyche*, the commonest genus, is usually found in fast-running water.

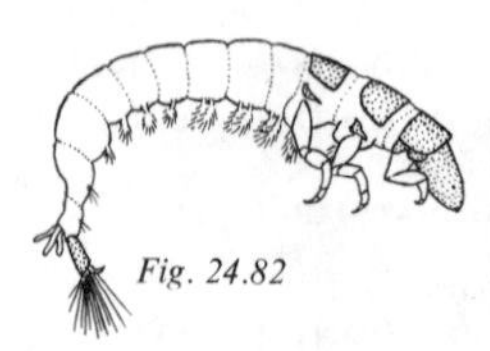

Fig. 24.82

Family Ecnomidae Larvae up to 8 mm long. Little is known of their habits, although according to some authorities they are associated with freshwater sponges. The sole genus *Ecnomus* is widely, but locally distributed in still or slow-flowing waters.

Family Lepidostomatidae *Pl. 242, Fig. 24.83* Larvae up to 11 mm long; mesonotum sclerotized, metanotum with two or more small sclerites. Abdomen with a few pairs of simple filamentous gills, its first segment with two small lateral tubercles and no dorsal tubercle. Case up to 17 × 2.5 mm, usually made of cut vegetation, sometimes including sand grains, typically square in section, at least anteriorly. Several genera, e.g., *Lepidostoma*, are widespread but rather local, mostly in running water.

Fig. 24.83

Family Brachycentridae *Pl. 243, Fig. 24.84* Larva up to 12 mm long; mesonotum sclerotized, metanotum with four small sclerites. Abdomen green or cream, with small bunches of filamentous gills dorsally on segments 2–7, and a well-developed lateral line on 3–6. Case up to 12 × 2.5 mm, cylindrical, slightly tapering, made of translucent,

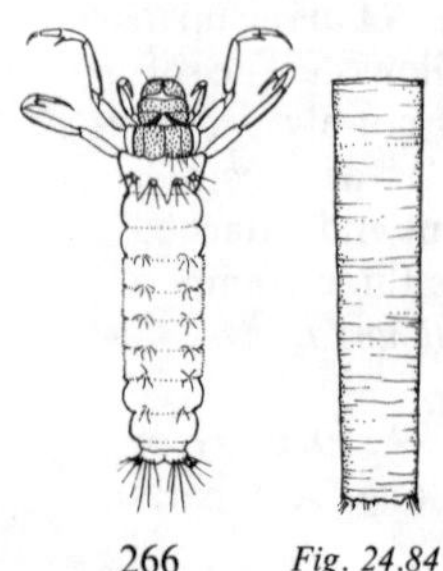

Fig. 24.84

brownish, chitin-like material, usually anchored to stones or weeds, but sometimes free. The sole species, *Brachycentrus subnubilis*, is common in clean, running waters.

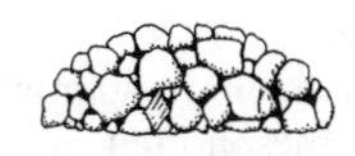

Fig. 24.85

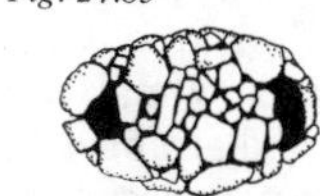

Family Glossosomatidae *Pl. 244, Fig. 24.85* Larvae 4–10 mm long; meso- and metanota membranous (*Glossosoma*) or with sclerites (*Agapetus*); abdomen lacking gills and lateral line. Cases up to 10 × 5 mm, made of small stones, convex with a flat ventral surface in which are two apertures. *Glossosoma* and *Agapetus* are common in fast-running stony streams.

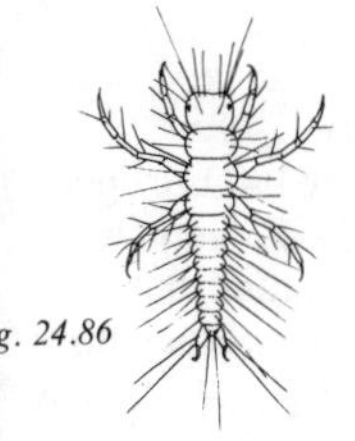

g. 24.86

Family Hydroptilidae *Pl. 245, Figs. 24.86, 87, 88* Young larvae up to the final (fifth) instar are free-living and markedly different in form (*Fig. 24.86*) from the latter stage. Fifth instar larvae (*Fig. 24.87*) live in cases, and are small, always less than 5 mm long, with sclerotized thoracic nota and fat, deep abdomens with deep intersegmental grooves and no gills. The tiny cases, never more than 8 mm long and usually much smaller, are typically compressed with narrow apertures, seed- or flask-shaped, and made entirely of secreted, translucent, chitin-like material (*Ithytrichia, Fig. 24.88, Oxyethira*), bean-shaped and covered with fine sand (*Hydroptila*), or oval and incorporating algal filaments (*Agraylea, Pl. 245*). These four genera are all common in either still or flowing waters.

Fig. 24.87

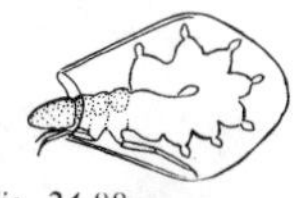

Fig. 24.88

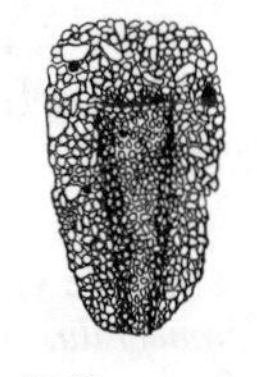

Family Molannidae *Fig. 24.89* Larva up to 17 mm long; mesonotum with several fused sclerites forming a hard patch, metanotum membranous. Abdomen with filamentous gills on segments 1–8, lateral line on 3–8. Case very characteristic, up to 25 × 13 mm, made of sand grains and shaped as in *Fig. 24.89*. *Molanna* is the only common genus; it occurs in still or slow-flowing waters.

Fig. 24.89

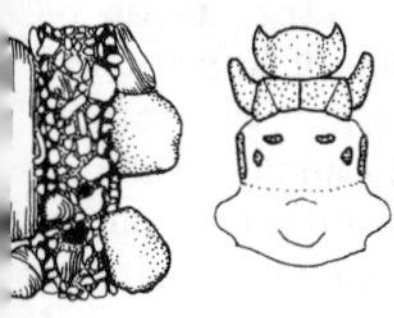

Fig. 24.90

Family Goeridae *Fig. 24.90* Larvae up to 11 mm long; pronotum with anterior corners extended forwards; mesonotum with four large dorsal sclerites, and two lateral coxal sclerites which project forward; metanotum with small sclerites (*Fig. 24.90*). Abdomen with small gills on segments 3–7, lateral line on 3–8. Cases up to 15 × 10 mm, of very characteristic form – a tube of coarse sand grains with a few large stones on either side for ballast. *Goera* and *Silo* are common in small streams and occasionally in still waters. Members of this family are sometimes parasitized by the wasp *Agriotypus* (*p. 295*).

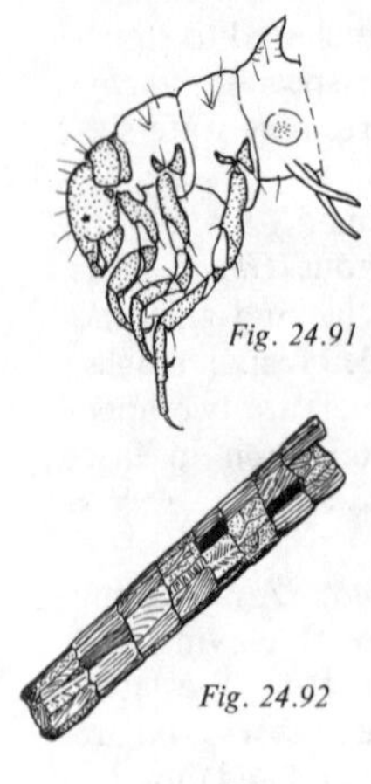
Fig. 24.91

Family Phryganeidae *Pl. 246, Figs. 24.91, 92* This family contains the largest caddis flies, with larvae up to 40 mm long. Thorax with a prosternal horn; mesonotum and metanotum membranous; abdomen stout and cylindrical with filamentous gills on segments 1–7, and a well-developed lateral line. The cases, up to 50×9 mm, are made of vegetable material, cut to shape and arranged in a spiral pattern. *Phryganea*, with the large species *P. grandis* and several others, is the only common genus, found in still or slow-flowing waters.

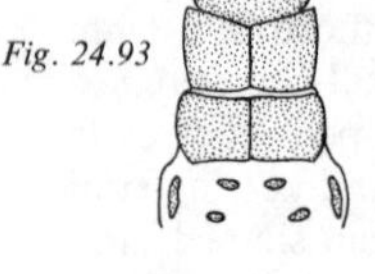
Fig. 24.92

Family Limnephilidae *Pls. 247–248, Figs. 24.75, 93, 94* Easily the largest family of caddis with over 100 European species; the larvae range from about 12–25 mm long. Pronotum and mesonotum sclerotized, metanotum with six sclerites arranged as in *Fig. 24.93*; prosternal horn present. Abdominal features shown in *Fig. 24.75*. The cases are manufactured from a wide variety of materials – stones, sand grains, twigs, leaves and other vegetable debris, even snail shells – a few examples are shown in *Fig. 24.94*.

Fig. 24.93

Fig. 24.94 Cases of various limnephilids

Limnephilus

Glyphotaelius

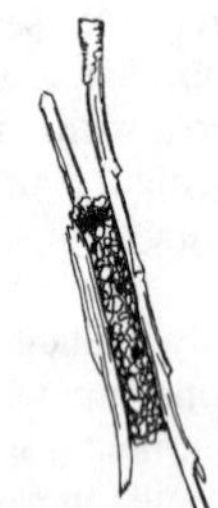
Anabolia

Stenophylax

Some of the commoner genera build distinctive cases but others, including the largest genus *Limnephilus* and others such as *Stenophylax*, use a variety of materials and designs.

Glyphotaelius cases are broad and flattish, made of whole small leaves or rounded portions of larger ones. *Anabolia* cases are tubes of coarse sand-grains with several relatively long twigs arranged longitudinally. *Halesus* cases are similar to *Anabolia* but the central tube is partly or entirely of vegetable matter. *Potamophylax* constructs smooth, parallel-sided, slightly curved cylinders of small stones, but some species of *Limnephilus* and *Stenophylax* (*Pl. 249*) do this too.

Limnephilid larvae occur in all sorts of habitats from fast-running streams to large lakes, and few localities do not hold at least one species.

Fig. 24.95

Family Odontoceridae *Fig. 24.95* Larva about 18 mm long; mesonotum sclerotized, metanotum with four sclerites arranged as in *Fig. 24.95*, prosternal horn absent. Abdomen whitish with filamentous gills on segments 2–7, lateral line weakly developed. Case made of coarse sand grains, slightly tapered and curved, up to 20 × 4 mm. The sole species is *Odontocerum albicorne*, which occurs in fast-running streams.

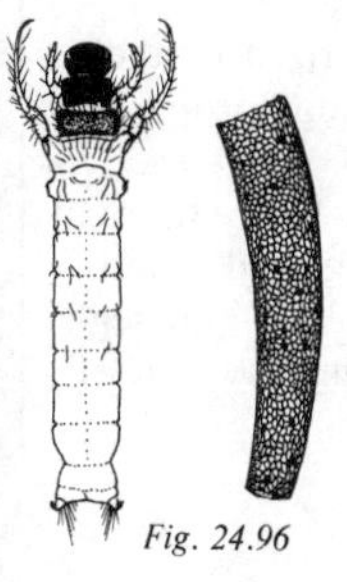

Fig. 24.96

Family Sericostomatidae *Fig. 24.96* Larvae up to 18 mm long; mesonotum sclerotized, metanotum membranous, thorax and legs noticeably hairy. Lateral tubercles on first segment of abdomen rather flat and slightly hardened (semi-sclerotized); a few small filamentous gills are present but variable in distribution. Cases made of coarse sand grains, curved and tapering, the surface with a characteristically smooth finish, up to 15 × 4 mm. *Sericostoma personatum* is the only common species, occurring in streams, more rarely in lakes.

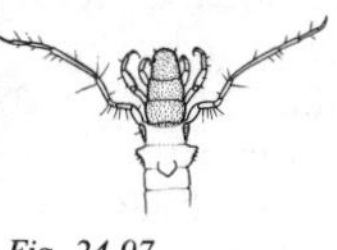

Fig. 24.97

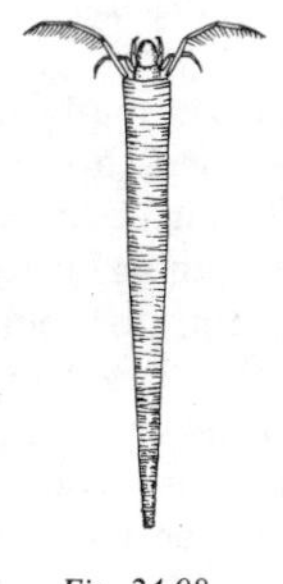

Fig. 24.98

Family Leptoceridae *Pls. 250–252, Fig. 24.97, 98* Larvae ranging from 6–15 mm in length, rather slender compared with other caddises. Mesonotum sclerotized, metanotum membranous; hind legs at least three times length of fore-legs, projecting forward well beyond the head. Abdomen with or without gills, lateral tubercles often with small sclerites. Cases typically slender, usually curved, and up to 18 × 3 mm in most species (up to 35 mm long in *Triaenodes*), made of various materials: sand grains, cut vegetation arranged in a spiral like Phryganeidae but more slender (*Triaenodes*) secreted translucent chitin-like material (*Leptocerus, Pl. 250, Fig. 24.98*), or a mixture of vegetable and mineral materials. There are many common genera found in still or flowing waters: *Athripsodes, Mystacides, Setodes, Triaenodes, Leptocerus*; the latter two can swim, using their hind legs, which are fringed with strong hairs, as paddles.

Family Beraeidae *Fig. 24.99* A small family with poorly-known larvae which possess no common distinguishing features. All are less than 10 mm long and inhabit strongly

Fig. 24.99

curved and tapering cases made of fine sand grains, in small streams and marshes. *Beraea* and *Ernodes* have forward-pointing lateral projections on the pronotum: *Beraeodes* lacks these but its hind claw is very long, exceeding the tarsus.

References: Hickin, 1967. Mosely, 1939. Edington & Hildrew, 1981.

Moths and Butterflies
(Order Lepidoptera)

Recognition features
Adult butterflies and moths are familiar insects characterized by their broad wings, which are covered with tiny scales, and the long proboscis which is carried rolled up beneath the head (*Fig. 24.100*).
Larvae are caterpillars, with a sclerotized head and soft, fleshy thorax and abdomen; thorax with three pairs of short stumpy legs and usually with a pronotal sclerite; abdomen with five pairs of prolegs – on segments 3–6 and 10. Aquatic larvae often construct simple cases out of pieces of cut vegetation or small whole leaves.

Several genera of small moths in the family Pyralidae have aquatic larvae which feed on various water plants. They are locally common, though seldom noticed, in shallow, still, or slow-flowing waters throughout Britain and Europe. Except for *Paraponyx statiotata* the larvae are ordinary-looking caterpillars up to 30 mm long, and apart from their habits and slight differences in colour and size are not easily distinguished. Adult aquatic moths are generally pale, usually whitish, in colour, but the china-mark moths (*Fig. 24.102*) are delicately-patterned with brown and white.

The life history follows a similar pattern in most species and is exemplified by *Nymphula nympheata*. Adult moths, which live for only a few days, appear from June to September and, after mating, eggs are laid on the underside of the floating leaves of various aquatic plants. When hatched, the young larvae burrow into the host plant, feeding as they go (mining) and live thus for some time. Later they break out of their tunnels and construct a flattish case from two oval portions cut from a leaf and spun together with silk. They live in this case through autumn and winter, when they hibernate. As the water warms in spring they resume feeding and, in early summer, spin a silken cocoon, attached to a plant stem near the water line, in which they pupate (*Fig. 24.103*).

The presence of oval holes, up to 50 mm long, or discoloured patches in the leaves of plants such as waterlilies and pondweeds are often signs that aquatic caterpillars are at large in a pond.

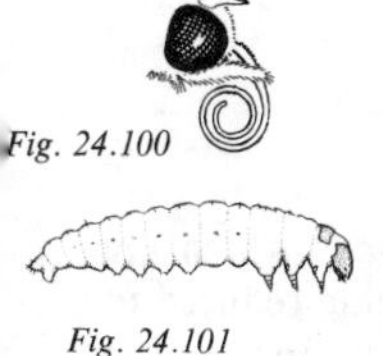

Fig. 24.100

Fig. 24.101

Nymphula nympheata (Brown China-mark Moth) *Pl. 254, Figs. 24.101, 102, 103* Larva up to 25 mm long and light brown in colour; its habits are described above. The case may be free or temporarily attached to the foodplant – typically water lilies, frogbit or pondweeds (*Potamogeton*).

Paraponyx stagnalis A smaller caterpillar than *Nymphula*, up to 20 mm long, yellowish-brown, with a buff head. The preferred food plant is *Sparganium* (Bur-reed) but *Nuphar* or *Potamogeton* may also be used. In *Sparganium* the larva may remain a miner throughout its life, but usually it emerges and builds a case of leaf portions.

Fig. 24.102

Paraponyx stratiotata *Pl. 255* The larva of this species is very distinctive, up to 30 mm long, whitish or pale green, with up to eight, slender, branching gill filaments on each abdominal segment (*Fig. 24.104*). A number of plants are used as food: *Stratiotes* (Water Soldier), *Potamogeton*, *Elodea* (Canadian Waterweed) and *Ceratophyllum* (Hornwort), the larva living free or in a loose tube of silk spun around the stems and leaves; it does not build a case and pupates in a cocoon underwater.

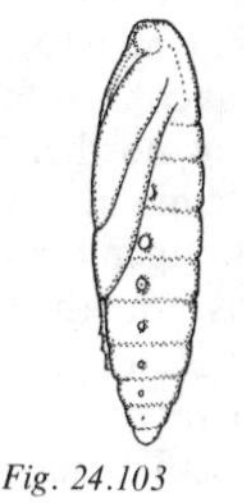

Fig. 24.103

Cataclysta lemnata The caterpillar is up to 18 mm long, dark green or blackish in colour, and lives amongst the floating duckweed *Lemna*. It makes a case of silk and duckweed leaves in which it lives and hibernates, pupating in a separate cocoon of silk and leaves.

Fig. 24.104

Acentria (Acentropus) nivea Adults are small moths with white wings and green bodies, but some females possess only vestigial wings and can swim underwater, paddling with their legs, to deposit their eggs on submerged plants. The larvae also have green bodies, except for the brown, sclerotized head and pronotum, and are up to 12 mm long. They occur on *Elodea, Chara* (stonewort), *Potamogeton* and other plants, around which they spin silken nets (they do not build cases); pupation takes place underwater in a cocoon.

Schoenobius spp. The larvae of the two or three European species in this genus are not truly aquatic but mine the stems of marginal plants such as *Phragmitis* (common reed) and *Carex* (sedge). They attain a length of 35 mm.

True-flies*
(Order Diptera)

Recognition features
Adults: insects with only one pair of membranous wings which are usually held flat over the back at rest, the hind wings being reduced to club-like halteres; with sucking (and often piercing) mouthparts. The best-known dipterans are probably the familiar house-flies and blue-bottles; forms with aquatic larvae include craneflies, midges, gnats, mosquitoes, horse-flies, hover-flies and others less well known.
Larvae: worm-like or maggot-like, without true (jointed) legs but often with short, unjointed prolegs; head sclerotized or not, other parts of the body rarely sclerotized; often with prominent spiracles, breathing tubes, or short filamentous gills at the posterior end. Most feed on decaying organic matter but some are carnivorous.
Pupae: vary from active free-swimming forms to immobile, featureless hard cocoons.

Adult dipterans exhibit a wide variety of form but are easily recognized by the possession of only two wings. No adults are aquatic. Many flies bite man (or, more correctly, pierce his skin and suck his blood) and in some areas constitute a considerable nuisance or even danger to him. Fortunately, in northern Europe, malaria and other fly-borne diseases are no longer endemic but in their season, certain species can make outdoor life insufferable. Biting midges (Ceratopogonidae) and black-flies (Simuliidae) are notorious for their irritating bites and appearance in enormous swarms; mosquitoes require little introduction as they often enter our houses and attack us there; and the menacing silent approach, usually behind the line of sight, of horse-flies adds unpleasant surprise to the pain of their bite. Even the harmless chironomid midges, which swarm in great numbers around water margins, can prove a distraction and constant irritation, no doubt because their buzzing flight is all too reminiscent of their biting cousins.

The species of Diptera possessing aquatic larvae probably outnumber all other aquatic insects together, and larval structure encompasses a wide range of forms. The body segmentation is usually distinct but the thoracic and abdominal regions are rarely differentiated. Abdominal segments are often subdivided. Most forms are slender and worm-like or fat, fleshy maggots; the bristly mosquito larvae constitute a third type. Only a few larval forms can swim and this is achieved with a violent wiggling action, the body bending through U or S shapes. Some pupae are active swimmers and descriptions of these are given where appropriate.

* Many insects of other orders are also called 'flies', so we here follow the system of Oldroyd (in Chinery, 1972) in hyphenating the names of dipteran flies but not those of others.

Dipteran larvae occur in most habitats, from truly terrestrial – in the soil, in or on plants, in animal carcases, in dead timber, etc. – through a wide range of transitional (damp or marshy) conditions to fresh or salt water. Only a minority live in 'normal' aquatic habitats, most living in temporary or polluted waters – puddles, rot holes in trees, sewage beds, saline or strongly alkaline lagoons, various farmyard habitats, drains, gutters, water-butts and so on. It is impossible to draw a line separating aquatic species from others so we have provided descriptions of those groups, and typical genera, most likely to be found in conditions that are definitely wet.

The larvae of many dipterans are little or not known, and this applies particularly to those which are associated with water and therefore may have aquatic larvae. It is therefore probable that the diligent collector will occasionally discover larval dipterans which cannot readily be identified. In such cases rearing the larvae through pupae to the adult stages may be the only means of discovering their identity. With a little forethought this is unlikely to be difficult, but bear in mind that the larvae may require provision to leave the water to pupate. Use a shallow, covered, container with a little of the natural substratum as a food source. Predatory forms must be fed regularly and various other dipteran larvae can be used as a convenient and often abundant food supply.

Variation between genera is often as great as between families and so the construction of a useful key to all families is not possible. The following key separates the three suborders of aquatic dipteran larvae.

---

1. a) Head very small or apparently absent, never sclerotized; mouthparts hook-like or absent; generally maggot-like larvae (never free-swimming). Suborder CYCLORRHAPHA *(p. 280)*
   b) Head always distinct and at least partly sclerotized (but may be retracted into the prothorax); worm-like or maggot-like larvae. ... 2
2. a) Head entirely sclerotized or nearly so; mandibles opposed (pincer-like) or adapted for filter-feeding (brush-like); antennae often distinct (larvae often free-swimming). Suborder NEMATOCERA *(p. 273)*
   b) Head only partly sclerotized, usually small, often retracted into the prothorax; mandibles form parallel hooks, not opposed; antennae small or hidden (larvae never free-swimming). Suborder BRACHYCERA *(p. 279)*

Reference to whole group: Stubbs & Chandler, 1978.

Suborder NEMATOCERA

Adults are mostly slender-bodied, long-legged flies with relatively long antennae, e.g., crane-flies and mosquitoes. Larvae have well-developed, sclerotized heads with the jaws opposed, for biting, or else brush-like and adapted for filter-feeding. They are generally worm-like in form and fairly active; many can swim (no other dipteran larvae swim voluntarily).

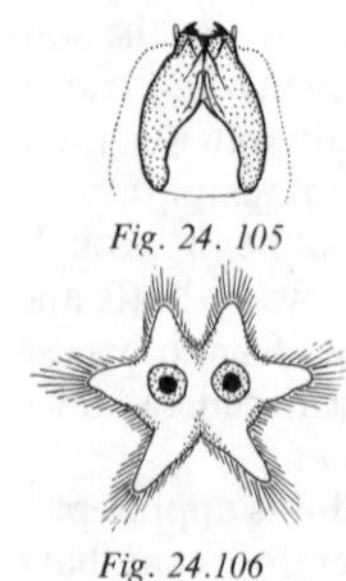

Fig. 24. 105

Fig. 24.106

Fig. 24.107

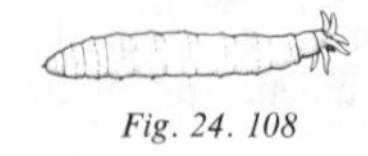

Fig. 24. 108

Fig. 24.109

Fig. 24.110

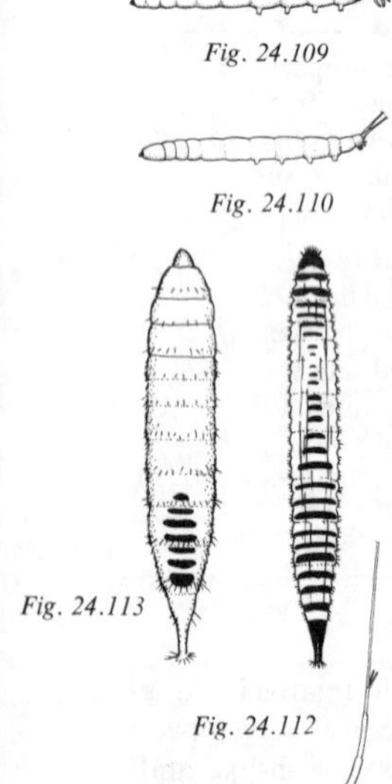

Fig. 24.113

Fig. 24.112

Fig. 24.111

Family Tipulidae (Crane-flies, Daddy-long-legs) *Pl. 256, Figs. 24.105–110* These are common and well known, often large flies, with larvae that live in many terrestrial or aquatic habitats. Aquatic larvae are mostly fat and fleshy grubs up to 50 mm long, with a well-developed head (*Fig. 24.105*) that can be withdrawn into the prothorax, and dirty whitish to dull brown in colour. The posterior end of the body forms a transverse lobed plate, usually fringed with water-repelling hairs, on which are two prominent spiracles (*Fig. 24.106*). Up to eight lobes may be present, the two ventral ones often being elongated. Paired ventral prolegs are often present on the abdomen and the penultimate segment may bear filamentous gills. One genus, *Phalacrocera* (*Fig. 24.107*), has long filamentous processes on the body.

Tipulid larvae are common and the numerous aquatic genera – e.g., *Tipula* (*Fig. 24.108*), *Pedicia* (*Fig. 24.109*), *Dicranota* (*Fig. 24.110*) – occur in all types of waters from stagnant pools to fast-running streams, where they live under stones, buried in the sediment, or amongst vegetation; most aquatic forms are fierce carnivores.

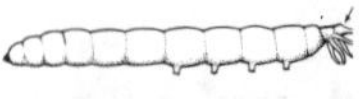

Family Ptychopteridae (Phantom Crane-flies) *Pl. 257, Fig. 24.111* Adults are similar to tipulids but the larvae differ in possessing a long, telescopic breathing tube at the posterior end; the body is up to 35 mm long. The larvae are locally common at the margins of small, shallow ponds and marshy places, buried in mud or decaying vegetation with the slender breathing tube reaching up to the surface. The pupa usually lies on the mud surface but will swim free if disturbed. It possesses two slender tubes projecting from the head; one is vestigial but the other is greatly elongated and functions as a breathing tube.

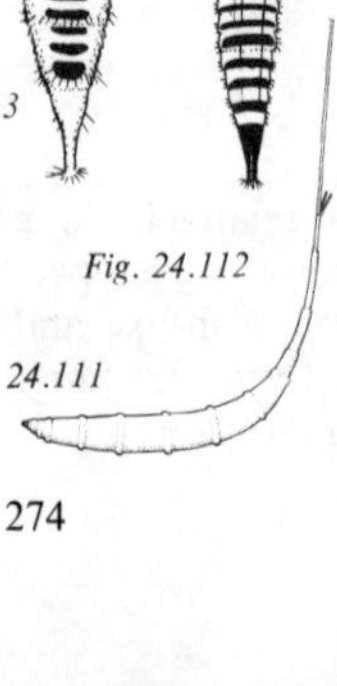

Family Psychodidae (Owl-midges, Moth-flies) *Figs. 24.112, 113* Adult flies are 2–4 mm long, with broad wings and bodies, both of which are covered with a dense pubescence. Larvae are up to 8 mm long, with slightly flattened bodies in which most segments are transversely subdivided. Prolegs are absent and the larvae are generally slow-moving. In *Pericoma* (*Fig. 24.112*) each subdivision bears a narrow sclerite; in *Psychoda* (*Fig. 24.113*), only the posterior segments bear sclerites; in both, the body is often encrusted with debris. These larvae typically occur in foul habitats – sewage beds, drains, etc., but are sometimes found in

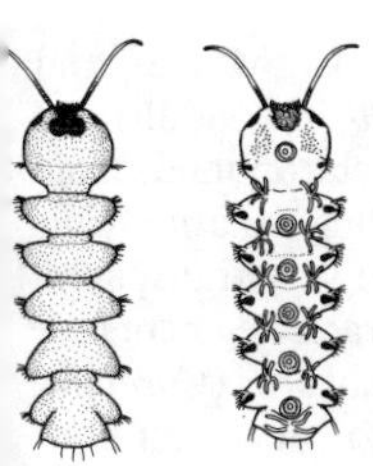

Fig. 24.114

ditches and gutters choked with dead leaves or other rubbish.

Family Blepharoceridae *Pl. 258, Fig. 24.114* Adults resemble small crane-flies. The unusual and very distinctive larvae have flattened bodies up to 10 mm long consisting of seven divisions: the first formed by fusion of head, thorax and the first abdominal segment; the next five by abdominal segments 2–6; and the remaining segments together forming a small caudal region. The dorsal surface may be hard, spiny or bristly. Ventrally each division bears lateral tufts of filamentous gills, and all but the last have a powerful median sucker for adhesion to the substratum. They live attached to the upper surfaces of rocks in torrential mountain streams, often in large groups with Simuliidae. This family is widespread in Europe but absent from Britain.

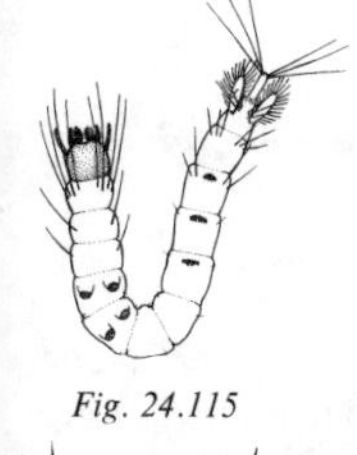

Fig. 24.115

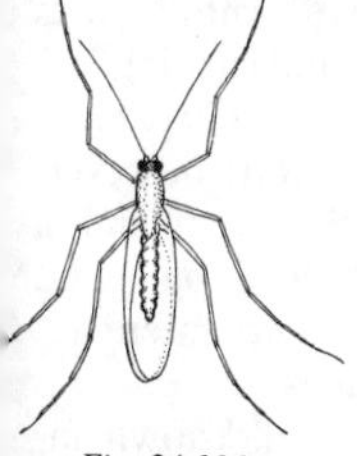

Fig. 24.116

Family Dixidae (Meniscus Midges) *Pl. 259, Figs. 24.115–117* Adults are up to 5 mm long with very long antennae and legs (*Fig. 24.116*); at rest the wings overlap and extend beyond the body. The larvae of this small family, with two European genera *Dixa* and *Dixella*, are exclusively aquatic. They are free-swimming and found at the water surface, usually in the meniscus amongst emergent vegetation, or where other objects break the surface, in ponds, rivers and marshes. The 4–8 mm body is characteristically bent *sideways* into a U-shape; one or two pairs of prolegs are present on the anterior abdominal segments; and the head and tail bear numerous dark bristles. Pupation occurs just above the water level on emergent plants.

Reference: Disney, 1975.

Fig. 24.117 Wing venation of adult midges

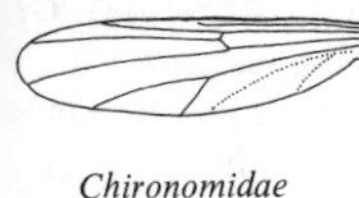

Chironomidae

Culicidae and *Chaoboridae*

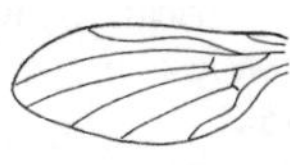

Thaumaleidae

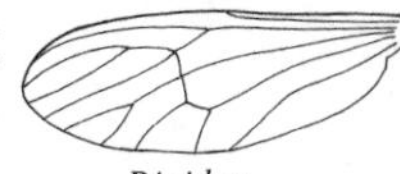

Dixidae

Family Chaoboridae (Phantom or Ghost Midges and Larvae) *Pls. 260–261, Figs. 24.117–120* The adults are small, pale-coloured midges resembling chironomids from which they are distinguished by their wing venation (*Fig. 24.117*). The free-swimming larvae have very transparent bodies, prolegs are absent and the thoracic segments are fused together. The common genus *Chaoborus* (*Corethra*) (*Fig. 24.118*) has two air-sacs which allow it to rest horizontally in mid-water, a characteristic attitude, and lacks a breathing tube; it is found in open water in ponds and lakes. *Mochlonyx* (*Fig. 24.119*) lacks air-sacs and possesses a short breathing tube; it lives in marshes and small overgrown ponds. The active free-swimming pupae of this family (*Fig. 24.120*) are similar to those of Culicidae but tend to rest hanging vertically, with straight bodies, in the water.

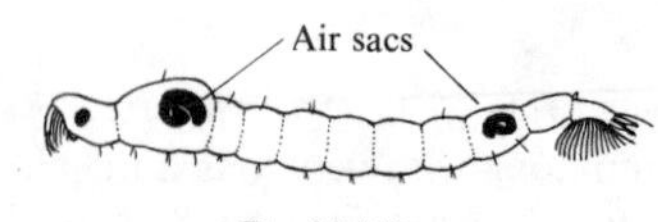

Fig. 24.118

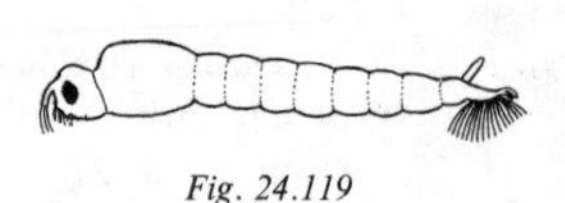
Fig. 24.119

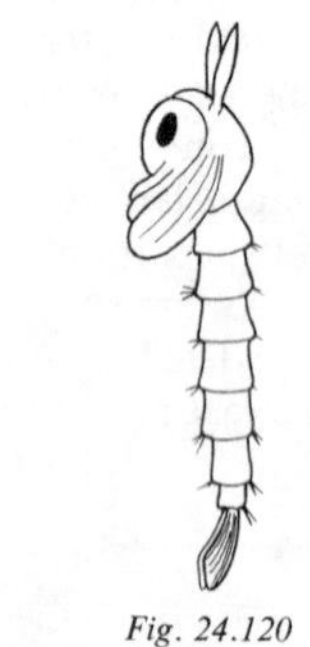
Fig. 24.120

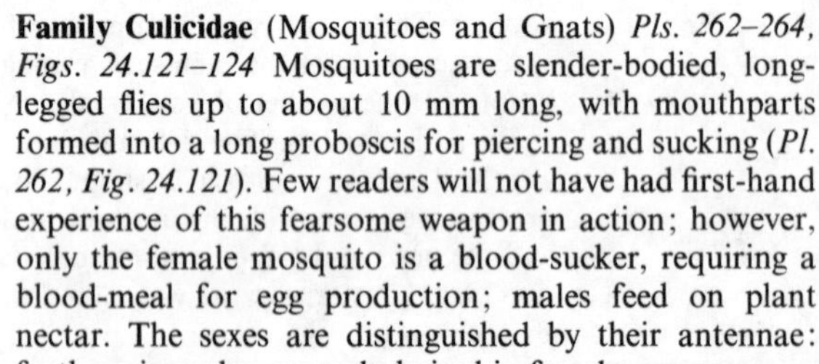

Family Culicidae (Mosquitoes and Gnats) *Pls. 262–264, Figs. 24.121–124* Mosquitoes are slender-bodied, long-legged flies up to about 10 mm long, with mouthparts formed into a long proboscis for piercing and sucking (*Pl. 262, Fig. 24.121*). Few readers will not have had first-hand experience of this fearsome weapon in action; however, only the female mosquito is a blood-sucker, requiring a blood-meal for egg production; males feed on plant nectar. The sexes are distinguished by their antennae: feathery in males, sparsely-haired in females.

The larvae (*Pl. 263, Fig. 24.123*) are well-known inhabitants of water-butts, gutters and other small stagnant water bodies, as well as bogs, marshes, ponds and ditches. They rarely survive long where fishes are present. The cylindrical, opaque, bristly body lacks prolegs and is broadest in the thoracic region where the segments are fused together. Most genera have an oblique caudal breathing tube and hang at an angle from the surface when breathing. *Anopheles* (*Fig. 24.122*) lacks a breathing tube and larvae of this genus lie horizontally just beneath the surface.

Fig. 24.121

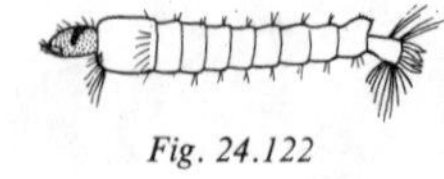
Fig. 24.122

Culicid larvae spend much of their time at the surface,

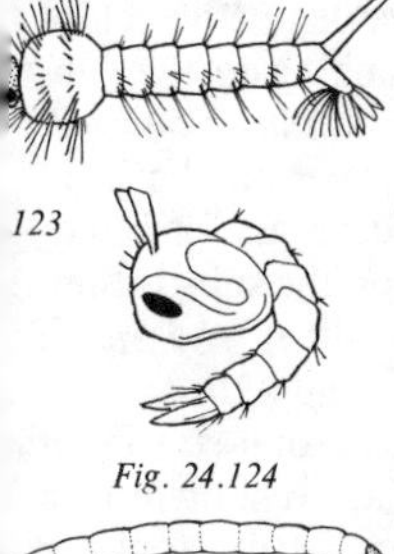

123

Fig. 24.124

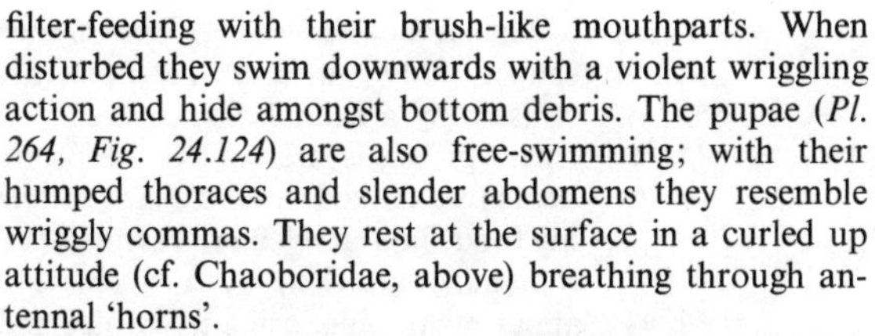

filter-feeding with their brush-like mouthparts. When disturbed they swim downwards with a violent wriggling action and hide amongst bottom debris. The pupae (*Pl. 264, Fig. 24.124*) are also free-swimming; with their humped thoraces and slender abdomens they resemble wriggly commas. They rest at the surface in a curled up attitude (cf. Chaoboridae, above) breathing through antennal 'horns'.

Culex and *Anopheles* are probably the best-known genera, although about twenty genera and fifty species with aquatic larvae occur in north-west Europe.

Fig. 24.125

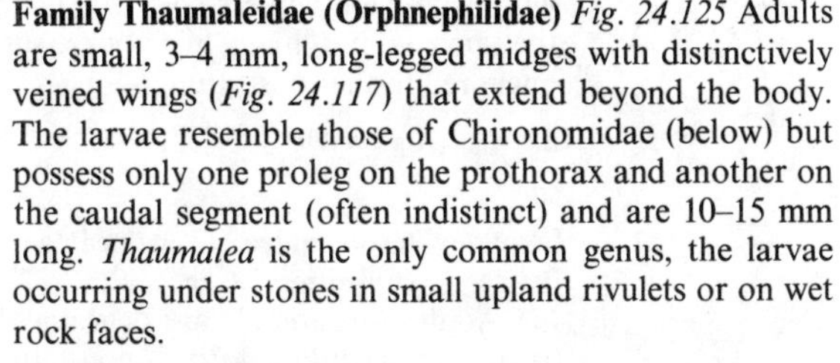

Family Thaumaleidae (Orphnephilidae) *Fig. 24.125* Adults are small, 3–4 mm, long-legged midges with distinctively veined wings (*Fig. 24.117*) that extend beyond the body. The larvae resemble those of Chironomidae (below) but possess only one proleg on the prothorax and another on the caudal segment (often indistinct) and are 10–15 mm long. *Thaumalea* is the only common genus, the larvae occurring under stones in small upland rivulets or on wet rock faces.

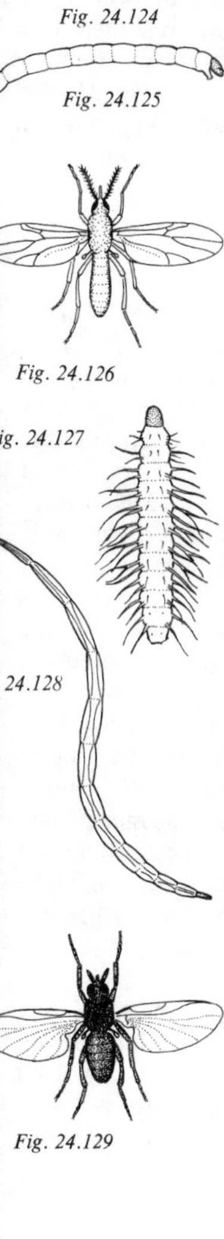

Fig. 24.126

ig. 24.127

24.128

Family Ceratopogonidae (Heleidae) (Biting Midges) *Pl. 265, Figs. 24.126–128* The adults are small midges, less than 5 mm long, with an extremely irritating bite; at rest the wings overlap and extend beyond the body (*Fig. 24.126*).

The larvae are mostly aquatic and fall into two, quite distinct categories. *Culicoides, Probezzia* and others occur in small pools and ditches, often in mats of filamentous algae. They are up to 12 mm long, with slender, transparent bodies (*Fig. 24.128*). They lack any obvious features such as prolegs, but there is a bunch of fine hairs at the posterior end; they swim weakly by lashing the body. *Atrichopogon*, *Forcipomyia* and others are more terrestrial but sometimes occur in shallow waters or damp places. They are less than 5 mm long, with single (sometimes divided) prolegs on the prothorax and caudal segment, and with spines or hairs on the body (*Fig. 24.127*).

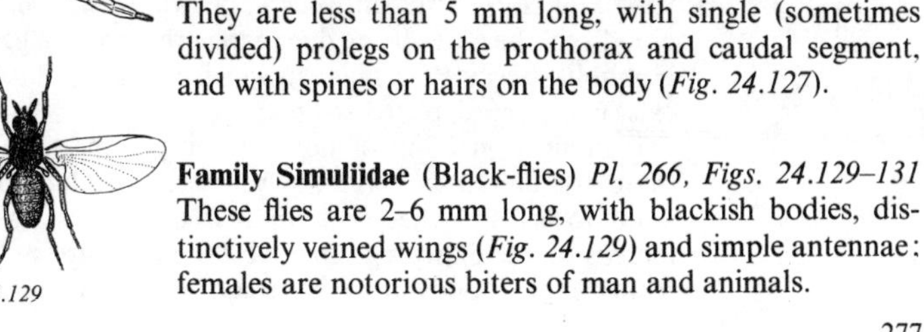

Fig. 24.129

Family Simuliidae (Black-flies) *Pl. 266, Figs. 24.129–131* These flies are 2–6 mm long, with blackish bodies, distinctively veined wings (*Fig. 24.129*) and simple antennae: females are notorious biters of man and animals.

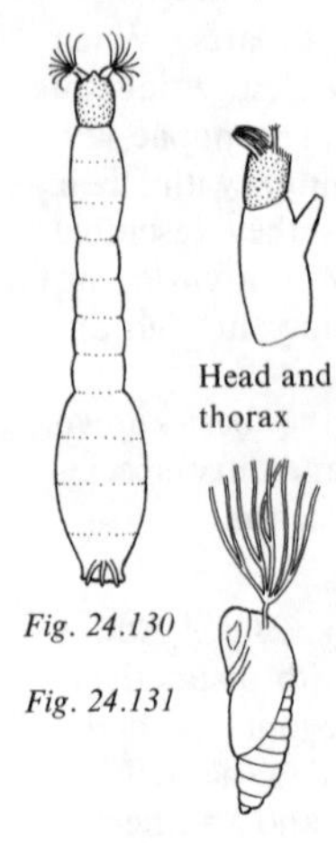

Fig. 24.130

Fig. 24.131

The larvae (*Pl. 266, Fig. 24.130*) have distinctive, club-shaped bodies up to 10 mm long, attaching to the substratum by a caudal disc covered with minute hooks. A single proleg is present on the prothorax and the head bears two stalked, foldable fans of bristles used for catching food. The larva can release its hold on the substratum, remaining attached by a silken safety-line, and move about by looping, using the proleg and caudal disc.

Simuliid larvae are common and often abundant in all types of flowing waters, being particularly characteristic of small, fast-flowing streams where they live on stones or plants, often in large aggregations. The pupae (*Fig. 24.131*), with prominent white respiratory filaments, are found in the same habitat, living in a pocket-shaped cocoon of silk anchored to the substratum.

Reference: Davies, 1968.

Fig. 24.132

Family Chironomidae (Tendipedidae) (Plumed Gnats; Buzzers, Non-biting Midges) *Pls. 267–270, Figs. 24.132–135* Adult chironomids are delicately-built, long-legged, frequently brightly coloured, non-biting midges up to 14 mm long. The distinctively-humped thorax overhangs the head, often concealing it, and the wings do not extend beyond the body (*Fig. 24.132*); males have elaborate, feathery antennae. These midges commonly form large swarms hovering above water margins, usually at dusk, and make themselves conspicuous by their continuous loud buzzing noise. Egg-laying females fly low over the water to drop prominent egg masses carried at the tip of the curved abdomen.

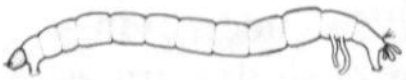

Fig. 24.133

Nearly all chironomid larvae are aquatic. They have worm-like cylindrical bodies up to 30 mm long, with a pair of prolegs on the prothorax, another pair on the caudal segment, and sometimes the penultimate segment bears filamentous gills (*Figs. 24.133–134*). They occur in a wide variety of colours: green, yellow, brown, white, pink and blood-red, the latter being due to the presence of a haemoglobin-like respiratory pigment which enables the larvae to exist in oxygen-depleted sediments.

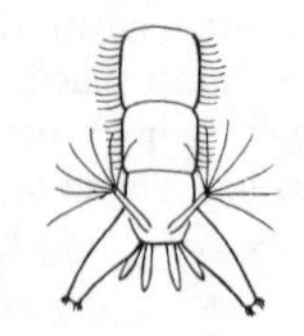

Fig. 24.134

Identification of most larvae is at present unreliable (many of the larvae have not yet been correlated to an adult form) and the great number of species involved increases the difficulty. One of the best-known larval forms is the 'bloodworm' that frequently occurs in water-

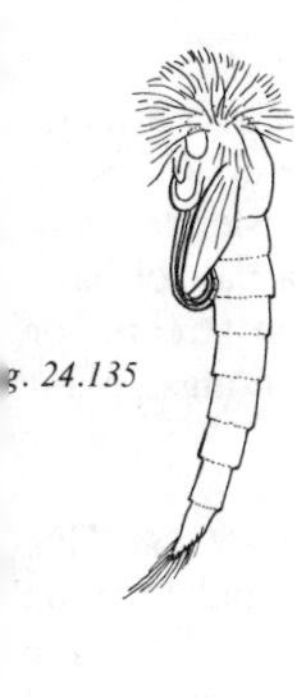
Fig. 24.135

butts and garden ponds (*Pl. 268*) and swims with a laboured, figure-of-eight looping motion. These red larvae are almost certainly *Chironomus* itself.

The pupae (*Fig. 24.135*) resemble those of other midges and gnats. Some species can be distinguished by the presence of tufts of fine white hairs on the head, but others have a pair of respiratory horns, as in the Culicidae.

This is one of the most important families of freshwater insects, with over 1500 European species, some of which occur in enormous numbers – larvae have been counted at up to 50,000 per sq m of lake bed. These larvae are often the dominant faunal element in muddy sediments, and are a major food source for fishes and other animals in otherwise unproductive lakes. They occur in all types of habitat, from water-butts and gutters to large lakes, although less common in fast-flowing waters. Some species are characteristic of water polluted by sewage or farmyard effluent.

Many species burrow in sediments, even in anoxic mud, often constructing a flimsy tube of silk and debris. Others are found in fixed tubes of various materials – silk, debris, sand, faecal pellets, filamentous algae, etc. – attached to the surface of stones or plants; a few uncommon species build tubular cases which they carry about with them like caddis larvae (Trichoptera).

Reference: Cranston, 1982.

Suborder BRACHYCERA

Adult brachyceran flies are generally intermediate in form between the typically slender and long-legged nematocerans and the house-fly-like cyclorrhaphids; their bodies are strongly-built and often colourful, but the legs are never conspicuously long. The larvae have small, but distinct, partly-sclerotized heads; the mandibles are never opposed, usually forming parallel hooks.

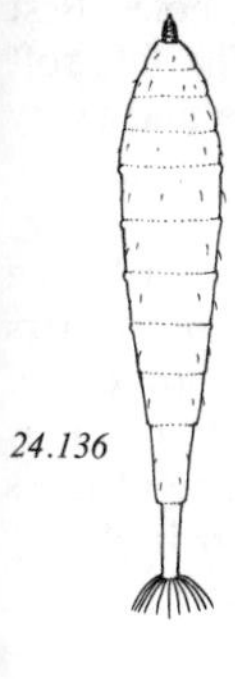
24.136

Family Stratiomyidae (Soldier-flies) *Pl. 271*, *Fig. 24.136*
A few genera in this family have aquatic larvae. They are generally spindle-shaped, slightly depressed, without prolegs, up to 50 mm long when fully extended, and not very active. The skin often contains a gritty deposit of calcareous particles. A slit-like cavity on the caudal segment encloses the spiracles and is surrounded by a prominent bunch of water-repelling bristles arranged like the spokes of an umbrella. *Stratiomys* and *Odontomyia* larvae are sometimes encountered in shallow ponds and ditches, typically at the surface amongst floating plant debris.

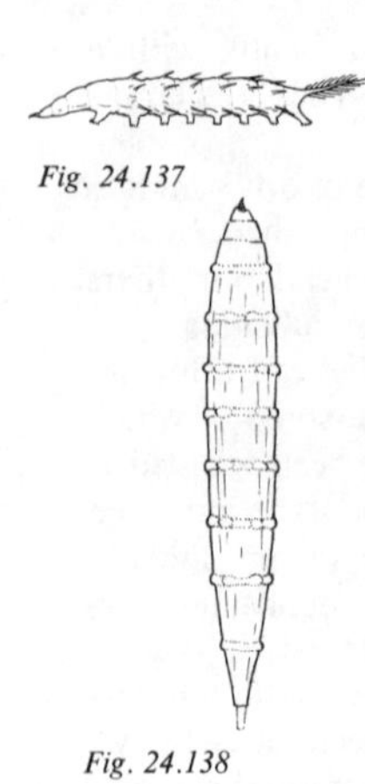

Fig. 24.137

Fig. 24.138

Family Rhagionidae (Snipe-flies) *Fig. 24.137 Atherix* and *Atrichops* are the only genera in this family with aquatic larvae. These are distinctive and very active creatures up to 20 mm long, with well developed prolegs, slender, filamentous processes on most segments, and a pair of hair-fringed caudal processes. They are found in streams or damp waterside places under stones or amongst plant debris.

Family Tabanidae (Horse-flies) *Pl. 272, Fig. 24.138* This family of well-known and universally-unpopular biting flies includes several genera with aquatic larvae. These are spindle-shaped, up to 25 mm long, and easily recognized by a raised ring, often giving rise to about eight short radially-arranged prolegs, on each body segment. *Tabanus* and *Chrysops* typically occur amongst debris or vegetation at the margins of woodland pools, in bogs and marshes, or occasionally in streams.

Other Brachyceran Larvae It is likely that some flies belonging to other brachyceran families have aquatic or semi-aquatic larvae (e.g., Empididae and Dolichopodidae) but these are little- or unknown and cannot, at present, be characterized.

Suborder CYCLORRHAPHA

The adults are mostly house-fly-like with small antennae. The larvae, 'maggots', have tiny or vestigial heads, never sclerotized, which can be retracted into the body.

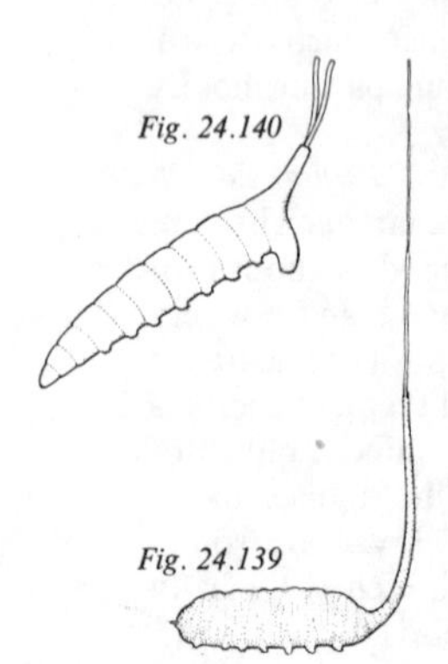

Fig. 24.140

Fig. 24.139

Family Syrphidae (Hover-flies) *Pl. 273, Fig. 24.139* These are common, easily-recognized flies which often mimic bees and other insects. Most larvae are terrestrial but some live in wet or aquatic habitats, typically in foul conditions or small temporary waters such as tree-holes. Best known is the rat-tailed maggot, *Eristalis*. The fat, soft body bears several pairs of small ventral prolegs, and a very long telescopic breathing tube at the rear.

Family Ephydridae (Shore-flies) *Fig. 24.140* Most larvae in this family are terrestrial, living in damp soil; others live in salt marshes, brine pools or on the sea-shore; only a few, e.g., *Ephydra*, *Notiphila*, are known in fresh water. They are recognized by the presence of about eight pairs of prolegs, the last often being enlarged, and a pair of long caudal processes, each with a terminal spiracle.

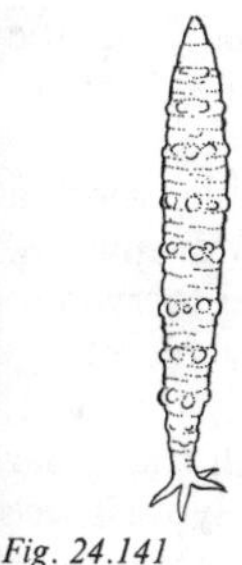

Fig. 24.141

Family Sciomyzidae *Fig. 24.141* The larvae of this family feed on aquatic or terrestrial molluscs, usually starting life as internal parasites but eventually devouring the whole host and then moving on to another. Pupation often takes place in the mollusc shell. The best-known aquatic genus is *Sepedon*, which is sometimes common amongst rafts of floating vegetation and attacks snails such as *Lymnaea* and *Succinea*. The larvae are spindle-shaped, up to 12 mm long, and covered with wart-like tubercles, with a lobed caudal disc surrounding the spiracles.

Family Scathophagidae The larvae of this family are distinguished by a ring of conical papillae surrounding the spiracles on the caudal segment. *Hydromyza* is known to mine the leaves of water-lilies; *Spaziphora* is free-living, occurring amongst shore debris and in sewage beds.

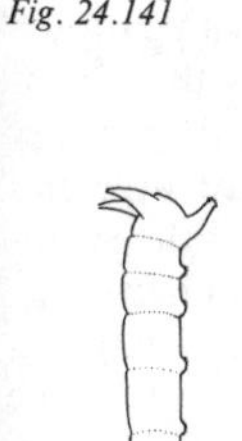

Fig. 24.142

Family Muscidae *Fig. 24.142* This family includes the familiar house-flies. Several genera have aquatic larvae, of which *Limnophora* is the best known. It has a carrot-shaped body with paired short prolegs, enlarged on the caudal segment, and two spiracles on short conical caudal processes; it is found under stones or amongst mosses or filamentous algae in running waters.

Beetles

(Order Coleoptera)

Recognition features

Adult beetles, with their hard, tough bodies, must be the most readily recognized of all aquatic insects. Superficially they might be confused with bugs (Hemiptera) but are easily distinguished by their biting, not sucking, mouthparts, and the form of their wing cases which meet along the centre line of the body and do not overlap.

Larvae are mostly active predators but some are more-or-less sedentary grubs. The head is always sclerotized and usually bears powerful, pincer-like jaws. In all but a few forms, the body is at least partly sclerotized, and legs are present in all types except weevil (Curculionidae) larvae, which are not strictly aquatic. Mature larvae are generally about 25–50% longer than the respective adult forms.

The beetles form the largest of all insect groups and include many aquatic species. Some are aquatic in both the adult and larval stages, others only as larvae.

The most obvious feature of an adult beetle is the hard body, which forms a strong, rigid box, beautifully streamlined in many of the specialized swimmers. The wing cases (elytra) are formed from the strongly sclerotized forewings, and fit over the second and third thoracic segments and the whole of the abdomen, meeting along the centre line. Thus from above, only the head, pronotum and elytra are visible, although in some groups a small triangular plate, the scutellum, can be seen at the junction of pronotum and elytra. These dorsal surfaces, particularly the elytra, are often ornamented with grooves (striae) or perforations (punctures) or are occasionally covered with a fur of fine soft hairs (pubescence). Nearly all beetles can fly and, during the warmer months, may leave the water for long periods. The hind wings are large and membranous and fold away neatly beneath the elytra. In many of the swimming species, the oar-like legs are fringed with a row of swimming hairs.

All species which live submerged breathe from a bubble of air which is trapped under the elytra. This is usually renewed during periodic visits to the surface, most species exchanging air via the tip of the abdomen, but hydrophiliids (see *p. 289*) use their antennae for this purpose. Some tiny beetles (e.g., Elminthidae) do not need to surface, as their air supply is sufficiently large in relation to their volume to act as a lung (plastron breathing).

Adult beetles are long-lived and most survive through the winter in a state of hibernation. Eggs are laid on or in aquatic plants in the spring, and the larvae grow rapidly during the warmer months; most of them leave the water to pupate in damp soil, but some spin a cocoon under water in which they pupate.

Ten families of beetles are either wholly aquatic or contain aquatic species. In six of these both adults and larvae are aquatic (although some species of Hydrophiliidae are more-or-less terrestrial). The other four families include species which have aquatic larvae and adults which are often associated with water, typically being found on emergent water plants. Beetles belonging to other groups sometimes turn up in wet or damp habitats but these are mostly accidental and are not taken into account here. The families of habitually aquatic adult and larval beetles are defined in the following keys:

Key to the Families of Adult Aquatic Beetles

1. a) Surface dwellers; each eye is divided horizontally into two separate parts. Family GYRINIDAE (*p. 285*)
   b) Beetles living *in* the water; eyes not divided, normal. . . . 2

2. a) Beetles that take in air at the surface via the antennae, which are club-shaped (*Fig. 24.162*) and typically shorter than the conspicuous mandibular palps.
   Family HYDROPHILIIDAE (*p. 289*)

b) Beetles that take in air at the surface via the tip of the abdomen; antennae thread-like and longer than the inconspicuous palps. ... 3

3. a) Coxae of hind legs forming large plates (*Fig. 24.146*); beetles less than 5 mm long.
Family HALIPLIDAE (*p. 285*)
b) Hind coxae normal (not as a) above); beetles 1–38 mm long. ... 4

4. a) Tiny, strongly-built and angular beetles usually about 1–3 mm long, found clinging to stones or plants, usually in running waters.
Family ELMINTHIDAE (*p. 293*)
b) Active, usually streamlined, swimming forms; often more than 5 mm long; usually found in still or slow-flowing waters. ... 5

5. a) Eyes noticeably protuberant; beetle capable of emitting a loud rasping squeak; hind legs move alternately when swimming.
Family HYGROBIIDAE (*p. 286*)
b) Eyes not protuberant; silent beetles; hind legs move together when swimming.
Family DYTISCIDAE (*p. 286*)

Key to the Families of Aquatic Beetle Larvae

Whilst this key will work satisfactorily for all common and fully-aquatic larvae, it should be used with caution for specimens from damp (e.g., marshy or bankside) habitats as other, not strictly aquatic, forms may occur here. Also the larvae of some forms, especially some Hydrophiliidae, are not well known and may not conform to the key characters.

1. a) Inactive and scarcely mobile, fat, soft-bodied grubs; legs vestigial or absent; head very small; typically found attached to or inside the roots, stems, or leaves of aquatic plants. ... 10
b) More-or-less active and mobile larvae (some crawl very slowly), rarely soft-bodied and usually with some degree of sclerotization, at least on the thorax; legs well developed, even if small; head well-developed, usually with prominent pincer-like jaws; found in most habitats and often on aquatic plants, but never attached to them or inside their tissues. ... 2

2. a) Legs consisting of five segments and terminating in a double claw (*Fig. 24.143*). ... 3
b) Legs consisting of four segments and terminating in a single claw. ... 7

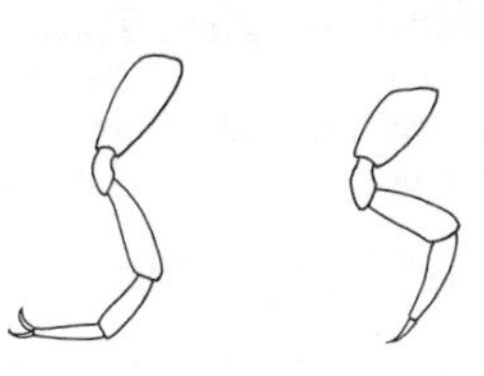

Fig. 24.143 Beetle larva legs

3. a) Abdomen consisting of nine or ten segments. . . . 4
   b) Abdomen consisting of eight segments. . . . 5

4. a) Caudal segment with four small hooks; lateral filamentous gills often present on abdomen; dorsal sclerites without spines.
   Family GYRINIDAE *(p. 285)*
   b) No hooks on caudal segment and abdomen never with lateral gills: posterior edges of dorsal sclerites on thorax and abdomen bearing spines which are usually short, but very long in one genus.
   Family HALIPLIDAE *(p. 285)*

5. a) Legs short and stout, used for digging; abdomen roughly cylindrical.
   Family DYTISCIDAE, subfamily NOTERINAE *(p. 287)*
   b) Legs moderate or long, used for crawling and swimming; abdomen usually tapering. . . . 6

6. a) Caudal segment elongated to form a third 'tail' as long as the two lateral tails; filamentous gills on underside of body.
   Family HYGROBIIDAE *(p. 286)*
   b) Caudal segment shorter than the two lateral tails; no filamentous gills on underside of body.
   Family DYTISCIDAE (all subfamilies except NOTERINAE, 5a) *(p. 286)*

7. a) Jaws usually large and prominent when viewed from above; abdomen soft, sometimes flattened, rather fat, sometimes with lateral or dorsal filamentous gills. Family HYDROPHILIIDAE *(p. 289)*
   b) Jaws inconspicuous when viewed from above; abdomen and thorax more-or-less strongly sclerotized; never with dorsal or lateral filamentous gills. . . . 8

8. a) Antennae much longer than head, consisting of twenty or more bead-like segments. Family HELODIDAE *(p. 292)*
   b) Antennae not longer than head, with less than twenty segments (usually only three or four). . . . 9

9. a) Legs short to moderate in length and strong, used for clinging to stones or plants in fast-flowing streams. Family ELMINTHIDAE *(p. 293)*
   b) Legs short and weak, not used for clinging; larvae found in mud, under stones, amongst debris, etc.
   Family DRYOPIDAE *(p. 292)*

10. a) Legs absent; larvae typically 'mining' inside plant tissues.
   Family CURCULIONIDAE *(p. 294)*
   b) Legs vestigial; aquatic larvae attached to submerged plants or their roots, by two posterior spines. Family CHRYSOMELIDAE *(p. 294)*

References to whole group: Linssen, 1959. Balfour-Brown, 1940, 1950, 1958; Balfour-Brown 1953 (first five families only).

Beetles

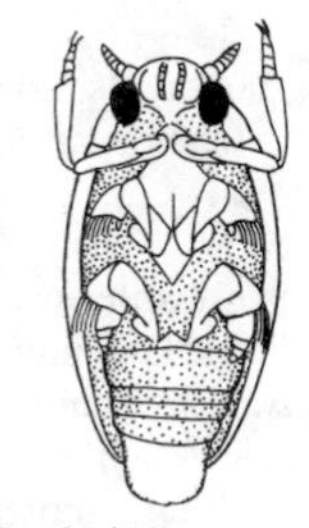

Fig. 24.144

Family Gyrinidae (Whirligig Beetles)
Small or medium-sized beetles, 3–8 mm long, that live on the water surface, commonly in groups. Each eye is divided into two components, one above and one below the water line, for aerial and aquatic vision. The middle and hind legs form short, flattened paddles used exclusively for swimming; the forelegs are 'normal' (*Fig. 24.144*). Whirligigs are very common and well-known beetles that zoom around on the surface like miniature power-boats, leaving a conspicuous wake. They can also dive in pursuit of prey or when alarmed.

The larvae are slender-bodied and somewhat flattened; each abdominal segment bears a pair of lateral gill filaments and the caudal segment has four inconspicuous chitinous hooks. They can run rapidly over the substratum and, at first glance, are very reminiscent of centipedes. There are two common genera:

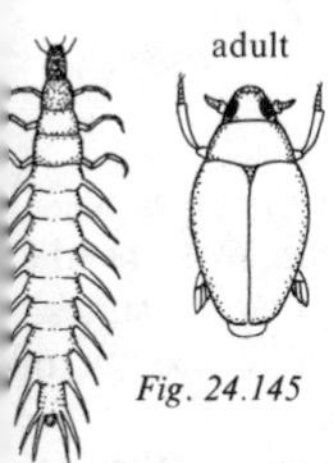

Fig. 24.145

Gyrinus *Pl. 274, Fig. 24.145* Smooth, shiny, blackish beetles, 3–8 mm long. About a dozen species occur in north-west Europe, mostly in still waters. The larvae are found amongst debris or marginal vegetation.

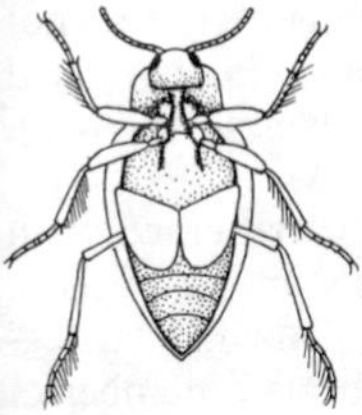

Orectochilus *Pl. 275* The sole European species is *O. villosus*, 5 or 6 mm long, distinguished from *Gyrinus* by its pubescent dorsal surface. It is found only in running waters and is not often seen due to its nocturnal habits, hiding under the banks by day. The larvae live under stones or amongst cushions of filamentous algae.

Fig. 24.146

Family Haliplidae
All the beetles in this family are about 2–4 mm long and easily distinguished by the enlarged plate-like coxae of the hind legs (*Fig. 24.146*). They swim rapidly for their size, moving the hair-fringed legs alternately. Most species are yellowish-brown with rows of dark punctures on the elytra. The slow-moving larvae have slender cylindrical bodies with several backward-pointing spines on each body segment, and short, weak legs. Adults and larvae are common in ponds and slow streams, usually being found amongst the filamentous algae on which they feed. Three genera are common:

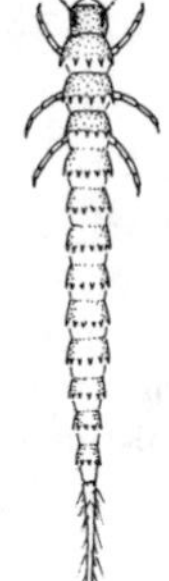

Fig. 24.147

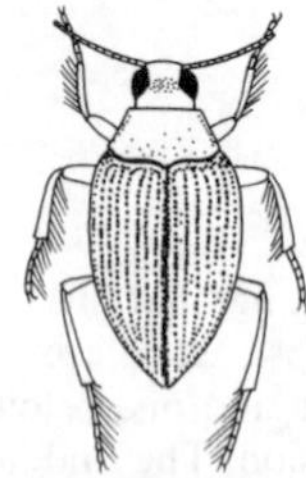

Fig. 24.148

Haliplus *Pls. 276–277, Figs. 24.147, 148* Pronotum narrowest anteriorly, with weak punctures. A large genus with more than twenty species in the area covered by the guide; common and widespread.

Brychius *Figs. 24.146, 149* Pronotum squarish, its sides not tapering. One species, *B. elevatus*, occurs rather locally in ponds and ditches.

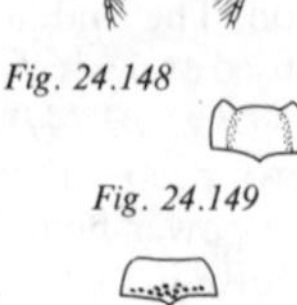

Fig. 24.149

Fig. 24.150

***Peltodytes* (*Cnemidotus*)** Pronotum narrowest anteriorly, with a transverse row of prominent well-marked punctures (*Fig. 24.150*). One species, *P. caesus*, is frequent in stagnant pools. The larvae are remarkable: each body segment bears two or three pairs of very long, jointed spines arising from the dorsal sclerites.

Family Hygrobiidae

This family contains a single distinctive species:

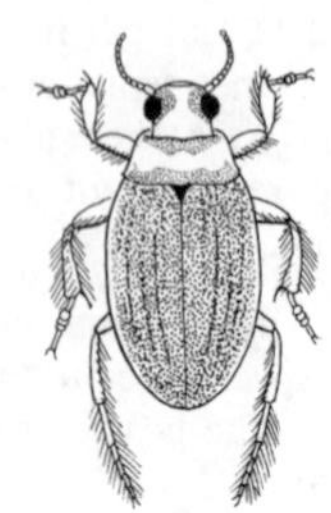

Fig. 24.151

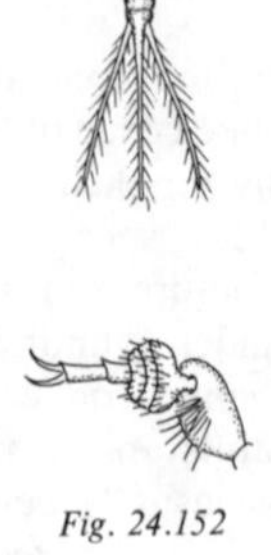

***Hygrobia hermanni* (*Pelobia tarda*)** (Squeak or Screech Beetle) *Pl. 278–279, Fig. 24.151* A medium-sized beetle, 8–12 mm long, which often emits a loud rasping squeak when handled or otherwise excited (a few species of Hydrophiliidae and some bugs, Hemiptera, also make stridulating noises but rarely as loud as *Hygrobia*; they are easily distinguished). The body is deep, with a pronounced ventral keel, and the head, with its bulging eyes, does not merge smoothly with the thorax as in the closely-related Dytiscidae (below). The upper surface is reddish-brown with dark markings.

The stout-bodied larva is strikingly patterned with cream and dark brown. It has three long tails – two cerci, and an even longer rodlike extension of the caudal segment – and the underside of its body bears a number of filamentous gills. Both adults and larvae feed avidly on various annelid worms, especially tubificids. The species is widely distributed but rather local in muddy ponds and ditches.

Family Dytiscidae (Diving Beetles)

A large family of aquatic beetles, 2–38 mm long, beautifully adapted for swimming: the head fits smoothly and snugly into the thorax and the whole body is sleek and streamlined. The legs are fringed with strong swimming hairs and move together like oars. Males are distinguished by their modified fore-tarsi (*Fig. 24.152*) and in *Noterus*

Fig. 24.152

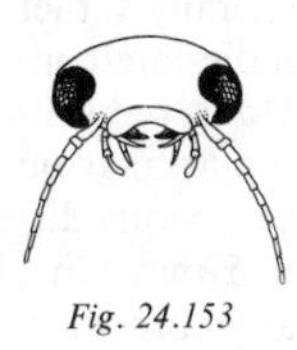
Fig. 24.153

have swollen antennae. Most species are voracious carnivores, devouring any invertebrates, amphibian tadpoles, and even young fishes.

Dytiscid larvae are characteristically swimming and crawling predators with powerful pincer-like jaws, swimming hairs on the legs, eight well-sclerotized abdominal segments without filamentous gills, and two caudal cerci ('tails'); but *Noterus* larvae are different (see below).

With over 200 species in northern Europe, this is the most important family of water beetles. The following key defines adult beetles of the five subfamilies (larvae cannot always be readily distinguished).

1. a) Scutellum hidden (body not more than about 5 mm long).		. . . 2
b) Scutellum visible (body often more than 5 mm long – up to 38 mm).		. . . 4
2. a) Fore tarsi of four segments.	Hydroporinae	(*p. 287*)
b) Fore tarsi of five segments.		. . . 3
3. a) Hind tarsi terminating in a pair of equal claws.	Noterinae	(*p. 287*)
b) Hind tarsi terminating in a single strong claw.	Laccophilinae	(*p. 287*)
4. a) Anterior margin of eye notched (*Fig. 24.153*).	Colymbetinae	(*p. 288*)
b) Anterior margin of eye not notched.	Dytiscinae	(*p. 289*)

Fig. 24.154
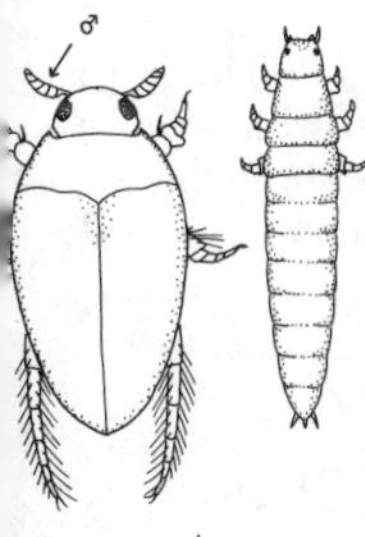

Subfamily Noterinae

Noterus *Pl. 280, Fig. 24.154* Adult beetles are 3–5 mm long, with bodies that are strongly convex above and almost flat beneath; dorsal surface brassy orange/brown; males have swollen antennae. The larvae are poorly known, with cylindrical bodies and short stout legs adapted for burrowing; they apparently spend all their time buried in mud. This is the only European genus in the subfamily, with two species which are locally common in ponds and ditches.

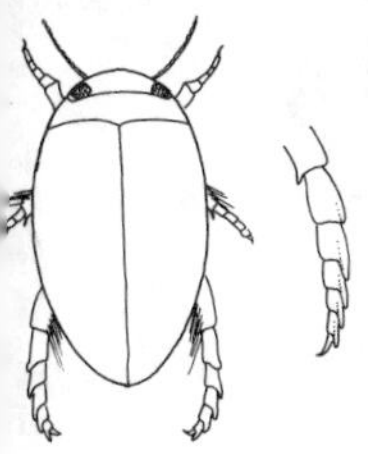
Fig. 24.155

Subfamily Laccophilinae

Laccophilus *Pl. 281, Fig. 24.155* The only common genus, with about three species. Body somewhat depressed, 3.5–4.5 mm long; hind tarsi lobed; greenish or brownish, sometimes mottled. The larvae are similar to those of Colymbetinae (below).

Subfamily Hydroporinae *Pls. 282–284*

This group contains several large genera and upwards of

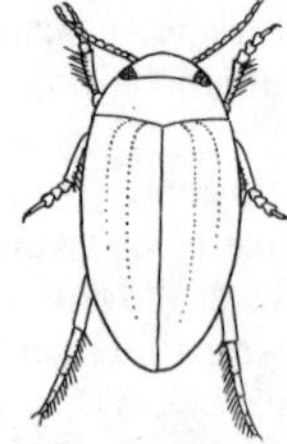

Fig. 24.156

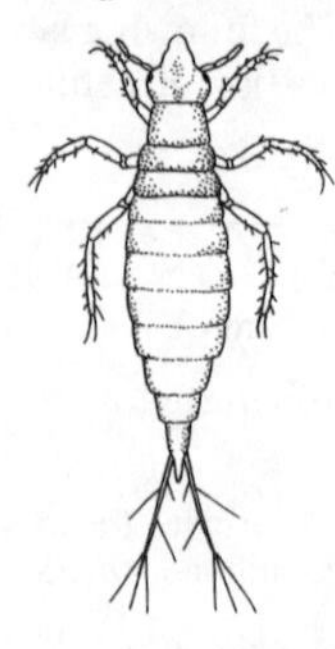

Fig. 12.157

100 species, all about 2–5 mm long, and generally rather similar in appearance (*Fig. 24.156*). Their bodies are convex above and below and usually dark in colour, reddish-brown to black; some have red or yellowish markings on the dorsal surface. Only one common genus is readily distinguishable: *Hyphydrus* (*Pl. 282*) is about 5 mm long, very deep-bodied and rotund, and dull orange/brown in colour. The other common genera can only be distinguished by study of minute anatomical details; they include *Hydroporus* with more than fifty species, *Hygrotus*, *Deronectes*, *Bidessus* and *Hydrovatus*.

Hydroporid larvae can be distinguished from other dytiscids by a snout-like frontal projection on their heads (*Fig. 24.157*).

Subfamily Colymbetinae

Medium-sized beetles, 6–18 mm long, all rather similar in shape. There are six common genera of which *Colymbetes* and *Platambus* are easily recognized, but the remainder are all rather similar in appearance; they include numerous species, some of which are very common.

The larvae of this subfamily are typically 'dytiscid' (see below) in form, distinguished from those of Hydroporinae by lacking a frontal projection, and from Dytiscinae by lacking a fringe of hairs on the last two posterior abdominal segments.

Colymbetes *Pls. 285–286, Figs. 24.153, 158* The largest beetle in the subfamily, 15–18 mm long, with yellowish elytra very finely cross-striated with blackish, so that they appear dusky to the unaided eye; common and widespread.

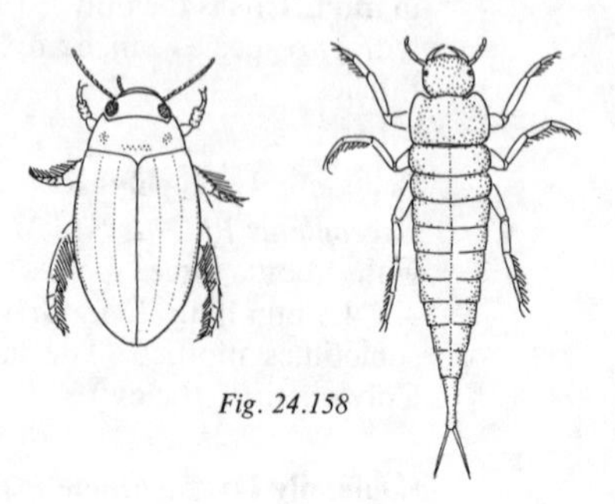

Fig. 24.158

Fig. 24.159

Platambus *Pl. 287* 7–8 mm long, with a distinctive pattern of yellow and blackish on the dorsal surface (*Fig. 24.159*).

Agabus, Copelatus, Ilybius and ***Rantus*** *Pls. 288–289* These genera are all very similar in appearance and very difficult to separate. They are mostly blackish, often with a greenish or purplish sheen, and some species are marked with red or yellow.

Agabus and *Copelatus* range from 6–9 mm in length and have two equal claws on the hind tarsi; *Ilybius* and *Rantus* are 8–14 mm long and have two unequal claws on the hind tarsi (*Fig. 24.160*).

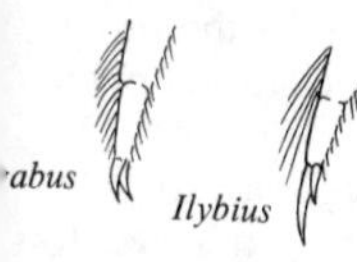

Hind tarsal claws
Fig. 24.160

Subfamily Dytiscinae
Medium to large-sized beetles. The subfamily includes several genera, but only two are common. The larvae have short cerci and a lateral fringe of hairs on the last two abdominal segments (*Fig. 24.161*).

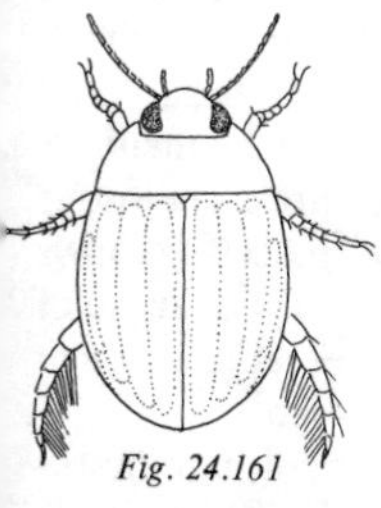
Fig. 24.161

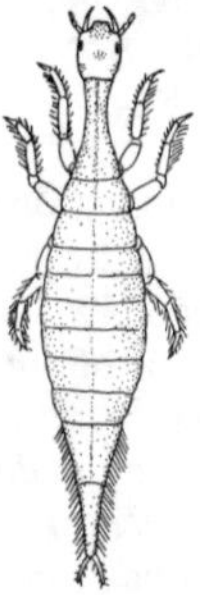

Dytiscus Great Diving Beetle *Pls. 290–292* This genus, with about six species, contains the best known of our water beetles. They are easily identified by their size – at 24–38 mm they exceed all others except *Hydrophilus* (*p. 290*) – and the well-marked yellow margins of the pronotum and elytra. Males are shiny black or green above, with smooth elytra; females are duller with conspicuously grooved elytra. The larvae are formidable beasts, up to 60 mm long, and well capable of tackling small fishes as well as tadpoles, any invertebrates and the fingers of careless collectors.

Acilus *Pls. 293–294, Fig. 24.161* Two species of this genus are fairly common. They are 14–18 mm long, broad-bodied, and greyish- or yellowish-brown with striking dark markings on the head and pronotum; sexual differences are as in *Dytiscus*. The larvae are distinctly 'long-necked' – with an elongated pronotum.

Family Hydrophiliidae (Scavenger Beetles)
Most of the beetles in this family are aquatic with aquatic larvae, but the adults frequently leave the water, especially when confined in aquaria. The terrestrial species inhabit rotting vegetation or animal dung.

Fig. 24.162

Hydrophiliids have club-like antennae (*Fig. 24.162*) and prominent mandibular palps which serve a sensory function, like the antennae in other beetles. The pubescent

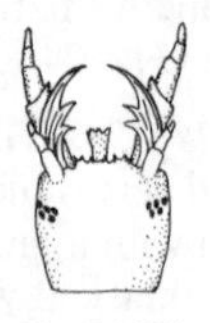
Fig. 24.163

undersurface traps a layer of air which supplements that carried under the elytra – hence their typical silvery appearance when submerged. The antennae are used to channel air to these reservoirs and the head-up breathing attitude at the surface is characteristic of the family. Although some species have swimming hairs on their legs most are poor swimmers and spend much of their time crawling around on vegetation or debris. In size they range from less than 1 mm to nearly 50 mm and include one of the largest of all British beetles.

Many species produce a distinctive egg-case of silk, which may be free-floating at the surface, attached to emergent or submerged plants, or carried around by the female.

The larvae are soft-bodied with little sclerotization except on the head and sometimes the thorax, and short, weak legs. The abdomen is plump, often broad and flattened, and in some species, bears lateral filamentous gills on the first seven segments. All larvae are carnivorous as suggested by their large and powerful, pincer-like jaws (*Fig. 24.163*). They feed on snails, worms or the larvae of other insects. Unlike the adults, larvae breathe through spiracles at the posterior end of the abdomen.

The family contains many genera of small, unremarkable beetles less than 3 mm long (e.g., *Hydraena*, *Ochthebius*) that are not easily identifiable, but some of the commoner larger forms are fairly distinctive.

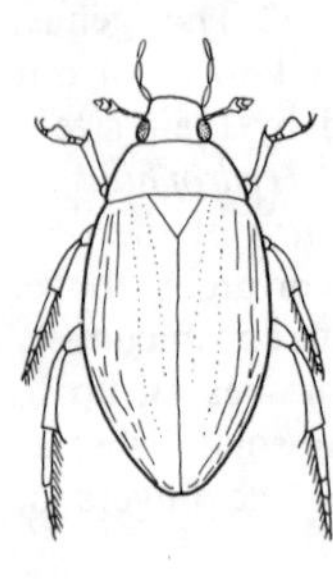

Hydrophilus (Hydrous) piceus (Great Silver Diving Beetle) *Pl. 295, Fig. 24.164* This is the best-known hydrophiliid and is easily distinguished by its great size alone: 40–48 mm long. The dorsal surface is blackish with a pronounced green sheen; middle and hind legs with swimming hairs on the tarsi but the beetle is a poor swimmer; a central ridge on the underside is produced into a sharp spike which is best avoided when handling the animal.

Larva up to 60 mm long and slightly flattened; with inconspicuous lateral gill filaments on the abdomen. It feeds on water snails.

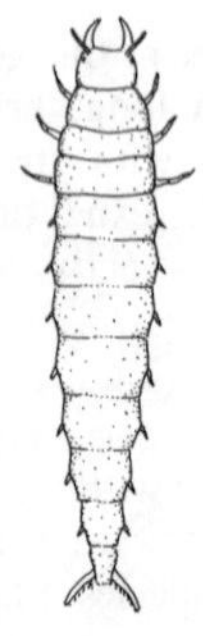
Fig. 24.164

Although once locally common, this species has declined in numbers in recent years, possibly due to its popularity as an aquarium pet – a beetle of this size is difficult to overlook and always tempting to collect. It is usually found in small, still waters.

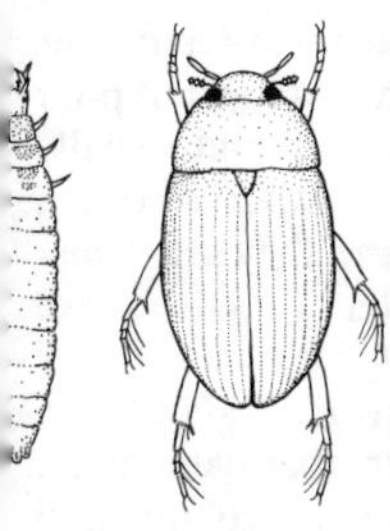

Fig. 24.165

Hydrocara caraboides Very similar to *Hydrophilus* but smaller, 14–18 mm; locally distributed in weedy ponds and slow rivers. The larva has conspicuous filamentous abdominal gills.

Hydrobius fuscipes *Pls. 296–297, Fig. 24.165* A blackish beetle with a more-or-less strong greenish sheen; 6–8 mm long; with eleven finely-punctured striae on each elytron; legs dark reddish-brown, with swimming hairs on mid and hind tarsi (but the beetle is a poor swimmer). Locally common in ponds and ditches.

Limnoxenus niger Very similar to *Hydrobius* but a little larger, up to 9.5 mm, black with a purplish sheen; local and rather uncommon.

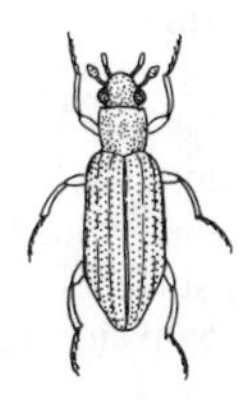

Fig. 24.166

Hydrochus *Fig. 24.166* Small, slow-moving beetles up to 4.5 mm long; eyes protuberant; pronotum markedly narrower than bases of elytra; head and pronotum usually metallic green or brown; elytra usually blackish but may be metallic too, with strong, punctured striae separated by ridges. Several species are fairly common in stagnant waters.

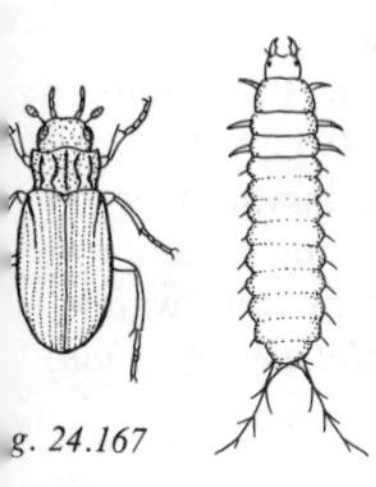

g. 24.167

Helophorus *Pl. 298, Fig. 24.167* A large genus of distinctive beetles up to 7 mm long. Head and pronotum metallic bronze; pronotum broadest anteriorly and bearing five deep longitudinal grooves; elytra with rows of close-set punctures (sometimes in striae), usually gold or greenish. Larva with two long caudal cerci which bear a few stiff hairs. Various species are very common in ponds and ditches, usually amongst dense vegetation.

Berosus *Fig. 24.168* Distinctive beetles with protruding eyes and swimming hairs on mid and hind legs – quite strong swimmers compared with other hydrophiliids. Head usually metallic bronze; elytra markedly convex and with distinct punctured striae, typically yellowish-brown; underside often black; up to 6 mm long. Larvae have very long filamentous abdominal gills. Fairly common in small stagnant waters; several species occur in the area covered by the guide.

Fig. 24.168

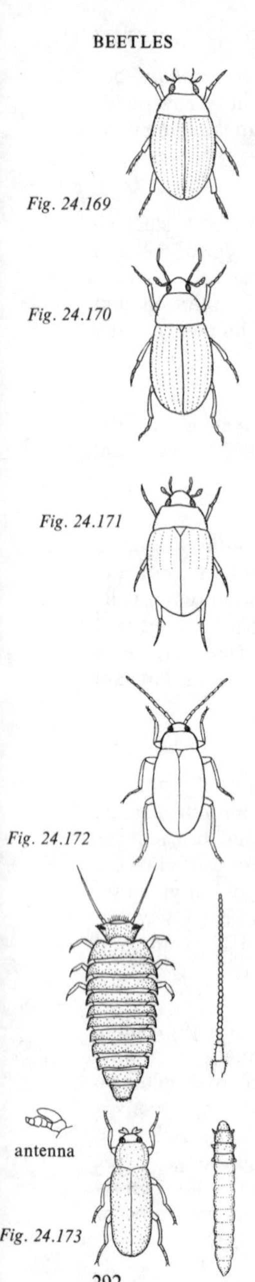

Fig. 24.169

Fig. 24.170

Fig. 24.171

Fig. 24.172

Fig. 24.173

Laccobius *Fig. 24.169* Plump little beetles up to 4 mm long, with weak swimming hairs on mid and hind tarsi, but poor swimmers. Upper surface strongly convex and usually yellowish; elytra with rows of punctures but not striate. Several species are widely distributed and they are sometimes found in brackish or running water.

Helochares *Fig. 24.170* The sole species is *H. lividus*, 6 mm long. Mandibular palps much longer than antennae; elytra punctured but not striate; overall colour dull brownish; female often carries a lens-shaped egg-sac attached to her abdomen; fairly widespread in still waters.

Enochrus (including ***Philydrus***) *Fig. 24.171* A large genus with species up to 7 mm long. Elytra usually with three short rows of punctures, sometimes with very fine striae between them; typically yellowish-brown with dark markings on legs and antennae; widespread in still waters.

Family Helodidae *Pl. 299, Fig. 24.172*
These are small terrestrial beetles with aquatic larvae. Adults of all species are less than 5 mm long, with long thread-like antennae and pubescent dorsal surfaces. They are common in damp places or amongst bankside vegetation.

The distinctive larvae are strongly sclerotized on all segments, with broad, flattened bodies and small but strong legs; a small bunch of white, retractile, filamentous gills protrudes from the tip of the abdomen. The antennae are very long and composed of about twenty to forty bead-like segments; other aquatic beetle larvae have no more than eleven antennal segments. Several genera are common in streams and occasionally in still waters, clinging to stones or vegetation, e.g., *Helodes*, *Cyphon*, *Scirtes*.

Family Dryopidae *Pl. 300, Fig. 24.173*
Small beetles about 5 mm long, with distinctively-shaped antennae somewhat rectangular in shape, and covered with a greyish or yellowish pubescence. Some species occur at the edges of streams, or occasionally in the water clinging to stones. The larvae resemble meal-worms, with cylindrical bodies and weak legs; they burrow into the banks or bottom sediments in the shallow water margins.

Family Elminthidae (Helmidae) (Riffle Beetles)
Small blackish beetles, 1–4 mm long, which are common in running waters or on exposed lake shores, clinging tenaciously to stones or amongst aquatic mosses or other plants; they cannot swim. Their bodies are angular and usually strongly sculptured, with deep punctures, but the raised ridges on the elytra, which are the main external identification characters, can only be seen adequately with strong side lighting and at least a good × *10* hand lens; the legs are strong with large, hooked claws. The larvae are up to 6 mm long and are found in the same habitats. They have well sclerotized bodies, usually more-or-less cylindrical in shape, but broad and flattened in *Elmis*. The legs are short but strong, and a tuft of white, retractile, filamentous gills protrudes from the tip of the abdomen.

These beetles are truly aquatic and neither adults nor larvae need to breathe at the surface. Probably the only time they leave the water is when the larvae pupate in crevices just above the water line. The commonest species are described below:

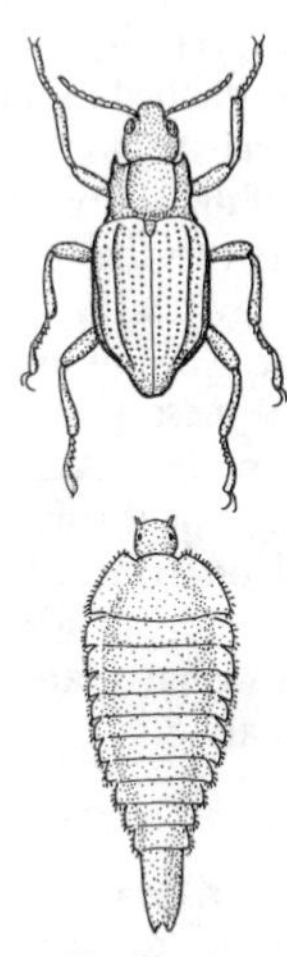
Fig. 24.174

Elmis aenea *Pls. 301–302, Fig. 24.174* Pronotum with a U-shaped ridge; elytra each with two strong ridges; black; about 2 mm long. Larva very broad-bodied with lateral flanges; sometimes with a pale band in the middle of the body, up to 4 mm long. Probably the most common and widely-distributed species.

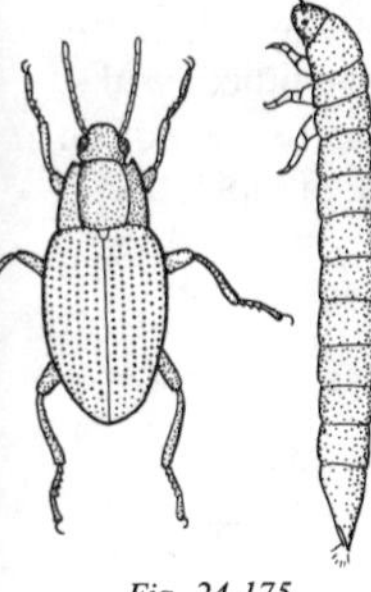
Fig. 24.175

Limnius volckmari *Pl. 303 Fig. 24.175* Pronotum with two separate longitudinal ridges; no ridges on elytra; black, about 3 mm long. Larva cylindrical, up to 5.5 mm long.

Esolus parallelepipedus *Fig. 24.176* Pronotum with two longitudinal ridges; elytra each with one ridge; brown or black; about 1.5 mm long. Larva triangular in cross-section with slight lateral flanges, tapering a little posteriorly, up to 2.5 mm long.

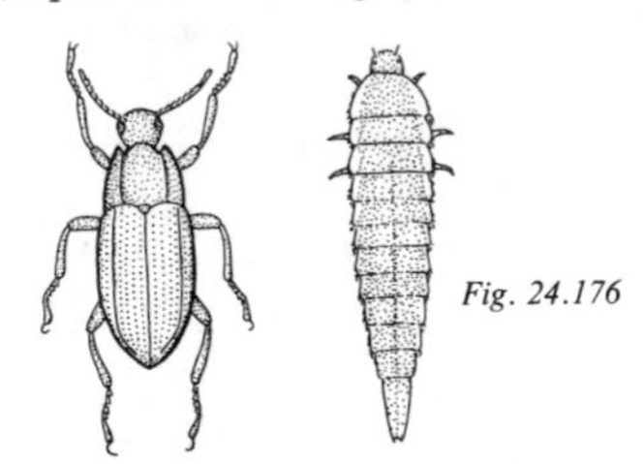
Fig. 24.176

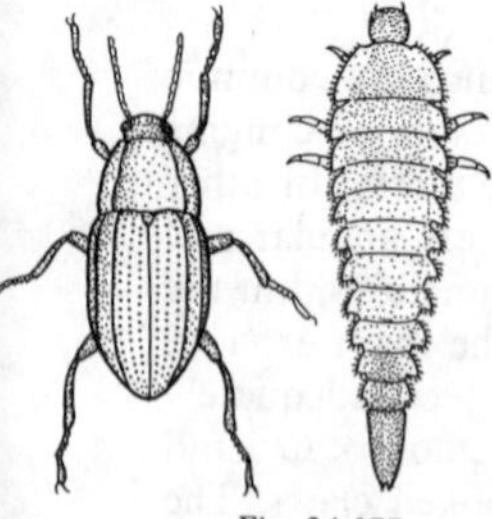

Fig. 24.177

Oulimnius *Fig. 24.177* Pronotum with two longitudinal ridges; elytra each with three ridges; dark brown; about 2 mm long. Larva similar to *Esolus*, but a little broader and up to 3 mm long. The two species *O. tuberculatus* and *O. troglodytes* are virtually indistinguishable externally.

Reference: Holland, 1972.

Family Chrysomelidae *Pls. 304–305, Fig. 24.178*
Three genera of this large, otherwise terrestrial, beetle family have aquatic larvae: *Donacia*, *Macroplea* and *Plateaumaris*. The larvae are fat, whitish or greenish grubs with very short legs, 10–12 mm long, found attached to the roots or stems of aquatic plants such as *Sagittaria*, *Potamogeton* and *Nymphaea*. Near the end of the abdomen are two sharp spikes which anchor the larva to the host plant and, remarkably, extract air from cavities in the plant tissues, enabling the larva to breathe without surfacing. Unlike other aquatic beetle larvae, these pupate underwater in cocoons attached to the plants. Adult beetles are often found on the floating or emergent leaves of water plants; many are brightly coloured, often metallic green. Other genera, such as *Galerucella*, are associated with emergent marginal or bankside plants such as water-pepper (*Polygonum hydropiper*) but are not aquatic.

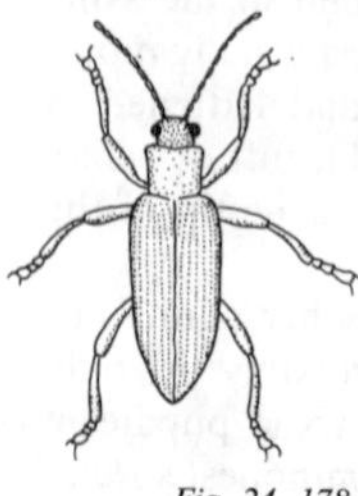

Fig. 24. 178

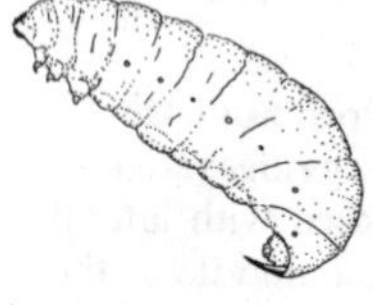

Family Curculionidae (Weevils) *Fig. 24.179*
Many species of these exclusively herbivorous beetles are found on the emergent parts of aquatic plants. Adults are small, from less than 1–5 mm long, often brightly coloured or metallic; the head is elongated to form a rigid, tubular proboscis with the antennae about half-way along its length. The larvae are small, legless, soft-bodied grubs, whitish or greenish, which bore into plant tissues, often occurring in the submerged parts although not strictly aquatic; they are small and seldom seen.

Fig. 24.179

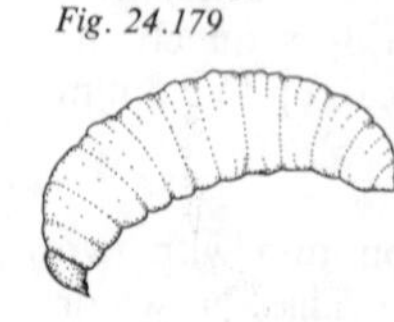

Wasps, Bees and Ants
(Order Hymenoptera)

This is the second largest order of insects and its principal members, wasps, bees and ants, are all very familiar insects. Less well known are the few aquatic species, all of which are parasites on either the egg or larval stages of other aquatic insects; most are microscopic. The larvae are relatively featureless grubs, with small heads and soft fat bodies, which are best identified by their highly-specialized habitats. Adults of all species leave the pupal case underwater but emerge from the water and fly about to find a mate, the females finally re-entering the water to lay their eggs.

About thirty species of aquatic Hymenoptera are known in Europe, and undoubtedly many more await discovery. The three examples below are the best known.

Superfamily Ichneumoniodea These are wasp-waisted insects with long, many-segmented antennae; the female has a slender ovipositor at the end of her abdomen.

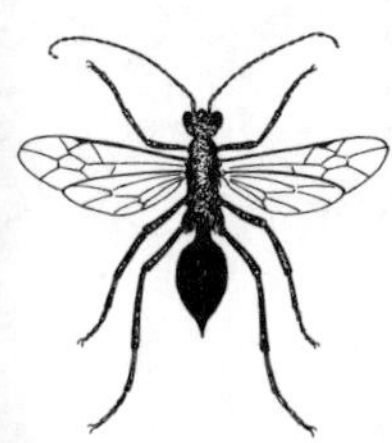
Fig. 24.180

Family Agriotypidae

Agriotypus armatus *Fig. 24.180* This species parasitizes caddis larvae of the genera *Goera* and *Silo* (family Goeridae). The adult is a blackish insect about 6 mm long. After mating the female crawls down into the water and seeks out a caddis larva in its case in which she lays a single egg. The larva feeds on the caddis grub but does not injure it fatally until it has sealed its case prior to pupation. The *Agriotypus* larva then consumes the whole caddis and pupates in its case. Parasitized cases at this stage can be recognized externally by a long (2–3 cm) ribbon-like filament protruding from one end; this probably functions as a gill for the pupa. This species is widely distributed but rather local.

Superfamily Chalcidoidea

This is, numerically, an enormous insect group but its members are all tiny and include some of the smallest-known insects (0.25 mm long). Many aquatic forms are known, most of which have narrow, hair-fringed wings. All of them parasitize the eggs of aquatic insects – beetles, bugs and damselflies in particular. Only one egg is laid in each host-egg but, as a remarkable adaptation to a parasitic way of life, this single egg can develop into several larvae (polyembryony).

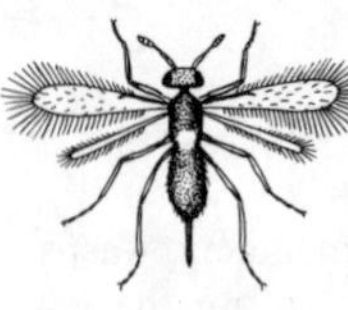
Fig. 24.181

Family Trichogrammatidae (Fairy Flies)
Prestwichia aquatica *Fig. 24.181* This species is dark brown with pale extremities, 0.6–1 mm long; the female has a prominent ovipositor; males and some females have only vestigial wings. The adult female spends some time underwater, swimming with her legs, in search of suitable host eggs, mostly those of water beetles. Although fairly common on the continent, this species is apparently rare in Britain.

Fig. 24.182

Family Mymaridae (Fairy Flies)
Caraphractus cinctus *Fig. 24.182* Fairy flies of this family are distinguished from Trichogrammatidae by their stalked wings. *C. cinctus* is a blackish insect about 1–1.3 mm long; females, identified by the long ovipositor, can swim underwater using the legs and wings for propulsion. It is a locally common species which typically attacks the eggs of various water beetles but may also infest those of bugs or damselflies.

CHAPTER 25 Fishes

Fishes are gill-breathing vertebrates, 'cold-blooded' in that their temperature varies according to that of their surroundings. They are divided into four main classes, each equal in status to the reptiles and amphibians, but only two of these – the Pisces (true or bony fishes) and the Marsipobranchii (lampreys) occur in European fresh waters. Anglers have their own classification into game fish – salmon, trout and grayling – and coarse fish – all the rest.

Before describing and identifying fishes, it is necessary to have some idea of their anatomy. The limits of the head, body and tail, and the names of the various fins are indicated in *Fig. 25.1*

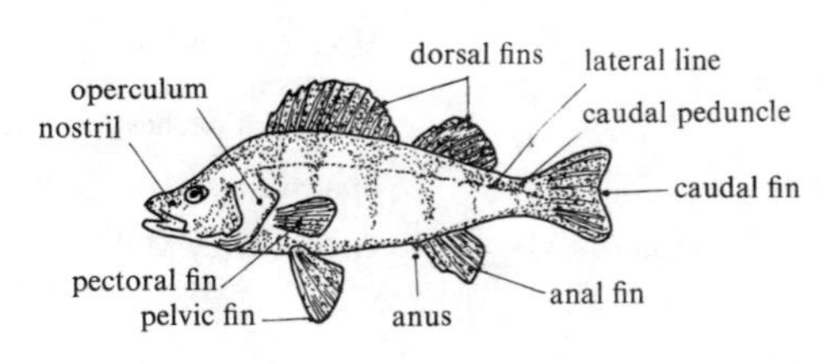

Fig. 25.1

Tabular Key to Fishes

Shape Features	
Big-headed:	Bullheads
Bream-shaped (deep-bodied):	Breams, Bitterling, Carps, Perch, Pumpkinseed, Sunfish
Elongated and snake-like:	Lampreys, Eels
Elongated but not snake-like:	Sturgeon, Pike, Loaches, Catfishes, Streber
Flattened:	Flounder
Herring-shaped:	Shads, Whitefish, Carp, Bass, Zander, Ruffe, Black Basses, Grey Mullets, Sticklebacks
Hump-backed:	Humpback Salmon
Trout-shaped (somewhat blunter-snouted than herring-shaped):	Salmon, Trout, Char, Smelt, Grayling, American Mud-minnow, Soufie, Asp, Burbot

Head Features

Barbels present – 2:	Tench, Gudgeon
3:	Burbot
4:	Sturgeon, Barbel, Carp
6:	Stone Loach, Spined Loach, Wels
8:	American Catfish
10:	Pond Loach
Eyes red:	Roach, Bitterling
yellow:	Rudd, Dace, Ide
Four yellow warts:	Four-horned Bullhead
Lower jaw hooked:	Adult Salmon
projecting:	Vendace, Smelt, Pike, Rudd, Roach, Asp, Black Basses, Pumpkinseed, Sunfish, Sticklebacks
No jaws:	Lampreys
Snout pointed:	Houting, Powan, Pike,
projecting:	Sturgeon, Nases, Barbel, Breams, Zaehrte, Loaches
Spines below eye:	Spined Loach, Perch, Ruffe
Yellow spot on gill covers:	Golden Grey Mullet

Body Features

Band along side blue-green:	Bitterling
purple:	Rainbow Trout, Soufie, Schneider
yellow:	Minnow
Belly red:	Char, American Brook Trout, Minnow (male breeding), Zaehrte (breeding), Three-spined Stickleback (male (breeding)
Bony plates present:	Sturgeon, Sticklebacks
Chin red:	Cut-throat Trout, Zaehrte (breeding), Three-spined Stickleback (male breeding)
Colour blackish:	Zaehrte (breeding), Nine-spined Stickleback (breeding)
reddish-orange:	Golden Ide or Orfe, Tench, Goldfish, Golden Carp
Longitudinal stripes:	American Mud-minnow, Pond Loach
Vertical bars:	Perch, Zander, Ten-spined Stickleback; sometimes Parr, Pike, Minnow
No lateral line:	Shads

No scales:	Lampreys, Burbot, Eel (apparently no scales), Catfishes, Bullheads
Spotted:	Salmon and Trout spp., Char, Powan (Schelly), Gudgeon, Barbels, Spined Loach, Ruffe
Uniformly dark colour:	Tench

Fin Features

Adipose fin present:	Salmon and Trout spp., Huchen, Char, Whitefish, Smelt, Grayling, American Catfish
Anal fin elongated:	Bleaks, Breams, Zope, Zaehrte, Catfishes, Burbot, Sticklebacks, Flounder
No paired fins:	Lampreys
Dorsal fin elongated:	Carps, Goldfish, Burbot, Pumpkinseed, Sunfish, Flounder
near tail:	Pike, Sticklebacks
very short:	Catfishes
2 separate:	Burbot, Bass, Perch (not quite separate), Zander, Grey Mullets, Bullheads
2 joined:	Ruffe, Black Basses
with dark spot:	Perch
with spines:	Perch Family, Grey Mullets, Sticklebacks (spines in front of fin)
Lower fins red:	Char, American Brook Trout, Roach, Rudd, Dace, Chub, Ide, Asp, Nases, Zaehrte (breeding), Perch, Minnow (breeding)
red-based:	Schneider, White Bream
white-edged:	Char, American Brook Trout (with black margin)
Pelvic fin with spine:	Bullheads, Sticklebacks
Tail fin forked, scarcely:	Trout, Char, American Brook Trout, Sticklebacks
forked, slightly:	Salmon, Trout, Rainbow Trout, Cut-throat Trout, Char, American Catfish, Black Basses
forked, well:	Sturgeon, Humpback Salmon, Chum Salmon, Whitefish, Smelt, Grayling, Carp Family, Bass, Perch, Zander, Ruffe, Pumpkinseed, Sunfish, Grey Mullets
pointed:	Burbot (sometimes)
rounded:	American Mud-Minnow, Pond Loach, Bullheads

square:	Tench, Stone Loach, Spined Loach, Wels, Flounder
spotted:	Gudgeon, Barbels
upper lobe longer:	Sturgeon
shorter:	Zope

Where Likely to be Seen from the Bank

(i.e., excluding primarily nocturnal fishes and those normally frequenting deep or turbid water)

Lakes/ponds, surface (basking):	Rudd, Tench, White Bream, Common Bream, Goldfish, Common Carp
(not basking):	Perch
(taking flies):	Trout, Roach
mid-water:	Trout, Roach, Minnow, Perch, Ruffe, Sticklebacks
shallows (spawning):	Char, Whitefish, Roach, Minnow, Tench, White Bream, Common Bream, Crucian Carp, Goldfish, Common Carp, Burbot, Perch, Ruffe, Sticklebacks
(feeding):	White Bream, Common Bream, Perch, Ruffe
River surface (migrating):	Lampern, Shads, Eel, Smelt
Barbel zone, surface (taking flies):	Trout, Dace, Ide, Bleak, Chub, Roach
mid-water:	Trout, Dace, Ide, Asp, Nase
bottom only:	Barbel
shallows (spawning):	Dace, Ide, Burbot
Bream zone, surface (basking):	Tench, Common Bream, Crucian Carp, Goldfish, Common Carp, Perch
(taking flies):	Roach, Rudd
mid-water:	Pike, Roach, Rudd, Perch, Ruffe
shallows (spawning):	Pike, Roach, Rudd, Tench, White Bream, Common Bream, Crucian Carp, Goldfish, Common Carp, Perch
(feeding):	White Bream, Common Bream, Perch, Ruffe
estuaries:	Grey Mullet (surface and mid-water), Bass (mid-water)
Streams	
Trout zone, surface (taking flies):	Trout, Minnow, Dace
mid-water:	Trout, Minnow
bottom:	Lampreys (attached to stones), Trout, Stone Loach (under stones), Miller's Thumb (under stones at dusk)
shallows (spawning):	Lampreys, Trout, Char, Minnow

Grayling zone, surface (taking flies):	Trout, Grayling, Chub, Bleak, Dace
mid-water:	Trout, Grayling, Chub, Bleak
bottom:	Lampreys (attached to stones), Grayling, Gudgeon
shallows (spawning)	Lampreys, Salmon, Huchen, Whitefish, Grayling, Chub, Asp, Nase, Barbel, Bleak, Ruffe
small ditches:	Sticklebacks, Umbra
Waterfalls, Weirs (migrating):	Salmon, Eel, Sea Trout, Shad

CYCLOSTOMES (Cyclostomata)
These are the only extant members of the Marsipobranchii, and are so primitive that nowadays they are often not considered to be fishes at all. They are indeed the most primitive of all living vertebrates, with no jaws, but horny teeth in their circular sucking mouths.

Lamprey Family (Petromyzonidae)
The European freshwater cyclostomes are four species of lamprey. They are eel-like, elongated and slimy, with fins only along the back and tail. Two species, the Lampern and the Sea Lamprey, migrate to the sea in late summer and autumn to spend their adult years in the sea, returning in a subsequent autumn or winter to the rivers to spawn in spring and then die. Adult lampreys feed mainly by sucking blood from live fish. Young lampreys, called ammocoetes, take three to five years to develop, and lie buried in mud, feeding on diatoms and other minute organisms.

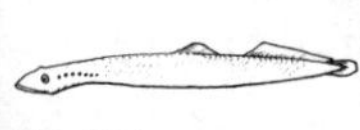
Fig. 25.2

Lampern or River Lamprey *Lampetra fluviatilis* (*Fig. 25.2*) Silvery-white, tinged blue or green on the back, with a gap in the middle of the dorsal fin: finger-thick, 30–40 cm; females larger than males. Sucker with three pairs of fairly stout teeth in an outer ring surrounding an inner group of two, widely-separated pointed teeth and one comb-like tongue-tooth. Rivers, brooks and lakes, most adults migrating to the sea, but a few overwinter fasting in fresh water, especially those in lakes. Can live out of water for days. Edible, and still eaten in Baltic countries.

Brook Lamprey *Lampetra planeri* Shorter and thinner than the Lampern, 12–16 cm, with a notch rather than a gap in the dorsal fin and the two separate teeth blunt. Non-migratory in streams, brooks, ditches and the headwaters of rivers. Used only for bait.

Sea Lamprey *Petromyzon marinus* Longer, up to 1 m, and stouter than the Lampern, of a grey, green or yellow ground colour, heavily marbled with black, brown or olive. Dorsal fin with a marked gap; teeth numerous. Rivers and streams, migrating to the sea when adult. Edible and more highly-esteemed than the Lampern; a surfeit of it killed King Henry I of England.

Tabular Key to Lampreys and Eels

Colour silvery:	Lampern, Brook Lamprey, Silver Eel
yellowish:	Yellow Eel
mottled:	Sea Lamprey
Dorsal fin with a gap:	Lampern, Sea Lamprey
with a notch:	Brook Lamprey
continuous:	Eels
Pectoral fins present:	Eels
absent:	Lampreys
Mouth with jaws:	Eels
with suckers:	Lampreys
Size more than 50 cm:	Sea Lamprey, Eels

Sturgeon Family (Acipenseridae)

Fig. 25.3

Sturgeon *Acipenser sturio* (*Fig. 25.3*) The largest fish found in European rivers, which it ascends to spawn. It can live for a century and grow to 6 m in length and 400 kg in weight. Its five rows of bony plates and shark-like tail are unmistakable. In Britain it is a 'royal fish', the prerogative of the Crown when caught.

Herring Family (Clupeidae) (*Fig. 25.4*)

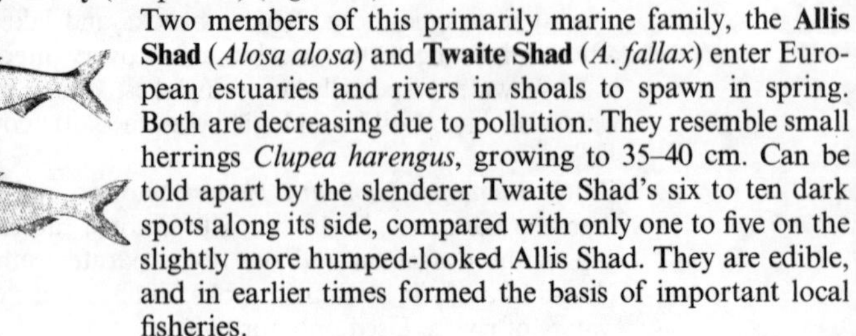

Fig. 25.4

Two members of this primarily marine family, the **Allis Shad** (*Alosa alosa*) and **Twaite Shad** (*A. fallax*) enter European estuaries and rivers in shoals to spawn in spring. Both are decreasing due to pollution. They resemble small herrings *Clupea harengus*, growing to 35–40 cm. Can be told apart by the slenderer Twaite Shad's six to ten dark spots along its side, compared with only one to five on the slightly more humped-looked Allis Shad. They are edible, and in earlier times formed the basis of important local fisheries.

Salmon Family (Salmonidae)
An important family of freshwater fishes, containing the species most valued by fishermen, both commercially and for sport – salmon, trout and grayling. Salmonids can be told from all other freshwater fish by their adipose fin, a small extra fin between the dorsal and tail fins. Many species are anadromous, i.e., visit fresh water from the sea only to spawn.

Fig. 25.5

Atlantic Salmon *Salmo salar* (*Fig. 25.5*) The largest European salmonid, males attaining 150 cm and 36 kg and females 120 cm and 20 kg (both can be even heavier). Its name and appearance change with each stage in its life history. The young hatch as Alevins, but once the yolk-sac is absorbed they become Parr, distinguished by their eight to ten so-called 'fingermarks' on their flanks, each separated by a red spot, and a grey-green adipose fin. After two years, most Parr become Smolts, turning silvery and migrating to the sea. Some travel thousands of miles in the ocean, as far as the coast of Greenland. After a year or more they return to their home river as silvery Grilse (*Fig. 25.6*) and ascend far upstream to spawn in gravelly beds known as redds. As the spawning time approaches, the males become redder and the females darker, both eventually spotted and mottled with black, red and orange (sometimes known as 'red fish' and 'black fish' respectively). At the same time the snout and lower jaw of the males elongate, the lower jaw becomes markedly hooked. After spawning they go back to the sea as spent fish or Kelts, some dying but others returning to breed again. A few salmon stocks are non-migratory, inhabiting land-locked lakes such as Vänern in Sweden. Salmon fresh from the sea can be told by their sea-lice *Lepeoptheirus salmonis* (*p. 221*), parasitic copepods up to 2 cm long, which die in fresh water and so drop off after a few days.

Fig. 25.6

Two Pacific species of salmon have been successfully introduced into northern European rivers. The Humpback Salmon (*Oncorhynchus gorbuscha*), which becomes markedly hump-backed as well as red when it spawns, and the Chum Salmon (*O. keta*), which has a line of reddish and blackish markings on its side. Both are silvery before they breed.

Trout *Salmo trutta* Another important game fish variable in both appearance and habits. One form, the River, Lake or Brown Trout, spends its whole life in fresh water.

Fig. 25.7

Another, the Sea Trout, Salmon Trout or Sewin, now considered to be conspecific with the Brown Trout, resembles the Salmon, in spending part of its life in the sea, though remaining much nearer the coast and sometimes hardly leaving the estuaries (*Fig. 25.7*). Some populations in lakes have local names, such as the Gillaroo of Lough Neagh and other Irish loughs.

Young trout have the same names as young salmon. Parr have numerous white-ringed red spots on their flanks and an orange adipose fin. Smolts, found only in Sea Trout, are silvery. Adults are exceptionally variable, the back colour ranging from silvery or bluish-grey through purple, olive, green, yellow and brown to almost black, and spangled with dark and reddish spots. Fully-grown Sea Trout can attain 80–100 cm and 10–15 kg, but Brown Trout usually only reach 50 cm (in smaller streams only 20 cm) and 1.5 kg.

It is not always easy to distinguish between salmon and trout, especially the larger Sea Trout, but if a fish is more than 100 cm long, and particularly if it has a hooked lower jaw, the presumption is salmon. The salmon is also somewhat less thick-set, its head more pointed, its gill covers usually less heavily-spotted, and its tail fin usually more forked. The old fishermen's rule is that a salmon does not slip through your fingers when you hold it by the tail, but a trout does. This is because the salmon's tail is so constricted that the base of the fin forms more of a shoulder. In the hand too, the slenderer and more pointed gill-rakers on the upper limb of the first gill arch of the salmon can be distinguished from the blunter and thicker ones of the trout (*Fig. 25.8*). The parr are the easiest to separate, trout parr having more numerous red spots, all with pale rings, and an orange instead of a grey-green adipose fin; their 'finger marks' are also smudgier. The great majority of European lakes only hold trout, and so do the smaller rivers and streams. In Britain, a distinction is usually made between a 'trout stream' and a 'salmon river'.

Fig. 25.8

Rainbow Trout *Salmo gairdnerii* Much the most frequent of several North American trout species introduced into European rivers, being a mixture of sedentary Rainbow and migratory Steelhead stocks from the Pacific coast. It can be distinguished by the broad purple or violet band along its flanks and the black spots on the tail fin. Rainbows can tolerate more polluted water than Brown Trout and this sometimes leads to feral breeding populations. The Cut-throat Trout (*S. clarki*), also from western North America, and easily told by the red on its throat, is bred on European fish farms, but does not seem to have established itself anywhere.

Fig. 25.9

Char *Salvelinus alpinus* (*Fig. 25.9*) A trout-sized cold-water fish, maturing at 60 cm and 3 kg, or rarely more, which inhabits Arctic rivers, where it migrates to the sea, and northern and alpine lakes, where it does not. In Britain some fifteen lake populations differ slightly and have acquired both subspecific scientific names and local vernacular ones, such as the Torgoch *S.a. perisii* of Snowdonia and the Haddy *S.a. killinensis* of Loch Killin in Strath Errick, by Loch Ness. Char vary as much as trout in ground colour and spotting, though the spots are rarely blackish and may be confined to below the lateral line. They are best told by the pure white edges of the pectoral, pelvic and anal fins, which in the breeding season, along with the belly, turn bright red. In the hand Char can also be recognized by the vomerine teeth being confined to the head of the vomer.

American Brook Trout *Salvelinus fontinalis* A North American relative of the Char introduced in Europe and established ferally in a number of, mainly cold-water, streams and lakes. It matures smaller than the native Char, and also differs in the pale mottling on its back and in the black margin to the white-edged lower fins.

Whitefish Family (Coregonidae)

Whitefish *Coregonus* A group of shoaling, silvery, herring-shaped fish distinguished from the Herring (*Clupea harengus*) by their salmonid adipose fin, and from other salmonids by their lack of spotting, rather small mouth and well-forked tail. All except the Houting have sedentary forms inhabiting cool-water lakes, and migratory forms that ascend rivers to spawn, especially around the Baltic and further north. Most species mature at 50 cm or so, but have dwarf forms not exceeding 20–30 cm. The numerous described European forms have now been reduced to the following six species, distinguished mainly by the shape of the mouth and the number of rakers in the first gill-arch.

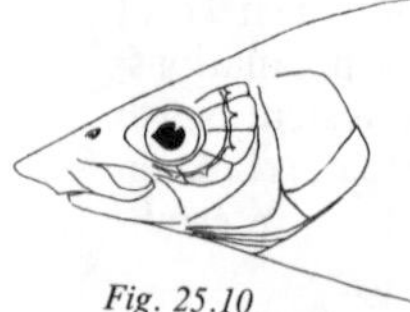

Fig. 25.10

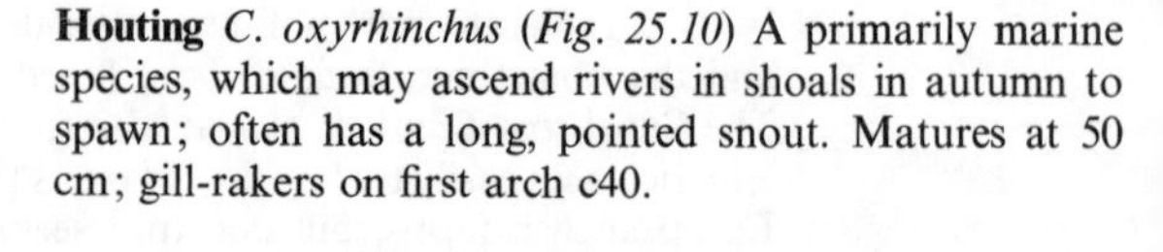

Houting *C. oxyrhinchus* (*Fig. 25.10*) A primarily marine species, which may ascend rivers in shoals in autumn to spawn; often has a long, pointed snout. Matures at 50 cm; gill-rakers on first arch c40.

Fig. 25.11

Vendace *C. albula* (*Fig. 25.11*) Easily told by its projecting lower jaw; matures at about 40 cm; gill-rakers on first arch 40–50. In Britain confined to two lakes in the Lake District and two in south-west Scotland; also in several loughs in Ireland, where it is known as Pollan.

Fig. 25.12

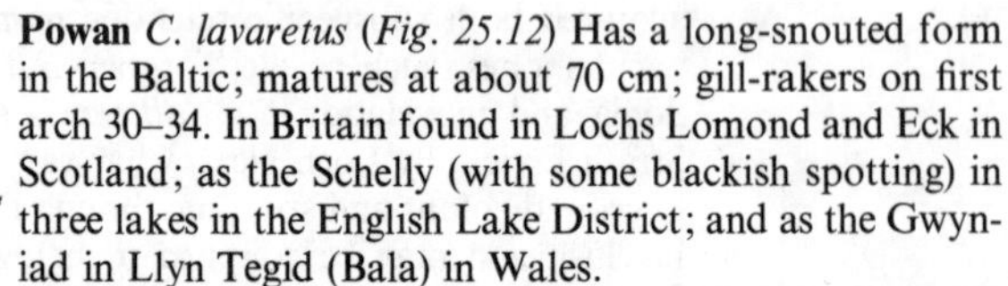

Powan *C. lavaretus* (*Fig. 25.12*) Has a long-snouted form in the Baltic; matures at about 70 cm; gill-rakers on first arch 30–34. In Britain found in Lochs Lomond and Eck in Scotland; as the Schelly (with some blackish spotting) in three lakes in the English Lake District; and as the Gwyniad in Llyn Tegid (Bala) in Wales.

There are three more very similar species in northern Europe: *C. pidschian*, maturing at about 50 cm with about 20 first-arch gill-rakers; *C. nasus*, about 50 cm and about 24 gill-rakers; and *C. peled*, about 70 cm and 50–60 gill-rakers.

Smelt Family (Osmeridae)

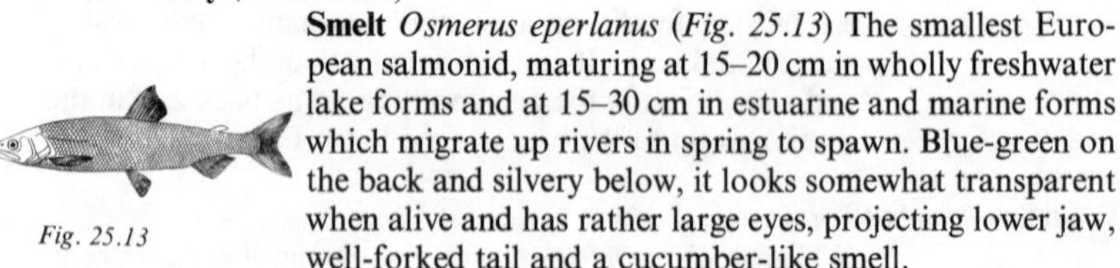

Fig. 25.13

Smelt *Osmerus eperlanus* (*Fig. 25.13*) The smallest European salmonid, maturing at 15–20 cm in wholly freshwater lake forms and at 15–30 cm in estuarine and marine forms which migrate up rivers in spring to spawn. Blue-green on the back and silvery below, it looks somewhat transparent when alive and has rather large eyes, projecting lower jaw, well-forked tail and a cucumber-like smell.

Grayling Family (Thymallidae)

Fig. 25.14

Grayling *Thymallus thymallus* (*Fig. 25.14*) A river fish, preferring fast-flowing hill streams below the char zone, but sometimes, as in the Gulf of Bothnia, living in brackish water. Named from its generally silvery-grey appearance, though with a greenish or purplish tinge on the back. Smaller than trout, maturing between 30 and 50 cm, and differing especially in being almost unspotted and having a much larger and longer dorsal fin, a more deeply-forked tail and a thymy scent.

Pike Family (Esocidae)

Fig. 25.15

Pike *Esox lucius* (*Fig. 25.15*) One of the most easily recognized fishes, with its elongated body, long, broad and rather flat snout, and dorsal fin set well back above the anal fin. Generally greenish in colour, it matures at 25–100 cm and 5–8 kg; exceptionally old fish, up to thirty years, may even reach 150 cm and 35 kg. Primarily a fish of lakes and slow-moving rivers, it can also live in brackish water, especially in the Baltic. A voracious predator, it can eat prey up to half its own weight.

Mud-minnow Family (Umbridae)

American Mud-minnow *Umbra pygmaea* A small greenish-brown, minnow-like fish introduced from North America to several places in north-west Europe. It matures at 7–12 cm and differs from the true Minnow especially in its rounded tail, pale flank stripes, dark lower jaw and dark spot at the base of the tail.

Carp Family (Cyprinidae)

The largest European freshwater fish family, whose members are mainly herring-shaped, with one dorsal fin a little behind the middle of the back, one anal fin and a well-forked tail. Breeding males often have white or yellowish tubercles on the head. The jaws are toothless, but the teeth on the curved lower pharyngeal bones (see *p. 297*) inside the throat are important in identification, as is the number of scales in the lateral line. Though few are considered good to eat, many cyprinoids are favourite anglers' fish.

Roach *Rutilus rutilus* (*Pl. 307*) A common shoaling fish of lakes, often eutrophic (rich in nutrients), and slow-flowing rivers. It matures at 10–25 cm, and is silvery, often greenish on the back and tinged red on the eyes and lower fins; 42–45 scales along the lateral line; lower jaw slightly protruding; pharyngeal teeth curved, in one row and slightly serrate. For distinctions from Rudd, see below.

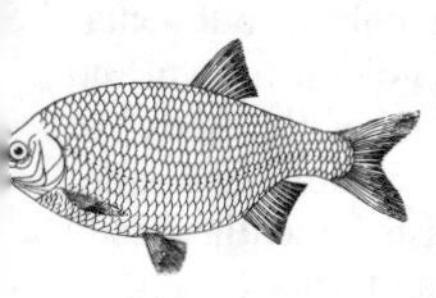

Fig. 25.16

Rudd *Scardinius erythrophthalmus* (*Fig. 25.16*) Similar in habitat and in many other ways to the Roach (above), but has a deeper body, sharply-keeled at the vent, and can be slightly larger, with the lower fins a brighter red; the dorsal fin further down the back, originating behind instead of just above the pelvic fins; the eyes yellow or orange; 40–43 scales along the lateral line; and the lower pharyngeal teeth in two rows and more deeply serrate. Usually found in still waters.

Fig. 25.17

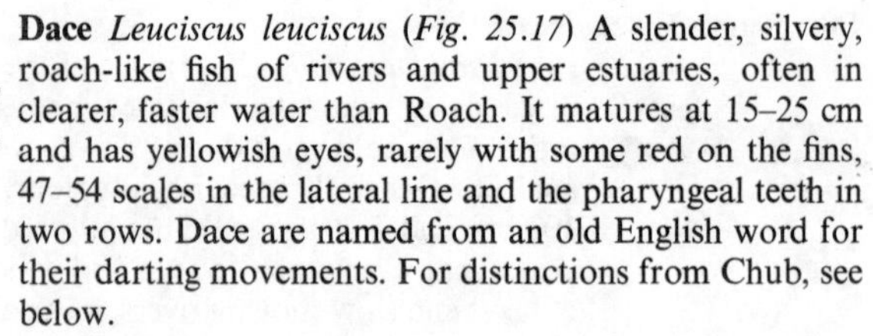

Dace *Leuciscus leuciscus* (*Fig. 25.17*) A slender, silvery, roach-like fish of rivers and upper estuaries, often in clearer, faster water than Roach. It matures at 15–25 cm and has yellowish eyes, rarely with some red on the fins, 47–54 scales in the lateral line and the pharyngeal teeth in two rows. Dace are named from an old English word for their darting movements. For distinctions from Chub, see below.

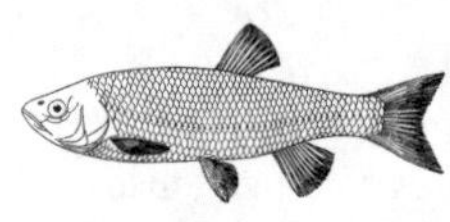

Fig. 25.18

Chub *Leuciscus cephalus* (*Fig. 25.18*) More of a river fish and less often in lakes than Dace (above), maturing larger, to 40 cm, and differing also in its black-edged scales, 44–46 in the lateral line, the lower fins being more often reddish, and especially in its convex anal fin.

Ide *Leuciscus idus* Another silvery, roach-like fish, of rivers, estuaries, from which it migrates upstream to spawn, and a few lakes. Matures at 30–40 cm, and has yellow eyes, reddish lower fins, the anal one straight-edged; two rows of pharyngeal teeth and 55–61 scales along the lateral line. It has a reddish-orange form, the Golden Ide or Golden Orfe, which may escape from ornamental waters into the wild and can be told from escaped Goldfish (*p. 311*) by its much shorter dorsal fin.

Soufie *Leuciscus souffia* A small, somewhat trout-shaped, shoaling fish, maturing at 12–25 cm, living in the fast hill streams also favoured by Grayling (*p. 306*). It is bluish on the back and silvery below, with a violet-purple band (brighter in the males) along the sides, and the lateral line orange-yellow.

Minnow *Phoxinus phoxinus* (*Pl. 308*) A small shoaling fish (maturing at 9–10 cm) of the upper parts of the rivers with cool, clear water and sandy or stony bottoms also favoured by trout; less often in lakes. Variable in colour, but generally dark on the back, usually either golden-brown and green with dark bars on the dorsal half of the body, or silvery tinged green, with a yellow line along the side; breeding males in summer turn reddish beneath. Easily told from sticklebacks (*p. 315*) by blunt snout, lack of spines, forward position of dorsal fin and forked tail. Often shoals with salmon and trout fry (*pp. 303* and *304*), whose adipose fin it lacks.

Motherless Minnow *Leucaspius delineatus* Differs from the true Minnow both in habitat and in being much more silvery, especially on the sides, and not reddening in summer; also has larger and looser scales, and the anal fin longer than, not shorter than the dorsal fin. Inhabits small ponds and the more slow-moving parts of rivers favoured by bream; less often in lakes. Its curious name, the German *Moderlieschen*, derives from its habit of suddenly appearing in small ponds.

Fig. 25.19

Asp *Aspius aspius* (*Fig. 25.19*) A solitary fish of the middle reaches of rivers, resembling an outsize Roach or Rudd – it matures at 50–55 cm – but with eyes not coloured and a sharp keel behind the pelvic fins. Breeding males have numerous distinctive white tubercles.

Tench *Tinca tinca* (*Pl. 309*) One of the most distinctive European fishes, with its deep body, dark blackish or brownish-green colour, almost square tail and the other fins rounded; also has a small barbel at each corner of its mouth. Matures at about 25 cm. Lies at the bottom of weedy ponds and slow-flowing rivers; in the Baltic sometimes in slightly brackish water. Has a golden variety that occasionally escapes from ornamental waters.

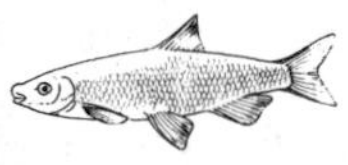

Fig. 25.20

Nase *Chondrostoma nasus* (*Fig. 25.20*) A silvery, herring-shaped, shoaling fish with lower fins tinged red, snout slightly projecting, mouth transverse and knife-shaped pharyngeal teeth, maturing at 25–40 cm. It feeds by scraping algae and their associated small animals off rocks and wood with its sharp-edged lips. Inhabits the middle, barbel zone, reaches of rivers, mainly in fairly deep water with a stony or sandy bottom; spawns higher upstream in the trout zone on gravel bottoms. Two very similar and very local species are the smaller (to 20 cm) *C. genei* in the Rhine, and *C. toxostoma*, with a smaller, bowed mouth, north to the Loire.

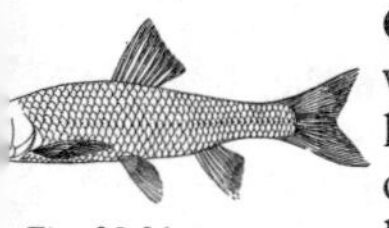

Fig. 25.21

Gudgeon *Gobio gobio* (*Fig. 25.21*) A small (8–14 cm) fish with a single pair of barbels at the corners of its mouth; lives in a wide variety of fresh and even brackish water, but chiefly in grayling-type streams with sandy or gravelly bottoms. It is brownish or greenish above, and silvery or golden on the sides, often with a row of blackish spots along the sides and smaller, scattered brownish ones; the fins are also dark-spotted.

Fig. 25.22

Barbel *Barbus barbus* (*Fig. 25.22*) A fairly large, olive-green, shoaling fish, maturing at 30–50 cm or even more, living on the bottom of the middle reaches of fast, clear rivers. Easily distinguished by its four barbels and projecting snout, also has lower fins red-tinged, and sometimes small dark spots both on the back and sides and on the dorsal and tail fins. *B. meridionalis* of central Europe has a more arched back and larger and redder spots.

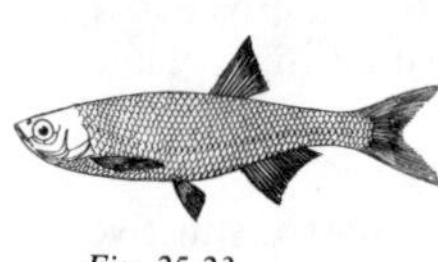
Fig. 25.23

Bleak *Alburnus alburnus* (*Fig. 25.23*) A smallish, slender, silvery, shoaling fish, greenish on the back, that matures at 12–15 cm, occasionally to 20 cm; anal fin rather long. Usually seen near the shores of lakes or slow-flowing rivers, often leaping after flies. The less slender **Schneider** (*Alburnoides bipunctatus*) has a deeper body, reddish bases to the lower fins and in the breeding season, a purple band along the side. It inhabits faster clear water.

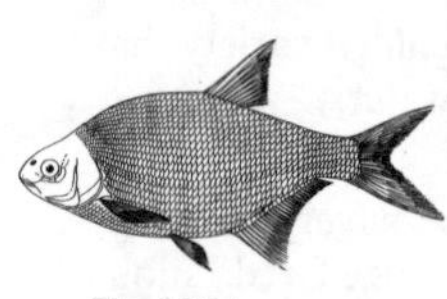
Fig. 25.24

Common Bream *Abramis brama* (*Fig. 25.24*) A thick-set, deep-bodied, brownish, shoaling fish with a slightly projecting snout and a long anal fin (23–28 rays), maturing at 30–40 or even 40–50 cm, up to 3 kg or more. Diameter of eye less than distance from snout. It inhabits slow-moving and stagnant waters, especially rivers with a muddy or clayey bottom, and in the Baltic, also brackish water. The **Zope** (*A. ballerus*) is smaller (20–40 cm) with the snout not projecting and 39–46 rays in the anal fin. It inhabits lakes and the lower parts of rivers, but not estuaries.

White or **Silver Bream** *Blicca bjoerkna* Differs from the Common Bream especially in being silvery with a greenish back, and having red-based and grey-tipped lower fins, 22–26 rays in the anal fin, the lower lobe of the tail fin longer than the upper and the diameter of the eye equalling its distance from the snout. Inhabits still or slow-moving water, often shoaling with Roach or Rudd.

Zaehrte *Vimba vimba* Like one of the smaller bream, but somewhat slenderer, often maturing at no more than 20–23 cm, and with projecting snout and 20–23 rays in the shorter anal fin. Usually silvery, but when spawning turns almost black, with underparts and paired fins reddish.

Inhabits lakes and the lower parts of rivers – the bream zone; in the Baltic in brackish water after spawning.

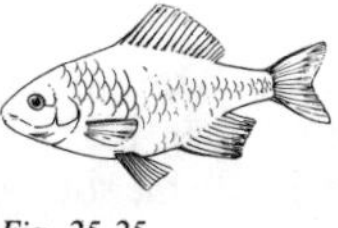

Fig. 25.25

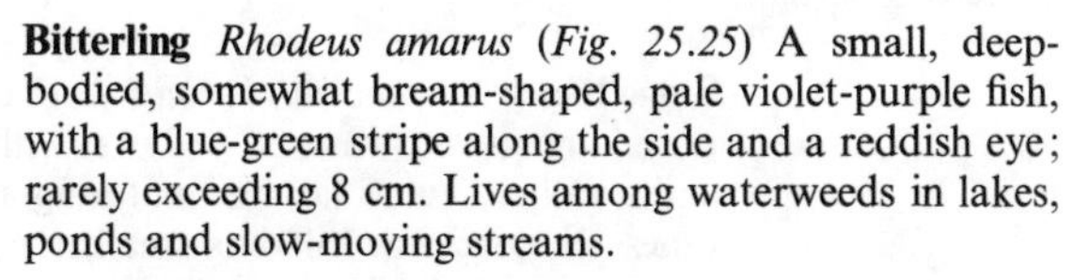

Bitterling *Rhodeus amarus* (*Fig. 25.25*) A small, deep-bodied, somewhat bream-shaped, pale violet-purple fish, with a blue-green stripe along the side and a reddish eye; rarely exceeding 8 cm. Lives among waterweeds in lakes, ponds and slow-moving streams.

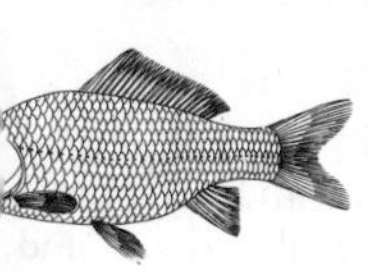

Fig. 25.26

Crucian Carp *Carassius carassius* (*Fig. 25.26*) Smaller and often less deep-bodied than the Carp, with which it hybridizes, maturing at 15–45 cm and with no barbels and a single row of pharyngeal teeth on each side of the mouth. Prefers heavily vegetated, almost stagnant waters. The Prussian Carp is a slender variety. Its name is anglicized from the German *Karausche*, which was also latinized by Linnaeus to *Carassius*.

Fig. 25.27

Goldfish *Carassius auratus* (*Fig. 25.27*) A well-known ornamental fish with many varieties combining golden-orange, black and silver and fins that have been bred into grotesque shapes. It reverts when wild into a fish not unlike the Prussian Carp form of the Crucian Carp, though usually silvery rather than greenish-yellow in colour, and inhabiting similar waters. It differs also in having 27–31 instead of 31–35 scales along the lateral line, and 35–38 instead of 23–33 gill-rakers in the first gill-arch.

Carp *Cyprinus carpio* (*Pl. 306*) A large, rather heavy and deep-bodied, shoaling fish of still and slow-moving waters, that matures at 20–40 cm and can attain 100 cm and 30 kg. Generally greenish-brown in colour, with golden highlights; dorsal fin elongated. Easily told from Crucian Carp and feral Goldfish by its four barbels, two long and two short, as well as by its three rows of pharyngeal teeth. Varieties bred in captivity and sometimes escaping include the Golden Carp, a golden-orange form, the Mirror Carp with larger scales, and the Leather Carp with almost no scales. Carp populations breeding in the wild develop more slender bodies than those reared in captivity, and resemble the Prussian Carp form of Crucian Carp.

Loach Family (Cobitidae)
Rather small, elongated, bottom-living fishes, with projecting snout and numerous barbels.

Stone Loach *Neomachilus barbatulus* In colour a mosaic of greys, greens, browns and yellows, with six long barbels and a rather square tail fin; maturing at 8–12 cm, sometimes up to 15 cm. Inhabits small, clear streams, with a sandy or gravelly bottom, usually hiding under stones; also along shallow lake shores and in slightly brackish water.

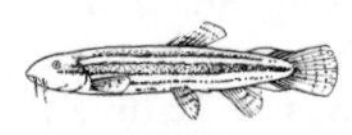

Fig. 25.28

Pond Loach *Misgurnus fossilis* (*Fig. 25.28*) The largest European loach, maturing at 20–25 cm and sometimes more; generally brown with longitudinal stripes, and distinctive for its ten barbels, and rounded tail fin. Shallow muddy ponds and lakes. Known as *Wetterfisch* in Germany from its habit of frequently rising to the surface before stormy weather.

Spined Loach *Cobitis taenia* Smaller than Stone Loach, maturing at 5–10 cm, occasionally to 12 cm; generally greenish-brown and easily told by its six shorter barbels, double spine below each eye and row of dark spots along each side. Clear streams or brooks with a sandy or gravelly bottom, in which it often lies buried up to the head; also in lakes.

Catfish Families (Siluridae and Ictaluridae)
Rather large, bottom-living fishes with stout, elongated, slimy, scale-less bodies, flat heads, broad mouths equipped with long barbels, long anal fins and small dorsal fins located well forward. They inhabit muddy lakes, slow-flowing rivers in the bream zone, and even swamps.

Wels or **Giant Catfish** *Silurus glanis* One of the largest European freshwater fishes, maturing at 50–100 cm (2–10 kg) and sometimes up to 200 cm or more; the record, from eastern Europe, is 500 cm and 306 kg. Grey-brown, mottled paler; anal fin very long; dorsal fin very short; barbels, two long on upper jaw, four shorter below.

American Catfish *Ameiurus nebulosus* A somewhat greyer, introduced species maturing at 30–40 cm, and differing from the Wels in having eight barbels, an adipose fin, a longer dorsal fin and a shorter anal fin. Another North

American species, *A. melas*, has also been introduced in a few places.

Eel Family (Anguillidae)

Eel *Anguilla anguilla* Elongated, snake-like fish with a long dorsal and two small pectoral fins; males 30–50 cm, females 40–100 cm; for differences from lampreys, see key on *p. 302*.

All European eels spawn in the Sargasso Sea in the tropical western Atlantic, whence their transparent, flattened larvae, known as Leptocephalus, drift eastwards on the Gulf Stream. When three years old they change to eel-shape, darken, and in winter and spring, ascend rivers as Elvers. As Yellow Eels, all yellowish except for the grey-brown back, they then spend from five to twelve years in both still and flowing fresh water, for they can cross damp grassland to reach land-locked ponds. Eels eat many other freshwater animals; those which enlarge their jaws by eating larger prey are called Frog-mouthed Eels. Finally, they change colour again and, as Silver Eels, return to the sea. Whether they actually recross the ocean and spawn before they die is still uncertain.

Cod Family (Gadidae)

Fig. 25.29

Burbot *Lota lota* A large, bottom-living fish, maturing at 40–80 cm and in the east up to 25–30 kg. It is the only European freshwater relative of the Cod, Haddock, Whiting and other important commercial white fish. Generally marbled brownish, greyish, greenish and yellowish, it has a distinctive trio of barbels, one on the lower jaw, a long anal fin, a bluntly-pointed tail fin and, unlike other freshwater fish, two dorsal fins, one of them very long. Inhabits lakes, slow-flowing rivers and, especially in the Baltic, shallow brackish water; rare and possibly extinct in Britain.

Sea-Perch Family (Serranidae)

Fig. 25.30

Bass *Dicentrarchus labrax* (*Fig. 25.30*) A silvery, herring-shaped, shoaling, marine fish, which ascends the lower parts of rivers in summer, especially those with sandy, gravelly or muddy bottoms. It has two dorsal fins, the front one with spines, and the anal fin, too, has three spines, and matures at 30–60 cm.

Perch Family (Percidae)

Perch *Perca fluviatilis* (*Pl. 310*) A medium-sized, greenish, bream-shaped shoaling fish, maturing at 25–45 cm, and noted for its barred flanks and red lower and tail fins. It also has spines in the more forward of its two dorsal fins, which has a dark spot at the rear, and a spine on each gill-cover. Inhabits rivers, lakes and ponds in the lowlands, and shallow brackish water, especially in the Baltic.

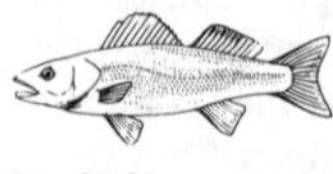
Fig. 25.31

Zander or **Pike-Perch** *Stizostedion lucioperca* (*Fig. 25.31*) Larger and slenderer than the Perch, maturing at 35–55 cm and occasionally to more than 100 cm (12 kg). Differs further in its less conspicuous barring, the lower and tail fins not reddish, the anal fins longer and no spines on the gill covers. Mainly in lakes and the lower reaches of rivers, especially in turbid water; introduced in Britain, in the Fenland.

Fig. 25.32

Ruffe or **Pope** *Gymnocephalus cernua* (*Fig. 25.32*) Smaller than the Perch, maturing at 12–25 cm, and with darker marbling rather than barring, the two dorsal fins joined together and the lower and tail fins not reddish. Shoals in slow-moving and fairly deep water in rivers, canals and lakes.

Another small, slender perch-ally, the **Streber** (*A. asper*), with dark bars, two separated dorsal fins and a rather pointed head, is nocturnal and occurs in the upper Rhône basin.

Sun Perch Family (Centrarchidae)

Two grey-green species of **North American Black Bass** (*Micropterus*), the Large-mouthed (*M. salmoides*) and the Small-mouthed (*M. dolomieu*), which mature at 40–60 cm, have been introduced by anglers into ponds, canals, and other still or slow-flowing European waters. Their dorsal fin is so sharply divided as to appear to be two fins, the front part being spiny. They can be distinguished by the size of the mouth, with its projecting lower jaw; the Large-mouth's upper jaw extends back behind the rear edge of its eye.

Two species of the more perch-shaped **Sunfish** (*Lepomis*) with a long single dorsal fin, have also been introduced from North America, the **Pumpkinseed** (*L. gibbosus*), which has a black-edged, reddish spot on the gill-cover, and *L. cyanellus*, which has become established near Frankfurt-on-Main, West Germany.

Grey Mullet Family (Mugilidae) *Fig. 23.33*

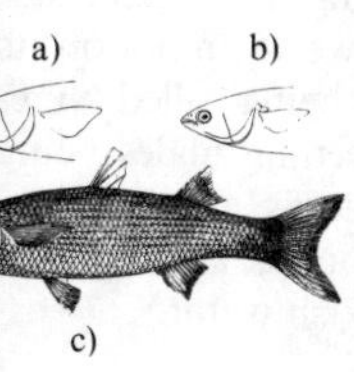

Grey Mullets are shoaling marine and estuarine fishes, maturing at 50 cm or more, that sometimes swim up rivers beyond the tide limit. The **Thick-lipped Grey Mullet** (*a*) (*Crenimugil labrosus*), the most frequent, is silvery-grey with two short dorsal fins, the front one spiny, and two rows of small warts on its distinctively swollen upper lip. The **Thin-lipped Grey Mullet** (*Liza ramada*) (*b*) has much thinner lips with no warts. The smaller and yellower **Golden Grey Mullet** (*Liza auratus*) (*c*) also has thinner unwarted lips, but is best told by the two golden-yellow spots on each side of its head, the larger on the gill covers and the smaller behind the eye.

Bullhead Family (Cottidae)

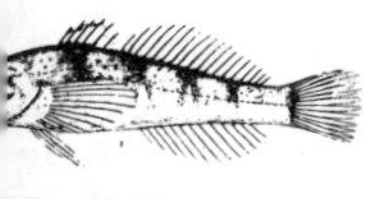

Fig. 25.34

Miller's Thumb or **Bullhead** *Cottus gobio* (*Fig. 25.34*) A small, scale-less big-headed fish, olive-brown or yellowish, marbled darker and with two dorsal fins, a spiny pelvic fin and a rounded tail fin; maturing at 10–18 cm. Inhabits clear water over gravelly or sandy bottoms, in lakes and the trout zone of rivers, where it is typically found under stones. The Siberian Bullhead (*C. poecilopus*) differs mainly in its pelvic fin being barred with the innermost ray less than half, instead of more than half, the length of the longest rays.

Four-horned Bullhead *Oncocottus quadricornis* An Arctic marine species that occurs in some lakes as a glacial relict, is darker and has four most distinctive greyish-yellow warts on its head and a taller rear dorsal fin.

Stickleback Family (Gasterosteidae)

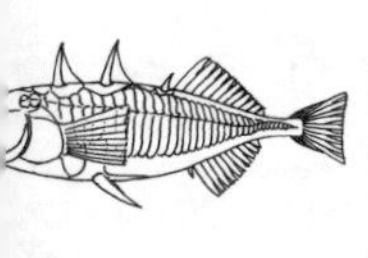

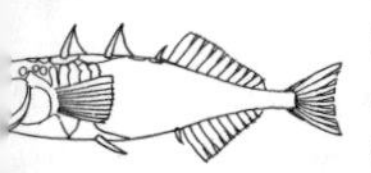

Fig. 25.35

Three-spined Stickleback *Gasterosteus aculeatus* (*Pl. 311, Fig. 25.35*) A diminutive shoaling fish, maturing at 5–8 cm, easily told by the three spines on its back and two on its belly (the pelvic fins). Generally dark olive-green, the breeding males with red underparts, and scaleless, but with a row of bony plates along each side. There are three intergrading forms: along the whole side, (*G. trachurus*); only along the front half (*G. leiurus*); and both along the front half and on the tail (*G. semiarmatus*), which is a hybrid between the other two. Inhabits both still and flowing water, especially small ponds and ditches, but also rivers; in the north often migratory, wintering in the sea. Builds a nest at the bottom in shallow water.

Ten-spined Stickleback *Pungitius pungitius* (*Pl. 312*) Often smaller than the Three-spined, maturing at 5–7 cm, and readily distinguished by the seven to twelve spines on its back; this variability has led to its also being called Nine-spined Stickleback. Differs also in breeding males being dark brown and the absence of bony plates, except sometimes on the tail. Prefers still water, ponds and ditches, but around the Baltic in shallow, brackish water.

Plaice Family (Pleuronectidae)

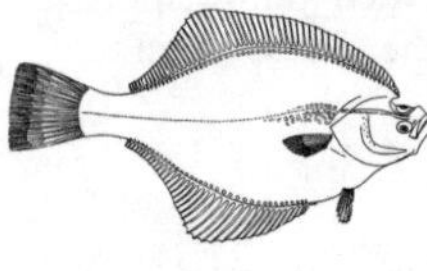

Fig. 25.36

Flounder *Platichthys flesus* (*Fig. 25.36*) The only flatfish that enters fresh water, migrating up rivers, sometimes considerable distances, after spawning in the sea in spring, and so hardly confusable with any other fish found in fresh water; matures at 30–50 cm.

CHAPTER 26 Amphibians

Amphibians, like fishes, are 'cold-blooded' vertebrates, but unlike them, breathe air through lungs as adults. They differ from reptiles particularly by their softer, moister and usually scale-less skin, and their need to resort to fresh water in order to breed. They lay eggs with a gelatinous covering, called spawn, usually in still water, often among aquatic vegetation. The resultant young, known as tadpoles – at first limbless blobs with a tail – are aquatic and have feathery external gills, but soon develop lungs and limbs and leave the water. Adult amphibians frequent fresh water primarily to spawn in the spring, though some species may be found in the water at almost any time. Tadpoles, of course, only occur in spring and early summer.

The two groups of British and European amphibians are the tailed, or Caudata (newts, salamanders), and the tail-less, or Anura (frogs, toads).

Newts and Salamanders (Caudata)

Newts and salamanders bear a general resemblance to lizards, but are always softer-skinned and without scales. Lizards do not enter the water.

Tabular Key to Newts and Salamanders

Adults	
Upper parts black with orange or yellow spots or stripes:	Fire Salamander
brown, tinged olive or yellow:	Smooth Newt, Palmate Newt
green, marbled darker:	Marbled Newt
grey:	Great Crested Newt, Alpine Newt
orange or yellow with black spots or stripes:	Fire Salamander
rough and wrinkled:	Great Crested Newt
Crest, jagged or toothed:	Great Crested Newt, Smooth Newt
smooth, barred or spotted darker:	Marbled Newt, Alpine Newt, Palmate Newt
Flanks spotted:	Alpine Newt
Belly grey, marbled darker:	Marbled Newt
orange, spotted darker:	Great Crested Newt, Alpine Newt, Smooth Newt (male)
yellow, spotted darker:	Smooth Newt (female), Palmate Newt

Chin grey, marbled darker:	Marbled Newt
warty:	Great Crested Newt
orange or yellow, spotted:	Smooth Newt
orange or yellow, unspotted:	Alpine Newt, Palmate Newt
Tail with pale midrib:	Marbled Newt (male), Great Crested Newt (male)
projecting filament:	Palmate Newt (male)

Tadpoles

Tail blunt:	Fire Salamander
pointed:	Smooth Newt, Palmate Newt
tapering to a fine point:	Marbled Newt, Great Crested Newt

Fig. 26.1

Fire Salamander *Salamandra salamandra* (*Fig. 26.1*) Adults are most distinctive, with broad heads, and very variable and conspicuous orange-red or yellow spots or stripes on a blackish background; up to 20 cm or even 28 cm long. Young may be born alive, especially at high altitudes, when they already look like miniature, unspotted, grey-brown adults, or as tadpoles, from eggs deposited singly on stones or plants in the water. Tadpoles can be up to 6.5 cm long, and also differ from newt tadpoles in their broader heads, a pale spot at the base of each leg, and the tail crest extending only a short distance on to the body. Ponds and streams with clear water, especially on wooded hills and mountains.

Newts (*Triturus*)
Five species in north-west Europe. Breed in still or slow-flowing water, attaching eggs singly to water plants or stones; sometimes remain in water out of the breeding season. Males develop striking breeding dress and use their tails for courtship display. Non-breeding males, females and tadpoles can be difficult to identify.

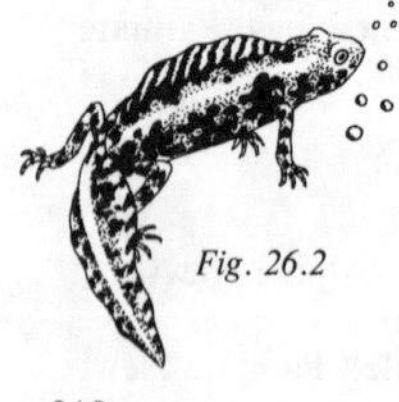
Fig. 26.2

Marbled Newt *Triturus marmoratus* (*Fig. 26.2*) The only green newt, marbled darker, the belly grey and the tail midrib pale; breeding males with a smooth, pale yellowish crest, barred darker; up to 14 cm. Tadpoles resemble those of Great Crested Newt, which it replaces in north-west France, but usually tinged green, and with only twelve to thirteen grooves between fore and hind limbs.

Great Crested Newt or **Warty Newt** *Triturus cristatus* (*Pls. 314, 315*) At 14–18 cm, the largest European newt, the upper parts rough and wrinkled, dark grey with no green tinge, the belly orange and dark-spotted, the male's tail with a silvery midrib. Breeding male's crest conspicuously jagged and not barred. Tadpoles with tail often dark-spotted and tapering to a long, fine point; crest extending more than half-way along back; up to 8 cm.

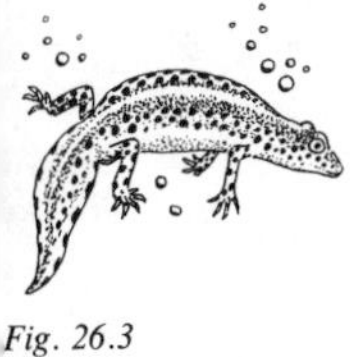
Fig. 26.3

Alpine Newt *Triturus alpestris* (*Fig. 26.3*) One of the smaller north-west European newts, to 12 cm, with belly yellow or orange-red and flanks distinctively dark-spotted. Breeding males have a rather inconspicuous crest, yellow with blackish spots or bars. Tadpoles, to 5 cm, are best told by their blunt tail.

Fig. 26.4

Smooth Newt *Triturus vulgaris* (*Pl. 313, Fig. 26.4*) Generally the commonest newt, to 11 cm, olive- or yellowish-brown, with numerous dark spots, sometimes fused into lines, and a yellow or orange belly and throat, also dark-spotted. Breeding males have a crest similar to, but much less jagged than the Great Crested Newt. Tadpoles to 4 cm, the tail pointed, but not to a fine point like the Great Crested Newt. Breeds in still water only.

Fig. 26.5

Palmate Newt *Triturus helveticus* (*Fig. 26.5*) The smallest north-west European newt, to 9 cm, differing from the Smooth Newt especially in its smaller feet, paler yellow belly, less conspicuous spotting, and unspotted throat. Breeding males are more distinct, being noticeably smaller than male Smooth Newts and having only a low, smooth crest, a filament projecting from the tail-tip, and the hind feet webbed. Tadpoles cannot be distinguished until a very late stage, when their belly becomes pink or brown.

Frogs and Toads (Anura)

There is no clear distinction between frogs and toads, except that the close relatives of the Common Frog are usually called frog, and those of the Common Toad are usually called toad! All of them lose their tails as they are transformed from tadpoles to adults.

Tabular Key to Frogs and Toads

Adults	
General Features	
Smelling of garlic:	Spadefoots, Parsley Frog
Spawn carried on back:	Midwife Toad
Body Features	
Skin fold down each side of back:	Common, Moor, Agile, Marsh, Pool and Edible Frogs
Skin warty:	Yellow-bellied, Fire-bellied, Common, Natterjack and Green Toads, Parsley Frog
Under parts orange:	Common Frog
orange variegated black:	Fire-bellied Toad
yellow variegated grey:	Yellow-bellied Toad
Upper parts blue:	Moor (breeding male), Pool and Edible (occasionally) Frogs
green with darker marbling:	Marsh, Pool and Edible Frogs
green and unspotted with dark stripe on flanks:	Tree Frog
with green marbling:	Green Toad
green spots:	Parsley Frog
orange spots:	Spadefoots
yellow stripe	Natterjack Toad, Pool and Edible Frogs
Head Features	
Eyes green:	Western Spadefoot, Green Toad
yellow or orange:	Common Spadefoot, Common and Natterjack Toads, Parsley, Tree, Common, Moor, Agile, Marsh, Pool and Edible Frogs
silvery:	Western Spadefoot, Natterjack Toads
pupils horizontal:	Common, Green and Natterjack Toads; Tree, Common, Moor, Agile, Marsh, Pool and Edible Frogs

pupils rounded:	Yellow-bellied, Fire-bellied Toads
pupils vertical:	Midwife Toad, spadefoots, Parsley Frog
Lump on head:	Common Spadefoot

Leg Features

Hind feet with 'spade':	Spadefoots
smaller hard tubercle:	Moor, Agile, Pool, Edible Frogs
smaller soft tubercle:	Common, Marsh Frogs
webbed:	Spadefoots, Common, Natterjack and Green Toads; Common, Moor, Agile, Marsh, Pool and Edible Frogs

Tadpole Features

Blackish:	Common and Natterjack Toads: Common Frog
Dark spotted:	Midwife Toad; Moor and Agile Frogs
Pale-spotted:	Natterjack Toad
Yellow above, white below:	Tree Frog
Belly white:	Marsh, Pool and Edible Frogs
white with yellow spots:	Agile Frog
Beak black:	Spadefoots
white edged black:	Parsley Frog, Common Toad
Tail sharply pointed:	Spadefoots, Tree, Moor, Agile, Marsh, Pool and Edible Frogs

Painted Frog Family (Discoglossidae)

Yellow-bellied Toad *Bombina variegata* A very small, warty toad, to 5 cm, easily told by its salamander-like under parts, bright yellow, variegated with dark grey, the yellow extending to the finger-tips. Sociable; frequenting and breeding in all kinds of shallow fresh water, down to cart-ruts. Call, often in chorus, a melodious 'poop, poop, poop', one to two times a second. Spawn in clusters, often on waterweeds – tadpoles with rather bluntly-pointed tail, and mouth oval.

Fig. 26.6

Fire-bellied Toad *Bombina bombina* (*Fig. 26.6*) Differs from Yellow-bellied Toad especially in its under parts being variegated bright orange and black (very variable and sometimes almost all black), but the orange not extending to the finger-tips, and often having some green colouring on its back. Calls slower, one every one-and-a-half to four seconds, and described as mournful. Tadpoles with triangular mouth.

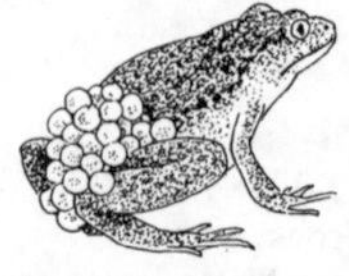

Fig. 26.7

Midwife Toad *Alytes obstetricans* (*Fig. 26.7*) Another very small toad, usually less than 5 cm; not unlike the Common Frog in colouring – some shade of brown, olive or grey. Primarily terrestrial, its name derives from the curious habit of the male carrying the eggs around, moistening them from time to time in fresh water, but only finally depositing them there when they are about to hatch. Tadpoles like *Bombina*, but larger and often with dark spots. Call higher-pitched and briefer than *Bombina*, and once in one to three seconds; confusable with Scops Owl in southern Europe.

Spadefoot Family (Pelobatidae)

Common Spadefoot *Pelobates fuscus* (*Pl. 316*) Very variable, brown, grey, olive or yellowish, with diverse patterns of darker spots, stripes and marbling, the spots sometimes orange; up to 8 cm. Has a prominent lump on the head behind the orange or yellow eyes, whose pupils are vertical; and a flattened, sharp-edged, pale tubercle ('spade') on each webbed hind foot. May smell of garlic. Breeds in ponds, otherwise terrestrial. Voice: submerged breeding adults have a staccato 'tock' call, the females harsher than the males. Tadpoles rather large, to 16 cm, with black beak and sharply-pointed tail.

Western Spadefoot *Pelobates cultripes* Larger than Common Spadefoot, to 10 cm, the marbling often greenish and rarely in stripes, no lump on the head, eyes silvery or greenish and the 'spade' black. Voice a fast 'co, co, co', likened to a hen clucking.

Fig. 26.8

Parsley Frog *Pelodytes punctatus* (*Fig. 26.8*) Much smaller than the spadefoots, only attaining 5 cm. Has a warty, greyish-olive skin, with lines of bright green, and sometimes also orange spots; eyes yellow with vertical pupils – and hind feet not webbed, with no 'spade'. May smell of garlic. Breeds in ponds, otherwise mainly nocturnal and terrestrial. Submerged males croak, the females have a softer response. Tadpoles with bluntly-pointed tail and beak white, but black-edged.

Toad Family (Bufonidae)
Three European species with dry, warty skins and horizontal pupils. Mainly terrestrial and nocturnal, they breed in ponds, the shallow parts of lakes, and slow streams. Spawn in strings, usually among waterweeds. Breeding males may call from the water, and both males and clasped pairs may be seen floating on or beneath the surface.

Common Toad *Bufo bufo* (*Pls. 317, 318*) The largest European toad, to 15 cm, typically brown or grey, in varying shades, but exceptionally more brightly coloured and/or spotted; eyes golden-yellow. Can begin to arrive in breedin grounds in a mild February or even January, but most in March and April. Male has a rather weak, high-pitched croak, recalling a small dog's bark. Tadpoles small, to 3.5 cm, blackish, with mouth as wide as space between eyes, and a black-edged, white beak.

Fig. 26.9

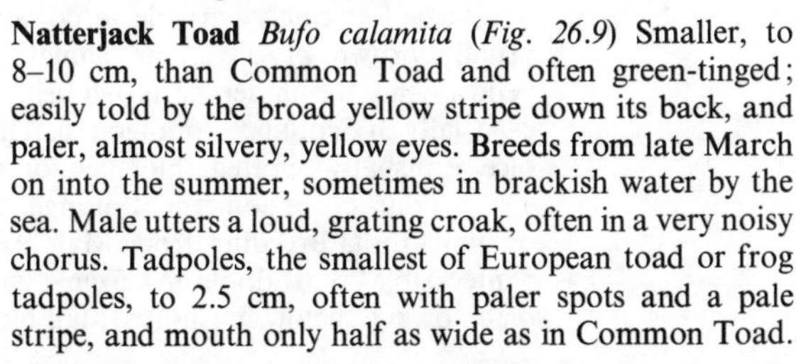

Natterjack Toad *Bufo calamita* (*Fig. 26.9*) Smaller, to 8–10 cm, than Common Toad and often green-tinged; easily told by the broad yellow stripe down its back, and paler, almost silvery, yellow eyes. Breeds from late March on into the summer, sometimes in brackish water by the sea. Male utters a loud, grating croak, often in a very noisy chorus. Tadpoles, the smallest of European toad or frog tadpoles, to 2.5 cm, often with paler spots and a pale stripe, and mouth only half as wide as in Common Toad.

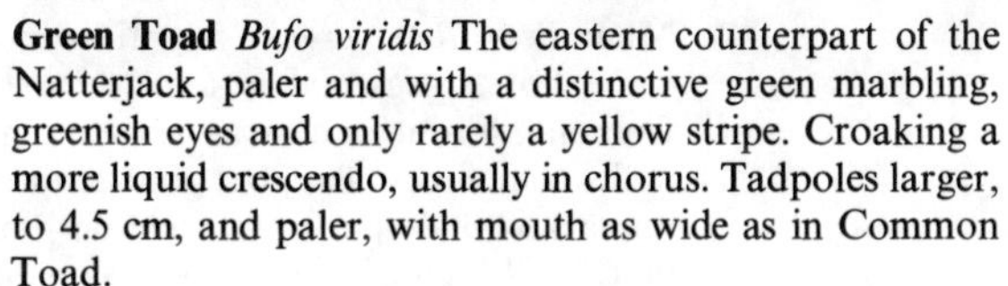

Green Toad *Bufo viridis* The eastern counterpart of the Natterjack, paler and with a distinctive green marbling, greenish eyes and only rarely a yellow stripe. Croaking a more liquid crescendo, usually in chorus. Tadpoles larger, to 4.5 cm, and paler, with mouth as wide as in Common Toad.

Tree Frog Family (Hylidae)

Tree Frog *Hyla arborea* (*Pl. 319*) The only all-green, small (to 5 cm) frog in north-west Europe, but can change colour to yellow or brown; has a distinctive dark, often cream-edged, stripe along the flanks and up to the eye, which is absent from its relative *H. meridionalis*, found only south of the Loire. Primarily terrestrial and arboreal, they only resort to the water to breed, at night. Voice, mainly at night on land, a harsh loud croak, three to six times a second. Tadpoles greenish-yellow above, white below and with a pointed tail. Absent from Britain, except for a small introduced population in the New Forest, Hampshire.

Frog Family (Ranidae)

Typical frogs differ from all other European frogs and toads in having a distinct fold of skin down each side of the back, and in laying their spawn in masses. From the typical toads (Bufonidae), they also differ in their smooth skins. Pupils horizontal; hind feet webbed. They breed in ponds and the shallow parts of lakes and slow-moving streams; breeding males often float in the water, either alone or clasping a female. Tadpoles in this family are hard to identify.

There are two distinct groups of these frogs in north-west Europe – the brown or grass frogs (*Rana temporaria, R. arvalis* and *R. dalmatina*) – and the green or water frogs (*R. ridibunda, R. lessonae* and *R. esculenta*). The brown frogs are largely terrestrial and diurnal, frequenting water only around the breeding period. The green frogs are much more aquatic, and spend much of their time close to water, into which they leap when disturbed.

Common Frog *Rana temporaria* (*Pls. 320–321*) Matures at 8–10 cm. Extremely variable in colour, the upper parts usually brown or olive, but sometimes grey, pink or yellow, and the under parts whitish, yellow, or even – especially in Scotland – orange-red. The whole body surface is usually spotted, blotched or marbled (but not striped) darker, or sometimes orange. Small soft tubercle on hind foot. Breeding from March (February in mild winters) to May. Tadpole, to 4.5 cm, black or brown above, black or grey below, unspotted; tail bluntly-pointed.

Moor Frog *Rana arvalis* Slightly smaller, to 8 cm, and with shorter legs and a somewhat sharper snout than Common Frog; almost as variable in colour (breeding males sometimes blue), but often striped on back and with unmarked underparts; tubercle on hind foot hard. Tadpole, to 4.5 cm, brown above, sometimes dark-spotted, and grey below; tail pointed.

Agile Frog *Rana dalmatina* Much less variable than Common or Moor Frogs, usually buffish-yellow or pinkish-brown, with fewer dark markings and flanks often yellow; maturing at 9 cm. Longer-legged than Moor Frog and with a more pointed snout than Common Frog; tubercle on hind foot larger than either. Tadpoles larger, to 6 cm, paler or rufous and sometimes dark-spotted above, white with yellow spots below, and tail more pointed than Moor Frog.

Marsh Frog *Rana ridibunda* (*Pl. 322*) The largest native European frog, to 15 cm, usually some shade of brown with darker markings, often tinged green, the back of the thigh whitish, grey or olive, marbled darker; external vocal sacs dark grey; tubercle on hind foot soft. Very noisy, making the harsh 'brek-ek-ek' croaking associated with the frogs of Aristophanes. May be seen resting on water-lily leaves or on submerged waterweeds with only head out of water; often hibernates submerged, emerging in April and spawning in June. Tadpoles to 9 cm, olive or green above, white beneath, with pointed tail.

Pool Frog *Rana lessonae* Markedly smaller than Marsh Frog, to 9 cm, more variable in colour and often mainly green, with the ground colour of the thighs orange or yellow, sometimes a pale (or green yellow) stripe down the back and the folds of skin down the back also often paler. Vocal sacs much paler and the hind legs shorter, their tubercle hard and sharp-edged. Breeding males may have yellow head and back. Voice softer than Marsh Frog. Tends to breed in smaller ponds. Tadpoles similar but smaller, to 7.5 cm.

Edible Frog *Rana esculenta* A hybrid between the Marsh and Pool Frogs, closely resembling the Pool Frog, from which it has only recently been definitely separated. Intermediate in size, to 12 cm, in length of hind legs and size of their hard tubercle, and in voice. Can be found in same ponds as either or both parents. Tadpoles not distinguishable from Pool Frog.

CHAPTER 27 Reptiles

Reptiles are marginal to the freshwater ecosystem in our area; only one terrapin and three snakes need be mentioned in detail.

Terrapin Family (Emydidae)

Fig. 27.1

Pond Tortoise *Emys orbicularis* (*Fig. 27.1*) The only terrapin found in northern Europe, extending into the south of our area in France and into the east in Germany, though aquarists often release them further north and west. Has a dark carapace, often spotted or streaked yellowish, 20–30 cm long. Prefers still or slow-moving water, including brackish, often poking its head above the surface or basking at the edge. Breeds on land.

Snakes (Serpentes)

Three snakes in our area are aquatic in the sense that they often hunt their prey in the water, the Viperine, Dice and Grass Snakes. Others, such as the Smooth Snake (*Coronella austriaca*) (Colubridae) and Adder or Viper (*Vipera berus*) (Viperidae) resort to the water to rest, travel to a lake or lake island, or for other purposes, and are not considered here. All snakes breed on land and are largely diurnal.

Typical Snake Family (Colubridae)

Grass Snake *Natrix natrix* (*Pl. 323*) The commonest and most widespread of the aquatic snakes, though frequenting water less often than the Viperine and Dice Snakes. All three are superficially rather similar, olive or greyish, often buffish along the flanks and belly, and with a series of dark blotches, but not with the marked zigzag blotching of the Adder, which can always be distinguished by its vertical, not rounded, pupils. The Grass Snake is much the largest snake likely to be seen in the water, attaining 120 cm, or exceptionally, 200 cm. (The Adder rarely goes above 90 cm, so the larger the snake the safer you are!) Grass Snakes often have a black-edged pale yellow collar on the neck. Feeds mainly on frogs and toads.

Fig. 27.2

Viperine Snake *Natrix maura* (*Fig. 27.2*) The most aquatic snake in the southern part of our area, in France. Smaller than Grass Snake, maturing at 70–100 cm, and very variable in colour and blotching, with no pale collar, but often two yellowish stripes down the back. Spends most of its time in or by ponds and streams. Feeds mainly on amphibians and fishes.

Dice Snake *Natrix tessellata* Even more aquatic than the Viperine Snake, from which it is hardly distinguishable in the field. Also even more marginal in our area, in only a few places in the east of it, and not overlapping with the Viperine Snake. Feeds mainly on fishes.

CHAPTER 28 Birds

Only birds which form part of the underwater aquatic ecosystem are discussed, and much less fully than other animals in the book, since there are so many good bird guides. Aquatic birds are classified by their diving and swimming habits. Waders, wagtails and other birds that feed only by wading in shallow water are omitted. See Fig. 28.1.

Fig. 28.1

Birds Which Dive and Walk on the Bottom

The only bird that not only dives, but actually walks along the bottom of a water body, is the Dipper* (*Pl. 324*), whose dumpy form, and white throat and breast, are a familiar sight by fast-flowing streams, especially in hill districts, often perched and bobbing on a rock in midstream. It dives from the surface, from rocks and from the air, and feeds mainly on the larvae of aquatic insects.

* Scientific names of birds will be found in the index.

Birds Which Dive from the Surface and Swim Submerged (*Pl. 325*)
Most of the birds that form part of the freshwater ecosystem dive from the surface only, and then hunt their prey or graze aquatic vegetation under water. They include fifteen diving ducks, five grebes, four divers, one cormorant and one coot. Some species resort to fresh water only to breed, wintering in estuaries or on the sea. Most of the 'up-enders', such as the dabbling ducks and moorhen also occasionally dive for their prey.

The fifteen diving ducks comprise four *Aythya* (Tufted and Ferruginous Ducks, Pochard, Scaup), three sawbills (Goosander, Red-breasted Merganser, Smew), two each of scoter and goldeneye, the Harlequin, the Red-crested Pochard and the Long-tailed and Ruddy Ducks. Of these the Red-crested Pochard feeds mainly on vegetation, the Harlequin, Long-tailed Duck, scoters, goldeneyes and sawbills on animal food, and the four *Aythya* and the Ruddy Duck on both. Only the Tufted Duck, Pochard, Ferruginous Duck, Goosander, Smew, goldeneyes, Red-crested Pochard and Ruddy Duck are found at all commonly on fresh water in winter.

The five grebes (Great Crested, Red-necked, Slavonian, Black-necked and Little) and the four divers (Great Northern, White-billed, Black-throated and Red-throated) are all primarily fish-eaters with some aquatic invertebrates. The grebes are much more frequent than the divers on fresh water in winter.

The Cormorant, which is primarily a seabird, feeds almost entirely on fish, but the Coot is omnivorous. Both occur on fresh water in winter; in most parts of Europe the Cormorant is more often found in fresh water at this time than in the breeding season, when only non-breeding birds are at all common inland.

Summary of Principal Prey
Most diving waterfowl prey or graze on a wide variety of vertebrate and invertebrate, and/or vegetable food. Only the primary and other principal prey items are listed here, and for species that winter at sea, only their breeding-season preferences.

Fishes:	Divers, Great Crested Grebe, Cormorant, Smew, Goosander, Red-breasted Merganser
Molluscs:	Little Grebe, Scaup, Long-tailed Duck, Scoters, Goldeneyes
Other Invertebrates:	Grebes, Pochard, Tufted Duck, Harlequin, Long-tailed Duck, Scoters, Goldeneyes, Smew, Goosander, Red-breasted Merganser, Ruddy Duck, Dipper
Waterweeds and Seeds:	Red-crested Pochard, Pochard, Ferruginous Duck, Tufted Duck, Scaup, Goldeneyes, Ruddy Duck

Tabular Key to Surface Divers

[Winter plumages of species that mainly winter at sea are not included. Most female diving ducks are some shade of brown, with few distinctive markings.]

Head Features

Black:	Great Northern Diver, White-billed Diver, male scoters, Ruddy Duck, Moorhen, Coot
Black with green sheen:	male Scaup, male Goldeneye, male Goosander, male Red-breasted Merganser
with purple sheen:	male Tufted Duck, male Barrow's Goldeneye
Blue, chestnut and white:	male Harlequin
Chestnut:	male Red-crested Pochard, male Pochard, male Ferruginous Duck, female and immature Goldeneyes, female and immature Sawbills
Grey:	Black-throated Diver, Red-throated Diver
White:	male Smew
Forehead red:	adult Moorhen;
white:	adult Coot;
peaked:	Goldeneyes
Pale chin:	juvenile Moorhen
White or pale cheeks:	breeding Red-necked Grebe, breeding Cormorant, female Red-crested Pochard, female and immature Smew, Ruddy Duck, juvenile Coot (partly)
White patch round eye:	breeding male Long-tailed Duck
White spots:	Little Grebe (breeding), Harlequin, breeding female Long-tailed Duck, male Goldeneye (rounded), Barrow's Goldeneye (crescentic)
White round base of bill:	female Tufted Duck, female Scaup (more conspicuous), Harlequin
Crests etc.: double-horned crest and rufous tippets:	breeding Great Crested Grebe
golden-chestnut ear-tufts:	breeding Slavonian Grebe (pointing upwards), breeding Black-necked Grebe (pointing downwards)
single crest:	male Tufted Duck, male Goosander (obscure)
double crest:	female Goosander (more conspicuous), Red-breasted Merganser

Bill Features

Flattened:	Red-crested Pochard, Pochard, Ferruginous Duck, Tufted Duck, Scaup, Harlequin, scoters, goldeneyes, Long-tailed Duck, Ruddy Duck
Pointed:	Divers, grebes, Moorhen, Coot, Dipper
Pointed and hooked at tip:	Cormorant
Pointed, hooked at tip and saw-toothed:	Sawbills
Pointed and slightly 'tip-tilted':	Red-throated Diver, Black-necked Grebe
Very short:	Long-tailed Duck
Blue:	breeding male Ruddy Duck
Red:	adult Moorhen
White:	White-billed Diver, adult Coot
With orange or yellow spot:	male Common Scoter (on top), male Velvet Scoter (on sides)

Neck Features

Black:	breeding Black-necked Grebe, male Tufted Duck, male Scaup, breeding male Long-tailed Duck, male scoters, Ruddy Duck, Moorhen, Coot (juvenile only at back)
Chestnut:	Little Grebe, breeding Red-necked Grebe, breeding Slavonian Grebe, male Pochard, male Ferruginous Duck
White:	Great Crested Grebe, goldeneyes, sawbills, juvenile Coot (front)
With alternating dark and pale bands:	breeding Great Northern Diver, breeding White-billed Diver
With black patch:	breeding Black-throated Diver
With red patch:	breeding Red-throated Diver

Body Features

Upper parts black:	male scoters, Coot
black and white:	male Smew
black and blue:	male Harlequin
chestnut:	Ruddy Duck
grey:	male Pochard, male Scaup

Breast black:	male Red-crested Pochard, male Tufted Duck, male Pochard, male Scaup, breeding male Long-tailed Duck, adult Moorhen, adult Coot
black, white and blue:	male Harlequin
chestnut:	male Red-breasted Merganser, Ruddy Duck
grey:	female and immature sawbills
white:	juvenile Coot, male Goosander, male Smew, Dipper
Flanks black:	Scoters, Coot
chestnut:	Ferruginous Duck, male Harlequin, Ruddy Duck
grey:	male Pochard, sawbills (except male Goosander)
white:	male Tufted Duck, male Scaup, goldeneyes, male Goosander
with white line:	adult Moorhen
White wing-bar conspicuous in flight:	Grebes (except Little), Tufted Duck, Scaup, Ferruginous Duck, Velvet Scoter, sawbills
Tail very long:	male Long-tailed Duck
held erect:	Ruddy Duck
Conspicuous white under tail coverts:	Ferruginous Duck, Moorhen
Chicks:	Ducklings can be told from Moorhen and Coot chicks by their broad bills. Downy young of Moorhen and Coot are black with orange or red bills, but Moorhen chicks have pale-tipped down on the head; the Coot chicks have red-tipped down.

Birds Which Dive from the Air, Seize their Prey and Emerge (*Pls. 326, 327*)
The only two birds which regularly feed in this way, the Kingfisher and the Osprey are very distinct from each other and from all other aquatic birds. Herons, especially the Grey Heron, also exceptionally forage in this way.

The Kingfisher is small, dumpy, and brilliantly-coloured with blue-green and warm chestnut. It perches on a branch or some similar vantage point above the water, and dives in, completely submerging to catch its prey – mainly fish but some invertebrates. Found mainly on rivers and streams.

The Osprey is a large bird of prey and the only one that actually submerges when diving to seek its prey, mainly fish, though rarely submerging more than 1 m. Like other birds of prey and also crows, it may also pick prey up off the surface of the water. Found mainly on lakes.

Birds Which 'Up-end' and Reach Downwards from the Surface (*Pls. 328, 329*)
Up-ending and stretching the neck downwards to forage beneath the surface mostly for aquatic vegetation, is practised especially by dabbling ducks, swans and the Moorhen, and to a lesser extent by geese. All these also sometimes dive, but more often to avoid a rival or a predator than to feed.

The seven common European dabbling ducks are the Mallard, Gadwall, Teal, Garganey, Wigeon, Pintail and Shoveler. The three swans are the Mute, Whooper and Bewick's. Both groups are set out in the keys below. The Moorhen is added to the diving birds key (*p. 330*), since the only bird it is likely to be confused with is the Coot.

Tabular Key to Up-Enders

---

1. Dabbling Ducks

[N.B. Females are brown, and mainly distinguished by general shape and the colour of the wing-patch (speculum). Mallard are the largest, and Teal and Garganey the smallest of the dabbling ducks.]

Head Features

Chestnut with buff crown:	male Wigeon
with green stripe:	male Teal
Dark brown with vertical white streak:	male Pintail
Dark green:	male Shoveler
with white neck ring:	male Mallard
Grey-brown:	Gadwall
White stripe over eye:	male Garganey
Forehead peaked:	Wigeon
Bill yellow:	male Mallard
short:	Wigeon
spatulate:	Shoveler

Wing Features

Forewing blue-grey:	Garganey, Shoveler
white:	male Wigeon
White line above:	male Teal
Speculum black and white:	Gadwall
black, white and green:	Teal
blue and white:	Mallard
bronze:	Pintail
green:	Wigeon, Shoveler
green and white:	Garganey

Other Features

Neck and breast white:	male Pintail
Flanks chestnut:	male Shoveler
Tail pointed:	Wigeon, female Pintail
long-pointed:	male Pintail
Under tail coverts buffish-yellow:	Teal

2. Swans

[NB Bewick's is smaller than the other two.]

All white:	adult swans
All or partly pale ash-brown:	juvenile and immature swans
Bill orange-red with black basal knob:	adult Mute (immature paler with no knob)
yellow at base, ending bluntly above nostrils:	adult Bewick's
yellow at base, extending beyond nostrils at acute angle:	adult Whooper

CHAPTER 29 Mammals

Nine European mammals, three of them introductions from America, play some part in freshwater ecosystems, either by grazing vegetation or by predation on other aquatic animals. Many others, of course, enter the water to bathe, or cross a stream, but these are not considered here. All the aquatic mammals, except for the two water shrews, have a pelage of some shade of brown, and breed in holes or other structures in stream and pond banks.

Tabular Key to Freshwater Mammals

Size large, 80–140 cm:	Beaver, Coypu, Otter
medium, 30–70 cm:	Muskrat, Mink
small, 10–20 cm:	Water shrews, Water Vole
Upper parts black:	Water shrews, Water Vole
dark brown:	Mink
warm brown:	Water Vole
Muzzle blunt:	Beaver, Water Vole, Muskrat, Otter
very blunt:	Coypu
pointed:	Water shrews, Mink
Chin white:	Mink
Throat and chest white:	Otter
Tail flattened horizontally:	Beaver
vertically:	Muskrat
Tail rat-like:	Water shrews, Water Vole, Coypu
stoat-like:	Mink
thick at base:	Otter
Feet webbed:	Beaver, Coypu, Otter
Underparts black:	Water Vole
white:	Water shrews
with white spots:	American Mink

INSECTIVORES (Insectivora)

Shrew Family

Water Shrew *Neomys fodiens* (*Pl. 330*) Like a small rodent, but with a sharper snout; 12–17 cm long including the tail, with slate-black upper parts sharply defined from the usually pale, or even white, underparts. Distinguished from the slightly smaller Miller's Water Shrew *N. anomalus* – which in western Europe occurs only in a few mountainous areas – by the keel of stiff, silvery hairs under its tail. Very active, by day and night, swimming, diving and walking on the bottom of streams, ponds and cress beds, seeking invertebrates, tadpoles and small fishes.

RODENTS (Rodentia)

Beaver Family (Castoridae)

European Beaver *Castor fiber* (*Pl. 331*) A very large rodent, to 120 cm, found in our area only on the Elbe, and distinguished from Muskrat, Coypu, Mink and Otter by its flattened, naked tail; feet webbed. Often breeds in lodges, piles of logs and branches built in the water. Only likely to be seen at dusk.

Vole Family (Microtidae)

Water Vole *Arvicola terrestris* (*Pl. 332*) A plump, warm brown (or black, especially in Scotland), blunt-nosed rodent, about 20 cm long, of still and slow-moving water. Sometimes called Water Rat, creating needless confusion with the larger (28–30 cm), sharp-snouted Common or Brown Rat, which often swims. Swims and dives actively by day, often heard plopping in; mainly grazing on grasses.

Muskrat *Ondatra zibethica* (*Fig. 29.1*) A large, to 70 cm, blunt-snouted North American rodent, with a vertically flattened tail and feet not webbed. Escaped from musquash fur farms, and established by still and slow-moving water in various parts of France, Germany and the Low Countries, but no longer in Britain. Most likely to be seen swimming and diving early in the day, feeding on water plants.

Fig. 29.1

Fig. 29.2

Coypu Family (Myocastoridae)
Coypu *Myocastor coypus* (*Fig. 29.2*) A large, to 100 cm, very blunt-muzzled, South American rodent, with a cylindrical, rat-like tail, and webbed feet. Escaped from nutria fur farms and established in several scattered localities in marshes and by lakes and rivers. Often seen swimming and diving by day; grazing many water plants.

CARNIVORES (Carnivora)
Stoat Family (Mustelidae)

Fig. 29.3

European Mink *Mustela lutreola* (*Fig. 29.3*) An elongated, ferret-sized and -shaped, dark brown predator, about 50 cm long, distinguished from the much larger Otter especially by its sharper muzzle, brown throat and chest, and scarcely webbed feet, as well as by the white patch on its chin, lower lip, and often also on the upper lip. For distinctions from American Mink, see below. Swims and dives in still and slow-moving waters. A catholic feeder, with such prey as water voles, birds and their eggs, frogs, fish and crayfish. Now found only in France and very locally in Germany.

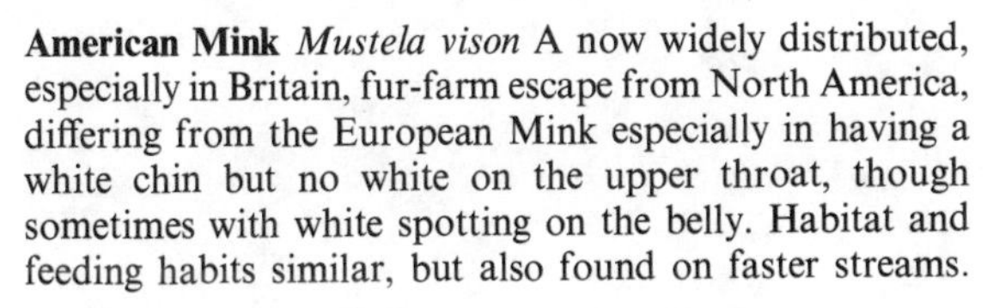

American Mink *Mustela vison* A now widely distributed, especially in Britain, fur-farm escape from North America, differing from the European Mink especially in having a white chin but no white on the upper throat, though sometimes with white spotting on the belly. Habitat and feeding habits similar, but also found on faster streams.

Fig. 29.4

Otter *Lutra lutra* (*Fig. 29.4*) The longest European freshwater mammal, 100–140 cm long, with a long tail, much thicker at the base than the darker brown mink, and with a distinctive blunt muzzle, white throat and chest, and webbed feet. Swims and dives in both still and flowing fresh and brackish water, preying especially on fish, but also on water voles, birds, frogs and crayfish. Breeding holts sometimes at base of trees by water.

Collection and Study

The study of aquatic life involves both direct observation at the waterside and collection of organisms for detailed study in home or laboratory. In the interests of conservation it is important that as few specimens as possible be taken from their natural habitat and that they should be returned unharmed whenever possible. If properly and responsibly carried out, collecting organisms should not conflict with sensible conservation policies: indeed, if a natural resource is to be accurately evaluated, and perhaps managed, it is important to know exactly what organisms are living in any particular habitat, and hence correct identification is essential. Potentially more damaging than simple removal of a few specimens, is the physical disturbance of the habitat caused by many collecting methods, and it is important that the collector's approach be carefully planned in order to keep this to a minimum.

Waterside Observation

Always approach the water slowly and quietly – stalk it! As any competent angler knows, fishes have excellent eyesight and are very sensitive to vibrations caused by footfalls; the same applies to many other aquatic animals. Scanning the area with binoculars first will suggest likely places to aim for and how best to get to them. Once at the waterside look around. The most obvious organisms – larger plants (macrophytes), flying insects and vertebrates that do not live permanently submerged, can usually be fairly easily identified by sight, or at least in the hand.

Fishes and some amphibians pose a problem because they live below the surface, are usually shy of observers, and are difficult to catch without resorting to angling or the use of large nets or traps. With patience, however, it *is* possible to discover most of the species present in a water by persistent observation, especially during the spawning season when their natural wariness is overcome by the urge to reproduce. At this time the intense activity, at or near the surface, of large species such as tench or carp is difficult to overlook! Small species, or the young of larger ones, are often captured in the water net when general collections are being made.

Simply looking into the water, through the surface, will reveal many of its secrets. In good light, a pair of polarizing sunglasses is invaluable for water-watching – they eliminate much of the glare reflected from the water surface and enable a clearer view of the goings-on beneath. The traditional glass-bottomed bucket or watertight box gives an even better view, though within a limited field. The average handyman should have no difficulty in making one: the main aim is to place a glass surface between the water (but in contact with it) and the air, which will give an uninterrupted view free of ripples and reflections – placing the viewer inside a fish-tank, as it were. This device can be used from a boat or kneeling by the water's edge – but take care that the bank is stable and safe, it is all too easy to over-reach! An even wetter way to achieve the same result is by using a divers' face mask, with a snorkel tube to breathe through. And, of course, qualified scuba divers can actually visit the organisms in their own habitat.

Collecting Tactics

Having surveyed the locality in general, decide on which habitats are worth further investigation and devise a sensible sequence for working them. If one is after aquatic invertebrates, for instance, start by studying the shallow margins – water only a couple of centimetres deep can be surprisingly rich in species. Search among the stems of marginal reeds or similar plants; sort through strand-line debris and any floating or stranded rafts of uprooted plant-life; look on and under the floating leaves of plants such as water-lilies, or amongst the trailing roots of willow and alder; these habitats provide shelter for many small organisms, including some of the most delightful subjects for microscopy such as sessile rotifers and protozoans. Turn over any stones or bits of timber and other large debris, above or below the shore-line, examining upper and lower surfaces to see what lives there. Be sure to replace them exactly as they were sited, otherwise many of the plants and animals that are adapted to live in just the positions they occupy will suffer and perish.

A simple way to get a good selection of micro-organisms is to fish out skeins of filamentous algae (those with a rough texture, e.g., *Cladophora*, are more productive than slimy ones such as *Spirogyra*) and squeeze them out over a container. This will provide a rich, concentrated 'soup' containing a wealth of specimens. Similar methods can be followed for loose drifts of larger plants, clusters of aquatic roots, etc., having first removed the larger life-forms clinging to or growing on them. Avoid uprooting growing plants, but dying leaves can be carefully removed for examination.

These immediately accessible areas can be searched with no special equipment other than waterproof boots and perhaps a hand lens; most of the specimens found can be collected with a small hand-net, forceps, fingers, or a wide-mouthed pipette (see *p. 345*). In deeper water, which can be reached by careful wading in boots or thigh-waders, or from steep banks, the long-handled water net, and other methods described in the following section, come into play.

In rivers, especially those with a strong flow, the current can be used to advantage. Holding a net across the current while turning stones or disturbing weed beds just upstream of it, will cause many organisms to be swept into the bag. Similarly, stirring the sediment with the foot or hand upstream of the net ('kick sampling') will provide a good selection of the burrowing 'infauna'.

In some rivers it is possible to find small patches of shell gravel, the accumulated dead shells of numerous snails and mussels. Apart from their value as indicators of the molluscan fauna, these patches are home to numerous burrowing creatures, particularly those with an aversion to fine sediment.

Apart from such 'obvious' aquatic habitats as ponds and lakes, rivers and canals, do not neglect such easily-overlooked places as the 'splash zone' of weirs, the permanently damp parts of lock-gates, jetties and sluices – the wet mosses, algae and liverworts that abound in such places harbour a rich variety of life. Tiny waters such as puddles, rain gutters, temporary pools in field-hollows or woods, tiny springs, or even abandoned crockery, all possess an interesting and often specialized fauna and flora, particularly for the microscopist.

The catch should be sorted on the bank to select specimens for further study, returning the remainder to the water. A shallow tray, ideally of white plastic or enamel, is best for this purpose, but not always easy to obtain. A gardeners' seed tray can be used, providing one can be found without drainage holes in the bottom! Even the lid of a plastic bucket can be used (these also make useful 'kneelers'). A pair of forceps and a wide-mouthed pipette, as well as such mundane items as teaspoons and tea-strainers, are useful tools to have at this stage, and of course a hand lens (magnifying glass) is virtually essential. A small gadget called a 'nature viewer', consisting of a clear plastic vial with a lens built into the screw lid, also has its uses.

Most specimens can be transported in a plastic bucket with a lid to prevent spillage. It is best to fill the container almost to the brim to prevent the water 'sloshing' about too much. If the journey is long it will be necessary to aerate or change some of the water periodically to prevent oxygen-depletion. Aeration can be simply carried out by rapidly working a large bulb-pipette in the upper half of the water. Overcrowding containers is another source of trouble that is avoided with a little forethought.

Small, easily lost, or delicate specimens should be carried individually in small containers. Be sure not to mix large carnivores with potential prey: dragonfly nymphs, beetles and their larvae, water bugs, large stonefly larvae and others will inevitably make short work of your most precious specimens, even during a short journey home. Plastic bags are good for carrying plants in a moist condition, but offer little protection against crushing.

It is vital that both specimens and water be kept cool; overheating, even by a few degrees, will kill most organisms very quickly. Do not leave containers in direct sunlight or in any warm place (such as a closed vehicle). If they have to be left for a while, stand them in the water in a shady margin, or in the shade *under* your car.

Collecting Equipment

The equipment described below is typical of what is available to the amateur naturalist or student. Most of these items can be obtained from biological suppliers (see *p. 355*), but are not cheap. Many can easily be made at home and a friend with workshop facilities, who could manufacture net frames, is a boon. The keen enthusiast will inevitably invent nets and other equipment to suit his own particular requirements; the list below should be regarded as a guide and in no way definitive. Professional gear, used by full-time scientists and water supply engineers, is generally concerned with quantitative work and is beyond the scope of this book.

Collecting freshwater organisms can be reduced to three basic methods: pushing or pulling a net more or less randomly through the water, in the open, or amongst vegetation, silt or detritus; sorting through bunches of weed, aquatic roots, debris or other substrata by hand; and picking out individual organisms with a small net, pipette or the fingers. Methods will obviously vary according to the nature of the habitat and the type of organisms sought, and the methods and equipment described below will have to be adapted to suit individual requirements and circumstances.

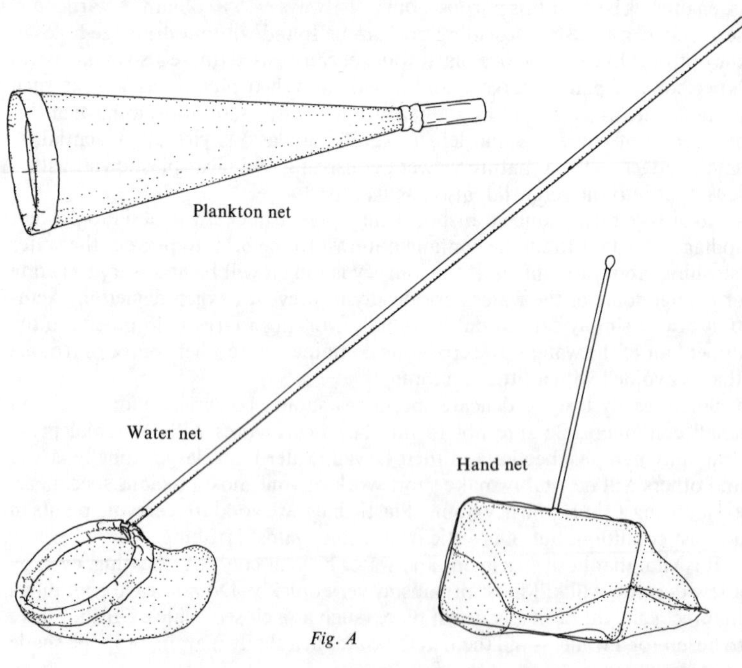

Fig. A

Nets A good water net is a basic essential. Its design, within broad limits, is largely a matter of personal choice or making the best of what is available. Ideally it should have a stout metal frame about 15–25 cm in diameter; a circular one is convenient but the corners of a square or diamond shape can be more useful in certain circumstances. The frame should be attached, either permanently or by a screw-threaded ferrule, to as long a pole as can be managed – 1.5–2 m is a good length. The net bag itself should be strong and rotproof, preferably nylon or some other synthetic material. For general purposes a mesh size of 1–3 mm is suitable. The bag should be rounded at the bottom, without corners where the catch could be concentrated and crushed, and a little deeper than the width of the frame. Durability is greatly increased by reinforcing the rim with tough webbing. Perhaps the ideal system is to have several nets and frames which can be screwed into the same pole, or to have interchangeable net bags of different mesh sizes, which press-stud over the frame.

The net can be pushed through the water just clear of, or just touching the bottom, to catch free-swimming animals and disturb bottom-dwellers and chase them up into the mouth of the net. Small samples of rooted vegetation, bottom sediment, or various debris can be dredged up and examined for the many organisms that live in them. Dense weedbeds should be worked through with a 'sawing' motion to dislodge the inhabitants.

A scraper net is a smaller version of the water net with a sharp metal edge to the frame, and is useful for sampling firm level surfaces such as brick supports, concrete retaining walls, wooden piling or 'camp-sheathing'. The scraper need only be 10–15 cm long – larger sizes are more difficult to manipulate. Do use these nets with care and discretion. It is always a good idea to study such surfaces by eye first, kneeling low over the water to establish what is there. One can then often scrape off just the specimen required and no more; if it can be reached by hand, so much the better.

Some authorities recommend the use of a grappling hook or heavy metal dredge which is thrown out on a line and dragged back to the bank. Such practices are, however, very unselective and damaging to the environment. They have no place in the armoury of the responsible naturalist.

Small hand nets (*Fig. A*), available from aquarist suppliers, are useful for sampling at short range in the margins, or in small streams, puddles, etc., and have sundry other uses. Sometimes they can be obtained in a variety of mesh sizes, which increases their potential value.

Collecting flying insects – the adults of aquatic nymphs and larvae, for example – calls for an entomologists' 'butterfly' net. This consists of a light wire frame (tennis-racket-size) on a 1 m-long handle. The bag is light and gauzy, and deep enough to allow it to be gathered in the hand to prevent escapes. Black netting seems to be best for visibility of the catch.

Plankton sampling in open water, particularly for microscopic forms, requires a special plankton net. This is typically a long conical net bag mounted on a circular frame, with a small glass (or plastic) bottle or vial fastened into the narrower (trailing) end of the bag by a drawstring (*Fig. A*). This arrangement funnels the catch into the container and prevents the delicate planktonic organisms from being crushed when lifted from the water. It also concentrates the catch and permits an immediate, if superficial, examination of the sample fresh from the water. Mesh size should be chosen according to the type of organisms to be caught. A very fine mesh may be necessary for some microscopic planktonic plants, but will clog rapidly; 0.3–0.5 mm mesh is a good general size, with a second net of 0.1–0.3 mm mesh for some of the smaller organisms.

A circular frame of 10–15 cm diameter is suitable for a plankton net mounted on a pole. The water resistance of the fine mesh causes larger nets to be almost unworkable. A towed net, attached to the main line by a 'tripod' of strong cord, can be larger, if the manpower is available to pull it, but there is no real advantage in larger sizes. A small net will provide sufficient specimens for weeks of study in a very short time. A strong, pole-mounted plankton net can serve equally as a water net, but the fine mesh is more expensive to replace if torn.

The composition of the plankton will vary according to depth, season and even time of day. Depth is easy to control with a pole net: twisting the pole will 'close' the net until the desired depth is reached, and a reverse twist will open it to make the catch. It can be closed again before being brought to the surface. Trailed nets need a weight on the line ahead of the net in order to 'fish' properly. Depth can be controlled by the speed of retrieve, which in any case should be as slow as possible or a 'pressure front' of water will prevent any organisms entering the net.

In some conditions any net will clog quickly with loose debris, sticks, rafts of algae, or other plant life. A coarse, wire-mesh 'mask' fixed over the mouth of the net will prevent much of this flotsam from entering and make later sorting easier.

Handling and Examining the Collection

Workers with access to laboratory facilities tend to take for granted a ready supply of suitable equipment. These notes are therefore biased towards those who study at home, often limited to a few hours' work on the kitchen table, and who have to supply their own equipment.

The first requirement is for a good selection of containers of different shapes and sizes, such as plastic kitchen-ware, in which the specimens can be housed, in water, while being studied. Clear plastic 'sandwich boxes' with close-fitting lids are particularly good and they make superb small aquaria for prized specimens. Covers are essential to prevent evaporation, contamination, or the escape of the occupants. Petri-dishes – shallow, circular, glass or clear plastic dishes 50–80 mm diameter and about 10 mm deep – are invaluable for use under the low-power microscope. Odd containers from many unlikely sources, e.g., plastic bottle-lids, are often surprisingly useful. If secondhand food containers are used be sure to rinse away thoroughly any trace of detergent as this is lethal to most small organisms. Never use tap water for your specimens unless it comes from an untreated natural source. Water from the original site is best, but clean rain or spring water will do.

Handling small aquatic organisms requires some expertise, and certain pieces of equipment are indispensable. One or two pairs of forceps ('tweezers') are essential: a medium-sized pair with fairly fine tips (*Fig. B*) and a light, positive action is best for general purposes; for fine work watchmakers'

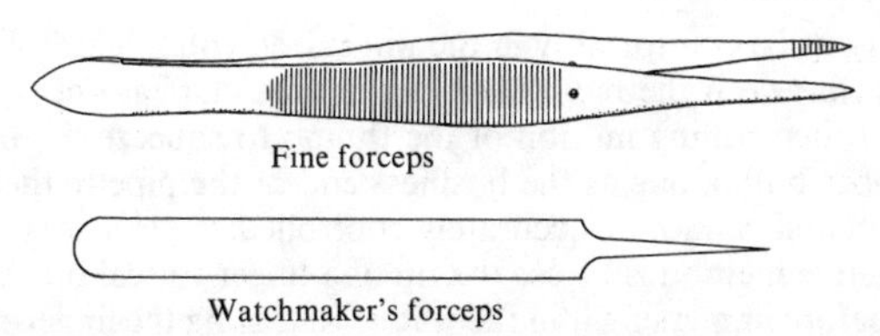

Fig. B

forceps, with needle-like points, are invaluable but rather expensive. Seekers – needles or short lengths of stiff wire mounted in a pencil-sized handle – are constantly useful and easily made at home. And a pair of stout scissors will be needed from time to time.

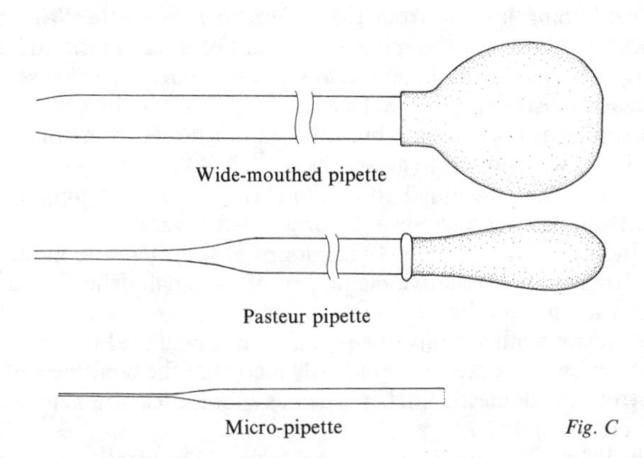

Fig. C

Probably the most important of all 'handling' apparatus is a selection of glass pipettes. These are virtually essential for transferring small volumes of water containing tiny or delicate organisms, and for sundry other tasks. A pipette is a simple glass tube drawn out at one end to form a narrow neck. Normally, only Pasteur pipettes (see below) are available from dealers, other sizes will probably have to be made by the user. They are not difficult to manufacture if one has access to a Bunsen burner or a small blowtorch, and a supply of glass tubing (see *p. 355* for method).

Three sizes of pipette cover most needs (*Fig. C*): a wide-mouthed pipette is useful for many purposes and will pick up animals up to the size of a freshwater shrimp (*Gammarus*); Pasteur pipettes are for general purposes requiring something smaller than a wide-mouthed pipette, and for larger micro-organisms; for handling micro-organisms a micro-pipette is required.

Water, together with its inhabitants, can be drawn up into the pipette by one of several methods. Wide-mouthed pipettes are most conveniently worked with a compressible rubber bulb of about 10–15 ml capacity. Many amateurs (and professional scientists) never seem to master the use of a bulb pipette.

Hold the glass tube comfortably in the fingers, as you might hold a pen, and grip the bulb between the ball of the thumb and the base of the forefinger, using a controlled rolling motion of the thumb to squeeze the bulb. Do not grip the rubber bulb alone as the business end of the pipette then acquires a life of its own and cannot be accurately controlled.

An alternative method is to use the tip of a finger to seal off the upper end of the tube before immersing it in the water. Releasing the finger-seal momentarily will allow a volume of water to rush into the tube; this can be caught at any desired level by smartly closing the tube again with the finger. Pasteur pipettes do not work satisfactorily by this method and require a small rubber bulb of 2–5 ml capacity.

Micro-pipettes can be very delicately controlled by using a length of thin, soft, rubber tubing leading from the mouth to the pipette. With practice, gentle sucking, biting and tonguing actions can be used to pick up individual organisms, however small, under the low-power microscope. Those who find such intimacies distasteful can make a small rubber bulb by knotting a short length of similar rubber tubing, but this is a much inferior method. Rubber tubing or bulbs will slide on to glass tubing easily if the latter is first lubricated with saliva. Forcing unwilling bulbs on to glass tubing is a common source of unpleasant and messy accidents in laboratories: beware!

'Irwin-loops' – tiny (less than 1 mm) loops of very fine wire mounted on a handle – provide an alternative method of lifting small delicate creatures as long as they are not too lively.

In time, various other items of equipment can be added to the collection. No doubt the home worker will gradually recognize the usefulness of utensils liberated from the domestic kitchen, such as spoons, tea-strainers and sieves, crockery, etc.

With the help of this book, many animals should be identifiable from living specimens. Some arthropods are an exception as they are often too lively to permit close examination and may need to be preserved. First they must be humanely killed by dropping them into hot (80°C, not boiling) water. They can then be preserved and stored in 70% ethyl alcohol; if this is unobtainable, methylated spirits diluted 3:1 with water will do. Lively micro-organisms can be subdued with methyl cellulose (see *p*. 353).

Obtaining laboratory equipment and materials is not always easy for the amateur working at home: friends who work in laboratories will take on a new significance! Sometimes the local chemist's shop can help, or at least suggest the nearest supplier. A list of British biological suppliers and microscopy dealers is given on *p*. 355.

Making a Culture

It is a common fallacy that a single drop of water contains tens of thousands of living organisms. However, the majority of water-drops that constitute a pond, lake or river contain disappointingly few micro-organisms, and because of their scarcity, these will be difficult to find and study. A reasonably concentrated collection of protozoans (which will inevitably include many other micro-organisms) can only be obtained by carefully selecting samples of suitable or likely substrata – bottom sediments, plant squeezings (*p.* 340), decaying vegetation, water enriched by animal droppings, etc. Protozoans rarely form blooms like the algae, but we have seen *Stentor coeruleus* covering every available substratum in a lake, including the surface film, with a dense 'fuzz' of deep blue-green; this only lasted for a couple of days.

Samples containing protozoans can be 'thickened up' by adding a quantity of suitable culture medium and waiting for a few days for the animals to multiply. A simple method is to add a small quantity of chopped hay or a few wheat grains to water. It is important that these have been thoroughly dried and do not carry any pesticides; do not use tap water unless untreated, clean rain or spring water is best, boiled pond water will do in an emergency. The exact proportions are not important for general purposes: a standard recipe gives 10 gm of hay (a good handful!) to 1 litre of water, or 5–6 wheat grains to 100 ml. These infusions should be brought to the boil and simmered for five minutes or so, then allowed to cool in a container stopped with a wad of cotton-wool. After standing it in a dark place for a day or two it is ready for use.

Cultures can be set up consisting of freshly-collected samples to which has been added about a 25% volume of medium. The protozoans (in theory) feed on the bacteria and fungi which appear in the medium and multiply to great densities. At first, a mixture of protozoans will be present, but usually one species becomes dominant and ousts the others. Sometimes the bacteria become dominant and the result is an evil-smelling black soup. Experimenting with culture methods can be an interesting pastime and gives hours of amusement. Different media – various plant matter, animal droppings, etc. – can be tried and inoculated with samples from any likely habitat: various aquatic plants, mosses, dead leaves, puddles, gutters, etc., often produce interesting finds. Leaving cultures in the dark or in daylight will give different results, the latter often producing rich cultures of algae. It is also possible to raise pure cultures of a single species by inoculating a small quantity of medium with a single animal. This has a low success rate but if several cultures are started at one time the chances of success are increased.

Searching collections or cultures with a low-power microscope for protozoans requires a little skill. A petri-dish is a suitable container. When the sample has settled first scan the bottom layer, focusing on the debris lying on the bottom. Most protozoans will be spotted by their movements, but amoebae are easily overlooked until the searcher has acquired a 'search-image' for them; they appear immobile under low powers of magnification. Use both direct and transmitted lighting if possible. Next, study the surface film, especially in the corners nearest to and furthest from the light; many species exhibit a preference for one or the other and congregate there. Finally, have a look with the unaided eye: it is surprising how many organisms can be seen when one knows what is present and where to look.

Understanding Microscopes

Microscopy is a subject full of delights and one which greatly enhances the study of aquatic life. Time spent at the microscope is never wasted: animals and plants take on a new dimension when the fine details of their structure, too small to be seen with the naked eye, are revealed for the first time. The amazing complexity of the insect eye and crustacean limbs, the delicate beauty of the 'wheel organs' of living rotifers, the clumsy gait and ridiculous faces of tardigrades, these are sights that never cease to fascinate. Some readers will already possess or have access to a microscope; others, we hope, may be inspired to invest in one for the first time. We do not have the space here to deal with microscopy comprehensively, indeed this would require a whole book to itself. Instead we will attempt to stir the reader's interest by describing the most useful types, their capabilities and a few simple techniques for use.

A microscope is basically an instrument for producing a magnified image of the subject or specimen and presenting it to the eye of the observer. A hand-lens or magnifying glass is just as much a microscope as the most sophisticated and expensive of instruments used in research laboratories. The light microscope enables the observer to view an image formed by the visible wavelengths of light and is the only type discussed here. Electron microscopes, which work by reflecting or transmitting electron beams (invisible to the eye) from the subject on to a viewing screen or photographic plate, are enormously complex and expensive and well beyond the scope of this book.

The Hand-lens

The hand-lens or magnifying glass is capable of giving a useful magnification of up to × *10* or more when used correctly. Hold the lens close to the eye and move the subject into focus; do not move the lens away from the eye towards the subject. Some types of hand-lens consist of two or more separate lenses, mounted separately, which can be pivoted into alignment to vary magnification.

The value of hand-lenses should not be underestimated, they are an important item in the equipment of the freshwater naturalist. Mounted on a simple stand they leave the hands free to work in much the same way as any other microscope.

Compound Microscopes

The 'typical' microscope. There are two different types – the Low-power and the High-power – which are each described in more detail later on. The optical parts of every compound microscope consist of two sets of lens elements – the objective lens and the eyepiece. Each of these compound lenses consists of several separate glass elements fixed in a tubular housing, which in turn is mounted on the main frame of the microscope. The subject is placed on a platform (stage) beneath the objective and focused by moving either the stage itself or the unit (head) of the microscope that contains the objective and eyepiece. The moving parts slide up and down a dovetailed groove, usually controlled by a rack-and-pinion mechanism, and worked by turning a focusing wheel.

The circular area seen through the lenses is the field (of view). When the field is magnified the quantity of light falling upon it (which is constant) is diminished because it is spread over a larger area. At most magnifications this is insufficient to give a worthwhile image. It is therefore necessary to use a supplementary (normally electric) light source for all but the lowest powers of magnification. Light can be directed in two ways: upward, through the subject from beneath the stage (transmitted light), or directly on to the subject (direct or reflected light). Transmitted light is, obviously, most useful for revealing internal details in translucent organisms, and it is the only means of illumination in a high-power microscope. Used in a low-power microscope it is useful for scanning collections of micro-organisms or small macro-organisms, showing up many creatures that would probably be missed by reflected light. Reflected light is more generally used in a low-power microscope and is necessary for illuminating opaque subjects such as insects and shelled animals.

Almost any electric lamp with a directional beam can be used for microscopy but a purpose-built light source with variable brightness and a beam focusser is far superior. These usually have an adjustable arm so that the beam can be directed at will. Many modern microscopes have a built-in light source for transmitted light; but the older types, with an angled mirror beneath the stage to direct light (even sunlight) upwards, should not be despised. They have the advantage of simplicity and can use the same light source that supplies reflected light.

The approximate total magnification of a compound microscope is calculated by multiplying the powers of the objective and eyepiece lenses together.

The exact size of a subject can be measured under the microscope by using an eyepiece micrometer. This is an ordinary eyepiece containing a scale of arbitrary length. This scale is calibrated against a stage micrometer, a glass slide upon which is engraved a series of parallel lines (sometimes a scale) having a known, precise spacing. From this the exact size of the eyepiece scale can be calculated for each objective. The subject is then measured against the eyepiece scale. The microscopist's standard unit of length is the micrometre or micron, μm, which is one thousandth of a millimetre.

scale

Low-Power Microscopes may also be called 'binocular', 'stereo', or 'dissecting' microscopes. They have the common features of a binocular head – with an eyepiece for each eye – and paired objectives, one aligned with each eyepiece, which provide a stereoscopic view of the subject. The range of magnification usually starts at around ×*5* or ×*10*, with the highest power being about ×*30* to ×*50*, depending on the model. Magnification can be varied by using different sets of eyepieces and/or objectives in combination Two or three sets of objectives are commonly mounted in a rotating turret beneath the head and should 'click' positively into position. The most pleasant models to use (and, naturally, the most expensive!) have a 'zoom' mechanism which allows magnification to be varied between two extremes, e.g., ×*1* to ×*5*. All binocular microscopes should have a working distance of about 10 cm between subject and objective lens to allow space for manipulation and orientation of the subject.

Illumination can be supplied by reflected or transmitted light. Many low-power microscopes have built-in direct lighting units mounted on a jointed arm for adjustment; others have a fixed light situated somewhere near the objectives – a very inferior arrangement. But a separate, external light source is often better than either of these as it avoids clutter around the working area and stage. Transmitted light is directed through a glass stage (preferably ground-glass), either from an integral source or reflected upward by an angled mirror.

A very basic, but good, low-power microscope is shown in *Fig. D*. In this model the power of the objectives is changed by unscrewing them and reversing them in their mounts. A stage is lacking, but any such model (which is considerably cheaper than ones with integral stages) can be placed on a glass platform, illuminated from beneath if required.

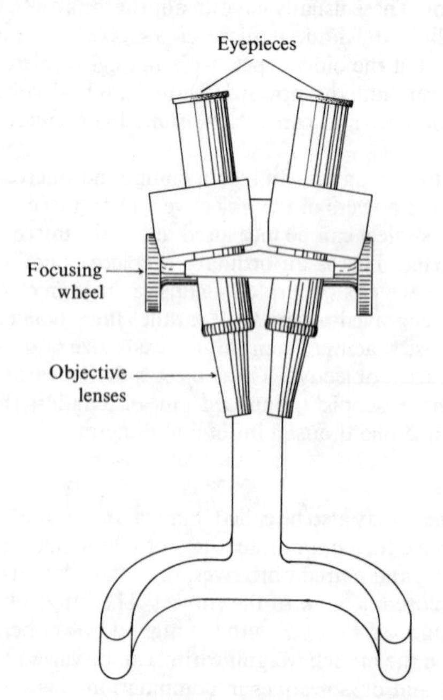

Fig. D Simple low-power binocular microscope

The low-power microscope is easy to use and requires no initial setting up. Living or preserved specimens require no preparation and are simply placed on the stage or in a conveniently shaped container, such as a glass or plastic petri-dish, if they are to be viewed in water. All the macroscopic groups described in this book (with very few exceptions) can be studied and identified with such an instrument, and most of the micro-organisms will be easily visible, if not in intimate detail, at least in overall view. As a general purpose microscope the low-power stereo binocular has no rival and the keen student of aquatic life will regard it as an almost essential aid to the study of these creatures.

The High-Power Microscope is an essential tool for the keen microscopist, to whom freshwaters are an almost endless source of first-class material for study. On the other hand, the field naturalist, even if he collects specimens for identification and study at home, will find it to be very limited in application compared with the low-power stereo microscope described above.

The main limitations are: Only relatively small objects can be viewed. Even the simplest specimen has to be mounted in some way on a microscope slide (see below) before it can be studied. Only transmitted light can be used, hence the subject must be naturally at least partly translucent, or rendered thus by various preparatory techniques. Manipulation of the specimen whilst under the microscope is very limited, often impossible, because of the short working distance between specimen and objective lens (e.g., about 5–6 mm for a normal × *10* objective). The microscope must be very carefully 'set up' – all the lenses and other optical components must be perfectly aligned and focused – in order to obtain the best results at the high magnifications involved, and this requires a certain amount of practice and expertise.

However, the high-power microscope offers the only means of viewing micro-organisms in sufficient detail to study their structure and workings, and hence to identify them; and some very beautiful effects can be achieved with it.

A typical modern instrument is shown in *Fig. F.* Most models have a single eyepiece but binocular heads (with two) are becoming increasingly popular; objective lenses, though, are always single, never paired as in a stereo-microscope. Objectives normally range in power from about × *5* up to × *100* (the latter are usually 'oil-immersion' lenses which require a drop of special oil between specimen and objective lens). The × *5* objective, in combination with an eyepiece of × *6*– × *10* magnification (which is usual) will give a total magnification roughly equivalent to the highest powers of a low-power microscope and is useful for searching the slide in order to find elusive specimens. Other usual sizes of objective are × *10*, × *20* and × *40*, which give correspondingly increased magnification and are the most useful sizes for viewing micro-organisms.

The light source, which may be external or built-in, is usually focused by a unit consisting of a lens and aperture iris, the condenser, which is mounted beneath the stage. In more sophisticated models the condenser includes provision for various different lighting effects in addition to 'normal' transmitted light (bright field). For example 'dark ground' shows the subject brilliantly lit against a black background (e.g., *Pl. 157*); 'phase-contrast' is a means of increasing the edge-contrast of specimens or details which are so transparent that they are almost invisible with normal illumination (*Pl. 135*); and polarizing the light by means of filters can produce some remarkably colourful and instructive effects from apparently dull and uninteresting subjects.

A wide range of preparatory techniques is available to the microscopist, mostly requiring elaborate equipment and a supply of chemicals, especially for dealing with preserved specimens. We will describe just a few simple methods which require the minimum of apparatus; the necessary skills can only be acquired with practice.

For most purposes the specimen is mounted in a drop of fluid mounting medium (which may set solid in permanent preparations) on a glass microscope slide. A slide is usually 3 in × 1 in in size and 1–2 mm thick. The preparation is completed by lowering a cover-slip – a small disc or square of paper-thin glass – onto the surface of the medium, to present a level surface to the lens and to protect the specimen from disturbance and evaporation (*Fig. E*).

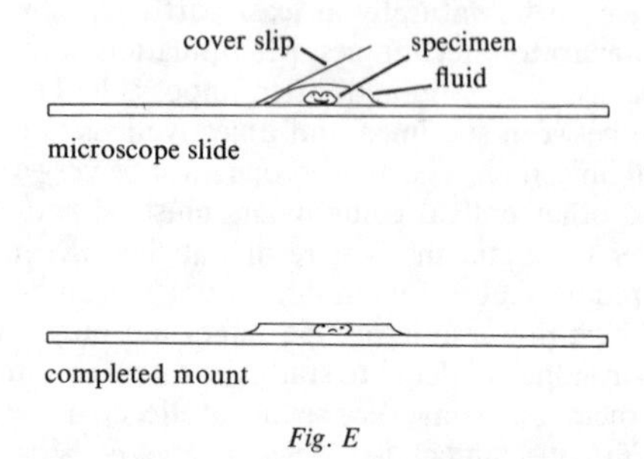

Fig. E

Small living plants or animals can be mounted for study using pond-water as the medium. The surface tension of the water tends to hold the cover-slip in place provided that the quantity of fluid has been judged correctly. Moving organisms can be held still by gently drawing off some of the water with the corner of a piece of blotting paper until the specimen is just trapped between slide and cover-slip. Larger subjects, up to about 1 mm thick (the practical maximum), can be accommodated by supporting the corners of the cover-slip with tiny balls of 'plasticene' or similar substance. This permits gentle controlled pressure to be applied to the cover-slip if required (use a piece of blotting paper placed over the cover-slip to spread the load and prevent breakage of the thin glass, and to soak up surplus water). This general method is perfectly satisfactory for the study of most living micro-organisms but the preparation only lasts for a short time as the focused heat from the light source, and accompanying evaporation, will soon cause deterioration and death of the specimen.

Movement in living specimens is a great problem as motion is magnified by the microscope as well as size. A good method of slowing micro-organisms down, literally stopping them in their tracks, is to place them in an inert, viscous medium such as methyl cellulose solution. Most chemists' shops stock methyl cellulose (as a dry powder or as tablets, which should be crushed before use) which can be dissolved in water to make a thick syrup. The syrup is used as a mounting medium in place of water and the specimen injected into it with a micro-pipette, or stirred in gently with a needle. (If the specimen is simply placed *on* the medium it will be forced to the edge when the cover-slip is added.) After a few minutes the specimen will be almost totally immobilized (but alive and unharmed) ready for detailed study.

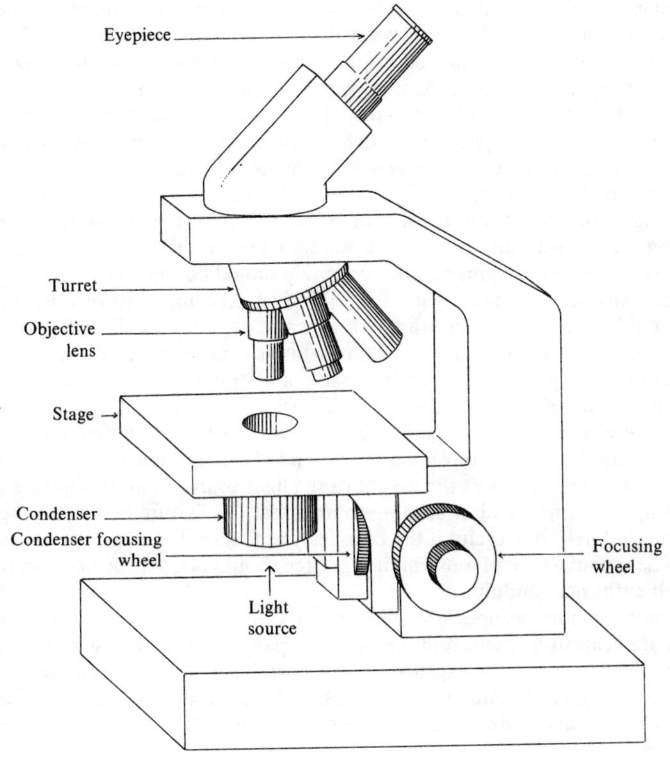

Fig. F High-power microscope

Specimens of arthropods which have been killed in hot water or alcohol (see *p. 346*) can be mounted, more-or-less permanently, in Berlese's Gum Chloral, obtainable from biological suppliers. This is used as a mountant, as above, but before adding the cover-slip the limbs should be splayed out or separated from the body using fine mounted needles; gum chloral is a viscous liquid, which aids this procedure. After a few hours the specimen will become translucent (this is called 'clearing') and reveal details of joints, hairs and other important structures. Such a preparation is only semi-permanent and will eventually deteriorate. It can be made virtually permanent by sealing the edge of the cover-slip to the surface of the slide with several coats of quick-drying varnish. If gum chloral is unobtainable, glycerine (usually available from chemists' shops) works as well but is more difficult to handle.

Buying a Microscope

As with most optical equipment the best is usually the most expensive and the final choice will probably be dictated by the depth of one's pocket. The quality can only really be judged, at the time of purchase, by comparing a number of models or makes. A high power of magnification is not the sole criterion: resolution – the ability to distinguish detail – is more important. A specialist microscope supplier should have on hand a suitable test-slide to demonstrate the resolving power of various models. Diatom frustules (*p. 54*) are often used for this purpose. If possible, take a model 'on approval' to test it thoroughly before buying. The moving parts, especially the rack-and-pinion focusing mechanisms and their slides, should work smoothly and positively, with no looseness or roughness, and preferably should be adjustable for wear. The vertically moving stage or head should never drop under its own weight and should be able to resist slight downward pressure. Rack-and-pinion parts should be made of brass or steel (quality is usually self-evident); some cheaper microscopes have these parts made of aluminium or soft alloys that wear quickly and should be avoided at all costs.

Lighting systems are legion and often a matter of personal preference. It is worth testing the ease of replacing bulbs; and checking their cost! Support arms which cannot support the weight of the light source in all positions are surprisingly common and infuriating to use. Test this feature before buying. For reflected light it is useful if the beam can be focussed to provide a bright spot. The manufacture of a useful light source should not tax the keen do-it-yourself enthusiast unduly.

Secondhand microscopes are often good value. Many 'antique' microscopes are beautifully made and their moving parts still function as new. The lens quality can vary from excellent to awful, so it is worth checking the size of the screw threads holding the objectives, and the eyepiece tubes, to see that they are of standard size so that modern replacement lenses can be fitted if desired.

References: Bradbury, 1984. Garnett, 1965. Hartley, 1979. Wells, 1957.

Making a Glass Pipette

Making a Glass Pipette
This is not a difficult skill to acquire but it does require practice before a satis-satisfactory result is achieved.

Cut a length of glass tubing, using a glass-cutting knife or diamond, twice as long as the thick part of the pipette required. Hold this 'blank' firmly at each end with just a short central part in the narrow flame of a bunsen burner or small blowtorch, directing the flame away from you for safety. Rotate the tube constantly and smoothly, avoiding lateral movements; the aim is to heat a portion of the tube about twice as long as its diameter evenly all round. When the heated portion softens (this can be felt before it is seen) remove it from the flame and immediately pull the two ends smoothly apart. If all goes well the softened region will be drawn out into a slender hollow 'neck'. This is the skilful part and the first few attempts, until the knack is acquired, will probably end in disaster; but glass tubing is cheap! The main problem is in judging just when to pull, how much force to exert, and pulling with smooth acceleration and deceleration in a straight line.

When the work has cooled, trim it to the desired length with the glass knife. Rough ends of tubing can be rounded off by holding and rotating them in a flame until the edge melts to form a smooth bead (but this does not work with fine tubes as they will seal off). Cutting and heating glass has its hazards and the use of safety glasses is recommended.

Suppliers

Bio-Science Supplies, 4 Long Mill North, Wednesfield, Wolverhampton, West Midlands WV11 1JD.

Brunel Microscopes, 113 Henbury Road, Henbury, Bristol BS10 7AA.

The Freshwater Biological Association:

Windermere Laboratory, The Ferry House, Ambleside, Cumbria LA22 0LP (Requests for publications to The Librarian at above address.)

River Laboratory, East Stoke, Wareham, Dorset BH20 6BB.

G.B. Nets, 50 Henshaw Road, Small Heath, Birmingham B10 0TB.

Hampshire Micro, The Microscope Shop, Oxford Road, Sutton Scotney, Hants SO21 3JG.

Micro-instruments (Oxford) Ltd, 7 Little Clarendon Street, Oxford.

Northern Biological Supplies, 3 Betts Avenue, Martlesham Heath, Ipswich, Suffolk IP5 7HR.

Science Studio (Oxford) Litd, 7 Little Clarendon Street, Oxford.

Watkins & Doncaster, Four Throws, Hawkhurst, Kent TN18 5ED.

Bibliography

General

Bradbury, S. 1984. *An Introduction to the optical microscope.* Oxford University Press, Royal Microscopical Society.

Brocher, F. 1913. *L'Aquarium de Chambre.* Payot, Paris.

Clegg, J. 1965. *The Freshwater Life of the British Isles.* Warne, London. (3rd Edn.)

Dowdeswell, W. H. 1966. *Keys to small organisms in soil, litter, and water troughs.* Nuffield Foundation, Penguin, London.

Edmondson, W. T. (Ed.) 1966. (Ward, H. B. & Whipple, G. C.) *Freshwater Biology.* John Wiley, New York & London; 1248 pp.

Garnett, W. J. 1965. *Freshwater Microscopy.* Constable, London.

Hartley, W. G. 1979. *Hartley's Microscopy.* Senecio, Charlbury, England.

Hutchinson, G. E. *A Treatise on Limnology.* Wiley, New York. Vol. 1, 1957; Vol. 2, 1967.

Hynes, H. B. N. 1970. *The Ecology of Running Waters.* Liverpool University Press.

Illies, J. (Ed.) 1967. *Limnofauna Europaea.* Fischer, Stuttgart.

Macan, T. T. & Worthington, E. B. 1951. *Life in lakes and rivers.* Collins (New Naturalist, No. 15), London.

Maitland, P. S. 1977. *A Coded Checklist of Animals occurring in Fresh Waters in the British Isles.* Inst. of Terrestrial Ecology, Edinburgh.

Mellanby, H. 1963. *Animal life in Freshwater.* Methuen, London. (6th Edn.).

Pennak, R. W. 1953. *Freshwater Invertebrates of the United States.* The Ronald Press Company, New York; 769 pp.

Quigley, M. 1977. *Invertebrates of Streams and Rivers, a key to identification.* Edward Arnold, London.

Thienemann, A. 1950. *Die Binnengewasser, Band 18. Verbreitungsgeschichte der Süsswassertierwelt Europas.* Schweizerbart, Stuttgart; pp. 809.

Wells, A. L. 1957. *The Microscope made easy.* Warne, London & New York.

Wells, S. M., Pyle, R. M. & Collins, N. M. 1983. *The I.U.C.N. Invertebrate Red Data Book.* I.U.C.N., Gland, Switzerland.

Wesenberg-Lund, C. 1939. *Biologie der Süsswassertiere; Wirbellose Tiere.* Springer, Vienna; 817 pp.

Works on Invertebrates and Lower Plants

(NB F.W.B.A. = Freshwater Biological Association)

Allen, G. O. 1950. *The British Stoneworts (Charophyta).* Haslemere Natural History Society.

Balfour-Brown, F. 1953. *Handbooks for the identification of British insects, VI, pt. 3. Coleoptera, Hydradephaga.* Roy. Ent. Soc., London.

Balfour-Brown, F. 1940. *British Water Beetles.* Ray Society, London. 3 Parts: Vol. 1, 1940; Vol. 2, 1950; Vol. 3, 1958.

Ball, I. R. & Reynoldson, T. B. 1981. *British Planarians.* Synopses of the British Fauna, N.S. No. 19, Linnean Society, Cambridge University Press.

Barber, H. G. & Haworth, E. Y. 1981. *A guide to the morphology of the Diatom Frustule.* F.W.B.A. Scient. Publ. No. 44.

Belcher, H. & Swale, E. 1978. *A beginner's guide to Freshwater Algae*. Inst. of Terrestrial Ecology, HMSO, London.
Belcher, H. & Swale, E. 1979. *An illustrated guide to River Phytoplankton*. Inst. of Terrestrial Ecology, HMSO, London.
Bierne, B. P. 1952. *British pyralid and plume moths*. Warne, London & New York.
Brauer, A. (Ed.) 1909. *Die Susswasserfauna Deutschlands*. Heft 10, *Phyllopoda*: Heft 11, *Copepoda, Ostracoda, Malacostraca*; Heft 13, *Oligochaeta, Hirudinea*: Heft 19, *Mollusca, Nemertini, Bryozoa, Turbellaria, Tricladida, Spongillidae, Hydrozoa*. Gustav Fischer, Jena.
Brinkhurst, R. O. 1971. *A guide to the identification of British Aquatic Oligochaeta*. F.W.B.A. Scient. Publ. No. 22. (2nd Edn.).
Brinkhurst, R. O. & Jamieson, B. G. M. 1971. *Aquatic Oligochaeta of the World*. Oliver and Boyd, Edinburgh.
Chinery, M. 1986. *Guide to the Insects of Britain and Western Europe*. Collins, London.
Corbet, P. S., Longfield, C. & Moore, N. W. 1960. *Dragonflies*. Collins (New Naturalist, No. 41), London.
Cranston, P. S. 1982. *A key to the larvae of the British Orthocladinae (Chironomidae)*. F.W.B.A. Scient. Publ. No. 45.
Curds, C. R. 1982. *British and Other Freshwater Ciliated Protozoa. Vol. 1*. Synopses of British Fauna, N.S. No. 22. Linnean Society, Cambridge University Press.
Curds, C. R., Gates, M. A. & Roberts, D. M. 1983. *British and Other Freshwater Ciliated Protozoa. Vol. 2*. Synopses of the British Fauna, N.S. No. 23. Linnean Society, Cambridge University Press.
Davies, L. 1968. *A key to the British species of Simuliidae (Diptera) in the Larval, Pupal, and Adult stages*. F.W.B.A. Scient. Publ. No. 24.
Disney, R. H. L. 1975. *A key to the British Dixidae*. F.W.B.A. Scient. Publ. No. 31.
Donner, J. 1966. *Rotifers*. Frederick Warne & Co., London & New York.
Edington, J. M. & Hildrew, A. G. 1981. *Caseless Caddis Larvae of the British Isles*. F.W.B.A. Scient. Publ. No. 43.
Edmondson, W. T. (Ed.) 1966. (Ward, H. B. & Whipple, G. C.) *Freshwater Biology*. John Wiley, New York & London; 1248 pp.
Elliot, J. M. 1977. *A key to the British Freshwater Megaloptera and Neuroptera*. F.W.B.A. Scient. Publ. No. 35.
Elliot, J. M. & Humpesch, U. H. 1983. *A key to the adults of the British Ephemeroptera*. F.W.B.A. Scient. Publ. No. 47.
Elliot, J. M. & Mann, K. H. 1979. *A key to the British Freshwater Leeches*. F.W.B.A. Scient. Publ. No. 40.
Ellis, A. E. 1969. *British Snails*. Clarendon Press, Oxford.
Ellis, A. E. 1978. *British Freshwater Bivalve Mollusca*. Synopses of the British Fauna. N.S. No. 11, Linnean Society, Academic Press.
Ewer, R. F. 1948. *A review of the Hydridae and two new species of Hydra from Natal*. Proc. Zool. Soc. 118: 226–244.
Flossner, D. 1972. *Krebstiere, Crustacea (Branchiopoda, Branchiura)*. Tierwelt Deutschlands, 60 Teil. Fischer, Jena.
Fox, H. Munro. 1949. *Triops in Britain*. Proc. Zool. Soc. 119:693–702.
Fryer, G. 1982. *The Parasitic Copepoda and Branchiura of British Freshwater Fishes*. F.W.B.A. Scient. Publ. No. 46.
Gibson, R. 1982. *British Nemerteans*. Synopses of the British Fauna, N.S. No. 24, Linnean Society, Cambridge University Press.

Gibson, R. & Moore, J. 1976. *Freshwater Nemerteans*. Zool. J. Linn. Soc., 58: 177–218.

Gisin, H. 1960. *Collembolenfauna Europas*. Museum D'histoire Naturelle, Geneve.

Gledhill, T., Sutcliffe, D. W. & Williams, W. D. 1976. *A key to British Freshwater Crustacea: Malacostraca*. F.W.B.A. Scient. Publ. No. 32.

Goodey, T. 1963. *Soil and Freshwater Nematodes*. Methuen, London. (Includes short appendix on Nematomorpha.)

Grayson, R. F. 1971. *The freshwater Hydras of Europe, 1. A review of the European species*. Arch. Hydrobiol., 68: 436–449.

Groves, J. & Bullock-Webster, G. R. 1920 & 1924. *The British Charophyta*. Vols 1 & 2, The Ray Society, London.

Hammond, C. O. 1983. *The Dragonflies of Great Britain and Ireland*. (2nd Edn.) Curwen, London.

Harding, J. P. & Smith, W. A. 1974. *A key to the British Freshwater cyclopoid and calanoid copepods*. F.W.B.A. Scient. Publ. No. 18. (2nd Edn.)

Harris, J. R. 1952. *An Angler's Entomology*. Collins (New Naturalist, No. 23), London.

Hickin, N. E. 1967. *Caddis Larvae*. Hutchinson, London.

Holland, D. G. 1972. *A key to the larvae, pupae, and adults of the British species of Elminthidae*. F.W.B.A. Scient. Publ. No. 26.

Hopkins, C. L. 1961. *A key to the water mites (Hydracarina) of the Flatford area*. Field Studies, 1, 3: 45–64.

Hynes, H. B. N. 1977. *A key to the British adults and nymphs of stoneflies (Plecoptera)*. F.W.B.A. Scient. Publ. No. 17. (2nd Edn.)

Ingle, R. W. 1980. *British Crabs*. British Museum, Oxford University Press.

Ingold, C. T. 1975. *Guide to Aquatic Hyphomycetes*. F.W.B.A. Scient. Publ. No. 30.

Jeffrey, C. 1977. *Biological Nomenclature*. Systematics Association; Edward Arnold, Southampton.

Karaman, G. S. & Pinkster, S. 1977. *Freshwater* Gammarus *species from Europe, North Africa, and adjacent regions of Asia (Crustacea – Amphipoda). Part 1.* Gammarus pulex *group and related species*. Bijd. Dierk. 47 (1): 1–97. *Part 2.* Gammarus roeseli *and related species*. Bijd. Dierk. 47 (2): 165–196.

Lacourt, A. W. 1968. *A monograph of the freshwater Bryozoa – Phylactolaemata*. Zool. verhand. 93: 1–159.

Laurent, P.-J. & Forest, J. 1979. *Données sur les écrevisses qu'on peut rencontrer en France*. La Pisciculture Française, 56: 25–37.

Levine, N. D. *et al.* 1980. *A new revised classification of the Protozoa*. J. Protozoology, 27: 37–58.

Lind, E. M. & Brook, A. J. 1980. *Desmids of the English Lake District*. F.W.B.A. Scient. Publ. No. 42.

Linssen, E. F. 1959. *Beetles of the British Isles*. 2 vols. Warne, London and New York.

Lucas, W. J. 1930. *The Aquatic (Naiad) stage of the British Dragonflies*. Ray Society, London.

Macan, T. T. 1965. *A revised key to the British water bugs (Hemiptera – Heteroptera)*. F.W.B.A. Scient. Publ. No. 16. (2nd Edn.)

Macan, T. T. 1977. *A Key to the British Fresh- and Brackish-Water Gastropods*. F.W.B.A. Scient. Publ. No. 13. (4th Edn.)

Macan, T. T. 1979. *A key to the nymphs of the British species of Ephemeroptera*. F.W.B.A. Scient. Publ. No. 20. (3rd Edn.)

Mackinnon, D. L. & Hawes, R. S. J. 1961. *An introduction to the study of the Protozoa*. Oxford University Press; 506 pp.
Miall, L. C. 1895. *The Natural History of Aquatic Insects*. Macmillan, London.
Michael, A. D. *British Oribatei*. Ray Society, London. Vol. 1, 1884; Vol. 2, 1888.
Morgan, C. I. & King, P. E. 1976. *British Tardigrades*. Synopses of the British Fauna, N.S. No. 9. Linnean Society, Academic Press.
Mosely, M. E. 1939. *The British Caddis Flies, Trichoptera*. Routledge, London.
Mundy, S. P. 1980. *A key to the British and European Freshwater Bryozoans*. F.W.B.A. Scient. Publ. No. 41.
Page, F. C. 1976. *An illustrated key to Freshwater and Soil Amoebae*. F.W.B.A. Scient. Publ. No. 34.
Pascher, A. (Ed.) *Die Susswasserflora Deutschlands, Osterreichs und der Schweiz*. Heft 1, 1914. *Zooflagellates*; Heft 2, 1913. *Chrysophyta, Cryptophyta, Euglenophyta*; Heft 3, 1913. *Dinoflagellata*; Heft 4, 1927. *Volvocales, Chlorophyceae* 1.; Heft 5, 1915. *Chlorophyceae* 2.; Heft 6, 1914. *Chlorophyceae* 3.
Penney, J. T. & Racek, A. A. 1968. *A Comprehensive Revision of a Worldwide Collection of Freshwater Sponges (Porifera: Spongillidae)*. Bull. U.S. Nat. Mus. No. 272.
Pontin, R. M. 1978. *A key to the Freshwater Planktonic and Semi-Planktonic Rotifers of the British Isles*. F.W.B.A. Scient. Publ. No. 38.
Reynoldson, T. B. 1978. *A key to the British species of Freshwater Triclads*. F.W.B.A. Scient. Publ. No. 23. (2nd Edn.)
Round, F. E. 1981. *The Biology of the Algae*. E. Arnold, London; 278 pp.
Rouseau, E. 1921. *Les larves et nymphes aquatique des insects d'Europe*. Bruxelles. Vol. 1 (? only) includes: Hemiptera, Odonata, Ephemeroptera, Megaloptera/Neuroptera, Plecoptera, Trichoptera.
Ruttner-Kolisko, A. 1974. *Plankton Rotifers, Biology and Taxonomy*. Die Binnengewasser, Supplementary Edition, English translation of vol. 26, part 1. pp. 1–146: Schweizerbart, Stuttgart.
Scourfield, D. J. & Harding, J. P. 1966. *A Key to British species of Freshwater Cladocera*. F.W.B.A. Scient. Publ. No. 5. (3rd Edn.)
Soar, C. D. & Williamson, W. *The British Hydracarina*. Ray Society, London. Vol. 1, 1925; Vol. 2, 1927; Vol. 3, 1929.
Stubbs, A. & Chandler, P. 1978. *A Dipterist's Handbook*. The Amateur Entomologist, Vol. 15. (Published by Am. Ent. Soc., 355 Hounslow Road, Hanworth, Feltham, Middlesex.)
Vickerman, K. & Cox, F. E. G. 1967. *The Protozoa*. Introductory Studies in Biology, John Murray, London.
Voigt, M. 1956–7. *Rotatoria, Die Radertiere Mitteleuropas*. Gebruder Borntraeger, Berlin-Nikolassee. Vol. 1, 1957; Vol. 2, 1956.
Wesenberg-Lund, C. 1943. *Biologie der Susswasser Insekten*. Springer, Berlin.
Young, J. O. 1970. *British and Irish Freshwater Microturbellaria: Historical Records, New Records, and a key to their Identification*. Arch. Hydrobiol. 67: 210–241.

Vascular Plants and Bryophytes

Cook, C. D. K. *et al.* 1974. *Water Plants of the World: a manual for the identification of the genera of freshwater macrophytes*. The Hague, Netherlands.
Fitter, R., Fitter, A. & Blamey, M. 1985. *The Wild Flowers of Britain and Northern Europe*. 4th edn. London.

Fitter, R., Fitter, A. & Farrer, A. 1984. *Guide to the Grasses, Sedges, Rushes and Ferns of Britain and Northern Europe*. London.
Hubbard, C. E. 1984. *Grasses*. 3rd edn. Harmondsworth, Middlesex.
Jermy, A. C., Chater, A. O. & David, R. W. 1982. *Sedges of the British Isles*. BSBI Handbook No. 1. London.
Nicholson, B. E. & Brightman, F. H. 1966. *The Oxford Book of Flowerless Plants*. Oxford.
Tutin, T. G. *et al*. 1964–80. *Flora Europaea*. 5 vols Cambridge.
Watson, E. V. 1963. *British Mosses and Liverworts*. Cambridge.

Vertebrates
Arnold, E. N. & Burton, J. A. 1978. *A Field Guide to the Reptiles and Amphibians of Britain and Europe*. London.
Corbet, G. B. & Southern, H. N. 1977. *A Handbook of British Mammals*. 2nd edn. Oxford.
Cramp, S. & Simmons, K. E. L. 1977. *A Handbook of the Birds of Europe, the Middle East and North Africa*. 7 vols. Oxford.
Frazer, D. 1983. *Reptiles and Amphibians in Britain*. NN 69 London.
Heinzel, H., Fitter, R. & Parslow, J. 1984. *The Birds of Britain and Europe with North Africa and the Middle East*. 4th edn. London.
Muus, B. J., Dahlstrom, P. & Wheeler, A. 1971. *Freshwater Fish of Britain and Europe*. London.
Regan, C. Tate 1911. *British Freshwater Fishes*. London.

Glossary

NB Words in **heavy type** refer to other glossary entries.

ABDOMEN The third and most posterior division of the body in **arthropods**; the belly region, containing the digestive organs, in **vertebrates**.

ACHENE A dry fruit containing one seed and not splitting open.

ALGAE An informal major group of relatively simple **eucaryote** plant organisms (*p. 42*).

ANABIOSIS Suspended animation during dry conditions, reviving when moisture returns, of certain aquatic organisms (e.g., Tardigrada, Nematoda, some Rotifera).

ANIMALCULE Quaint name for a **microscopic** animal.

ANNULI Structural rings, separated by circular grooves, around the surface of a cylindrical object or organism (e.g., an earthworm).

ANTENNA/ANTENNAE The first one or two pairs of appendages on the head of insects and crustaceans; 'feelers'.

ANTERIOR The 'front' or 'head end' of an organism; pertaining to that region.

ANTHER See **stamen**.

ANTHERIDIUM The male sexual organ in stoneworts (Charophyta) and other non-seed-producing plants.

ARTHROPODS An informal major group containing several phyla of animals characterized by jointed, paired limbs (e.g., Chelicerata, Crustacea, Uniramia). Regarded by some as a single phylum, Arthropoda.

ASEXUAL REPRODUCTION Reproduction by **division**, **budding**, or **parthenogenesis**, of a single parent individual. Asexual reproduction is faster and more energy-efficient than **sexual reproduction** and is employed by many organisms to increase their populations rapidly during favourable conditions.

AWN A bristle on a grass floret.

AXIAL Pertaining to the **axis**.

AXIS The longitudinal centre-line or core of any object or organism which exhibits symmetry other than spherical; the line about which such symmetry occurs.

BASIC WATER (Base-rich water) Water containing a high level of dissolved (basic) mineral salts which tend to make it alkaline.

BENTHIC Living on or near the bottom substratum, sedentary or mobile.

BIRAMOUS Two-branched.

BIVALVED Consisting of two equal **valves**, e.g., the shell of a mussel or diatom.

BOG An area of wet, markedly acid peat with a characteristic vegetation of *Sphagnum* mosses.

BRACT A small leaf immediately below a flower.

BUDDING A method of **asexual reproduction** by which a new individual is formed from an outgrowth of the parent body – e.g., 'runners' in plants, buds in hydras.

CALCAREOUS Containing calcium in the form of chalk or lime.

CALYX The outer ring of a flower, usually green, and often divided into three or more **sepals**.

CARAPACE The hard shell of many crustaceans and some other animals.

CAUDAL The most posterior part of the body; the tail.

CELL All the organisms described in this book consist of one or more cells. A cell is a microscopic quantity of living material (cytoplasm) bounded by a membrane.

Most kinds of cell have a very complicated internal organization, including various structures known collectively as **organelles** which are concerned with such functions as nutrition, respiration, reproduction, etc.

Many plants (some **algae**) and animals (**protozoa**) consist of just one cell. Some species of these **unicellular** organisms form aggregations or simple **colonies**, which usually have a regular structure that is characteristic of the species or genus. All other plants and animals are **multicellular**, consisting of numerous cells which tend to be specialized for different functions in the working of the whole organism. This is the main difference between colonial unicells and multicellular organisms.

Plant cells are distinguished from animal cells by the presence of certain pigments used in **photosynthesis**, the food-manufacturing process of plants. These pigments are typically green (**chlorophyll** is the best known), but can be red, brown, yellow, blue, blue-green, or purple. The cells of most plants have relatively thick and rigid walls made of cellulose: animal cells are typically enclosed by just a thin flexible membrane.

CEPHALOTHORAX A region of the body formed by fusion of the head and **thorax** in some **arthropods**.

CERCUS (*pl.* CERCI) A jointed tail appendage of **arthropods** and some other animals.

CHITIN A tough, hard, or slightly flexible substance forming the exoskeleton of arthropods and parts of some other animals.

CHITINOUS Made of **chitin**; any substance having similar characteristics to chitin.

CHLOROPHYLL The commonest and best-known of the **photosynthetic pigments** found in plants; always green.

CHROMOSOME A structure present in the **nucleus** of every **eucaryote** cell which contains the genetic material of the cell.

CHRYSALIS The pupal stage of an insect (Class Endopterygota) during which **metamorphosis** from larva to adult occurs.

CILIA (singular CILIUM) **Organelles**, structurally identical to **flagella**, functioning as a group (tract or field), and working in harmony (usually with each one beating slightly out of phase with its neighbour, like a field of corn rippling in the wind). Cilia may occur in numbers on the surface of a single cell (e.g., ciliated protozoans) or as an aggregation of adjacent ciliated cells (e.g., on flatworms). They are usually used for locomotion, or the generation of feeding or locomotory currents.

CILIARY FEEDING **Filter feeding** by utilizing **cilia**-generated water currents to carry suspended food particles to the feeding organism. A common method of feeding in many rotifers, protozoans, and other small animals. The freshwater mussels use a field of cilia to draw water currents over their **gills**; these currents are vital for both respiration (stale water being replaced by fresh) and feeding, particulate matter being filtered out of the water by the cilia, sorted, and carried to the mouth on a 'conveyor belt' of cilia.

CILIARY GLIDING A characteristically smooth, gliding motion of some small **soft-bodied invertebrates** over a substratum by means of fields of **cilia** (e.g., in flatworms, gastrotrichs, and others).

CLASS A taxonomic rank between **phylum** and **order**.

CLASSIFICATION (BIOLOGICAL) The arrangement of living organisms into designated groups intended to reflect their relationships (see *p. 13*).

CLEISTOGAMOUS (FLOWERS) Self-pollinating and not opening.

COLONY A group of organisms of the same species living connected together in a common mass. Colonies usually arise by **asexual division** or **budding**, the daughter individuals remaining joined and so building up into the colonial body. It is also common to refer to numbers of separate organisms living closely together, but not physically connected, as colonies, but **aggregation** is then a more accurate term.

COMPRESSED Flattened laterally, so that the body is distinctly taller than wide (*cf.* **depressed**).

CONJUGATION The physical joining, temporary or permanent, of two separate organisms, or their sexual cells, during which their **genetic material** is exchanged: a part of **sexual reproduction**.

In many simple organisms, that do not have separate sexes, the two individuals separate after conjugation and then divide into two or more 'daughter' individuals, each of which possesses genetic elements from both 'parents'.

CONTRACTILE Capable of being shortened, typically by muscular action, and therefore capable of gross changes in shape or size, particularly in soft-bodied invertebrates such as flatworms (*cf.* **retractile**).

COROLLA The most conspicuous part of most flowers, lying inside the **calyx**: in aquatic plants usually either absent or divided into three or more petals.

CUTICLE An outer skin, consisting of secreted material or dead cells, found in many animal groups (*e.g.*, nematodes, **arthropods**).

DEPRESSED Flattened dorso-ventrally, so that the body is distinctly wider than tall (*cf.* **compressed**).

DETRITUS Sediment formed by abrasion or other natural causes: may be inorganic or organic in origin, e.g., remains of decaying plants.

DISTAL The region furthest from the point of attachment (*cf.* **proximal**).

DIVISION (ASEXUAL) **Asexual reproduction** by dividing the parent body into two or more parts, each of which grows into a complete individual. (Also called **fission**.)

DORSAL The upper side or 'back' of an organism's body; pertaining to that surface (*cf.* **ventral**).

ECOSYSTEM A unit of vegetation with all the animals associated with it, and all the physical and chemical components (soil, microclimate) which make it a recognizably self-contained habitat.

EPIZOOIC (EPIZOOITE) A **sessile** animal that uses another animal as a substratum for attachment.

EUCARYOTE (or EUKARYOTE) Organisms with eucaryotic cells: all kingdoms other than Monera. Eucaryotic cells, compared to **procaryotic** cells, are relatively large and complex, containing **organelles** and a discrete **nucleus** bounded by a membrane. Eucaryotic plants contain their **photosynthetic pigments** in organelles called **plastids**.

EUTROPHICATION The over-enrichment of an aquatic environment (usually artificial, e.g., by run-off of agricultural fertilizers) which may result, for instance, in a sudden and catastrophic growth of **algae** and bacteria (especially blue-green algae), which reduces the oxygen content of the water, rendering it uninhabitable by other organisms (*cf.* **water bloom**).

EYESPOT The 'eyes' of many small organisms; a dark spot sensitive to light but not equipped to form an optical image.

FAMILY A taxonomic rank intermediate between order and genus.

FEN An area of wet peat that may be alkaline, neutral or slightly acid.

FILTER FEEDING A common method of feeding in aquatic animals by filtering small organisms or organic particles out of suspension in the water. Various devices are used to achieve this: tentacles, **gills**, nets made of silk or mucus, various **ciliated** structures, etc. Filter feeders are usually **sessile** but some are **planktonic**. Filter feeding is only possible in water, where there is abundant food material suspended in the dense, often moving medium.

FLAGELLUM (*pl.* FLAGELLA) A microscopic, motile, hair-like **organelle** protruding from the wall of a cell, typically occurring singly or in pairs or fours; they can be lashed rhythmically from side to side to generate movement of the organism or a localized water current (*cf.* **cilia**). (NB The flagella of **procaryotes** are different in structure from **eucaryote** flagella, although serving a similar function: they appear to revolve in a ball-and-socket joint and do not lash from side to side.)

FLORET A grass or sedge flower, aggregated into spikelets.

FLUSH A patch of wet ground, usually on a hillside, where the water flows diffusely and not in a fixed channel.

FRUITING BODY A cell or organ in fungi and many lower plants, which bears the reproductive cells, **spores**, etc.

GAMETE Sexual cell.

GENETIC MATERIAL The hereditary information (genes) contained in the **chromosomes** of a cell.

GENUS A taxonomic rank intermediate between **family** and **species**, analogous to the human 'surname'.

GILL A respiratory structure present in many aquatic animals, typically filamentous, comb-like, or plate-like.

GLIDING A form of locomotion in many monerans and **algae**. by which the organism glides smoothly over a substream with no obvious means of propulsion (probably related to **mucilage** secretion).

GLUME A small bract at the base of a grass or sedge spikelet.

HOST An organism that serves as the food or substratum of a **parasitic** or **epizooic** organism.

INFLORESCENCE The whole flowering part of the plant, including the flowers, stems, stalks and **bracts**.

INSTAR A single growth stage between moults in an **arthropod**.

INTERNODE Space between **nodes** on a stem.

INVERTEBRATES An informal major group including all animals that are not **vertebrates**.

LANCEOLATE Broad in the middle but tapering at both ends.

LARVA An immature but independent animal.

LATERAL The side parts of an organism's body (not **dorsal** or **ventral**): pertaining to those sides.

LEMMA One of the scale-like bracts in a grass floret.

LIGULE A small strap-like piece of tissue at the base of a leaf.

LONGITUDINAL AXIS See **axis**.

LOOPING Method of locomotion in worm-like animals by alternately attaching and detaching the extremities of the body, thus:

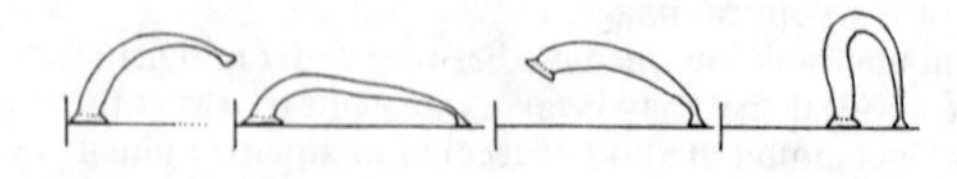

LORICA A protective case or shield, typically formed by hardening of the skin or **cuticle**, in various **invertebrates**.

MACROSCOPIC Easily visible to the unaided eye, defined in this book as longer than 1 mm.

MANDIBLE One of a pair of appendages forming the jaws in **arthropods** and some other **vertebrates**; the lower jaw of a **vertebrate**.

MARSH An area of wet ground on mineral, as distinct from peaty, soil.

MAXILLA A mouth appendage posterior to the mandibles in **arthropods**: the upper jaw of a **vertebrate**.

MEDIAN Aligned with the **axis** of the body; centrally situated.

MENISCUS The raised concave 'rim' formed where the water surface meets an emergent object.

METAMORPHOSIS A change from one physical form to another, e.g., larva to adult.

METAZOA An informal major group containing all **multicellular** animals except sponges (phylum Porifera).

MICROCLIMATE The special climate of any habitat where some factor or factors differentiate it from its surroundings.

MICROMETRE 1/1000th of a millimetre or 1/1,000,000 of a metre: 1 μm.

MICRON A micrometre (μm).

MICROSCOPIC Not easily visible, or too small to be visible, to the unaided eye: defined in this book as being less than 1 mm long.

MUCILAGE Plant mucus; a slimy or gelatinous substance secreted by many **algae** and monerans.

MULTICELLULAR An organism composed of more than one **cell**: typically with various cells or groups of cells (tissues) specialized to serve specific functions – digestion, reproduction, locomotion, etc. (See also **colony** and **unicellular**.)

NECTAR A sugary substance secreted in flowers by a gland – the nectary.

NODE Junction of leaf and stem.

NOMENCLATURE The application of names to species and higher taxa (of organisms) according to the *International Codes of Nomenclature* (Jeffreys, 1977; see *p. 13*).

NUCLEUS An **organelle** present in all **eucaryote** cells containing the **genetic material** of the cell.

NYMPH An immature insect of the class Exopterygota.

OOGONIUM The female reproductive organ in stoneworts (Charophyta) and some other lower plants.

ORDER A taxonomic tank intermediate between **class** and **family**.

ORGANELLE A differentiated part of a **eucaryotic** cell which serves a specific function, analogous to organs and tissues in the body, e.g., **nucleus**, **plastid**, **flagellum**.

ORGANISM Any living animal or plant – an organized being or entity.

PALMATE Divided like the fingers of a hand.

PALPS Paired appendages arising from the mouthparts of **arthropods**.

PANICLE A branched **inflorescence**.

PAPILLA A small, nipple-like protruberance on the surface of an organism.

PARASITE An organism that habitually lives on or in another organism (**host**), from which it derives food and shelter, usually to the detriment of the host.

PARTHOGENESIS A form of **asexual reproduction** whereby an unfertilized egg develops into an **organism**.

PEAT A pure organic soil, the result of plant material accumulating in waterlogged conditions.

PETAL See **corolla**.

PHARYNX The tubular 'throat' of an animal, part of the alimentary canal between mouth and gut (sometimes large and protruded for feeding).

PHOTOSYNTHESIS The main method of food production in plants. Plants, like all living creatures, need food for energy and growth, and photosynthesis is the method by which they manufacture it. Simply, photosynthesis is the conversion of water and carbon dioxide gas (from the air, or dissolved in water) into carbohydrates, using the energy of sunlight. **Chlorophyll** and other **photosynthetic pigments** are vital to this process, as they convert light-energy into the chemical energy which fuels the process. The net result is that plants obtain carbohydrate food (in the form of starch and sugars) and release oxygen – a waste product of the process – into the atmosphere.

Producing food from simple inorganic chemicals in this way is known as autotrophy (self-feeding). Animals cannot do this; they must utilize ready-made food – plants or other animals – a feeding method known as heterotrophy (feeding on others). From this it can be seen that, ultimately, all animals depend on plants for their existence.

PHOTOSYNTHETIC PIGMENT A coloured substance, often green, in the **plastids** of plants, necessary for the occurrence of **photosynthesis**.

PHYLUM The highest taxonomic rank containing organisms that are demonstrably related to each other (see *p. 14*).

PINNATE Of a compound leaf, the leaflets arranged in opposite pairs along the midrib, like a feather.

PLANKTON A general term for all plant and animal organisms drifting or floating free in the water at the mercy of currents.

PLASTID (CHROMATOPHORE) An **organelle** in a **eucaryote** plant cell containing the **photosynthetic pigments**.

POSTERIOR The hind or 'tail end' of an organism; pertaining to that region.

PROBOSCIS A prolongation of the mouthparts, usually tubular and flexible, of many animals, especially some insects.

PROCARYOTE (or PROKARYOTE) Organisms with **procaryotic** cells, confined to kingdom Monera. **Procaryotic** cells lack discrete membrane-bound **organelles** and have no distinct **nucleus**.

PROLEG A small protruberance on the body of some insect larvae which functions as a simple leg; unjointed, single or paired, usually with bristles or hooks at the tip.

PROTOZOA An informal group containing all the phyla of **unicellular** animals.

PROXIMAL The region nearest the point of attachment (*cf.* **distal**).

PUPA A chrysalis; the stage of an insect's life (Endopterygota only) during which **metamorphosis** from larva to adult occurs.

RECEPTACLE The flat top of a stem from which a flower arises.

RETRACTILE Capable of withdrawing or inverting one part of the body into another (*cf.* **contractile**).

ROSTRUM A prolongation of the head or mouthparts, usually tubular and rigid, of some animals, especially insects.

SAPROPHYTE An organism that feeds on the dead remains of other organisms.

SCLERITE A single **sclerotized** element or plate.

SCLEROTIZED Hardened; especially pertaining to the **cuticle**, of **arthropods** and other animals, which is hardened to form a more or less rigid 'shell'.

SEPAL See **calyx**.

SESSILE (Habit) sedentary; stationary; fixed, temporarily attached, or resting on or in a substratum; (anatomical) without a stalk, attached directly to the main body or stem (e.g., eye, leaf).

SESSILE ANIMALS Although we are used to plants being stationary, immobile organisms, it is not generally realized that many invertebrate animals also live like this. Such animals are, almost invariably, **filter feeders** and many are colonial. These sessile animals may be permanently fixed to one place on the substratum (e.g., sponges, bryozoans, some rotifers and protozoans). Other animals may be mobile but live in a sessile manner, usually inhabiting fixed, typically tubular, houses or cases which they themselves construct, e.g., certain insect larvae and worms. Many sessile animals are encased in some way, and when disturbed or threatened, can retract rapidly into their houses for protection, emerging only with caution.

SEXUAL REPRODUCTION Reproduction involving two separate parent organisms, each of which contributes a copy of its genetic material to the offspring. Thus the offspring inherits characters from each parent, but is itself slightly different from both. It is this variation with each generation that enables organisms to evolve to suit changing conditions in the environment. Sexual reproduction is not limited to organisms that have recognizable male and female individuals, but can also occur in 'sexless' **protozoans**, **algae**, and others, by means of an activity called **conjugation**.

SINGLE-CELLED (UNICELLULAR) An organism consisting of just one **cell** which is capable of performing all the functions of life.

SOFT-BODIED Many invertebrates are soft-bodied, with few or no hard skeletal parts, and thus their body shapes are not fixed or constrained by a rigid framework. Most soft-bodied animals have a typical shape which is assumed when they are 'happy' – either resting or moving normally. But when they are 'unhappy' (such as when being prodded by a biologist) they are capable of contracting, stretching, twisting, pinching, or even knotting themselves through a wide range of contortions. The illustrations and descriptions in this book are of animals in their 'normal' shapes, but it is important to understand that they do not always look like this!

Examples of soft-bodied groups are hydras, flatworms, segmented worms, etc. Molluscs possess a hard shell but can protrude at least the foot clear of the shell, so it functions as a soft body. On the other hand arthropods are completely encased by a more-or-less hardened cuticle, and thus cannot vary their shape beyond movement (and slight contraction) of the joints.

SPATHE A sheath.

SPECIES The lowest taxonomic rank; a group or population of similar organisms that can or do interbreed amongst themselves.

SPICULE A crystalline mineral structure, usually of silica or calcium, typically needle-like, secreted by various animals and retained within the body for support, defence, etc. (e.g., in sponges, some protozoans).

SPIKELET A group of florets in grasses; a single floret in sedges.

SPORE A **microscopic** reproductive body, produced **sexually** or **asexually**, by many plants, fungi, and some protozoans; usually resistant to dessication, extremes of temperature, etc.

SPORE-CASE or SPORANGIUM Receptacle containing spores.

STAMEN The male organ of a flower, consisting of a stalk, the filament of which bears at its tip the pollen-producing anther.

STIGMA See **style**.

STYLE The stalk of the female organ of a flower, which bears the stigma at its tip.

SUB . . . Descriptive prefix meaning *a*) nearly, roughly, approximately, not quite, etc. – e.g., an apple may be said to be subspherical, a tree-branch subcylindrical, and so on; or *b*) 'beneath', often implying immediately beneath (adjacent), e.g., suborder, substratum.

SUBSPECIES (abbreviated ssp.) A subdivision of a species.

SYMBIOSIS A mutually beneficial association between two organisms (symbionts) of different species.

TAXON (*pl.* TAXA) A unit or rank of classification which may be a species or a group of related species, e.g., **genus**, **class**, **order**, etc. (see *p. 13*).

TAXONOMY The study and practical application of **Classification** (and **Nomenclature**).

THORAX The middle one of three major divisions of the body in many **arthropods**, usually bearing walking legs; the chest region in vertebrates.

UMBEL A compound flowerhead in which all the flowers are borne on stalks arising from a single point, like the spokes of an umbrella.

UMBO The apex of each valve of a **bivalve** shell (*pl.* umbos).

UNICELLULAR See **single-celled**.

VALVE A shell or hard covering, e.g., in molluscs, many crustaceans, diatoms.

VASCULAR PLANT A 'higher plant' that is provided with internal vessels for the transport of fluids (a vascular system).

VENTRAL The underside of an organism's body, normally the side facing the substratum; pertaining to that surface (*cf.* **dorsal**).

VERTEBRATE Any animal possessing a backbone (most of phylum Chordata).

VOMER (adjective VOMERINE) A bone in the nasal region of fishes.

WATER BLOOM A discolouration or opacity of the water caused by a superabundance of one species of organism (*cf.* **eutrophication**).

ZOOCHLORELLAE **Unicellular** green algae (**Chlorella**-like) living **symbiotically** within the cells or tissues of animal organisms.

ZOOID One of the individual animals constituting a **colony**.

Index

References to genera and common names are for descriptive text only. For invertebrates and algae names of major groups are in capitals, alternative names are in brackets.

INDEX

INDEX

INDEX